SEDIMENTARY PETROLOGY

by

W. v. ENGELHARDT, H. FÜCHTBAUER, G. MÜLLER

PART II

SEDIMENTS AND SEDIMENTARY ROCKS 1

by

HANS FÜCHTBAUER

Ruhr-Universität Bochum

with a contribution by

HANS-ULRICH SCHMINCKE

Ruhr-Universität Bochum

Second revised and enlarged edition

With 199 Figures and 39 Tables in the Text

1974

E. Schweizerbart'sche Verlagsbuchhandlung
(Nägele u. Obermiller) Stuttgart

Halsted Press Division
John Wiley & Sons, Inc., New York - Toronto - Sydney

Library of Congress Cataloging in Publication Data (Revised)
Engelhardt, Wolf, Freiherr von.
  Sedimentary petrology.
  Vol. 1 has pt. of illustrative matter in pocket.
  Translation of Sediment-Petrology.
  Distributed in U. S. A. by Hafner Publishing Co.
  Bibliography: v. 1, p.
  CONTENTS: pt. 1. Methods in sedimentary petrology, by G. Müller. — pt. 2. 1. Sediments and sedimentary rocks, by H. Füchtbauer with a contribution by H.-U. Schmincke.
  1. Rocks, Sedimentary. I. Füchtbauer, H., joint author. II. Müller, German, joint author. III. Title.
  QE471.E523   552'.5   67-28575
  ISBN 0-471-28500-5

First edition (in German) © E. Schweizerbart'sche Verlagsbuchhandlung (Nägele u. Obermiller) Stuttgart 1970
Second edition © E. Schweizerbart'sche Verlagsbuchhandlung (Nägele u. Obermiller) Stuttgart 1974
All rights reserved, including translations into foreign languages.
This book, or parts thereof, may not be reproduced in any form without permission from the publishers.
Published in the United States, Canada, Australia by Halsted Press Division:
John Wiley & Sons, Inc., New York
In all other countries: E. Schweizerbart'sche Verlagsbuchhandlung (Nägele u. Obermiller) Stuttgart.
ISBN 3-510-65007-7

Design of the jacket: Wolfgang Karrasch
Printer: Gebrüder Ranz, Dietenheim
Printed in Germany

PREFACE

A geologist concerned with sedimentary rocks must first master the methods of field and laboratory investigation. Then, while accumulating his own observations, he becomes interested in the work and interpretation of other investigators, and finally, in a synthesis of our present knowledge about sedimentary rocks. He will want to know, how sediments presently form and what this tells us about old sediments; and he will have to concern himself with diagenesis, the agent which destroys primary structures and porosity, but which also forms or modifies economically important deposits. He is now in a position to synthesize this information and to justify those extrapolations, generalizations and speculations which provide the working theories for purposes of applied geology and for further research. In doing so, he becomes more and more interested in a quantitative understanding of the laws and forces behind geology.

This book is part of a three-part textbook on sedimentary petrology, which tries to provide graduate students, applied, teaching and research geologists with the necessary background for the three steps of investigation mentioned above. Its first editions were published in German in 1964 (Part I), 1970 (Part II), and 1973 (Part III).

Part I, "Methods in Sedimentary Petrology" (G. MÜLLER) was published in English in 1967.

Part II, "Sediments and Sedimentary Rocks" (H. FÜCHTBAUER and G. MÜLLER) in the English edition will be split because of the increase of material to be incorporated.

The present volume (II/1) deals with sandstones and conglomerates, with pyroclastic rocks, carbonate rocks, and cyclic sedimentation.

Part II/2 (G. MÜLLER), dealing with fine-grained clastic sediments and with sediments found in special environments (e. g., evaporites, siliceous sediments, phosphate sediments, sedimentary ores, residual deposits, coals, lunar sediments) will appear presumably in 1975.

Part III, "Origin of Sediments and Sedimentary Rocks" (W. v. ENGELHARDT), is now being translated.

The present book is a revised and enlarged version of the corresponding chapters in the German edition. Revision was necessary in order to include new ideas and observations published since 1970. The chapters on modern sedimentation as well as the number of figures have been considerably enlarged. Moreover, keys to the identification of heavy minerals and of carbonate particles, which may help to make an educated guess, are included.

It was the intention to present a condensed synopsis of composition, fabric, sedimentary structures, origin, and diagenesis of sedimentary rocks. Since many excellent pictures of sedimentary structures and textures in sandstones and carbonate rocks have appeared during the last decade, it was decided to provide only one example for each of the different rock types. More photographs, however, were included in the chapter on pyroclastic rocks (by H.-U. SCHMINCKE).

The first chapter is a short review of the processes of formation of sedimentary rocks (which will be treated in detail in Part III). It includes general ideas concerning compaction and sedimentary balance. The last chapter deals with cyclic processes and their manifestation in sedimentary sequences. "Authors' index" and "references" were combined. Emphasis was placed upon more recent references which will also allow the reader to find his way back to important older work.

Thanks go to Dr. JOHN D. MILLIMAN (Woods Hole) and to Mrs. and Dr. MALMSHEIMER (Bochum) for their assistance in translation. Many valuable suggestions are gratefully acknowledged; they are quoted in footnotes at the beginning of the chapters. I also wish to thank the staff of the Geological Institute in Bochum who helped to type the manuscript and to prepare the drawings and some of the photographs. Dr. E. NÄGELE (E. Schweizerbart'sche Verlagsbuchhandlung, Stuttgart) provided generous support and many valuable suggestions during the preparation of this second edition.

Bochum, May 1974　　　　　　　　　　　　　　　　　　　　　　　　　　H. FÜCHTBAUER

TABLE OF CONTENTS, PART II/1

1. The Exogenous Cycle 1
   1.1 Introductory notes 1
   1.2 Weathering and soil formation 1
      1.21 Mechanical weathering 1
      1.22 Chemical weathering 2
      1.23 Soils 2
   1.3 Transport and deposition 3
   1.4 Diagenesis 4
   1.5 Sedimentary balance 6

2. Principles of Nomenclature 10

3. Sandstones, Conglomerates and Breccias 13
   3.1 Primary components 13
      3.11 Sandstones 13
         3.111 Composition and nomenclature 13
         3.112 Quartz 19
         3.113 Rock fragments 22
         3.114 Feldspars 24
         3.115 Phyllosilicates 25
         3.116 Heavy minerals 26
            3.116.1 Distinguishing characteristics and identification key ... 26
            3.116.2 Grain size of heavy minerals 33
            3.116.3 Stability of heavy minerals 39
               3.116.31 Resistance to weathering 39
               3.116.32 Resistance to transport 40
               3.116.33 Resistance to intrastratal solution 40
               3.116.4 Use of heavy minerals 42
      3.12 Conglomerates and breccias 46
   3.2 Texture, fabric and environment 51
      3.21 Grain size 51
         3.211 Grain size distribution 51
         3.212 Regional variations in grain size 56
         3.213 Grain size distribution as an environmental indicator ... 58
         3.214 The matrix problem 61
      3.22 Grain form 63
         3.221 Shape 63
            3.221.1 Shape of pebbles 64
            3.221.2 Shape of sand grains 64
         3.222 Roundness 65
            3.222.1 Roundness of pebbles 65
            3.222.2 Roundness of sand grains 67
         3.223 Grain surface texture 69

3.23	Orientation	70
	3.231 Orientation of pebbles	70
	3.232 Orientation of sand grains	71
3.24	Stratification	72
	3.241 Internal structures of sandstone beds	72
	3.241.1 Horizontal bedding	73
	3.241.2 "Flaser" and lenticular bedding	74
	3.241.3 Cross-bedding	74
	3.241.31 Origin and classification of cross-bedding	74
	3.241.32 Variability of cross-bedding direction	78
	3.241.4 Preconsolidation deformations	79
	3.241.41 Mottled structures and trace fossils (Spurenfossilien)	79
	3.241.42 Flow-structures (Convolute lamination, slumping and others)	80
	3.241.43 Shrinkage cracks, injection structures and sand dikes	82
	3.242 The bedding plane	85
	3.242.1 Rates of accumulation and formation of sediments	85
	3.242.2 Bedding surface marks	87
	3.242.21 On the upper surface (ripple marks, trace fossils)	87
	3.242.22 On planes within the beds (current lineation)	90
	3.242.23 On the lower surface (load casts, current-induced casts, trace fossils)	91
	3.242.231 Load casts	91
	3.242.232 Current-induced casts; turbidity currents	91
	3.242.233 Trace fossils	98
3.25	Sand bodies	99
	3.251 Geometric characteristics	99
	3.252 Sand facies and environment	100
	3.252.1 Continental environments	100
	3.252.11 Dunes	100
	3.252.12 Lakes	103
	3.252.13 Rivers	103
	3.252.14 Glaciofluvial and glacial sediments	107
	3.252.15 Red beds	108
	3.252.2 Transitional environments	109
	3.252.21 Deltas	109
	3.252.22 Estuaries	115
	3.252.23 Tidal flats	116
	3.252.24 Lagoons	118
	3.252.25 Marginal seas and Gulfs	118
	3.252.26 Beaches	119
	3.252.3 Marine environments	121
	3.252.31 Shallow sea	121
	3.252.32 Deep sea	123
3.26	Sandstone facies and tectonics	124
	3.261 Cratonic tectofacies	124
	3.262 Orogenic tectofacies	125
	3.262.1 Geosynclinal facies	125
	3.262.2 Leptogeosynclinal facies	125
	3.262.3 Flysch facies	127
	3.262.4 Molasse facies	128
	3.263 Facies maps	129

3.3 Diagenesis 130
3.31 Diagenetic processes in sandstones 130
3.311 Definitions 130
3.312 Mechanical compaction 130
3.313 Dissolution 133
3.314 Neoformation (cementation, in part) 135
3.314.1 Criteria for determining the sequence of precipitation . . 135
3.314.2 Quartz neoformation 138
3.314.3 Feldspar neoformation 141
3.314.4 Carbonate neoformation 142
3.314.5 Sulphate neoformation 143
3.314.6 Clay mineral neoformation 143
3.314.7 Other neoformations 144
3.315 Replacement and alteration 144
3.32 Examples of diagenetic sequences 146
3.321 Diagenesis of quartz sandstones 147
3.322 Diagenesis of calcareous sandstones 149
3.323 Diagenesis of feldspathic sandstones 149
3.324 Diagenesis of feldspathic, slightly evaporitic sandstones (red beds) . 152
3.325 Diagenesis of coal-bearing micaceous sandstones 153
3.33 The transition from diagenesis to metamorphism in sandstones 155
3.4 Porosity, permeability, and oil saturation 158

4. Pyroclastic rocks (H.-U. SCHMINCKE) 160

4.1 Grain size classification 160
4.2 Pyroclastic rocks produced under subaerial conditions 162
4.21 Pyroclastic fall deposits (tephra) 162
4.211 Origin of tephra 162
4.212 Transport and deposition of tephra 163
4.22 Pyroclastic flows 172
4.23 Base surges 177
4.24 Lahars 180
4.25 Volcanic breccias 181
4.26 Peperites (lava-sediment mixtures) 183
4.3 Pyroclastic rocks produced under subaqueous conditions 183
4.31 Hyaloclastites and subaqueous tuffs 183
4.32 Subaqueous pyroclastic flows 184
4.4 Alteration of volcanic glass 186

5. Carbonate Rocks 190

5.1 The primary sediments and their origin 190
5.11 Calcilutites 190
5.12 Clotted limestones 197
5.13 Pelletal limestones 199
5.14 Detrital limestones 201
5.15 Skeletal limestones 206
5.151 Ecologic and mineralogic considerations, and identification key . . 206
5.152 Calcareous algae and Schizophyta 214
5.153 Nannoplankton 225

                5.154 Protozoans 227
                5.155 Sponges (Porifera) 232
                5.156 Coelenterates 233
                5.157 Bryozoa 236
                5.158 Brachiopoda 239
                5.159 Worms 240
                5.1510 Mollusca 241
                5.1511 Arthropoda 246
                5.1512 Echinodermata 248
        5.16 Oölites 250
        5.17 Crustose limestones 257
        5.18 Biostromes, bioherms and their environment 260
                5.181 Nomenclature 260
                5.182 Biologic aspects 263
                5.183 Modern reefs 265
                5.184 Ancient biostromes and bioherms 268
    5.2 Nomenclature of carbonate rocks 275
    5.3 Diagenesis 279
        5.31 Introductory 279
        5.32 Isochemical diagenesis 279
                5.321 Cementation 281
                    5.321.1 Cement A and cryptocrystalline cement (micritization) . . 281
                    5.321.2 Cement B 288
                    5.321.3 Chemical aspects 290
                5.322 Neomorphism 290
                    5.322.1 Crystal enlargement 290
                    5.322.2 Transformation aragonite → calcite 294
                5.323 Stylolites 296
                5.324 Cone-in-cone 299
                5.325 Concretions 301
        5.33 Allochemical diagenesis 303
                5.331 Dolomitization 303
                    5.331.1 The processes 303
                    5.331.2 Occurrence of dolomite 314
                        5.331.21 Early diagenetic dolomites 314
                        5.331.22 Late diagenetic dolomites 318
                5.332 Mg-loss 320
                    5.332.1 Dedolomitization 320
                    5.332.2 Magnesian calcite → calcite (incongruent dissolution) . . . 321
                5.333 Isotopic composition of carbonate sediments and rocks 324
                5.334 Anhydritization, celestitization and other processes 325
                5.335 Silicification 327
                5.336 Authigenic silicates 333
                5.337 Authigenic pyrite 337
    5.4 Pore space 338
        5.41 Porosity of calcilutites 338
        5.42 Porosity of particle limestones and dolomites 340
                5.421 Pore types 340
                5.422 Examples of depositional porosity 347
                5.423 Origin of secondary porosity 348
                    5.423.1 Secondary porosity in limestones 348
                    5.423.2 Secondary porosity in dolomites 351

5.5 Environmental indicators 352
   5.51 General aspects 352
   5.52 Continental environments 353
   5.53 Transitional environments 355
   5.54 Marine environments 356

6. Cyclic Sedimentation 361

6.1 Definitions and methods 361
6.2 Possible causes 364
   6.21 Fluctuations in climate 364
   6.22 Large-scale tectonic movements of the ocean bottom 365
   6.23 Different subsidence rates of basins 366
   6.24 Fluctuations in the rise of the hinterland 366
   6.25 Episodic shifting of rivers and deltas 366
   6.26 Biological factors 366
   6.27 Cycles caused by steady processes 367
6.3 Examples of rhythms and cycles 367
   6.31 Deep sea 367
   6.32 Shallow sea 369
   6.33 Transition land — sea 374
   6.34 Lacustrine cycles 379
   6.35 Fluvial cycles 380
6.4 Cycles of higher order (megacycles) 380
   6.41 Symmetrical megacycles 380
   6.42 Asymmetrical megacycles 381

References and Authors' Index 383
Subject Index 454

CONTENTS OF PART II/2

7. Fine-grained clastic sediments (G. MÜLLER)

   7.1 Classification of fine-grained clastic sediments — Designation according to structural and textural characteristics — Textural classification based on the sand-silt-clay ratio — Classification based on the nature of the components — Classification based on the chemical composition

   7.2 Dust sediments
   Definition — Terrestrial dust — Cosmic dust

   7.3 Terrigenous (silicate) muds
   Definition — Origin and composition of muds — Depositional areas of muds; pH and eH in the zone of deposition — Primary composition of the terrigenous muds — Altered and newly formed components as a result of aquatolysis, halmyrolysis and hydrothermal processes — Porosity and fabric of the terrigenous muds — Examples of recent mud deposits

   7.4 Diagenetic alterations in silicate dusts and muds
   The formation of silt- and claystones
   Definition of the term diagenesis — Chemical and mineralogical changes during diagenesis — Changes of the porosity and the fabric during diagenesis — The relations of the mechanical and mineralogical alterations to the depth of burial, pressure, temperature and time

   7.5 Silt- and claystones
   Siltstones — Shales

   7.6 The boundary range between diagenesis and metamorphism: Anchimetamorphism

8. Sediments formed in particular environments (G. MÜLLER)

   8.1 Salt rocks (evaporites) including evaporitic silicates
   Marine evaporites — Continental evaporites

   8.2 Siliceous sediments
   Classification — Behaviour of silica in the sedimentary environment — Biogenic siliceous rocks — Siliceous rocks of different genesis

   8.3 Sedimentary phosphate rocks
   Phosphate minerals, classification of phosphate rocks — Recent phosphate deposits — Fossil phosphorite deposits — Guano

   8.4 Sediments and sedimentary rocks rich in iron and manganese
   Sedimentary iron and manganese minerals — Separation of iron and manganese in the sedimentary cycle — Sedimentary rocks rich in iron — Sedimentary rocks rich in manganese

   8.5 Ores in sedimentary rocks

   8.6 Residual rocks
   Bauxite — Residual clays

   8.7 Peat — Brown coal — Hardcoal (with anthracite) — Rocks composed of coaly and mineral substances

   8.8 Sediments and sedimentary rocks of the moon

References and Authors' Index

Subject Index

1. The Exogenous Cycle[1]

1.1 Introductory notes

Sedimentary rocks are the weathering products of pre-existing rocks. During early stages in the evolution of the earth, only igneous rocks existed, thus being the sole source for the earliest sedimentary rocks of the geologic record. Subsequently, metamorphic rocks formed by deep burial and deformation of both sedimentary and igneous rocks.

Thus, in a general way, sediments form by exposure of igneous, metamorphic, and sedimentary rocks to the hydrosphere and atmosphere. This exposure leeds to the mechanical disintegration and chemical decomposition of the rocks ("weathering"). The weathering products are carried away (transport) and laid down (sedimentation) to form sediments. Subsequent burial by later sediments and simultaneous increase in temperature and pressure, as well as equilibration with changing pore fluids transform the sediments into sedimentary rocks ("diagenesis"). In a loose way, the term "sediments" is commonly used to cover both sediments and sedimentary rocks.

BLATT et al. (1972) and v. ENGELHARDT (1973) have presented a more detailed account of processes leading to the formation of sedimentary rocks.

1.2 Weathering and soil formation

1.21 Mechanical weathering

Because of the small primary pore volume in magmatites, meteoric water can only enter a rock after the fabric has been partly disintegrated. Prior to this, weathering is restricted to the rock surface and to joints. The type and extent of mechanical disintegration depends partly upon climate. In cold climates frost breaking occurs; in arid areas small fissures may be enlarged by salt crystallization as well as by continual adsorbtion and evaporation of water. Moreover, the severe diurnal temperature changes also may cause disintegration. In humid climates the pressure of plant roots, which penetrate small joints and haircracks, will fracture rocks. The resulting disintegration is especially prevalent in coarse crystalline rocks composed of different minerals (e.g. granite).

---

[1] Several suggestions by H. WEDEPOHL as well as the contribution at the end of ch. 1.4 by D. MARSAL are gratefully acknowledged.

1.22 Chemical weathering

The chief agent in chemical weathering is CO_2-bearing rainwater. Its effect is demonstrated by the following examples: as CO_2-pressure increases, more CO_2 is dissolved in the rainwater and the pH falls. The following two reactions, the first being the model case of silica weathering, move more intensely from left to right:

1) $2\,KAlSi_3O_8$ (K-feldspar) $+\,2\,H^+ +\,2\,HCO_3^- +\,H_2O =$
 $= Al_2Si_2O_5(OH)_4$ (kaolinite) $+\,4\,SiO_2 +\,2\,K^+ +\,2\,HCO_3^-$
 (by weathering of plagioclases Na and Ca are set free instead of K)
2) $CaCO_3 + H^+ + HCO_3^- = Ca^{++} + 2\,HCO_3^-$

According to SIEVER (1968) chemical weathering in general may be described as the uptake of H^+, combined with the freeing of K, Na, Ca, Mg and Si. Diagenesis works in the opposite direction, thus closing the exogenous cycle.

The work of bacteria can be a very important part of chemical destruction (KRUMBEIN 1968, 1969).

Quartz is relatively stable and consequently is enriched in the weathering residue.

As long as the water is in equilibrium with the atmospheric CO_2 and does not adsorb additional organic CO_2 or humic acids (which lower its pH), carbonates will remain relatively stable also.

Within the feldspar group, the K, Na, and Ca ions are easily removed at low pH's, whereas Si and Al are more resistant and form thin residual layers engulfing the grains; kaolinite is formed at low pH. K, Na and Ca are less soluble at high pH's, and montmorillonite is formed in such environments.

Among the micas, biotite weathers more readily than muscovite, due to its iron content. Release of potassium and uptake of H^+-ions and water result in the following weathering sequences (JÖRGENSEN 1965):

Illite → mixed layer illite-montmorillonite → montmorillonite
Illite → mixed layer illite-vermiculite with regular alternation of the two types of layers
Chlorite → mixed layer chlorite-vermiculite → vermiculite.

In a traverse from the north pole southward to the equator we find the following sequence: mica and mixed-layer mica-montmorillonite → kaolinite → gibbsite. This clay mineralogy is an expression of the fact that leaching causes first the removal of cations and then silica, whereby the structures become simplified (CARROLL 1970).

The other rock-forming minerals (e. g. amphiboles and pyroxenes) are more soluble than quartz, feldspar and mica. Gypsum and a number of salts are extremely soluble.

1.23 Soils

Soils are composed of the products of weathering and of precipitation, as well as of the residue of decomposing organic materials. Thus soils can be distinguished by climate. Red "laterites", rich in iron but poor in silica, occur in the humid tropics;

"chernozem" is typical of the temperate prairies, and "brown earth" and "para-brown earth" of semi-humid climates. "Gray podzolic soil", with an upper "A-horizon" leached of iron by humic acids, is typical of the cool and humid to temperate-humid climates. The iron is found in the underground as a hard, cemented B-horizon, the so-called "Ortstein". Further down there is the weathered but solid rock (C-horizon).

1.3 Transport and deposition

Minerals and materials which do not remain in situ after disintegration are removed by a) traction, b) suspension, c) solution. The material is moved either by wind, river, or ice action and redeposited in desert, alluvial, lacustrine or marine environments. In accordance with the three modes of transport, three main types of sediments exist:

1) sandstones, conglomerates and breccias (traction)
2) clay- and siltstones (suspension)
3) carbonate rocks and salts (solution)

Rock groups one and two are called clastic sediments (from Greek klastein = to break s. th.). The materials within the third group are called chemical sediments. The term "detritus" (from Latin deterere = pulverize, grind s. th.) not only includes clastic but also redeposited chemical sediments (e. g. detrital limestones).

Transportation, deposition and reworking are governed by the general diagram shown in fig. 1-1.

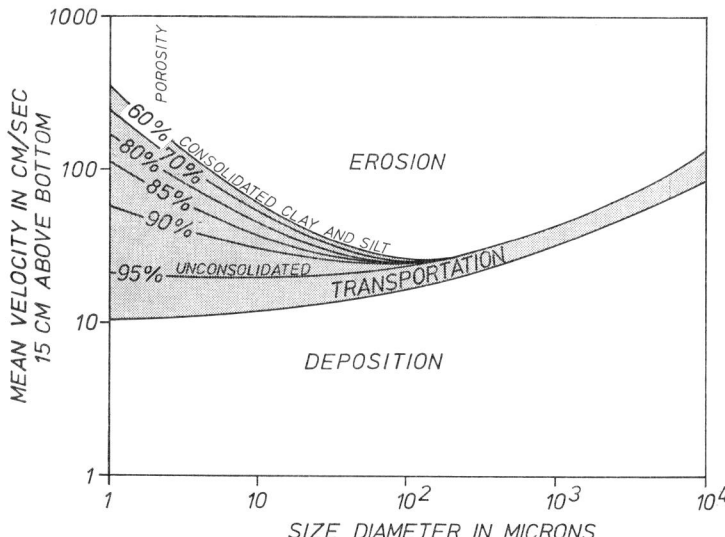

Fig. 1-1. Erosion, transportation, and deposition velocities for different grain sizes. The diagram indicates scour lag effects for various stages of consolidation (modified from POSTMA 1967).

Of the minerals and decomposition products, quartz occurs predominantly in group 1, clay minerals in group 2, and the dissolved ions K·, Na·, Ca··, Mg·· in group 3 (as well as in sea water and interstitial fluids). The precipitation of dissolved matter is frequently complicated by biological processes. Inorganic as well as biological processes result in sedimentary iron, manganese, and phosphate deposits.

1.4 Diagenesis

The transformation of sediments into sedimentary rocks is called "lithification". The term "diagenesis" (from the Greek dia = through; genesis = making s. th.) comprises not only this process but all alterations which occur within the sediment after deposition. Distinction must be made between mechanical diagenesis (reduction of porosity by pressure of the overlying sediments and by deformation) and chemical diagenesis. The latter results from increase in temperature and chemical changes of the pore fluids under increasing overburden of younger sediments, and is associated with an additional reduction of porosity. During subsidence, the conditions of diagenesis including temperature, pressure and composition of the interstitial fluids are changing continuously. For this reason, and due to low temperature, mineral equilibria are seldom attained during diagenesis.

The overburden pressure is important during mechanical as well as during chemical diagenesis. The following discussion applies to shallow depths, in which buoyancy is active because of the mobility of pore water.

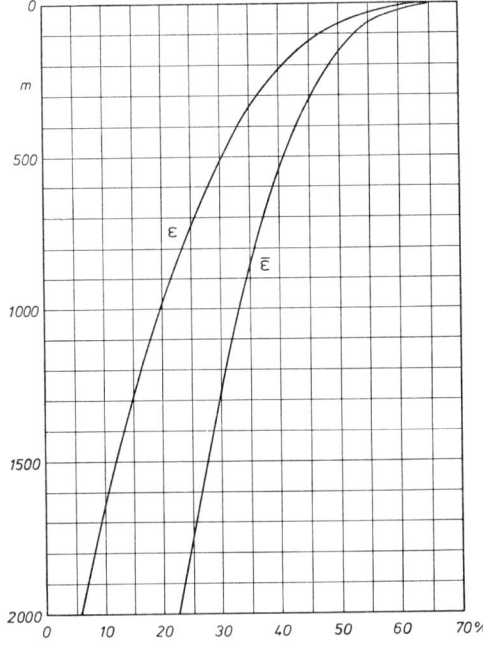

Fig. 1-2

ε = porosity of claystones, related to the depth below surface, from ATHY (1930).

$\bar{\varepsilon}$ = average porosity of the whole column of sediments above the depth h, based on ATHY. This diagram has been provided by Dr. D. MARSAL (Hannover).

Let $\bar{\varepsilon}$ be the average porosity (measured as a fraction of 1) of the whole column of porous sediments above the depth h [cm] (fig. 1-2), and assume that the density of the grains and the interstitial water is 2.7 and 1.1 g/cm³, respectively. Then the sediment presses onto each square centimeter of solid matter at the depth h below the sediment surface with an effective weight of 981 (2.7—1.1) (1 − $\bar{\varepsilon}$) · h [dynes] (981 cm/sec² = acceleration due to gravity). If ε is the porosity at the depth h (fig. 1-2), then only the fraction 1 − ε of each square cm of the horizontal interface at the depth h is in contact with the solid grains beneath. Hence the effective overburden pressure P at the depth h becomes

$$P = 1570 \cdot \frac{1 - \bar{\varepsilon}}{1 - \varepsilon} \cdot h \qquad \left[\frac{\text{dynes}}{\text{cm}^2}\right]$$

At greater depths, in which interconnected pore spaces no longer exist, buoyancy becomes unimportant. Instead, the pressure increase with increasing depth becomes more important, and in the limiting case of nonexistent buoyancy, the pressure at the bottom of the sediment column is

$$P = 981 \cdot \varrho_s \cdot h / (1 - \varepsilon) \qquad \left[\frac{\text{dynes}}{\text{cm}^2}\right]$$

ϱ_s being the average bulk density of the sediment, which, with increasing depth, finally reaches the density of the minerals. The pressure in the interstitial water can exceed the hydrostatic pressure of $981 \cdot \varrho_w \cdot h$ [dynes · cm⁻²] (ϱ_w = average density of the interstitial water; h = depth [cm]!),

(1) if the equilibrium, i. e. a self-supporting framework, is not reached, because the increase of overburden is too rapid due to high sedimentation rate, and the permeability of the sediments is too low to allow the release of pore fluids. Overhydrostatic pressure may be observed in sand lenses included in such clay sequences. The pressure depends on the (small) rate of escape of the fluid, on the elasticity of the overburden rock, and on the compressibility of the water in a complicated way and may exceed twice the hydrostatic pressure, e. g. in the Gulf Coast area.

(2) if the fluid pressure is increased by faulting and diastrophism accompanying the intrusion of salt domes into the soft and incompetent sediments (Gulf Coast; LEVORSEN 1967, p. 408), or by folding which is still active, e. g. N of the Alps (LEMCKE 1972).

Continuous sedimentation takes place only when the area of deposition is subsiding. By accumulation of material on the sediment surface, the overburden pressure increases within the deposits underneath. Reduction in thickness, i. e. compaction, results.

But, though the interstitial water moves upward through the sediment section, it does not move upward with respect to the sediment surface, because the total volume of pores increases in the sediment section during deposition. This is explained by the following model:

	stage I	stage II
bed 1	70% porosity	70% porosity
bed 2	40% porosity	40% porosity
bed 3	20% porosity	20% porosity
bed 4		10% porosity

The upper line indicates the sediment surface, the lower lines correspond to the interface of porous sediments and the pore-free basement. Stage II develops from stage I by addition of a new bed 1. Simultaneously, the previous bed 1 passes over into bed 2, with reduced porosity. The same applies to the underlying beds. Due to the reduction in porosity, bed 2 is thinner than bed 1, and so on. This model is based upon the following assumptions:

a) An equilibrium is reached immediately, i. e. together with deposition; water is released from the lower strata corresponding to the new overburden pressure.

b) In every bed (1-4) the thickness is equal for stage I and II.

c) The porosity in every bed is homogeneous.

Eventually, the pore volume of the newly-deposited upper stratum will be filled to a certain degree with compaction water which is pressed out of the lower strata. The thicker the porous sequence the lower is the increase of the pore volume of the entire sequence during sedimentation of bed 1 and the more pores of the newly deposited material will be filled with interstitial water from beneath. If a new layer of mud (bed 1) is superimposed on a very thick sedimentary sequence, nearly all of the interstitial water of bed 1 will be replaced by compaction water from below, and this expelled water will flow into the sea.

Though this is an idealized model, it leads to the conclusion that the compaction is not as effective in bringing deeply buried interstitial water back to the ocean as suggested by several workers. There is only an exchange from the sea water into the uppermost sediment layer and back to the bottom sea water. This idea is supported by the small differences in chemical composition between the ocean water and the interstitial water of the uppermost sediments. Only during erosion can large quantites of interstitial water be released to the earth surface and to the ocean.

Calculations by MARSAL & PHILIPP (1970, fig. 9) yielded the following figures: A column of clay sediment 670 m thick would be enlarged to 715 m ($= +45$ m), a column 1160 m thick would be enlarged to 1185 m ($= +25$ m), by superposition of a clay layer 165 m thick, with an initial porosity of 60%. This demonstrates that much sediment is needed to fill up basins already containing thick sequences of clay.

1.5 Sedimentary balance

The average chemical analyses of igneous and metamorphic rocks are similar to those of sedimentary rocks, except the excess of Na_2O in the former. This excess, however, is matched by salt deposits, by ocean salts, and by the salt dissolved in the pore fluids (GARRELS & MACKENZIE 1971).

In this chapter, the following topics will be discussed:

a) annual discharge of material into the oceans

b) mean erosion in selected drainage areas

c) relation between solid and dissolved matter in rivers

d) estimates of the mean thickness of the earth's sedimentary cover, including its distribution in terms of rock types in geosynclinal areas, platforms, shelf areas and ocean floors.

e) distribution of rock types through time

Table 1-1. Estimate of the annual discharge of material into the oceans, according to JUDSON (1968).

		[10^6 tons per annum]
by rivers		
	before human influence	9 300
	after human influence	24 000
by wind		60—360
by glaciers		100
from extraterrestric sources		0.35—140

For fluvial areas the following data may serve as examples:

Table 1-2. Transport and erosion efficiency of large rivers compared with the size of their drainage area (from Judson 1968).

River system	Drainage area (10^6 km^2)	Yearly transport (tons/km^2) dissolved	solid*	Erosion** (cm/1000 years)
Amazon	6.3	37	87	4.7
Kongo	2.5	39.5	13.5	2.0
Colorado	0.63	23	417	17
Mississippi	3.24	39	94	5
Columbia	0.68	57	44	4

* bed load (estimated at 10% of the measured suspension load) is included.
** calculated as pore-free material.

When considering the North American rivers of this table, one must take into account the increase in solid load due to human influence, i. e. destroying of forests and ploughing of the soil (see table 1-1, according to Judson 1968).

The material transported into the oceans by rivers can be subdivided as follows (Lopatin 1952): suspended load: bed load: solution load = 3.5 : 0.35 : 1.

The amount of clastic material deposited within alluvial areas locally can exceed the material delivered to the sea. For instance, the Tertiary rivers in the Molasse basin N. of the Alps delivered only about 10% of their total solid load to the delta and the open sea, as estimated from Füchtbauer's maps (1964; Aquitanian map).

According to Gibbs (1967) the world rivers carry an (weighted) average of 114 ppm (0.0114%) of dissolved material. In the Amazon river, the quantity of the dissolved material as well as the quantity and grain size of the suspension load increases with the relief of the drainage area (Gibbs, l. c.). The dissolved load is about 50% of the total load in the larger tropical tributaries and 30% in the mountainous tributaries (from the Andes). A detailed comparison of two mountainous and the four most important tropical tributaries of the Amazon follows (Gibbs, l. c.).

a) Dissolved load (salinity)
   in mountainous rivers: 112 ppm (wet season), 192 ppm (dry season)   at
   in tropical rivers: 9 ppm (wet season), 16 ppm (dry season)   Amazon mouth: 36 ppm*

b) Suspended solids concentration
   in mountainous rivers: 600 ppm (wet season), 70 ppm (dry season)   at
   in tropical rivers: 10 ppm (wet season), 5 ppm (dry season)   Amazon mouth: 90 ppm*

* discharge (weighted mean)

Feldspar is nearly completely missing in the suspended load of the tropical tributaries, whereas in the tributaries of the Andes it is common, indicating a predominance of mechanical weathering in the latter. On the other hand, kaolinite is the prevailing clay mineral in the tropical tributaries. In the humid season the medium grain size of the suspended material in the Amazon river is nearly twice as high as in the dry season. Such data are very important in interpreting the source areas and paleoclimates of fossil fluvial and deltaic sedimentary rocks.

According to RONOV (1968), the medium thickness of the sedimentary cover is 1.8 km. This value lies between the estimates of KUENEN (1950: 2 km) and POLDERVAART (1955: 1.4 km), and amounts to $9.2 \cdot 10^8$ km^3 of sediments upon the earth's surface.

Table 1-3. Estimation of the medium thickness (A), relative volumes (B) and composition of sediments on continents and seafloors, from RONOV (1968).

	A (km)	B (Vol.-%)	Sediments and volcanites (Vol. %)				
			Clay- and Siltstones	Sandstones	Limestones	Evaporites	Volcanites
Geosynclines	10	40	39.4	18.7	16.3	0.3	25.3
Platforms	1.8	13	49.5	23.6	21.0	2.0	3.9
Σ sediments on continents	4.5	53	42	20	18[1]	1	19
			Terrigenous Mud	Lime Mud	Silica	Brown Clay	Volcanic Mud
Shelves & Continental slopes	0.86[2]	20	56.7	29.5	7.8	1.0	5.0
Ocean floors[3]	0.4	27	7.3	41.5	17.0	31.2	3.0
Σ sediments below Oceans	0.52	47	28.3[4]	36.5	13.0	18.4[4]	3.8

A = Average thickness of the sediments (sedimentary rocks) on the earth.
B = Distribution of the sediments (sedimentary rocks) with respect to the geotectonic units (the calculation is based on compacted sediments; density 2.5).

[1] The total carbonate content may be higher because many claystones, siltstones and sandstones contain carbonate (SIEVER 1968). On the other hand, however, many carbonate sediments contain clay.

[2] The figures are based on the following area distribution: continents $14.7 \cdot 10^7$ km^2, shelves $9.3 \cdot 10^6$ km^2, oceans $27.0 \cdot 10^7$ km^2 (WEDEPOHL 1967).

[3] First seismic layer.

[4] The distinction between "terrigenous mud" and "brown clay" (which also is partly of terrigenous origin) is questionable.

Whereas sediments on platforms are relatively thin, those in geosynclines (i. e. in rapidly sinking basins) are very thick. Drilling and seismic refraction work in the ocean floors during the last few years (JOIDES) have shown that the sedimentary cover in all ocean basins is very thin because of the low age. It also thins from the continental slopes — where most material is accumulated — to the mid oceanic ridges where new basaltic crust constantly forms.

1.5 Sedimentary balance

The relative abundance of sedimentary rocks changed during geological times. Precambrian evaporites are nearly completely missing. Precambrian carbonate sediments are rare and are closely connected with blue-green algae. Not only did these blue-green algae produce rocks, but they also may have been instrumental in producing much of the free oxygen in the atmosphere. The CO_2 necessary for the production of carbonates was present already in the very earliest atmosphere, according to RUBEY (1951), whereas UREY (1951) claims that this early atmosphere contained carbon only in the form of methane. According to RUBEY (1951, cf. also WEDEPOHL 1963, 1967) a large portion of the CO_2 came from magmatic exhalations.

Systematic variations in sediment type occur throughout the geologic record. With certain reservations it can be shown that for very large areas (sometimes for the whole earth) there were periods in which chemical sediments are predominant, e. g. Silurian, Late Devonian, Early Carboniferous, Late Jurassic, Late Cretaceous, and some in which clastic rocks prevail, e. g. Precambrian, Early Devonian, Late Carboniferous, Triassic, Early Jurassic, Early Cretaceous, Late Tertiary. Moreover, one can distinguish between periods with prevailing reducing environments (e. g. Silurian) and those with oxidizing environments (e. g. Triassic). In Silurian times, large areas of the continents were flooded, whereas in Triassic times, continental sediments were abundant.

Vertical and lateral tectonic movements (epeirogenies, orogenies, plate tectonics) influenced the distribution of land and sea and may be responsible for the shifting of sea level as well as for short-termed rhythms and cycles (see ch. 6 and RONA 1973).

RONOV (1968) correlated the frequency of carbonate rocks and the percentage of water cover of the earth. Other important influences are climate and magmatic exhalations which may have influenced directly (e. g. by CO_2) the sedimentation. For instance, a cool climate mostly favors mechanical weathering producing clastic sediments, whereas warm and humid climates support chemical weathering, favoring the formation of carbonate sediments.

2. Principles of Nomenclature

The nomenclature of the rocks discussed in this book is based on the following principles:

1. There are different possibilities of defining rock names: (a) artificial names, (b) names taken from localities, as common with magmatic rocks, (c) names of common use. The latter are chosen in this book, because, in the writer's opinion, rock names should give an immediate impression of the rock without using diagrams.

2. Moreover, the names should give an approximate idea of the quantitative composition. For this purpose, a gradation 10—25—50 is suggested, which may be applied as follows: The group term (sand, silt, clay, limestone, dolostone) is formed by the main constituent, generally surpassing 50 percent. Minor components are added as attributes:

Sandstones containing   25—50% silt: silt-sandstones
Sandstones containing   10—25% silt: silty sandstones
Sandstones containing less than 10% silt: (slightly silty) sandstones

This gradation is general enough to enable a preliminary classification by field methods. Results of laboratory measurements, however, should be listed quantitatively, in order to avoid further confusion in the nomenclature.

A nomenclature for various mixtures of sand, clay and carbonate is shown in the triangle diagram of fig. 2-1. The components at the corners of the triangle can be replaced by others (e. g. sand-silt-clay, sand-silt-cements). The nomenclature, however, is applicable, independent of such triangles.

3. The main component can require further definition according to the principles described in ch. 3 through 5; e. g., calcareous *sandstone with rock fragments;* sandy lime *oolite*.

4. In chemical sediments (e. g., carbonate rocks, evaporites, chert), the rock names are based on chemical composition. Other properties which can be presented are grain size and particle type.

5. Clastic sediments, which are composed mainly of quartz and silicates, are subdivided according to grain size (Wentworth scale). The silt-clay limit has been modified according to Doeglas (1968), in order to achieve equal distribution of sand, silt, and clay on a logarithmic scale (fig. 2-2). The Phi grades ($-\log_2$ of grain diameter) are useful mainly for statistical treatment of grain size data. Rather than to give names, e. g. "medium sandstone" without referring to the nomenclature applied, the mean size (measured or estimated) in millimeters should be indicated, e. g. "sandstone (0.2)".

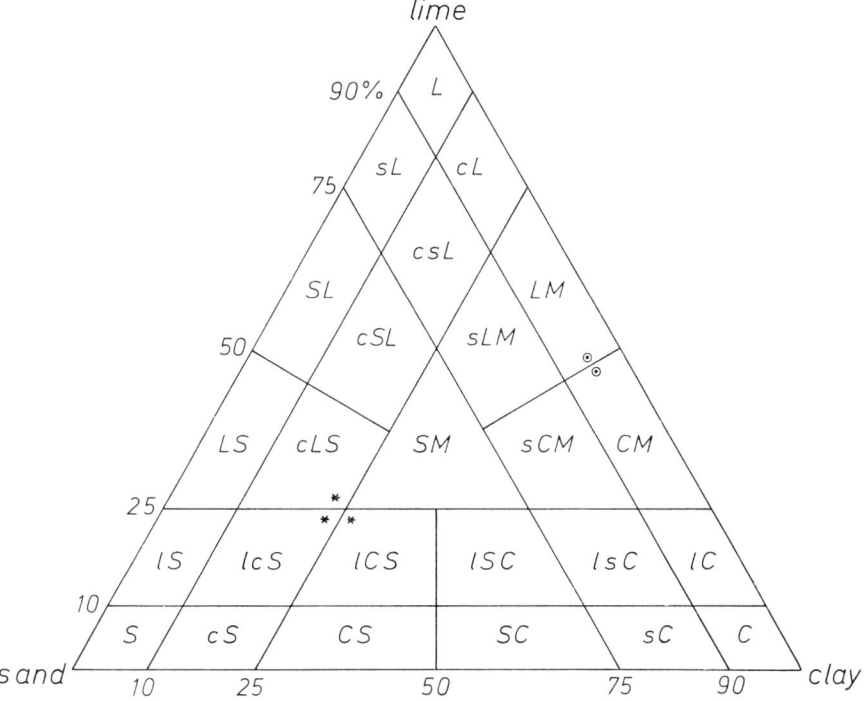

Fig. 2-1. Triangular diagram sand-clay-carbonate (from Füchtbauer 1959 and G. Müller, pers. comm.).

6. Three kinds of grain size names are used:

    1. Lutite (from Lat. lutum = mud)
       Arenite (from Lat. arena = sand)
       Rudite (from Lat. rudus = pebble)
    2. Clay-Silt-Sand-Pebble
    3. Pelite (from Greek pelos = clay)
       Psammite (from Greek psammos = sand)
       Psephite (from Greek psephos = pebble)

The first group is used mainly for carbonate rocks (calcilutite, calcarenite, calcirudite). — "Clay", a grain size term, is used for silicate minerals only. The same is suggested for "Silt" and "Sand". — Pelite and psammite could be useful as terms independent of composition, e. g., for marlstones or calcarenitic sandstones. (Pelite would include silt and clay).

7. These definitions raise difficulties: The term "Calcareous sandstone" is normally used for sediments composed of quartz or silicate grains cemented by calcite, because sandstones, according to paragraph 5, are rocks composed of quartz and silicate grains. But what about unconsolidated sediments, composed of $CaCO_3$-grains

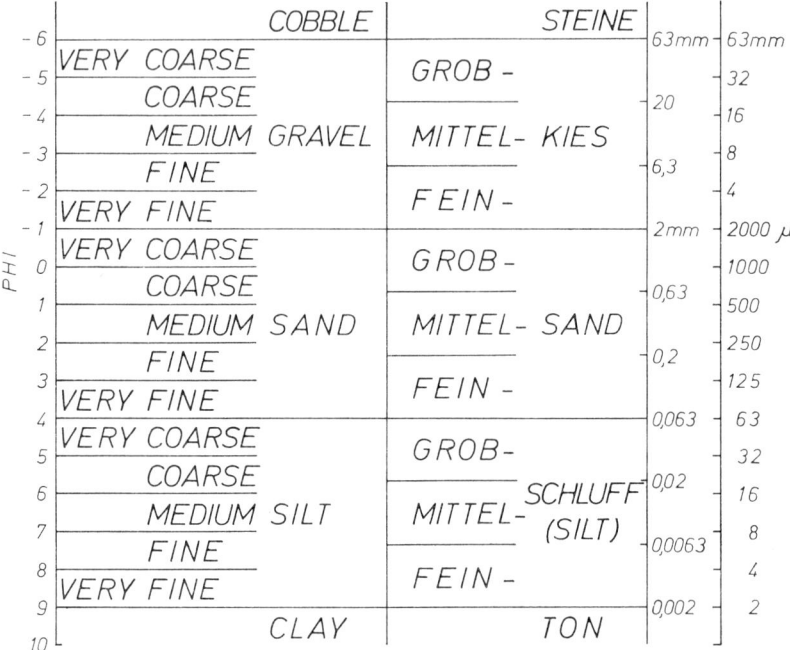

Fig. 2-2. Grain size nomenclature. At left the WENTWORTH scale with modifications by DOEGLAS (1968), at right the DIN 4022 scale used in Germany.

only? "Calcareous sand" is used frequently. This, however, is in contrast to the above definition of sandstone ... Sometimes the difficulty can be avoided by naming the rock by its particles (e. g. oolite).

8. Matrix is defined in grain size histograms by distinctly finer grains than the coarser grains composing the framework. The matrix of sandstones is mostly composed of silt and clay; the matrix of conglomerates can be composed of sand and silt. However, matrix is often defined in a more general way as the grain size fraction smaller than 40 or 20 microns in sediments of sand size or coarser. Matrix that formed by disintegration of unstable grains is discussed in ch. 3.214 and 5.13.

9. Cement designates chemical fillings in the pore space which occur after sedimentation; these therefore are diagenetic.

3. Sandstones, Conglomerates and Breccias[1]

This chapter deals mainly with sands and sandstones, that is, sediments composed of quartz and silicate grains ranging from 0.063 to 2 mm in size. Several sub-chapters and separate paragraphs, however, also discuss conglomerates and breccias, rocks with grain sizes larger than 2 mm.

The primary components and the fabric of rocks and unconsolidated sediments are treated in the first two sub-chapters. The third sub-chapter deals with diagenesis, the post-depositional alteration of sands, gravel and angular debris and their subsequent consolidation into sandstones, conglomerates and breccias. Alterations during erosion are called weathering and are briefly discussed in ch. 1.2 and 3.116.31. Alterations during and after transport and before burial are called aquatolysis (in freshwater) and halmyrolysis (in haline waters). They are discussed in the chapters dealing with silt and clay (pt. II/2); such alterations are minor in sandstones and conglomerates and can be neglected, for practical purposes.

3.1 Primary components

3.11 Sandstones

3.111 Composition and nomenclature

Many classifications of sandstones have been proposed, and have been summarized by HUCKENHOLZ (1963 a, b), KLEIN (1963) and others. The large number of classifications in part is the result of various definable parameters present within sandstones:

A. Chemical constituents

Chemical composition has not been used as a basis for classification of sediments due to its difficulty of predicting mineralogy. A summary of chemical analyses of sandstones is shown in table 3-1.

B. Mineral phases

A classification based on mineral phases such as quartz, feldspar, phyllosilicates, carbonates is unsuitable because of the length of time required for identification of

---

[1] I am indebted to H. J. DRONG, D. MARSAL, W. PLESSMANN, and H. E. REINECK for valuable suggestions and insights. J. D. MILLIMAN gave much valuable advice during the preparation of the English edition.
For reading and correcting ch. 1—3 and 6, grateful acknowledgement is made to F. B. VAN HOUTEN. For further corrections I am indebted to S. M. CASSHYAP and H.-U. SCHMINCKE.

the mineral components of rock fragments and fine grained material by microscopic and X-ray diffraction methods.

C. Component grains

1. grouped according to "maturity". PETTIJOHN (1957) defined two types of maturity:

a) compositional maturity $= \dfrac{\text{quartz} + \text{chert}}{\text{feldspars} + \text{rock fragments}}$

b) textural maturity $=$ sorting and roundness of sand grains

According to this definition, the "compositional maturity" of a sandstone is inversely proportional to the frequency of those components that are sensitive to transport, abrasion and weathering. Compositional maturity therefore increases with each successive sedimentary cycle. The higher the "textural maturity" (FOLK 1951) the better the sorting and roundness of the grains. Roundness increases during each sedimentary cycle provided diagenesis does not interfere too severely; under certain conditions sorting also improves. However, sorting depends on different factors including (1) the sorting of the source material, (2) the hydrodynamic regime, (3) the length of transport, and (4) the rate of sedimentation (STRAKHOV 1969).

Compositional and textural maturity therefore can reflect the number of times the grains have been "recycled".

The compositional maturity is derived from a descriptive compilation of the components. The textural maturity reflects the fabric properties and therefore should not be used in a petrographic nomenclature of sandstones; however, it can be considered separately (FOLK 1954).

2. grouped according to the parent rocks (grains of volcanic, plutonic, metamorphic and sedimentary origin). As a useful modification, PETTIJOHN (1957) defined a

$$\text{Source index} = \dfrac{\text{feldspar}}{\text{rock fragments}}$$

A classification by parent rocks is rarely possible if it is based on single mineral grains. Even with rock fragments it is often difficult to distinguish whether they originate directly from crystalline rocks or from older sediments ("recycled"). Therefore such a grouping although important in principle is unsuitable for sandstone classification.

3. descriptive (quartz, feldspar, rock fragments, phyllosilicates $> 63\,\mu$, silt, clay, cements).

The descriptive compilation of the sand particles is the best basis for sandstone nomenclature.

Before introducing quantitatively defined rock terms, two well-used but often mis-used terms of special sandstone types will be discussed: "graywacke", a term used by the miners of the Harz Mountains in the 18th century, and "arkose", a name introduced by BROGNIART (1826) (HUCKENHOLZ 1963b).

As the original definitions of these terms were not quantitative, their use in any exact classification is restricted. Though typical graywackes are rich in rock fragments and typical arkoses are rich in feldspars, these observations cannot be used as

3.11 Sandstones

Table 3-1. Chemical analyses of typical sandstones (from PETTIJOHN 1963).

	sandstone quartzose A	subgray-wacke B	gray-wacke C	arkose D	average sandstone E
SiO_2	98.91	56.80	68.85	79.30	77.6
Al_2O_3	0.62	8.48	12.05	9.94	7.1
Fe_2O_3	0.09	1.67	2.72	1.00	1.7
FeO			2.03	0.72	1.5
MgO	} 0.02	1.24	2.96	0.56	1.2
CaO		15.25	0.50	0.38	3.1
Na_2O	0.01	1.31	4.87	2.21	1.2
K_2O	0.02	1.46	1.81	4.32	1.3
H_2O+		0.50	2.30	0.55	1.7
H_2O-			0.77	0.41	0.4
TiO_2	0.05	0.10	0.74	0.22	0.4
P_2O_5		Sp.	0.06	0.05	0.1
MnO			0.05	0.02	0.1
CO_2		12.95	0.08		2.5
S; SO_3			0.08		0.1
Cl; F					Sp.
BaO; SrO			Sp.		Sp.
ignition loss	0.27		0.07		Sp.
sum	99.99	99.76	99.94	99.68	100.0

A St. Peter Sandstone (Ordovician), Mennota, Minn.
B Aquitan-Molasse, Lausanne, Switzerland (calcareous)
C Tanner graywacke (Devonian/Carboniferous) Harz Mts., Germany
D Jotnian Sandstone (Precambrian), Köyliö, Finland (44% feldspar)
E 34% of average quartzose sandstone + 25% of average subgraywacke + 26% of average graywacke + 15% of average arkose

universal criteria. HUCKENHOLZ (1963a) showed that the graywackes from the type localities in the Harz Mountains and the arkoses from the type localities in the Auvergne differed significantly only in the composition of their phyllosilicates: mica and chlorite are typical in graywackes, and kaolinite is typical in arkoses. A possible explanation is that the source areas of graywackes are relatively extensive, therefore contributing the common phyllosilicates mica and chlorite. Arkoses, on the other hand, develop from disintegration of granitic rocks or other plutonites, and are typical only as long as the mechanical and chemical weathering products (feldspar and kaolinite, respectively) are neither separated from each other nor mixed with other detritus. Therefore, arkoses mostly are deposited relatively close to their parent rocks, in many cases in continental environments. Kaolinite, however, may be replaced by chlorite during diagenesis, and feldspar may change into mica, so that diagenesis may blur the limits between graywackes and arkoses. On the other hand, the use of phyllosilicates and clay minerals as criteria for sandstone classification is meaningless by itself. The names graywacke and arkose however, should be maintained as field terms without quantitative implications:

Graywackes are dark-(green-) gray sandstones, mostly rich in rock fragments, with variable feldspar content, and with a matrix consisting of illite and chlorite or of montmorillonites and zeolites in younger rocks. They are very poor in sorting and roundness, and generally are strongly lithified. The matrix frequently is derived from unstable rock fragments during diagenesis (p. 62). Varieties of graywacke are lithic graywackes (rich in rock fragments) and feldspathic graywackes, pointing to a variety of source rocks including plutonites, volcanites and sediments. A condition for production of graywackes is lack in the sorting processes such as reworking immediately after deposition, typical of high-energy environments. Steadily sinking mobile belts with deep-water turbidite sequences are a dominant realm of graywackes.

Arkoses are light-gray to reddish sandstones, rich in feldspars, frequently with kaolinite and varying content of rock fragments. In most cases they are poorly sorted also. Thick accumulations proximal to granitic source areas are typical, thus explaining the lack in separation of feldspars and their alteration products (kaolinite). Conditions for production of arkoses are either high relief with consequent rapid erosion or glacial or arid climate lowering the rate of weathering for feldspars.

3.11 Sandstones

Matrix-poor "subgraywackes" (or "lithic arenites") and "subarkoses" (or "feldspathic sandstones") are transitional to "quartz arenites" (or "quartzose sandstones" or formerly "orthoquartzites") which contain more than 90% quartz grains, chert and quartzitic rock fragments not included. Quartzose sandstones frequently were formed during marine transgressions on stable platforms. Sorting and maturity improved by reworking, before final deposition occurred. For thin section photographs of these types see fig. 3-1.

"Hybrid sandstones" include "greensands", "phosphatic sandstones", "calcarenaceous sandstones" (calcarenites and biocalcarenites of this book), and "tuffaceous sandstones". PETTIJOHN et al. (1972) collected and described many examples of these varieties in detail with excellent photographs.

Nearly half of all sands belong in the category of lithic sandstones and graywackes, whereas the rest are feldspathic sandstones and arkoses (15 to 16%) and quartzose sandstones (34%; PETTIJOHN et al. 1972).

Though it is certainly appropriate for a genetical understanding of sandstones to discuss in detail these distinct sandstone types, a strictly descriptive system shall be used as a basis of sandstone classification in this book.

◀————————————————————————

Fig. 3-1

A. Sandstone rich in dolomite grains ("hybrid sandstone"). Upper Marine Molasse, Miocene, Hörnli/St. Gallen, Switzerland (no. 2.47). Md 0.18 mm. 26% quartz, 5% feldspar, 10.5% rock fragments + chert, 5% mica + chlorite, 32% dolomite grains, 4.5% calcite grains, 5% clay + silt matrix, 12% calcite cement. Chert grain in the center, dolomitic rock fragment left from center (width of photograph 0.32 mm; + nicols).

B. Micaceous sandstone with rock fragments ("subgraywacke"). Upper Carboniferous "molasse", Itterbeck-Halle Z 5, 2376 m (Emsland area). Md 0.15 mm, porosity 10.5%. 49% quartz, 9% feldspar, 11.5% rock fragments, 12% mica, 10% clay + opaque matrix, 8.5% kaolinite. Right side: several muscovite crystals with pressure solution of adjacent quartz grains, at right border and left from center, kaolinite neoformation (width of photograph 1.3 mm; + nicols).

C. Feldspathic sandstone rich in anhydrite cement ("subarkose"). Lower Triassic "Buntsandstein", Linsburg Z 1 (Weser) (no. 3-15). Md 0.2 mm. 42.5% quartz, 14% albite, 9% potash feldspar, 1.5% rock fragments, 1% clay + silt matrix, 32% anhydrite cement and replacements. Right from center (dark gray): plagioclase with albite overgrowth, upper left: potassium feldspar (dark gray, large, round) with albite overgrowth (white and gray), left border: anhydrite cement (width of photograph 0.32 mm; + nicols).

D. Conglomeratic sandstone rich in rock fragments ("graywacke"). Upper Devonian/Lower Carboniferous, Harz Mts., Germany (no. 23-48). Md of the sand portion 0.3 mm; pebbles up to 5 mm; very poor sorting. 25.5% quartz, 8.5% feldspar (pla > or), 27% rock fragments, 6% mica + chlorite, 33% clay and silt matrix. Center: plagioclase with albite twins, upper left: several rock fragments (width of photograph 1.3 mm; + nicols).

E. Sandstone rich in feldspars, with laumontite cement ("arkose"). Cuchara fm.; Eocene, 1/2 mile N of La Veta (Color.), contact-metamorphic. Md 0.5 mm. 36.5% quartz, 35.5% feldspar (pla > or), 4% rock fragments, 1% pyroxene?, 0.5% calcite, 22.5% laumontite cement. Many potassium feldspars and plagioclases are visible, laumontite (dark gray) is not well visible (width of photograph 1.3 mm; + nicols; cf. VINE & TOURTELOT 1973).

F. Rock fragment-sandstone ("graywacke"). Chinle fm. (Upper Triassic, Cameron, Ariz.). Md 0.5 mm. 16% quartz, 8.5% feldspar (pla > or), 50% rock fragments (mostly montmorillonitic), 1.5% mica, 24% clay mineral cement. Angular quartz grains (light), rock fragments (gray), desiccation cracks of the thin section (width of photograph 1.3 mm). 200 grains were pointcounted, in each of the above thin sections.

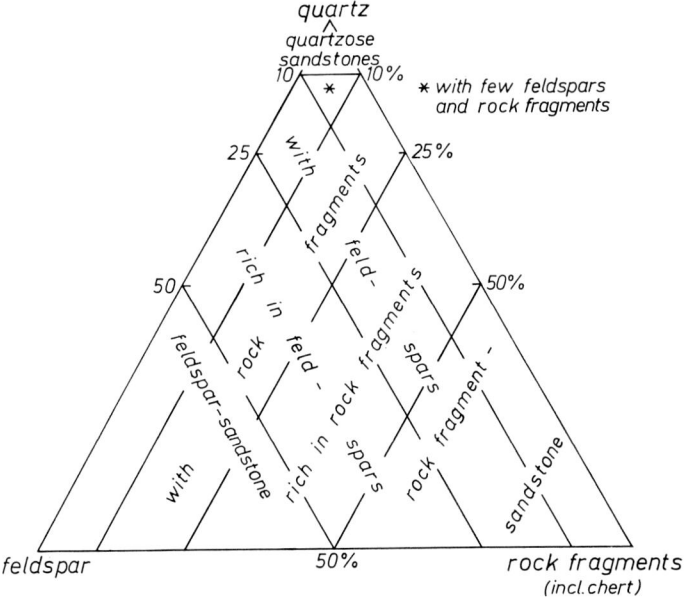

Fig. 3-2a. Diagram of sandstone nomenclature. Example: 55% quartz, 30% feldspars, 15% rock fragments: "sandstone, rich in feldspar, with rock fragments".

The quantitative classification of sandstones is based on the 10—25—50% system (fig. 3-2a) as follows:

More than 50% feldspar: feldspar-sandstone

More than 50% rock fragments: rock fragment-sandstone

25—50% feldspar: feldspar-rich sandstone (or sandstone rich in feldspars)

25—50% rock fragments: rock fragment-rich sandstone

10—25% feldspar: feldspar-bearing sandstone (or sandstone with feldspars)

10—25% rock fragments: rock fragment-bearing sandstone

Less than 10% feldspars or rock fragments: feldspar and rock fragment content can be dropped in the term, or may be indicated as "sandstone with few feldspars or rock fragments".

Less than 10% feldspars + rock fragments
(i.e. more than 90% quartz): quartzose sandstone

The triangular diagram of fig. 3-2a is not as symmetrical as the triangle sand-clay-carbonate (fig. 2-1), because, contrary to the latter, the frequency of the components is highly unequal in sandstones, quartz usually predominating. It would not make much sense, therefore, to call a rock with more than 50% quartz a "quartzose sandstone"; 90% is a more realistic limit. Sandstones from areas poor in quartz, such as the sandstones in New Zealand indicated in the lower right part of fig. 3-2b with 10—25% quartz, can be named "rock fragment-sandstone" with quartz grains.

Other constituents such as mica, chlorite, glass, and heavy minerals, as also clay and silt, can be included in this nomenclature in a similar manner.

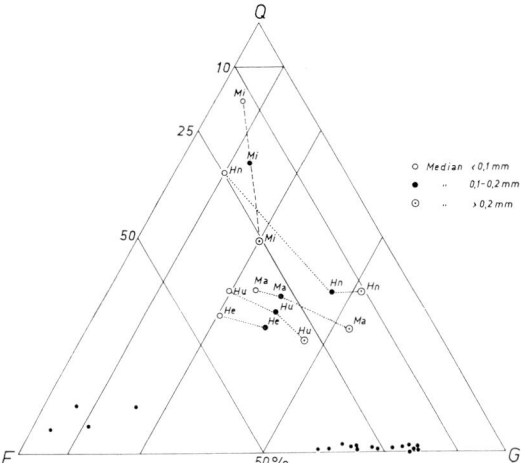

Fig. 3-2b. Triangular diagram quartz-feldspar-rock fragments (cf. fig. 3-2a). Upper half shows graywackes. The mean compositions for coarse, medium and fine sandstones, which are shown separately and are connected by lines, demonstrate the dependence of rock names upon grain size. He = HELMBOLD (1952), Hu = HUCKENHOLZ (1959), both Devonian/Carboniferous in the Harz Mountains. Hn = HENNINGSEN (1961), Ma = MATTIAT (1960), both Lower Carboniferous, Hessen and Harz Mountains, respectively. Mi = MIZUTANI (1957), Permian, Mugi/Japan. — Lower half: Nearly quartz-free arkoses (left side, with An_{26-34}-plagioclase) and graywackes (right side, with An_{2-4}-plagioclase and volcanic rock fragments) from the Paleozoic of New Zealand, according to CROOK (1960) (from FÜCHTBAUER 1967a).

The limited value of a rigid sandstone nomenclature is demonstrated by the dependency on grain size shown in fig. 3-2b. Even small changes in grain size at the same locality may change the composition considerably, especially the rock fragment content.

Locally used rock terms include:

Ganister (England): Silicified fine-grained sandstone, grains not rounded (lower Upper Carboniferous).

Grit (England): Relatively coarse-grained sandstone, grains not rounded, well lithified, with silica or carbonate cementation (MILNER 1962, CUMMINS 1959).

Itacolumite (Brazil): Elastic sandstone (DAPPLES 1967a), due to strong indentation and subsequent partial dissolution of the grains.

Sparagmite (Norway): Sandstone with many multi-colored feldspars (arkose) in the Caledonian geosyncline, especially near the Precambrian/Cambrian boundary.

3.112 Quartz

Quartz is the main mineral component of sandstones. Different characteristics of quartz grains may give clues as to the source rocks (KRYNINE 1946a). Such characteristics are:

        a) inclusions
        b) mode of extinction
        c) color

d) trace elements
e) shape
f) roundness   } cf. ch. 3.22
g) surface texture

a) **Inclusions** may be divided in the following groups
1. liquid and gas inclusions
2. mineral inclusions

   isometric
   needles
   bladed

According to MACKIE (1896) quartz from igneous rocks has more needle-shaped inclusions, whereas quartz from mica schists shows more isometric-euhedral mineral inclusions (PETTIJOHN 1957). This generalization, however, is not always applicable. Quartz with chlorite inclusions often originate from vein fillings, whereas quartz grains with rounded secondary enlargements which are separated from the detritic core by clay seams originate from older sandstones. Inclusions of hydrocarbons indicate a very special environment (MURRAY 1957). Generally, there is only a possibility of a direct comparison of quartz grains listed according to their inclusion types with the source rocks in question.

b) Many quartz grains show **undulose extinction**. Areas with more or less uniform extinction meet in blurred planes which run nearly parallel to the c-axes (4—5°, BLATT 1967). According to CARTER et al. (1964) clear deformation bands ∥ c and sharp deformation lamellae ⊥ c can be distinguished.

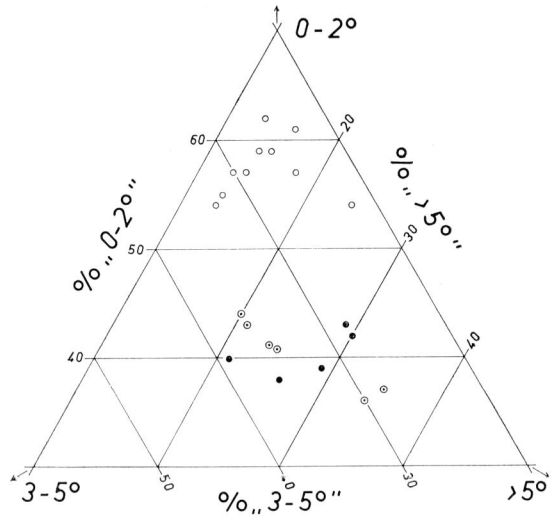

Fig. 3-3a. Undulose extinction of quartz in medium sandstones (Md about 0.25 mm) of the Buntsandstein (circles), Lower Permian (circles and dots) and Upper Carboniferous (dots) from boreholes in Northwestern Germany. The range of extinction is indicated in degrees.

Undulose extinction is found in quartz grains from rocks which were deformed after crystallization. Therefore, it is frequent in many, but not in all, metamorphic rocks. Research work by BLATT & CHRISTIE (1963) showed that metamorphic and plutonic rocks generally cannot be distinguished by their contents of undulose quartz. Such grains are found even in granites in folded areas or near larger thrusts. They are rare in volcanic rocks, which, however, do not contribute significantly to the sedimentary quartz balance. Quartz grains of volcanic origin commonly have embayed outlines, the stout prismatic and bi-pyramid habit of high-temperature quartz, or are associated with volcanic rock fragments (REIMER 1971).

In graywackes and arkoses, the contents of undulose quartz grains are similar to the supposed crystalline source rocks. Very few undulose grains, however, are found in frequently recycled quartzose sandstones (BLATT 1963). This means, that undulose quartz grains are more strongly affected by mechanical and chemical degradation than are grains with uniform extinction.

Undulose quartz can be used to characterize sandstones (GREENSMITH 1963, FÜCHTBAUER 1964a). In order to gain reproducible results, the range of extinction (in degrees) is measured for each grain and shown in histograms or combined into groups and plotted in triangular projections. For example, fig. 3-3a shows the undulosity of 21 samples from different formations within a small area. There is quite a difference between the quartz grains of the Buntsandstein and of the Lower Permian + Upper Carboniferous. Measurements of the undulosity on the universal stage would be more meaningful, but such measurements are too time-consuming for most cases.

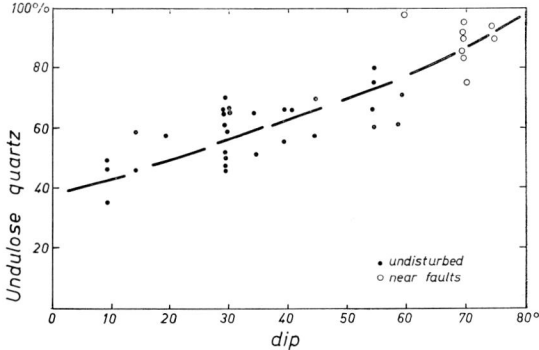

Fig. 3-3b. Relation between percentage of undulose quartz grains and dip, in sandstones of the Upper Devonian of New South Wales (from CONOLLY 1965).

Undulosity may increase considerably if the sedimentary rocks are subjected to folding (fig. 3-3b). — The lithostatic pressure of an overburden of 2000 m can produce wavy extinction of single grains. Generally, however, lithostatic pressure alone has only a minor influence on the percentage of undulose quartz grains, as revealed by the samples shown on fig. 3-3a: Though their maximum subsidence depth varies considerably (2500—6000 m), undulosity does not correlate with depth.

c) The color determination of the quartz grains is a new method introduced by SCHNITZER (1957, 1963).

The quartz grains are cleaned by boiling with concentrated hydrochloric acid and subsequently separated by heavy liquids. They are then covered by a liquid of the same refractive index as quartz to enable comparison with the color chart independently of grain size. KRÄMER (1961) refined this method by color notation of the individual grains. The most frequent colors are the following:

Red   (by hematite or biotite inclusions or by the uptake of metallic ions)
Brown   (by goethite inclusions)
Yellow (by inclusions of iron hydroxids or by the uptake of metallic ions)
White   (by inclusions of liquids or gas)
Gray   (by inclusions of mica, carbonaceous material, ores, graphite or "color centers" caused by the irradiation of AlO_4 tetrahedra within the quartz lattice)
Green   (by chlorite, biotite or amphibole inclusions)

Less frequent are the following colors:
Blue   (inclusions of rutile)
Violet (by "color-centers")
Pink   (by rutile, hematite or Mn-compounds)

The color of quartz grains has been successfully used for stratigraphic mapping in the Buntsandstein and Upper Triassic (PATZELT 1964). This technique, however, may be biased by dust rims of variable color which are included between grains and overgrowths, and which cannot be removed by acids.

d) Trace elements in quartz may serve as indicators of provenance (DENNEN 1967). According to SIPPEL (1968) luminescence is stronger in quartz grains from crystalline rocks than in diagenetic quartz. This can help in differentiating detrital quartz grains from diagenetic overgrowths.

e) Shape, roundness and surface texture are discussed in subsequent sections (in ch. 3.22 and 3.23).

3.113 Rock fragments

Rock fragments are defined as sand grains consisting of more than 2 distinct crystals of the same mineral species (twinning not included) or at least two crystals of different mineral species exclusive of inclusions. The probability of a polycrystalline grain increases with increasing grain size. Therefore, rock fragments are more frequent in coarse sandstones than in finer ones (fig. 3-2 b).

HELMBOLD (1952), HUCKENHOLZ (1959), MATTIAT (1960) and HENNINGSEN (1961) in describing Devonian and Carboniferous graywackes from the Variscian mountains distinguished the following groups of rock fragments:

1. Igneous Rocks

a) plutonic rocks

grains consisting of large isometric quartz, feldspar and phyllosilicate crystals. Such rock fragments are classified by feldspar composition. Needle-like apatite-inclusions are frequent in quartz and feldspar. A special type is the eutectic intergrowth of quartz and feldspar.

b) volcanic rocks

lamellar, acidic to intermediate plagioclases with flow textures are a main component. The interstices are filled with chlorite or mica which sometimes are replacements of smaller feldspars. Ore particles and carbonates are frequent.

2. Metamorphic Rocks

a) gneiss

spindle-shaped quartz and feldspar crystals surrounded by layers rich in mica and chlorite containing elongate quartz crystals.

b) mica schists, chlorite schists, and phyllites

flat rock fragments consisting mainly of mica and chlorite with lense-like elongate quartz grains.

c) metaquartzites and quartzite schists

predominantly intensively indented quartz crystals, frequently with undulose extinction, partly with a lamellar appearance ("quarzite schists"), sometimes with chlorite or mica, and with feldspar in varying amounts.

The distinction between 2 a—c is not sharp. Moreover, the grains sometimes are too small for an exact identification of the source rock. There are also indistinct transitions between 2c and 3a.

3. Sedimentary Rocks

a) sandstones and quartzites

The sand grains in these rock fragments are more or less isometric. They are connected with each other by clay matrix or cement (quartz, carbonate and others), or they are slightly interlocking due to pressure solution. Different types can be distinguished on the basis of the feldspar, rock fragment and mica content, as well as by the types of grain boundaries.

b) claystones and slates

Small chips of redeposited claystone (shale) can be found in the sand fraction. Often they cannot be distinguished from the matrix (examples have been published by BOSELLINI 1967a). Slate fragments may be distinguished by shear-planes sometimes covered with recrystallized mica.

c) chert, flint, radiolarite et al.

micro- to cryptocrystalline chalcedony and quartz. Often dark-pigmented grains, sometimes with inclusions of radiolaria, often with small crevices filled with quartz; generally strongly undulose.

d) limestones occur sporadically.

In many instances rock fragments can define petrographic correlations. In Northern Germany, the sandstones of the Lower Permian differ from those of the Carboniferous by the presence of dusty quartz aggregates which can be traced to lower Permian porphyries (DRONG 1959). In the Alpine Molasse, differences in the percentage of chert fragments are typical (FÜCHTBAUER 1964a). Ambiguous rock fragments are frequent, for example devitrified glass from older eruptive rocks can be mistaken for chert or fine-crystalline metamorphites.

One has to be very careful in deducing the rock distribution in a source area from the rock fragment assemblages in a sandstone, because resistant rock fragments will be enriched during transport and even during diagenesis.

Sands and sandstones in the circumpacific volcanic belt area are rich in feldspars and unstable rock fragments derived from extrusive basic igneous rocks (DUNCAN & KULM 1970). Such rock fragments can be altered chemically during diagenesis and will be squeezed between more resistant grains thus forming a secondary matrix (see ch. 3.214).

3.114 Feldspars

The feldspar content is probably the most important primary property of a sandstone. It generally diminishes during diagenesis by kaolinitization or sericitization. The best way to identify feldspars in thin section is by staining (see part I). In smear slides feldspars can be identified by embedding in oil with refractive indices of 1.527 and 1.540, in order to count the percentages of potassium feldspars, sodium-rich plagioclases, and quartz. Isolation of the feldspars by flotation (VAN DER PLAS 1966) and separation of the different plagioclases by heavy liquids facilitate further differentiation, e. g. by twin types which are indicative of the provenance (PITTMAN 1970).

Granites contain about 50% feldspar, whereas the mean in sandstones is about 10% (PETTIJOHN 1957). Assuming that the primary source of the sandstone material is granite and granite gneiss, these figures point to a lower stability of feldspars compared with quartz. The decrease in feldspar content which is due to decomposition and abrasion, therefore can serve as a useful measure of the maturity of a sandstone.

Because of distinct cleavage planes, feldspar grains disintegrate more easily than quartz grains during transport. Thus they contribute to the finer grain fractions. As a consequence, coarser sand fractions generally are impoverished in feldspar. Longshore transport is more effective in reducing the feldspar content than fluvial transport (RUSSELL 1937, RITTENHOUSE 1943).

In the arkose of Vic-le-Comte the feldspars are larger than the quartz grains (HUCKENHOLZ 1963 a), presumably because the grains were transported only for a short distance (see also p. 16).

Chemical weathering generally is more important than abrasion in reducing feldspar content. Detritus rich in feldspar originates predominantly from crystalline areas with steep relief in which abrasion predominates over chemical weathering (KRYNINE 1935). Moreover, because chemical decomposition is low in arid areas and in extremely cold climates, detritus rich in feldspars is common in these areas.

Among the feldspars, potassium feldspar and albite are the most resistant to weathering (GOLDICH 1938). Similarly, these two feldspars are diagenetically more stable than Ca-plagioclases. For both reasons, they are the most frequent feldspars in sandstones. Their higher diagenetic stability is underlined by the fact that they are the only feldspars formed authigenically.

In the Devonian and Carboniferous graywackes of Northern Germany the mean composition of plagioclase grains is about An_{3-10}. If the feldspars enclosed in rock fragments are included, the mean is about An_{10} (HELMBOLD 1952, HUCKENHOLZ 1959, MATTIAT 1960). In the Alpine Molasse, albite is by far the most frequent feldspar, especially in the finer sizes (FÜCHTBAUER 1964a). In the Permian graywackes of Japan, however, andesine (An_{30-34}) predominates (MIZUTANI 1959).

Table 3-2. Plagioclase and orthoclase percentages of various sandstones (free feldspar grains; feldspars in rock fragments were not considered).

	plagioclase	orthoclase	Author
88 graywackes, Harz (Devon/Carbonif.)	31 %	3 %	HUCKENHOLZ 1963 b
21 graywackes, Mugi (Permian)	1.4%	12.4%	MIZUTANI 1957
1 arkose, Saint-Sauve (Permian)	45.4%	0.7%	HUCKENHOLZ 1963 b
2 arkoses, Royat and Vic-le-comte (Oligocene)	0.2%	18.7%	HUCKENHOLZ 1963 b
17 subarkoses, S. Germany (Buntsandstein, Triassic)	7.6%	14.0%	VALETON 1953

The chemical analyses of PETTIJOHN (1963) show that plagioclase predominates mostly in graywackes, whereas orthoclase predominates in arkoses (tab. 3-1, also tab. 3-2). The high Na_2O/K_2O ratio in graywackes compared with the high K_2O/Na_2O ratio in contemporaneous shales and slates led PETTIJOHN et al. (1972) to the conclusion of a diagenetic exchange of K and Na caused by an albitization of K-feldspars in the graywackes. Textural evidence (albitized perthites) has been given by MIDDLETON (1972). K-feldspar in general becomes unstable near the diagenesis-metamorphism transition. At the same time, illite regenerates into muscovite by K-uptake. Potassium feldspar normally exceeds plagioclase in acid plutonic rocks, whereas they are subequal or plagioclase is more frequent in volcanic or less acid plutonic rocks (FOLK 1965 c). Potassium feldspars generally predominate in the coarser grain fraction, whereas plagioclases predominate in finer sizes (ZIMMERLE 1963, FÜCHTBAUER 1967 b). Perthitic feldspar is indicative of slow cooling and hence characteristic of the plutonic sources.

Zoning of feldspars can indicate the source rock: PITTMAN (1963) studied 45 volcanic, 30 plutonic and 15 metamorphic rocks, mostly from North America. He showed that an "oscillatory" zoning, consisting of thin bands with alternating extinction, occurs almost exclusively in volcanic plagioclases; an average of 30.6% of the volcanic plagioclases showed this zoning. Plagioclases with normal zoning were found in all igneous rocks. Plagioclases in metamorphic rocks, however, showed hardly any zoning (see also BOTTINGA et al. 1966). By statistical treatment of twinning MIZUTANI (1959) speculated on the origin of greywackes from certain metamorphic source rocks: According to TURNER (1951) and Ross (1957), complex combinations of twinning (such as albite-Karlsbad A) are missing in metamorphites.

Plagioclases in the approximative range between An_5 and An_{25} are absent in low-grade metamorphic rocks (below the amphibolite stage). In high grade metamorphic rocks and in granites they contain "peristerite" unmixing structures. These feldspars thus are indicative of the thermal history. Provenance studies based on plagioclase extinction have been published by ERMANOVIES (1967).

3.115 Phyllosilicates

The content of mica and chlorite increases with decreasing grain size, as a general rule. Most of the claystones contain more than 50% illite, which originates from mica by the exchange of H_3O for potassium. In sandstones, the presence or absence of mica and chlorite can be an interesting parameter. For example, the mica content distinguishes the Upper Carboniferous from the Lower Permian sandstones of Northern Germany.

Primary green biotites are typical of the epizone and in part of the (para-) amphibolites of the mesozone of metamorphism, whereas brown biotites prevail in the meso- and katazone as well as in igneous and certain volcanic rocks (TRÖGER 1967, 525).

The optical character of mica changes with increasing decomposition. These effects are stronger in biotite than in muscovite. In surface samples of the Buntsandstein, RIMŠAITE (1957) found the following decrease of refractive indices:

	fresh	slightly weathered	weathered	strongly weathered
Biotite	1.63 —1.65	1.62—1.63	1.59—1.61	1.575—1.59
Muscovite	1.584—1.590	1.57—1.58	—	1.535—1.545

This feature may occasionally permit important deductions as to the manner of erosion, but the possibility of diagenetic decomposition also must be taken into consideration.

Due to their relative instability biotites are missing in most second-cycle (recycled) sandstones. In the Alpine Molasse, reddish-brown biotite is indicative of brackish and marine sediments. It is missing in fluvial beds of the same age. Apparently such biotites could not tolerate the oxidation during alternating drying and moistening of the continental environment. In surface samples this same "bleaching" has been observed (FÜCHTBAUER 1963).

Reddish brown oxychlorites from greenschist rocks may be mistaken for biotites due to their considerable birefringence (0.02—0.024; n_y = 1.645—1.650). X-ray data, however, indicate chlorite [d(0015) = 14.15 Å, d(060) = 1.549 Å] (CHATTERJEE 1966). Stilpnomelane, a mineral forming in the very low stage of metamorphism, can also be mistaken for biotite, though the cleavage is less perfect and the color of x is always yellow in stilpnomelane; z is dark green to dark olive brown (TRÖGER 1967, 555).

A primary clay matrix of a sandstone mostly corresponds to the composition of surrounding claystones. In cases where a primary clay matrix is scarce or absent, diagenetic minerals, especially kaolinite, chlorite and mica, may differ appreciably from the shale composition. Less frequent are corrensite, vermiculite, mixed layer and montmorillonite. All these minerals also have been observed in bore holes, although some of them may originate within weathering zones as well.

Early diagenetic chlorite ooids are frequent in several limestones and calcareous sandstones. They were deposited while still soft, as shown by their mutual deformation.

3.116 Heavy minerals

3.116.1 Distinguishing characteristics and identification key

The following descriptions of 33 heavy minerals may serve as a first orientation for the beginner. Identification is only possible by using optical tables (e. g. TRÖGER

1971). The number of common heavy minerals (as indicated by the notations) is surprisingly few, compared with the confusing number listed in the following textbooks and monographs: MILNER (1962): 147; PETTIJOHN (1957, p. 117): 45, 21 of them frequent; BAKER (1962): 44; HUTTON (1950): 52. On the other hand, rare minerals sometimes may be most important in differentiating a sandstone series. (If no color is indicated, the grains are predominantly colorless. The notation of the refractive index is related to an immersion of $n = 1.6$. The occurrences are mainly from v. ENGELHARDT 1973.)

A. Transparent heavy minerals (fig. 3-4)

Allanite (rare)
Mostly green or brown, pleochroic grains, markedly elongated, with oblique and high birefringence. Axial angle large. Radioactive. Occurrence: in pegmatites and occasionally in plutonites.

Amphibole (locally frequent)
Mostly green or brown, pleochroic grains, markedly elongated, with oblique extinction (less than 20°). Refractive index and birefringence are moderate. In sediments the brown basaltic amphiboles are rarer than green ones. Authigenic amphiboles are rare (ENLOWS & OLES 1966). Occurrence of amphiboles: Ordinary hornblende in granites, syenites, diorites and their volcanic equivalents, as well as in metamorphic rocks of the amphibolite and hornblende-hornfels facies. Basaltic hornblende in basalts, andesites, hornblende-gabbros and syenites.

Anatase (generally rare)
Colorless to yellowish-brown grains with dark rims due to high refractive index. Authigenic rectangular crystals are sometimes frequent. Anatase is not indicative of special parent rocks.

Andalusite (rather rare)
Pleochroism colorless to bright red. Refraction and birefringence relatively low. Occurrence: low-pressure metamorphic rocks of the greenschist facies, especially in contact metamorphism.

Anhydrite (locally frequent)
Cleavage faces at right angles mostly with straight extinction. Low refraction; medium birefringence positive. Axial angle $+ 2V = 42°$. In most cases diagenetic.

Apatite (very frequent)
Prismatic to oval, mostly small grains with low refraction and birefringence less than quartz; uniaxial negative. Grains with pseudopleochroism (caused by greyish inclusions) occur. Occurrence: ubiquitous in igneous and metamorphic rocks.

Barite (relatively frequent)
Irregular, slightly rounded grains. Optically similar to apatite, but biaxial positive, with slightly higher birefringence. Predominantly authigenic. Baritocelestite occurs forming slightly reddish sheaves.

Biotite (frequent)
Green or reddish plates. In smear slides, the high birefringence can be observed only at the edges and on kinks and fractures within the grains. Occurrence see p. 26.

Brookite (rare)
Dark, tabular, irregular grains with high refraction and birefringence. Usually with high dispersion of the optical axes showing red and blue colors at alternating extinction positions. This observation is frequently possible in smear slides, because

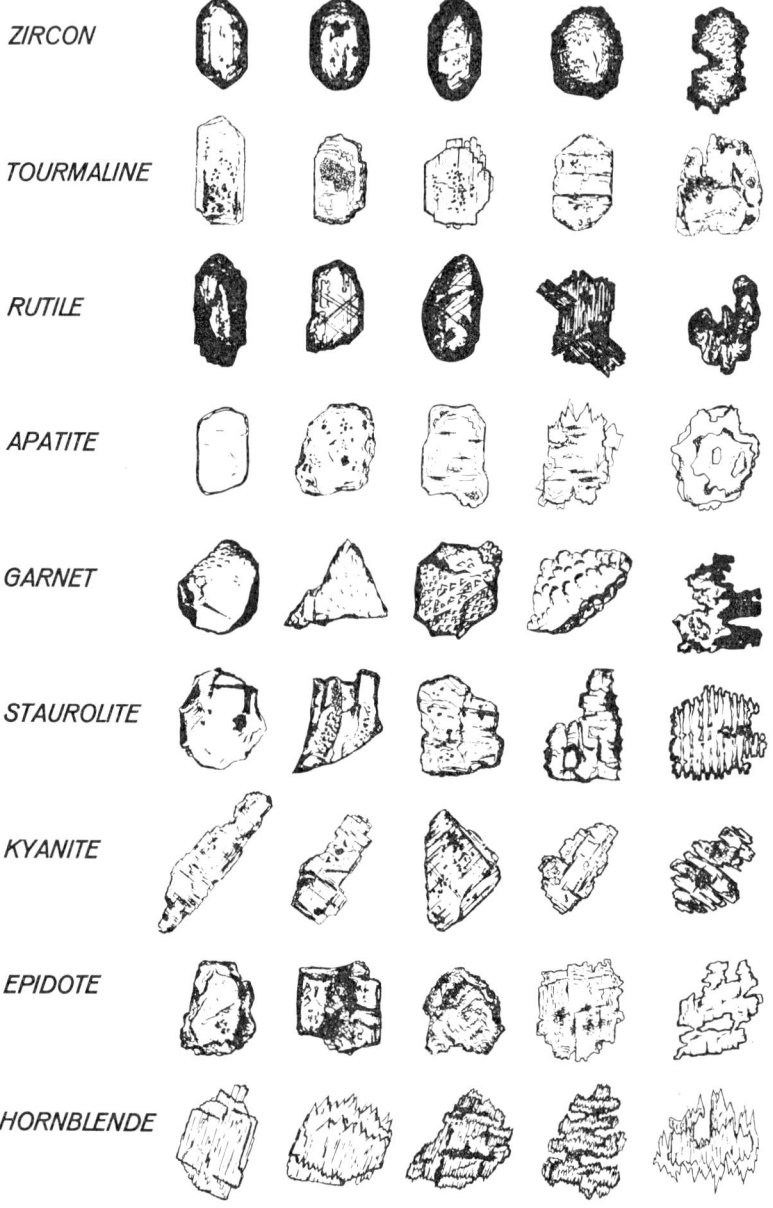

Fig. 3-4. Common heavy minerals. Five grains of each species with weathering increasing to the right. Fluvial Upper Miocene Molasse, Eastern Bavaria; selected drawings from GRIMM (1973). Less weathered samples = Vollschotter; strongly weathered samples = Quarzrestschotter and Decksande (see fig. 3-12).

the grains usually are tabular parallel to (100). Occasionally authigenic (ZIMMERLE 1963), not indicative of special parent rocks.

Chlorite (very frequent)
Bluish-green plates with very low, frequently anomalous birefringence. Occurrence: mostly metamorphic, indicative of the lower greenschist facies; also in hydrothermal veins.

Celestite (occasionally frequent)
Distinction from barite is difficult; refraction and birefringence are lower. Mostly authigenic (especially in evaporites).

Dolomite (frequent)
Dissolves only partially in hot acetic acid. Mostly diagenetic.

Epidote (quite frequent)
Yellowish-green, faintly pleochroic grains, partly prismatic, partly rounded, surface often like an aggregate ("alterite", VAN ANDEL 1950). Moderately high birefringence. Frequently one optical axis is visible near the rim of the visual field. Within prismatic grains the axial plane is about perpendicular to the elongation of the grain. Clinozoisite has a weaker birefringence and is colorless. Occurrence: in metamorphic rocks of the upper greenschist and the almandine-amphibolite facies; frequently replacing plagioclases in metamorphic granites: "saussuritization".

Fluorite (rare)
Cleavage pieces, partly etched, isotropic, very low refraction. Vein fill, cement, and replacement mineral in carbonate rocks.

Garnet (frequent)
In most cases relatively large, irregular, isotropic grains, corroded or covered with etch figures. High refraction. The most frequent garnet, almandine (FÜCHTBAUER 1964a), often is slightly red. Occurrence: almandine, grossular, spessartine and andradite in metamorphic rocks of the amphibolite facies; pyrope in the granulite facies and in eclogites; grossular and spessartine also in contact-metamorphic rocks.

Glaucophane (rare)
Alkali-amphibole; Pleochroism blue and violet. Axial plane parallel to the direction of fibres. Crossite: Axial plane perpendicular to the fibres. Occurrence: in crystalline schists of the glaucophane facies, i. e. at high pressure and low temperature.

Hematite (occasionally very abundant)
Thin plates are red translucent. High refraction and birefringence. Detrital; in red beds authigenic or replacements of magnetite.

Kyanite (commonly low quantities)
Rodlike grains, crossed by cleavage cracks at right angles. Refraction moderate, birefringence low. Most typical is the oblique extinction. Occurrence: in metamorphites of the middle almandine-amphibolite facies.

Leucoxene (present, but seldom common or abundant)
Generally an opaque mineral, see below.

Monazite (rare)
Yellowish grains, refractive index higher than zircon. Moreover, a distinction can be made by a spectroscope head-piece: Whereas in most cases zircon does not show any lines, monazite has two strong (Nd-) absorption lines close to 525 mμ and 3 weaker ones (Nd, Pr) between 570 and 590 mμ (HERING & ZIMMERLE 1963). Occurrence: in SiO_2-rich igneous and metamorphic rocks.

Olivine (very rare in pre-Recent sediments)
Slightly greenish grains. High refraction and birefringence. Occurrence: in ultramafic rocks, gabbros, and basalts.

Pumpellyite (rare)
Similar to epidote and clinozoisite, but colorless grains, have lower refraction and birefringence compared with latter. The colored grains show blue-green absorption in n_y. Anomalous interference colors. $2 V_z$ between $26°$ and $85°$. In Quaternary sands of the Netherlands (LANGENBERG & DE ROEVER 1955). Occurrence: in metamorphites of the pumpellyite facies, between zeolite and greenschist facies.

Pyroxene (rare; only common in modern sediments)
Yellowish-green to brownish, rounded prisms or cleavage fragments; low pleochroism, inclined extinction. Occurrence: Orthopyroxene and diopside: in andesites, igneous and metamorphic rocks of the granulite and pyroxene-hornfels facies. Augite: in syenites, diorites, gabbros and the corresponding volcanic rocks.

Rutile (frequent)
Depending on the iron content, yellowish-brown or reddish-brown to opaque. Irregular, prismatic crystals with high refraction and birefringence, slight pleochroism. Occurrence: in pegmatites, crystalline schists, and contact-metamorphic rocks.

Sillimanite (rare)
Rounded, elongated colorless grains, frequently composed of irregular fibres with undulatory extinction. Moderate refraction and birefringence, low axial angle. Occurrence: in metamorphic rocks of the amphibolite and pyroxene-hornfels facies.

Sphalerite (rare)
Yellowish- to brownish-grey, isotropic grains of high refraction. From ore veins and authigenic (less colored).

Sphene (not frequent)
Brownish color due to high refraction; high birefringence and strong dispersion of the axes causing metallic irridescence. Very often conoscopic pictures containing both optical axes. Smaller grains may be mistaken for zircon. Occurrence: in granites, diorites, syenites and their volcanic equivalents, as well as in metamorphic rocks.

Spinel (sometimes, in trace amounts)
Dark red, irregular, sharp-edged, isotropic grains with high refraction; in most cases, very likely picotite or chromite (WOLETZ 1963, GASSER 1967). Occurrence: in mafic, ultramafic, mafic metamorphic, and contactmetamorphic rocks.

Staurolite (frequent)
Yellow, pleochroic grains, often sharp-edged or corroded, similar to a rooster's comb. Refraction relatively high. Birefringence low. Occurrence: in metamorphic rocks of the almandine-staurolite subfacies of the mesozone.

Tourmaline (very frequent)
Very often well rounded, drab, brown, green or blue grains with low refraction, moderate birefringence and high pleochroism. Occasionally tourmalines show bottle-green secondary enlargement rims (KRYNINE 1946b). Well-rounded grains, and especially grains with rounded authigenic seams, are recycled. Occurrence: in diorites, syenites, granites, especially when pneumatolytically influenced, as well as in schists.

Xenotime (very rare)
Reddish-brown grains of high refraction and birefringence. Distinguished from zircon and monazite by lower refraction, and by two weak absorption lines close to 524 (Er) and 521 mµ, and also by the missing of monazite lines between 570 and 590 mµ (HERING & ZIMMERLE 1963). Occurrence: in granites and pegmatites.

Zircon (very frequent)
Globular or oval, sporadically zoned grains, and doubly terminated prisms. Refraction and birefringence high. Zircon is the most frequent heavy mineral. Its varieties as to length : width ratio, inclusions and color (rose to colorless) may be used for paleogeographic reconstructions and stratigraphic correlations.

Zircon is mainly associated with acid intrusive rocks, gneisses and quartz porphyries, the latter being distinguished by the greater number of growth impediments (large inlets) (Hoppe 1962). Occasionally the origin can be deduced from the habit, provided there is a chance of comparison with the possible source rock (Hoppe 1951, 1963). Statistical work on granites (Piller 1951) and observations on zoned zircons (Claus 1936, Piller l. c.) proved the length/width ratio to be a function of the crystal size: The growth velocity of the pyramidal planes overtakes the one of the prismatic planes. The result is that larger crystals become more oblong than the smaller ones. According to Claus (1936) the length/width ratio also depends on the cooling velocity of the magma: Longer zircons occur within (faster-cooling) dikes. Finally, zircon can resist remelting due to its high melting point. Therefore in granites rounded zircons of the former sediments can be found (Poldervaart 1950, Piller 1951). Overgrowths or facets on rounded zircons therefore are no proof of authigenic formation; they could have regenerated within granites. Refraction and birefringence of the "Malacon" variety are reduced due to radiation activity; a reddish-brown crust can result (Hutton 1950).

Zoisite (rare)
Moderate refraction, very low birefringence and strong dispersion of the optical axes, causing variable color in the grain (yellowish or blueish). Relatively often shows the conoscopic picture the 1st bisectrix; $2V_z$ is small. Occurrence: in metamorphic rocks of the greenschist facies.

This enumeration shows that many sandstones contain only 7 out of the 33 heavy mineral species listed above: zircon, tourmaline, apatite, garnet, staurolite, epidote and rutile. (Carbonates, sulphates, chlorite and biotite are not considered as heavy minerals "sensu stricto").

For many purposes it is unnecessary and tedious to note the exact percentage of heavy minerals. Milner (1962, p. 394) suggested for such cases the use of abbreviations or the numbers 1—9 to indicate increasing frequency. — It is even easier to combine the initial letters of the heavy mineral's names in the order of their frequency in one "formula" which can be put easily on maps (Füchtbauer 1964a). Minerals of more than 10% are marked in capital letters, those between two and 10% in small letters (with reference to v. Moos 1935). If a letter occurs more than once, single or double apostrophes may be used.

Key to the identification of common non-opaque heavy minerals
1. isotropic
   colorless or slightly reddish, etch figures: G a r n e t (A l m a n d i n e)
2. anisotropic
   2.1. colored
      2.1.1. blue
         P colorless - bl., −1 a, well-rounded or elong. // x: T o u r m a l i n e
         P bl.-violet, $2V_x$ low, axial plane // elongation: G l a u c o p h a n e
                          axial plane ⊥ elongation: C r o s s i t e
      2.1.2. green
         P light. gn.-dark gn., −1 a, well-rounded or elong. // x: T o u r m a l i n e
         P yell. gn.-blue gn., $2V_x$ high, elong. at low angles with z: A c t i n o l i t e, gn. Hornbl.
         P weak, C bn.-gn., $2V_z$ med., elong. mostly at high angles with z: P y r o x e n e s
         P weak, C colorless to light gn., 2V high, B high: O l i v i n e

P and color weak, aggregate appearance freq., 2 V high, B red+gn., AP ⊥ elongation: Epidote
P light gn.-dark gn., plates, B med., elong. // z: Biotite
P colorless to light gn.-bl. gn., plates, B very low, often anom., elong. // z or x: Chlorite

2.1.3. brown

P br.-nearly opaque, $-1a$, well-rounded or elong. // x: Tourmaline
P med. br.-dark br., $2V_x$ high, elong. at low angles with z: Hornblende, mostly volc. rocks
P light br.-dark br., plates, B med., elong. // z: Biotite

2.1.4. yellow

P light ye.-med. ye., B and R low: Staurolite
P weak, C yel. br./red. br., B and R very high: Rutile

2.1.5. red

P colorless-reddish, B low: Andalusite

2.2. colorless

2.2.1. B very low

$-1a$: Apatite
$2V_z$ low, anomal. Interfer. blue/yellow: Zoisite

2.2.2. B low ($\sim$ Quartz) to medium

$2V_z$ 37°, B 0.012: Barite
$2V_z$ 51°, B 0.009: Celestite
$2V_z$ 42°, B 0.044, rectangular cleavage plains, parall. ext.: Anhydrite
$2V_x$ 82°, B 0.017, rectang. cleav. pl., oblique extinction: Kyanite
check also amphiboles, clinozoisite, and quartz!

2.2.3. B medium ($\sim$ 0.02—0.05) to high (0.1)

Irreg. grains, $2V_x$ high, excentr. axis, (greenish), B-colors red/green: Epidote
Rounded grains or elongate idiom. cryst., $+1a$, high B-col., R high: Zircon
Rounded grains, 2 V high, high B-colors: Olivine
$2V_z$ low, high dispers., both axes often in conosc. pict.: Sphene

2.2.4. B very high

Carbonates.

Abbreviations: AP = axial plane, B = birefringence, C = color, E = extinction, P = pleochroism, R = refractive index. $-1a$ = negative uniaxial.

B. Opaque heavy minerals

Some heavy minerals like anatase, hematite, rutile, spinel, tourmaline, and zircon can occasionally or in certain positions appear opaque. Such **grains**, however, can be recognized by crushing them and embedding the translucent fragments (HUTTON 1950). The ore minerals occurring in sandstones can only be identified in translucent light if the grains have typical forms, e. g. square and hexagonal cross-sections of authigenic pyrite cubes or pentagondodecahedra, or grains of ilmenite with crusts or pockets of leucoxene. According to X-ray work by NELSON & NIGGLI (1950), a great part of the opaque fraction in sands can consist of rutile, in part as leucoxene.

Leucoxene forms fine crystalline aggregates, white or of various colors, in most cases opaque. It is a reaction (transformation?) product of ilmenite and predominantly consists of rutile, anatase, brookite, titanite and goethite or hematite.

Generally, the identification of opaque minerals is only possible in sectioned and polished grain mounts by using reflected light. Only by this technique is it possible to identify the diagnostically important intergrowths. The following intergrowths are characteristic, according to STUMPFL (1958):

a) for acid plutonic rocks: magnetite with fine ilmenite-exsolution lamellae. Pure magnetites. Ilmenite-hematite exsolutions.

b) for mafic plutonic rocks (richer in titanium): magnetite with coarse exsolution lamellae of ilmenite. Pure ilmenite.

c) for volcanics: pseudo-brookite Fe_2TiO_5 and high-temperature martite Fe_2O_3, formed from titanomagnetite $(Fe,(Ti))_3O_4$, ilmenite $FeTiO_3$, or magnetite Fe_3O_4 by heating and oxidation.

STUMPFL (in FÜCHTBAUER 1964a, p. 261) was able to distinguish the Tertiary molasse east of the Peruvian Andes from the Tertiary Molasse north of the Alps by use of opaque minerals: pseudo-brookites and high-temperature martites occur only in the South American Tertiary, whereas the translucent heavy minerals are similar in both areas.

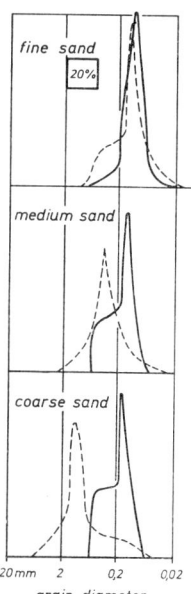

Fig. 3-5. Diagrams showing an almost complete independence of the grain size distribution of heavy minerals (straight line) compared with the grain size distribution of the entire sediment (dotted line). Samples with about 0.1% heavy minerals; from the Lower Permian south of the Harz Mountains (from G. LUDWIG 1955).

3.116.2 Grain size of heavy minerals

If heavy mineral analysis is not restricted to samples of uniform grain size, attention should be given to the influence of the sample grain size upon the composition of the heavy minerals fraction. As it will be shown below, this can be done in two ways:

1. by separating the heavy minerals from a large grain size range in which the variation of heavy mineral composition versus size can be observed directly,

2. by splitting the sand sample into grain size grades and separating from each the heavy minerals.

The relationship between grain size of the whole sediment and the grain size of the heavy minerals depends on several factors including the original grain size of these minerals, the specific sorting during transport, as well as differences in shape and specific gravity.

The Lower Permian sandstones shown in fig. 3-5 have different grain size: the heavy minerals, however, are similar in grain size. In this case, the influence of transport sorting on the heavy minerals is small, compared with the original size distribution of heavy minerals.

In order to study the behavior of the different minerals, LUDWIG (1955) suggested the conversion of abundance percentages into weight percentages. After determination of 100—300 grains per size grade, the expression $f\varrho W/\Sigma f_i \varrho_i$ is calculated for every heavy mineral in each grain size fraction, with f = number of grains, ϱ = density of heavy mineral, W = weight of the heavy mineral grain size grade, and $\Sigma f_i \varrho_i$ = sum of the product of grain number and density of all heavy mineral types in this grade. These expressions summed up for all grades and converted into percentages, lead to a grain size distribution for each heavy mineral, as shown by fig. 3—6 for samples no. 2 and 5 from table 3-3.

Table 3-3. Heavy minerals (weight percent) of 5 samples of the Valendis sandstone, Rühlermoor (Emsland area), grade 0.02—1.0 mm.

Sample	Median grain size of the sample in µ	Percent transparent heavy minerals	Percentage of individual heavy minerals				
			zircon	tourmaline	rutile	staurolite	garnet
1	96	0.0100	62.1	29.0	5.2	1.2	2.5
2	105	0.0364	67.2	16.8	5.3	0.9	3.8
3	124	0.0349	51.0	35.8	6.3	2.1	3.9
4	190	0.0165	55.4	36.7	3.4	2.9	1.4
5	280	0.0089	47.9	43.0	2.9	4.9	0.7

The curves of the heavy minerals are shifted towards the finer grains compared with the total sediment (S in fig. 3-6) since the hydraulic equivalents of any heavy mineral are considerably larger in light minerals (such as quartz). The barite of sample 2 (ca. 6% of the heavy minerals) is an exception; it cannot, therefore, have been transported mechanically together with the sediment, but must be diagenetic in origin. This is supported by the fact that barite was found only in one of the samples in table 3-3, all of which belong to the same heavy mineral province (barite is not shown in the above table).

Fig. 3-6. Cumulative curves of individual heavy minerals of two samples from the Valendis sandstone of Rühlermoor (Emsland area); abscissa: grain size (millimeter and phi scale); ordinate: weight percent; probability net. S = total sediment; heavy minerals: Ba(rite), Gr = garnet, Ru(tile), St(aurolite), Tu = tourmaline, Zi(rcon). Dotted lines: cumulative curves of heavy minerals in source area (sample nos. 2, 3 and 5, s. tables 3-3 to 3-5), calculated under the assumption of suspension transport (Ls) and bottom transport (Lr); explanation in the text.

3.11 Sandstones

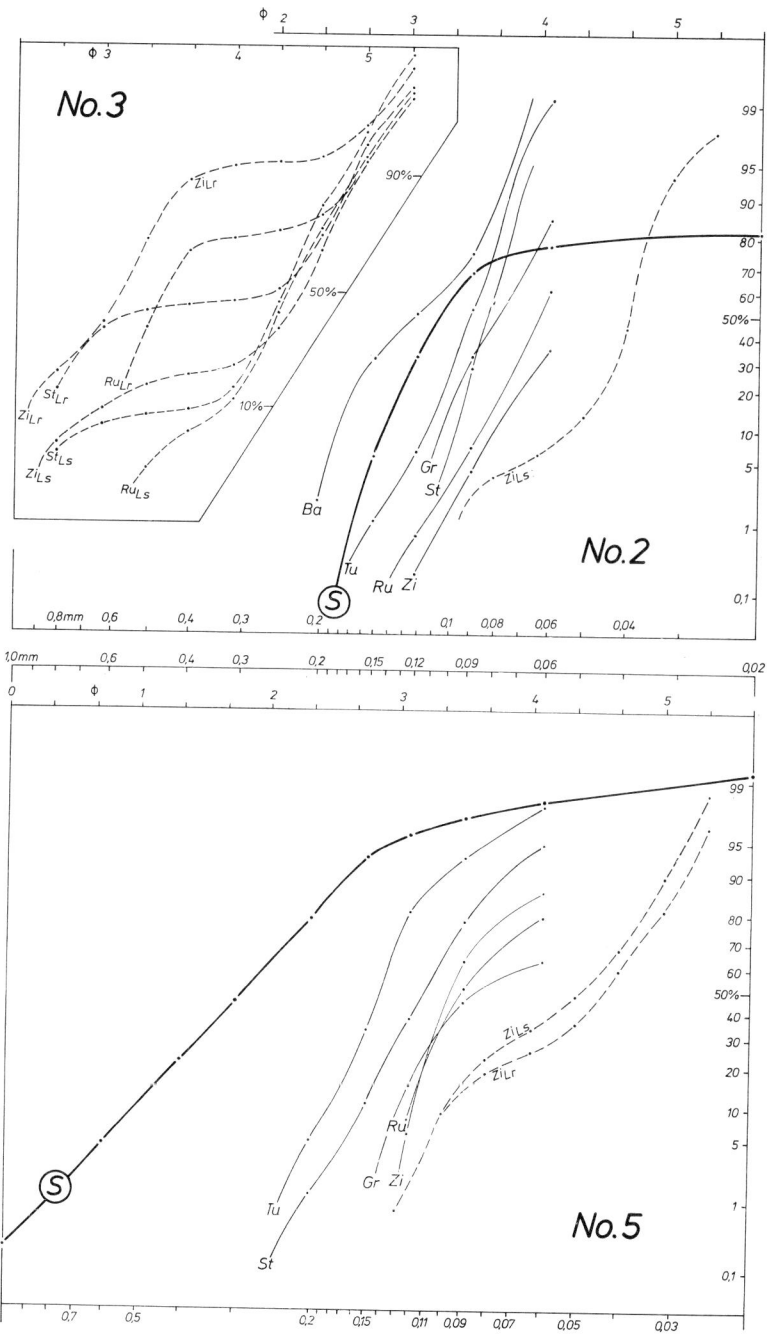

The pronounced differences in the distribution of various heavy minerals with size raise the problem of choosing the proper size grade for heavy mineral analysis. This problem can be discussed using the information presented in table 3-3, especially comparing two of the more frequently used grades, a broad one (0.06—0.4 mm) and a very narrow one (0.105—0.15 mm). Assuming these fractions were used instead of the entire sample (0.02—1.0 mm), then only the percentages of the individual heavy minerals shown in tab. 3-4 would have been identified. (This has been calculated by using graphs, part of which is shown in fig. 3-6.)

Table 3-4. Comparison of heavy mineral distributions which result from using different fractions (0.06—0.4 and 0.105—0.150 mm). Samples 1—5 as in tab. 3-3.

A. Weight percent of the different minerals identified in the fractions 0.06—0.4 and 0.105—0.150 mm compared with the original fraction (0.02—1.0 mm):

	fraction 0.06—0.4 mm					fraction 0.105—0.150 mm				
	zirc	tour	rut	stau	garn	zirc	tour	rut	stau	garn
1	32%	99%	46%	100%	72%	1.2%	22%	3%	29%	6.2%
2	35	99	60	100	85	1	19	2	4	10
3	73	96.5	82	75	78	9.5	37.5	3	19	7.4
4	90	99	88	96.5	81	19.3	46	8	21.5	10
5	86	98	79	94	62	30	55	26	48	30

B. With the limitations caused by counting the fractions 0.06—0.4 and 0.105—0.15 mm instead of 0.02—1.0 mm, the analyses of table 3-3 would change as follows:

1	36.8	53.2	4.5	2.2	3.3	9.6	82.0	2.0	4.4	2.0
2	46.7	32.9	6.3	1.8	6.3	14.4	68.4	2.3	0.8	8.1
3	45.3	42.1	6.3	1.9	3.7	25.1	69.5	1.0	2.0	1.5
4	53.5	38.9	3.2	3.0	1.2	37.3	58.8	1.0	2.2	0.5
5	45.2	46.1	2.5	5.1	0.5	34.6	56.9	1.8	5.6	0.5

C. Conversion of weight percent into grain number percent (for 2 minerals only, as an example):

1	34.1	55.9
2	45.1	34.5
3	45.6	41.8
4	60.4	32.0
5	57.5	33.8

As shown in part A of this table, the narrow grade (right hand side) comprises a very small part of the total heavy mineral population, e. g. in the fine-grained samples 1 and 2 only about 1% of the zircon. Moreover, the proportions are changed considerably (part B): In contrast to the total sample (table 3-3), tourmaline in the 0.105—0.150 mm fraction is more frequent than zircon. Finally, there is a considerable difference between coarse (4 and 5) and fine (1 and 2) samples, i. e. percentages are highly influenced by the sample grain size.

On the other hand, the wider grade (0.06—0.4 mm) changes the proportions too, but the differences compared with tab. 3-3 are much smaller than for the grade 0.105—0.150 mm. The elimination of the fraction 0.02—0.06 mm compared with

3.11 Sandstones

table 3-3 is in many cases an advantage because this fraction frequently contains only zircon. This elimination would then diminish the differences caused by grain size between the heavy mineral population of coarse and fine samples of the same occurrence. The broader grade, therefore, is preferred. Nevertheless, routine analyses of heavy minerals should not overlook the grain size differences.

In the samples used for fig. 3-6, the heavy minerals of the coarser sample (No. 5) are coarser than of the finer sample (No. 2) (opposite to fig. 3-5). Stronger currents in this case selected larger grains from the heavy mineral source area. These transport sorting phenomena can be eliminated by calculation. Two extreme cases shall be considered: transport by traction and by suspension:

Cumulative curves of the heavy minerals are first divided into logarithmically equal subfractions (e. g. in phi/3 sections, fig. 3-6). Then the percentage of the equivalent light mineral sub-fractions (equivalent with respect to rolling and sinking velocity, respectively; v. ENGELHARDT 1938) is taken from a graph (cf. fig. 3-6), without considering the form factor. Each heavy mineral sub-fraction is then divided by the weight of the corresponding light mineral sub-fraction. These figures, converted into percentages for each individual mineral, represent a grain size distribution of the heavy minerals within the detritus of the source area, which is independent of the transport sorting and which is called "availability distribution" (WALGER 1966). This calculation implies the assumption of equal percentages for all grain size grades of the whole sediment betwen 0.02 and 0.2 mm, in the source area. Since this assumption does not hold in nature, presumably this calculated heavy mineral grain-size distribution is only a relative distribution. Nevertheless, it is a characteristic property, because it is theoretically independent of the grain size of the sandstone, and therefore enables comparison of sandstones of different grain sizes. The independency of the "availability distribution" from grain size is demonstrated by table 3-5: Though the median grain size of each heavy mineral is different in samples 1—5 as shown in the columns "Sed.", nearly the same median size in the source area (availability grain size, columns "Source") has been obtained for each mineral in all samples. (The mean is shown in the last line.)

Table 3-5. Median grain size (μ) of the heavy minerals in the Rühlermoor samples (table 3-3) = "Sed". Arithmetic elimination of sorting effects for suspension (s) and traction (r) transport led to the 'availability grain size' in the supply area ("Source").

Sample No.	bulk sample	zircon Sed.	Source	tourmaline Sed.	Source	rutile Sed.	Source	staurolite Sed.	Source	garnet Sed.	Source
1	96 µ	52 r: s:	— 44	88 r: s:	— 90	58 r: s:	— 50	96* r: s:	— 125*)	70 r: s:	— 62
2	105 µ	50 r: s:	— 40	91 r: s:	— 82	63 r: s:	— 47	83 r: s:	— 79	79 r: s:	— 63
3	124 µ	76 r: s:	(112) 44	103 r: s:	94 (66)	80 r: s:	(90) 47	80 r: s:	(51) 49	74 r: s:	(110) 48
4	190 µ	93 r: s:	(78) 59	132 r: s:	94 90	97 r: s:	(56) 57	91 r: s:	75 74	81 r: s:	50 56
5	280 µ	96 r: s:	44 49	140 r: s:	90 92	90 r: s:	45 (55)	110 r: s:	66 72	82 r: s:	40 42
Means			ca. 45		ca. 90		ca. 50		ca. 70		ca. 50

* unreliable because of the low staurolite contents

In brackets are unlikely figures, according to the overall shape of the cumulative curves (dotted lines in fig. 3-6). For instance, in No. 3 (fig. 3-6), the curves for zircon, rutile and staurolite grains transported by traction at the bottom (subscript "Lr") indicate two distinct grain size maxima, which is unlikely. The assumption of traction transport for samples No. 1 and 2 leads to unreliably high medians (dashes). Some numbers in the table have been put into brackets because of their deviation from the other figures.

By repeating the above-mentioned calculation at different places along the path of transport, events during transportion such as abrasion, fracturing and chemical alteration can be reconstructed. More extensive work on this subject has been done by WALGER (1966).

Calculations such as shown in table 3-5 allow one to estimate the most probable mode of transportation of each mineral in each grain size: in the present example suspension seems to be preferred by grain size mixtures below 0.1 mm whereas grain size mixtures above 0.1 mm presumably are moved by traction.

In tab. 3-6, for purposes of simplification, only the median diameter is indicated. In the last two columns the hydraulic equivalent diameter for quartz compared with each individual heavy mineral was calculated, based on v. ENGELHARDT (1938): i. e. those grain sizes which are transported together with quartz grains (0.43 mm in diameter) by suspension and by traction. The mode of transport (suspension or traction) can not be inferred solely by these rough figures.

Table 3-6. Sands of the Rhine River near Wageningen (average of 6 samples from VAN ANDEL 1950).

	Density	measured median diameter	Diameters equivalent to quartz grains 0.43 mm in size	
			suspension transport	traction transport
quartz	2.65	ca. 0.43 mm		
augite	3.4	ca. 0.4	0.36 mm	0.30 mm
hornblende	3.2	ca. 0.3	0.37	0.32
epidote	3.3	ca. 0.2	0.37	0.31
garnet	4.2	ca. 0.18	0.31	0.22

The actual diameters of the augite and hornblende grains nearly attain the equivalent diameters. The grain size of the garnets and of the epidotes, however, are below these diameters, because they are supplied from the Alps and filtered by the Lakes of Constance, Zurich, and Thun allowing only small grain sizes to pass, whereas augite and hornblende come from Tertiary volcanic rocks of the Rheinisches Schiefergebirge. The variety of grain sizes in this particular example is great.

The importance of the primary grain size of the heavy minerals is shown also by the following example from the Rhone delta (VAN ANDEL 1959). Whereas the alluvial association contains pyroxene, hornblende, and epidote, the corresponding prodeltaic assemblage is devoid of pyroxene. This is explained by the different median diameters of the heavy minerals in the Rhone river:

Pyroxene (density 3.4): 0.31 mm
Hornblende (density 3.2): 0.165 mm
Epidote (density 3.3): 0.12 mm

Pyroxene, supplied from volcanics in central France, is larger in diameter than hornblende and epidote derived from the Alps and is therefore transported only by the river, but not in the low-energy shallow-sea environment in front of the delta.

Heavy mineral ratios may change during transport. For example, hornblende/garnet-ratios commonly increase with distance from source (GROBA & LUDWIG 1956, VAN ANDEL & POOLE 1960, SEIBOLD 1963). This may be due to the smaller primary grain size, lower density, and higher grain elongation of hornblende compared with garnet, resulting in a progressive relative enrichment of hornblende with distance from source.

The shape factor also has to be considered in calculating the equivalent diameters (BRIGGS & McCULLOCH 1962). Shape and roundness of heavy minerals have been successfully used as descriptive characteristics. For example, POTTER & GLASS (1958) compared tourmalines of three different sandstones:

	median of the sandstone	rounded tourmalines in % of all tourmalines
Ordovician quartzose sandstone	0.14 mm	97%
Pennsylvanian quartzose sandstone	0.17 mm	42%
Pennsylvanian subgraywacke	0.22 mm	2%

Based upon these figures they considered the Ordovician St. Peter Sandstone to be the most mature, and be representative of material which has been reworked and redeposited ("recycled") many times.

3.116.3 Stability of heavy minerals

3.116.31 Resistance to weathering

Different species of heavy minerals show various degrees of resistance to weathering. With increasing intensity of weathering, instable species are progressively eliminated while the absolute abundance of heavy minerals decreases. By such observations, GOLDICH (1938) and DRYDEN (1946) established sequences of stability of heavy minerals in soil profiles. WEYL (1951) used the degree of etching as an additional criterion of weathering.

The resulting sequences of stability differ widely which is not surprising in view of the various types of weathering. Minerals can be arranged into 4 groups:

(1) Most stable heavy minerals, i. e. zircon, rutile, tourmaline. Only under extremely alkaline weathering conditions has the etching of zircon been found in laterite profiles (CARROLL 1953).

(2) Moderately sensitive minerals: Kyanite, staurolite, sillimanite, andalusite, and epidote.

(3) Heavy minerals with relatively low stability and high sensitivity to chemical environment: Apatite is more sensitive to acid, carbonate-free soils than garnet (PILLER 1951); a carbonate-bearing environment has contrary effects (LEMCKE et al. 1953): The solubility of apatite is presumably lower in the presence of $Ca^{\cdot\cdot}$-ions.

(4) The most sensitive group with respect to weathering includes heavy minerals such as amphibole, and especially pyroxene and olivine (WALKER 1967a).

NICKEL (1973) was able to experimentally verify the observed order of chemical resistance.

3.116.32 Resistance to transport

The studies of RUSSELL (1936) on the Mississippi River and VAN ANDEL (1950) on the Rhine River sediments showed that the influence of abrasion during fluvial transport can be neglected even for minerals like augite and hornblende. Abrasion at the shore face may be more effective, because the grains normally stay longer in this environment than in an river channel. Eolian abrasion is much more effective, as shown also by the sand grain roundness (ch. 3.222.2).

FREISE (1931) investigated the weight loss of different heavy minerals during experimental abrasion, whereas DIETZ (1973) used the roundness increasing during tumbling bottle experiments to establish a sequence of increasing resistance to transport. DIETZ showed that zircon is most resistant among heavy minerals against mechanical abrassion, while, for example, tourmaline and hornblende are relatively easily rounded, thus veryfying the high degree of rounding characteristic of tourmalines in sandstones.

3.116.33 Resistance to intrastratal solution

Four different groups of observations also indicate solution of heavy minerals during d i a g e n e s i s, i. e. below the zone of weathering, a process called "intrastratal solution".

(1) PETTIJOHN (1941) established an order of dissolution stability based upon the distribution of the heavy minerals throughout the geologic sedimentary column. For each geologic period and for each heavy mineral he calculated the percentage of occurrences in which this heavy mineral was present. These values were then averaged for each heavy mineral in all periods prior to the Holocene, and divided by the percentage of occurrences of this heavy mineral in recent deposits. According to PETTIJOHN this ratio may serve as a standard for the diagenetic stability.

Table 3-7. Persistency sequence according to PETTIJOHN (1957) arranged in the order of decreasing stability.

— 3 Anatase	7 Ilmenite	15 Sphene
— 2 Muscovite	8 Magnetite	16 Zoisite
— 1 Rutile	9 Staurolite	17 Augite
1 Zircon	10 Kyanite	18 Sillimanite
2 Tourmaline	11 Epidote	19 Hypersthene
3 Monazite	12 Hornblende	20 Diopside
4 Garnet	13 Andalusite	21 Actinolite
5 Biotite	14 Topaz	22 Olivine
6 Apatite		

Negative numbers indicate that the mineral occurs more frequently in older rocks than in younger ones. With anatase this is due to diagenetic formation.

(2) Another indication of intrastratal solution is the presence of intensively etched heavy minerals in unweathered rocks. In many instances these could not have been transported to the place of deposition such as they are. Such corrosion on the surface of grains of pyroxene, amphibole, staurolite, kyanite, garnet, epidote and sphene was first described by EDELMAN & DOEGLAS (1934).

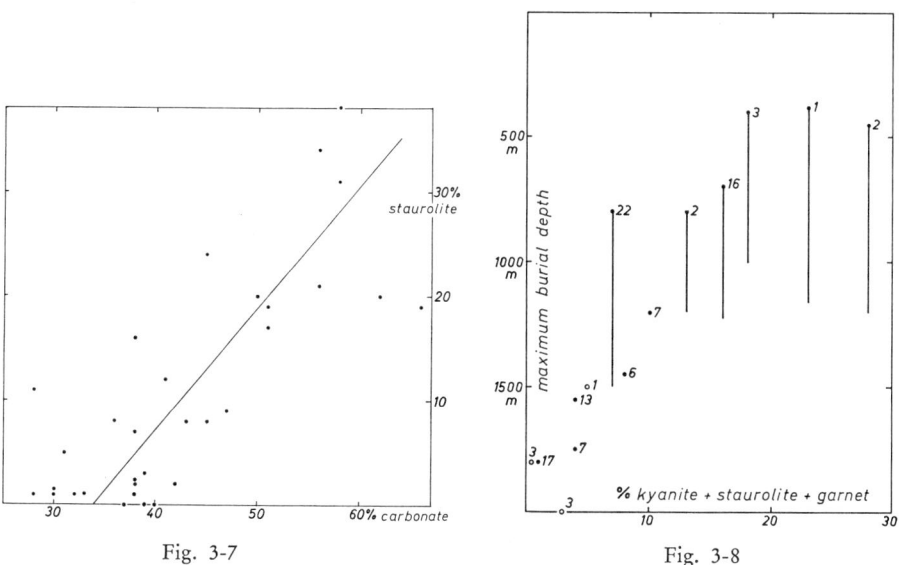

Fig. 3-7.   Fig. 3-8

Fig. 3-7. Intrastratal solution depending upon cementation. Molasse sandstones containing 30—40% dolomite grains. Carbonate in excess of 30—40% is calcite cement. Only cemented samples contain staurolite (from deep wells; FÜCHTBAUER 1964a).

Fig. 3-8. Intrastratal solution as a function of burial depth. Dogger-beta quartzose sandstones, N. Germany. The amount of unstable heavy minerals decreases with increasing maximum depth of burial. For oil-bearing sandstones (dots), the maximum depth before oil migration is indicated, according to structural geology and quartz diagenesis (PHILIPP et al. 1963). The lower ends of the vertical lines indicate the maximum depth which was attained after the oil migration. The number of samples is indicated. Compiled after DRONG (1965).

(3) The third series of observations on intrastratal solution is based upon the comparison of heavy minerals in porous sandstones with those in sandstones within the same sequence but which underwent early carbonate cementation: BRAMLETTE (1941) and WEYL & WERNER (1951) found hornblende to be preserved only in cemented parts of Tertiary sandstones. The same is shown by fig. 3-7 for staurolites from the Molasse rocks north of the Alps. Other examples were published by G. LUDWIG (1955), CARPENTER & SCHMIDT (1962), KNOBLAUCH (1963), and HUDSON (1964). According to observations by V. LUDWIG (1968) in Lower Carboniferous flysch graywackes, even high clay contents may preserve minerals as unstable as hornblende from intrastratal solution. This agrees with BLATT & SUTHERLAND (1969) who found that heavy minerals, especially hornblende and kyanite, are better preserved in shales than in the associated sandstones. Oil accumulations can preserve unstable heavy minerals

as well which consequently may be used to compare the age of oil migration into different pools (YURKOVA 1970) (fig. 3-8).

(4) The fourth group of investigations is based on drill cores in which heavy minerals decreased with increasing depth: In the Tertiary of the Vienna basin WIESENEDER (1953, 1958) found an increasing corrosion of garnet, staurolite, and epidote and a decrease of hornblende with increasing depth. In uncemented quartzose sandstones of the Dogger-beta in Northern Germany, the percentages of kyanite, staurolite, and garnet decrease with increasing maximum depth of burial (fig. 3-8).

Contrary to the multitude of weathering-stability sequences, stability with respect to intrastratal solution can be expressed in one scheme:

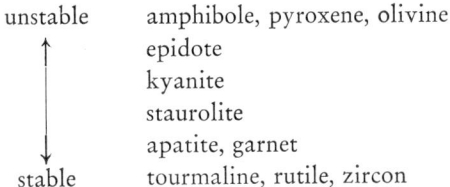

This sequence fits surprisingly well with the stability sequence of PETTIJOHN (table 3-7). The main difference in comparison with the weathering sequence is the diagenetic stability of the easily weathered minerals garnet and apatite. This difference is explained by the more acid environment of weathering compared with diagenesis. Experimental weathering of 12 heavy and light minerals at different pH levels revealed a relatively low resistance of garnet and apatite at low pH (5.6) and a relatively high resistance at high pH (10.6) (NICKEL 1973).

The high stability of tourmaline and zircon may be explained at least in part by the build-up of very thick surficial residual layers during the weathering experiments (NICKEL, l. c.).

This discussion shows clearly that both weathering and intrastratal solution must be taken into consideration when using heavy minerals for stratigraphic and paleogeographic purposes.

3.116.4 Use of heavy minerals

Due to their low abundances (0.01—1%), heavy minerals are called "accessory minerals". In spite of this they often are more useful in describing sandstones than the light minerals. In older sandstones, however, generally all but the most stable minerals, i. e. tourmaline, rutile, zircon, apatite, and garnet, are eliminated, unless an early cementation (e. g. calcite) has prevented intrastratal solution. Therefore, calcareous sandstones or concretions are preferably used when studying heavy minerals within older sandstones.

Heavy mineral analyses are mainly used for 3 purposes:
1) correlation of rock series barren of fossils, e. g. fluviatile sediments
2) determination of provenance and direction of transport
3) reconstruction of eroded rock series

A few examples will illustrate these applications.

ad 1) In the barren Upper Freshwater Molasse (Upper Tertiary) north of the Alps a large area was mapped structurally by means of heavy mineral analyses (LEMCKE et al. 1953). For this purpose counterflush wells of 300—500 m depth were placed at intervals of about 12 km. Samples were taken every 10 meters from the continuous core. To avoid grain size influences only fine-grained sandstones were analyzed when possible. The results were correlated as shown in fig. 3-9. The disappearance of zoisite towards the top was the only qualitative change noted, but this proved to be a most useful criterion though zoisite nearly never exceeded 2 percent of all heavy minerals present. The quantitative changes of the other heavy minerals, however, were helpful in the detailed local correlation.

ad 2) In order to determine the direction of supply of light and heavy minerals within a sedimentary basin one must designate groups which have similar sedimentological properties. Such populations define the sediment-petrological "provinces" as a group of sediments which constitute a natural unity by age, provenance, and distribution (EDELMAN 1933). The directions of supply and the source areas must be deter-

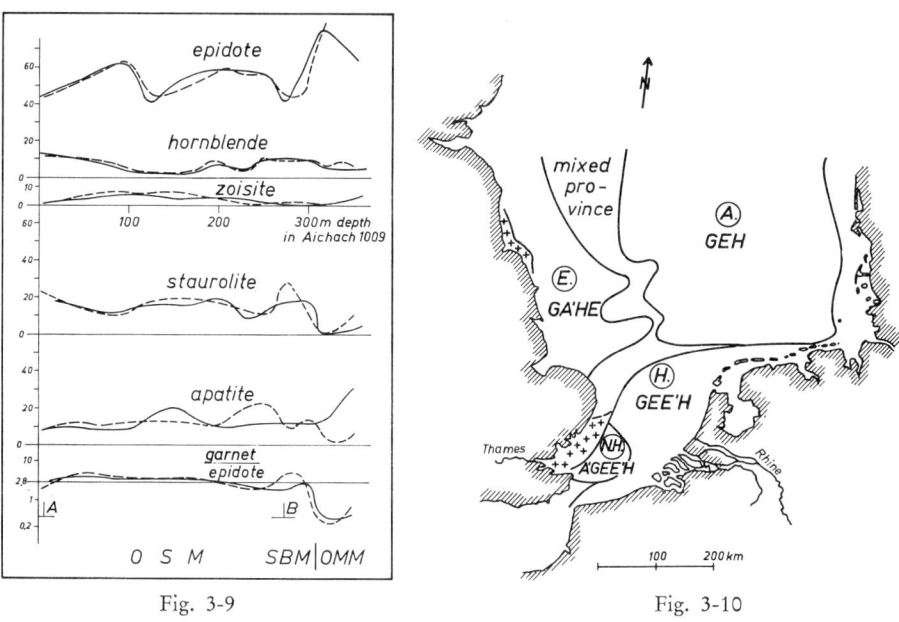

Fig. 3-9   Fig. 3-10

Fig. 3-9. Correlation of two well logs, 17 km apart (dotted and full lines, respectively), by means of heavy minerals (OSM = Upper Freshwater Molasse, SBM = Freshwater/Brackish Molasse, OMM = Upper Marine Molasse, A and B = correlation marks). Ordinates = percentage of heavy mineral grain numbers after elimination of garnet which is plotted separately below; after LEMCKE et al. (1953).

Fig. 3-10. Heavy mineral provinces within the North Sea according to BAAK (1936). The different provinces, A., E., H. and N.H. are characterized by different groupings of the minerals A = augite, E = epidote, E' = saussurite, G = garnet, H = hornblende. They are arranged in the order of decreasing frequency (heavy minerals with less than 10% have not been considered). Crosses = Tertiary and older.

mined for each of these provinces. These parameters frequently can be distinguished in modern sediments by direct petrological comparison with the possible parent rocks as well as by geographic considerations. In older sediments, however, such geographic criteria are lacking. Also the parent rocks frequently are eroded or difficult to find. In such cases fabric criteria like cross-bedding and changes in thickness must be taken into consideration.

The heavy mineral provinces of the North Sea are shown in fig. 3-10. According to EDELMAN (1933) the "A-province" material comes from Scandinavia. It is characterized by transparent epidote and zoisite. The H-province contains no zoisite, and, in addition to clear epidotes, dull epidotes which were formed as aggregates during saussuritization of plagioclases. These "alterites" (VAN ANDEL 1950) are frequent in the Rhine sediments and in the Molasse N. of the Alps. Consequently the H-province material is supplied by the Rhine River.

During or after the Niederterrasse (after Riss glacial stage) the Rhine eroded the volcanic rocks of the Eifel mountains and the Westerwald. This caused the appearance of green augites which were deposited in the small "North Hinder province" (N. H.) of the North Sea.

The minerals of the E-province are similar to those of the A-province, and they probably originate from Scandinavia too, but the E-province is also characterized by purplebrown augite. SINDOWSKY (1948) suggested a possible admixture of Oslo graben minerals. A thorough comparison of the varieties of heavy minerals could help to answer this question. Garnet and hornblende occur in all these provinces. They are therefore worthless for the present problem. These mineral provinces were dispersed mainly by fluvial action during the glacial lowering of the North Sea. At these lowered sea levels, rivers like Rhine and Thames drained much of the area, and reworked early Pleistocene and Tertiary sediments.

The heavy minerals within the Gulf Coast also can offer insights into the Quaternary history of the area. During the post-glacial transgression, the detritus from the Mississippi River spread over the shelf. A transport some 500 km towards the west has been delineated from heavy mineral analyses by VAN ANDEL & POOLE.

ad 3) Heavy minerals within the detritus of mountain chains can reflect periods of successive erosion, as shown by the following example (table 3-8).

Up to the Upper Cretaceous predominantly stable heavy minerals were deposited within the sediments of the Alpine geosyncline. These sediments originated from the surrounding continental areas or from older reworked geosynclinal sediments. Occasional spinel contents suggest a source of basic volcanic rocks (initial volcanism of the geosyncline).

Since Late Cretaceous time, garnet has dominated the heavy mineral suite. Garnet is abundant in the crystalline schists of the Variscian metamorphism. Due to their low resistance, these schists do not produce pebbles, but disintegrate and release the heavy minerals.

Beginning in the Tertiary, the feldspar content increases. During Early Oligocene, with the beginning of the Molasse sedimentation, staurolite, kyanite and apatite appear. They probably originate mostly from gneisses and schists of the Variscian (or older) metamorphism.

During the Late Oligocene, epidote and, in the eastern Molasse only, additional hornblende appear in the so-called "granitic sandstones". These are reworked Tertiary and older granites and gneisses altered during the Alpine metamorphism. Epidotebearing granites also occur as boulders within these strata. At the same time, barroisite,

Table 3-8. Mineralogical development of the Cretaceous and Tertiary in the area of the Alps (from VATAN 1949, HOFMANN 1960, WOLETZ 1963, FÜCHTBAUER 1967a).

	Calcareous Alps	Helveticum	Flysch	Molasse
Miocene				Gr. Ep. Ho. Ap. St.; f
Upper Oligocene				Gr. Ep. Ap. Ho.; F
Lower Oligocene			*Gr. Ap. St. Stab.; f	Gr. Stab. St. Ap.; f
Paleocene + Eocene	Gr. Stab. St.; (f)	Stab. Gr.; f	Stab. Ap. Gr.; f	
Upper Campan + Dan	Gr. Stab.; (f)	Gr. Stab.; (f)	Gr. Stab.; (f)	
Lower Cretaceous to Lower Campan	Stab. Sp.; (f)	Stab.; (f)	Stab.; (f)	

The abbreviations stand for Ap(atite), Ep(idote), F(eldspar): F = > 25%, f = 10—25%, (f) = < 10%, Gr = garnet, Ho(rnblende), Sp(inel), St(aurolite), Stab(le heavy minerals) = zircon, tourmaline, rutile.
* Deutenhausen formation.

a bluish-green hornblende, is found as an index mineral of the Alpine metamorphism (KARL 1959). During later stages of the Molasse and up to the present time, hornblende increases in the detritus of the Alps.

Thus, the following time sequence of heavy minerals is developed: tourmaline, zircon, garnet, staurolite, kyanite, epidote, hornblende. The same sequence was found by VAN ANDEL (1952) at numerous places of the world in connection with the Alpine orogenesis: The main filling of molasse foredeeps generally begins with the staurolite-kyanite association in all cases.

The preservation of the heavy minerals in these and the following examples is due to the high erosion rate in orogenic areas, preventing intensive weathering. Preservation is also related to the young age of these sediments as well as their frequent carbonate cementation which has reduced intrastratal solution (WIESENEDER 1953).

SARKISYAN (1949, 1958) described a similar sequence of heavy minerals from the molasse deposits west of the Southern Ural Mountains: Within the lower part of the Upper Permian, zircon, tourmaline and garnet are followed by epidote; later, hornblende, pyroxene and garnet. At the end of the Upper Permian, garnet, tremolite, hornblende, actinolite and glaucophane are found. In the extreme south, zircon, tourmaline, and garnet are followed by staurolite and kyanite only in the uppermost part of the Upper Permian, whereas epidote does not appear before the Triassic. It is notable that even such unstable minerals as pyroxene can be preserved over long periods of time. (Much care has to be put into the correct determination of heavy minerals like pyroxene and epidote. Zircon especially can be mistaken for these minerals.)

A third example is the French Central Massif, investigated by VATAN (1950). The detritus of this area is found in the three surrounding basins: the Aquitanian basin, the Rhone Valley and the Paris basin. In the Aquitanian and the Rhone area, the deposition of the debris from the Central Massif frequently was interrrupted by supplies from the Alps and Pyrenees. In the continental sediments of the Paris basin, however, deposition from the Central Massif has not been interrupted (table 3-9).

Table 3-9. Origin of the heavy minerals in the clastic rocks of the Paris basin (VATAN 1950).

age	minerals	parent rocks
Miocene	Zi, Ap, Mo, Br, Sp; F Si, Au; F	lowermost granites katametamorphic gneisses
Upper Eocene	St, Ky, Ad, An, Tu, Zi	mesometamorphic schists (6000 m)
Carboniferous-Triassic	St. Ky — Ch. Ep	Hercynian granite diapirs epimetamorphic schists

Abbreviations:
Ad = andalusite, An = anatase, Ap = apatite, Au = augite, Br = brookite, Ch = chlorite, Ep = epidote, F = feldspar, Ky = kyanite, Mo = monazite, Si = sillimanite, St = staurolite, Sp = sphene, Tu = tourmaline, Zi = zircon.

The successive erosion of epi-, meso-, and kata-metamorphic rocks is clearly shown. The absence of garnet compared with the Alps is striking.

In many rock sequences the heavy minerals association decreases downward, for two possible reasons:

a) The first debris reaching a basin frequently is derived from the sedimentary cover, which normally is poor in heavy minerals,

b) the decrease in unstable heavy minerals downwards reflects the influence of intrastratal solution increasing in the same direction (see above).

Vectorial analysis of heavy minerals has been introduced as a valuable aid in complex cases (IMBRIE & VAN ANDEL 1964).

The natural remanent magnetization of the rocks which mainly reflects properties of the opaque heavy minerals indicates at least two major periods of numerous reversals in polarity. One of them was in early Triassic (Scythian) time, the other began in the late Tertiary and has continued until present. The intermediate intervals were poorer in reversal events. These events may be helpful for exact time correlations in the two periods indicated above (BUREK 1970).

3.12 Conglomerates and breccias

A sediment consisting of more than 50% of pebbles (i.e. rounded pieces of rocks or minerals more than 2 mm in diameter), is called a gravel; if lithified it is termed a conglomerate. If the fragments are angular the sediment is called debris; if lithified, a breccia.

A distinction can be made between oligomictic and polymictic conglomerates and breccias. The first consist of only few rock species whereas the latter are rich in

3.12 Conglomerates and breccias

different rocks. In general, oligomictic conglomerates or breccias are classified according to their predominant components, e. g. limestone conglomerate, quartzite conglomerate, chert breccia. Polymictic conglomerates and breccias can be classified by the same system shown for sandstones in ch. 3.111. The determination of the components often indicates the parent rock. This method was especially well developed during the investigation of the Pleistocene ice movements [HESEMANN 1939, LÜTTIG 1954, 1958, 1964b (good summary and many references), K. RICHTER 1933, 1958, WOLDSTEDT 1950]. An example of a detailed investigation of older conglomerates is the paper by VOGLER (1956).

A transition exists between breccias and conglomerates. Boulder clays (glacial drift deposits; if lithified: tillites) for instance are composed of rounded and angular components embedded in a fine matrix. Similar rocks are found in arid areas (arid boulder clays; lithified: fanglomerates). Nonglacial conglomeratic mudstones, however, often are the product of submarine slumps or olisthostromes (CROWELL 1957). If such mudstones do not contain marine fossils, the distinction from glacial tillites may be difficult. However, nonglacial conglomeratic mudstones do lack the characteristic faceted and striated pebbles and are not associated with laminated claystones containing rafted blocks (PETTIJOHN 1957).

Solution-breccias of dolomite suspended in a clay-mud matrix have been described by MCCALEB & WAYHAN (1969). They are caused by the development and subsequent collapse of a solution cave system.

Flat clay pebbles generally are produced by the reworking of freshly deposited clay layers. They are found in fluvial (WILLIAMS 1966) and marine sediments (TREFETHEN & DOWN 1960) as well as in turbidites. The density of such pebbles during transportation is only about 1.6 (calculated from a porosity of 65%); clay pebbles 14 mm in diameter are equivalent in weight with 10 mm-quartz pebbles. Equivalent flat clay pebbles are much larger.

Pebbles in gravel worked by waves tend to be better segregated into discrete beds, and to be laterally more regular, or less lenticular, than in alluvial gravel (CLIFTON 1973).

The following local names are used in Europe:

Nagelfluh: conglomerates of the Alpine Molasse. Subtypes are calcareous Nagelfluh with predominant carbonate and sandy limestone pebbles (fig. 3-14; SCHIEMENZ 1960), and polygene or multicolored Nagelfluh with up to 50% granite and gneiss pebbles in addition to sedimentary rocks (KLEIBER 1937; SPECK 1953). The sand matrix is mainly cemented by calcite (RUTSCH 1968).

Verrucano: violet fanglomerates, unsorted breccias and feldspathic conglomeratic sandstones, all rich in volcanic components, interbedded with violet green and gray silty shales with volcanic intercalations, from the lower Permian and Permo-triassic of the western and southeastern Alps, respectively (TRÜMPY & RYF 1965).

The detritus from areas with high relief and rapid erosion (such as orogenic areas) often contains pebbles of low weathering and transport resistance, for example, schists, granites, gneisses and limestones. On the other hand, slow denudation of an area with low relief and deep weathering often produces pure quartz conglomerates.

The size distribution of different types of pebbles is influenced by the following factors:

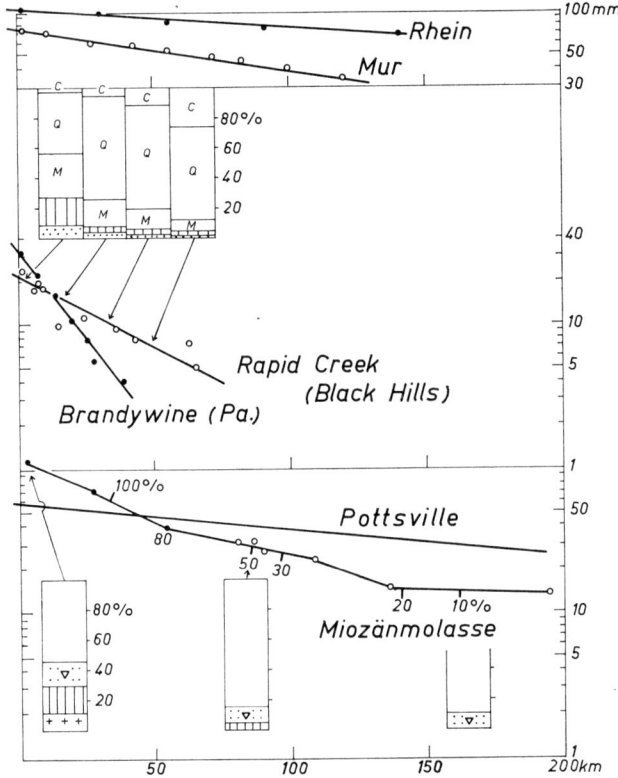

Fig. 3-11. Downstream decrease in maximum pebble size, in the percentage of conglomeratic layers (figures at 'Miozänmolasse', from FÜCHTBAUER 1954), and in the amount of pebbles sensitive to transport (histograms) are shown for modern rivers (upper two diagrams, PETTIJOHN 1960), and for older fluvial sediments (lower diagrams): Pottsville (PETTIJOHN 1957); Miocene Molasse, 0—100 m below the A-marker (circles from LEMCKE et al. 1953, dots from BLISSENBACH 1957). The average composition of the gravel in Rapid Creek from PLUMLEY (1948), for the Miocene Molasse from LEMCKE et al. (1953) and STIEFEL (1957). Legend: C = chert, Q = quartz and quartzite, M = metamorphic rocks, vertical hatching = carbonate rocks, dots = sandstones, dots with triangle = silicified sediments and metamorphic rocks, crosses = feldspar-bearing crystalline (igneous & metamorphic) rocks, white = quartz.

1. size distribution in the source area
2. sorting during transport
3. abrasion during transport

In the Miocene Molasse conglomerates shown in fig. 3-11 and 3-12 the problem arises as to whether sorting or abrasion is responsible for the decrease of the maximum pebble size in the direction of transport (see also ch. 3.212). The average composition of the gravel at various places is indicated for the Miocene Molasse in the lower part of fig. 3-11: Both the size and the percentage of pebbles as well as the relative percentage of the less resistant species decrease downstream. This can be caused by transport abrasion, but also by transport sorting. (The latter case

3.12 Conglomerates and breccias   49

would occur, for example, if the less resistant pebbles were coarser than the more resistant quartz pebbles and therefore, were deposited in the upstream areas.) The decrease in size of the resistant quartz pebbles over a short distance certainly points to the occurrence of transport sorting in this area. On the other hand, the composition of the size fractions shown in fig. 3-12 (corresponding to the very left point of the "Miocene Molasse" graph of fig. 3-11 which is about 100—200 km below the source of this river) indicates that the percentage of quartz pebbles is lower in the

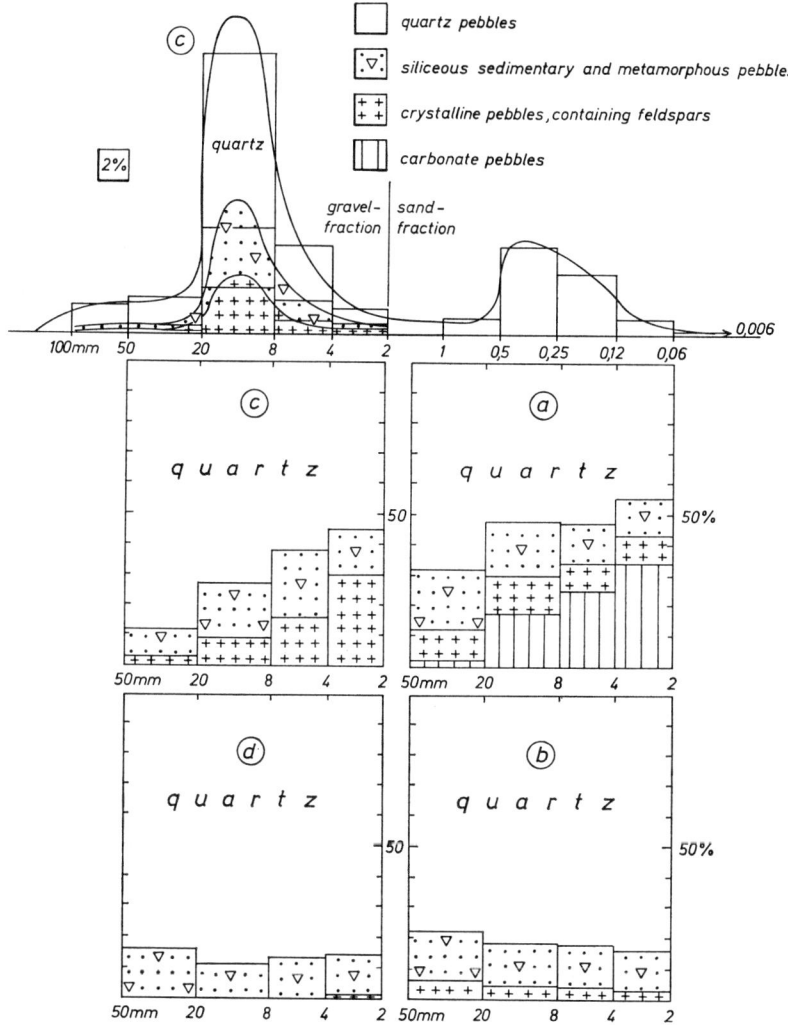

Fig. 3-12. Petrography and grain size distribution of conglomerates from the Miocene Molasse ("OSM") in Eastern Bavaria (modified from STIEFEL 1957); a and c are unweathered Northern and Southern "Vollschotter", b and d are weathered (d = "Quarzrestschotter") (3-11, 3-12, and 3-14 from FÜCHTBAUER 1967a).

finer than in the coarser fraction (fig. 3-12, a and c). If only sorting were active, the downstream increase in quartz pebbles with decreasing pebbles size (as shown in the lower part of fig. 3-11) would occur only if quartz percentages were higher in the smaller pebble sizes. The opposite is true (fig. 3-12) which suggests the transport abrasion of the less resistant pebbles. As a result, both transport sorting and abrasion were active in the distribution of the Molasse river pebbles.

As a general rule, prevailing transport sorting is indicated in cases, where the coarser fractions contain a higher percentage of less resistant pebbles than the finer fractions, and no downstream increase of the percentage of less resistant pebbles occurs in the finer fractions.

Prevailing transport abrasion generally results in an enrichment of resistant pebbles within the coarser fractions. This is because of the increase in shearing forces with increased weight of the pebble. This explains why quartz sandstones generally contain only quartz pebbles. On the other hand, quartz conglomerates are not always associated with quartz sandstones; nearly pure quartz- and quartzite-conglomerates have been found within feldspathic sandstones of the Buntsandstein (FORCHE 1935) and in the Precambrian Lorrain series in Ontario (PETTIJOHN 1957, p. 253) as well as in sandstones with rock fragments (claystone, siltstone, phyllite, and quarzite grains) of the Mississippian Pocono formation in Pennsylvania (PELLETIER 1958).

Each of the conglomerates shown in figs. 3-13 and 3-14 is composed of pebbles having similar abrasion resistivities. The conglomerate shown in fig. 3-13 has a mean grain size of 15—20 mm. It consists of (in order of decreasing frequency) quartzite,

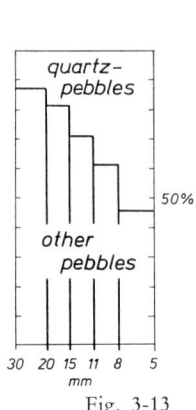
Fig. 3-13

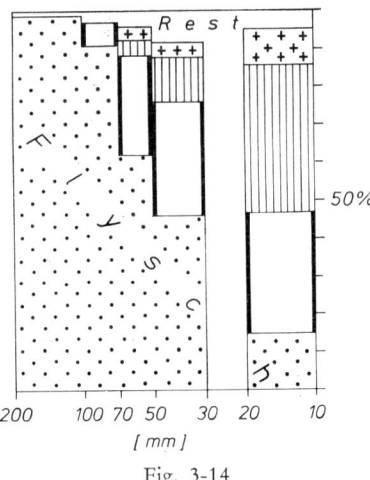
Fig. 3-14

Fig. 3-13. Decrease in the percentage of quartz pebbles with increasing grain size. Upper Carboniferous "Holzer Conglomerate" of the Saar district (modified from RÜCKLIN 1955, Loc. 28).

Fig. 3-14. Increase in Flysch pebbles (sandy limestones) with increasing grain size. Lower Freshwater Molasse of Southwestern Germany (modified from SCHIEMENZ 1960). Dots = Flysch, white = Jurassic marls, limestones, and radiolarites. Vertical hatching = Triassic dolomites and limestones. Crosses = crystalline rocks, predominantly gneisses.

quartz, "quartzite schists" + slates (RÜCKLIN 1955). Nearly equal transport distances can be assumed for all pebble types. The distribution shown in fig. 3-13 is supposed to be the result of primary size differences within the source area. The same may apply to the observation published by VALETON (1955) from the Main River area that vein quartz pebbles are less than 20 mm in diameter whereas pebbles of Mesozoic and Paleozoic rocks are as large as 50 mm: The size of the vein quartz pebbles is limited by the small width of the quartz veins in folded clastic sediments. Petrographic histograms of glacial gravel north of the Alps (BRUNNACKER 1965) show a decrease in quartz with increasing grain size of the sediments (between 2 and 40 mm), which also probably is inherited from the source area.

Contrary to transport abrasion, weathering preferentially attacks smaller pebbles, as demonstrated by comparing the less resistant non-quartz pebbles in fig. 3-12, samples b and d (weathered) with samples a and c (unweathered). Evidently, the finer fractions are changed drastically compared with the coarser fractions.

A marked increase in one species of pebbles (Flysch) with increasing grain size is shown in fig. 3-14. The resistance of these sandy limestone pebbles does not differ considerably from the other pebbles. SCHIEMENZ (1960) explained the increase by two assumptions:

(1) The source of the Mesozoic limestone pebbles was farther from the depositional area (Molasse) than the Flysch pebbles.

(2) The Flysch area was uplifted, which increased the slope for flysch pebbles, whereas the slope for the limestone pebbles decreased.

In other words, the Mesozoic limestone pebbles were removed from the river load by transport sorting.

The characteristics of coarse conglomerates in the proximal facies of turbidites are discussed by R. WALKER (1970).

3.2 Texture, fabric and environment

3.21 Grain size

Grain size, generally defined as the grain diameter, is one of the most important parameters used mainly in clastic sediments, because many properties depend on grain size (e. g. permeability, porosity, filtration properties, composition of primary minerals). On the other hand, grain size is currently used as a characteristic of the depositional environment though grain size is influenced by the source material and by previous transport mechanisms as well.

3.211 Grain size distribution

The different methods of grain size analysis are discussed in Vol. I of this series. Whereas sieve analyses separate grains according to their medium diameter (= maximum width b; ch. 3.221), sedimentation tubes measure settling velocity, which is a function of grain diameter, weight and shape. The latter type of analysis better approximates grain transport by suspension than does sieve analysis. Neither analytical method exactly simulates saltation-load and traction-load transport.

The grain size classification used in this book is shown by fig. 2-2. Sandstones are better described, however, by indicating both average grain size and sorting, e. g. "well sorted sandstone (0.2)" instead of "fine to medium grained sandstone ...".

MOSEBACH (1954) and WALGER (1965) dealt with the problem of plotting grain size analyses. Statistical problems connected with it, as well as the advantages of cumulative curves versus frequency curves are discussed by MARSAL (1967). Frequency curves and histograms are more clear than cumulative curves (fig. 3-19), but they are more difficult to use in determining statistical parameters. Moreover, they require arithmetical treatment if the grain size intervals differ. Therefore, such curves are used less frequently than cumulative curves.

Logarithmic scales are commonly used for grain size studies, for the difference between 0.01 and 0.02 mm for example, is as important as the difference between 0.1 and 0.2 or 1 and 2 mm. Using a linear scale, one would need to show the 1.01—1.02 mm interval if the .01—.02 mm interval were to be given. Several logarithmic scales have been proposed, but the Wentworth scale (based on the root of 2: 2, 1, $1/2$, $1/4$, $1/8$ mm, etc.) is most commonly used. If it is sufficient to describe a given grain distribution by a few parameters, those listed in table 3-10 are recommended. They use the Phi scale defined by

$$\Phi = -\log_2 d$$

where d is the diameter of the grains: $\Phi 0 = 1$ mm, $\Phi 1 = 0.5$ mm, $\Phi 2 = 0.25$ mm, see fig. 2-2.

Average grain size is measured in several ways: the median diameter or "median" is that diameter at which half the grains (by weight) are finer, and half

Tab. 3-10. Different grain size characteristics, listed on a Phi basis (Φ 50 for example is the grain size at the 50 percentile of the cumulative curve).

author	average grain size	sorting
TRASK (1932)	median = Φ 50	(Φ 75 − Φ 25)/2
INMAN (1952)	mean ∼ (Φ 16 + Φ 84)/2	(Φ 84 − Φ 16)/2
FOLK & WARD (1957)	mean ∼ ∼ (Φ 16 + Φ 50 + Φ 84)/3	(Φ 84 − Φ 16)/4 + (Φ 95 − Φ 5)/6.6

author	skewness	kurtosis
TRASK (1932)	$\Phi 75 + \Phi 25 - 2 \Phi 50$	—
INMAN (1952)	$\dfrac{\Phi 84 + \Phi 16 - 2 \Phi 50}{\Phi 84 - \Phi 16}$	$\dfrac{(\Phi 95 - \Phi 5) - (\Phi 84 - \Phi 16)}{\Phi 84 - \Phi 16}$
FOLK & WARD (1957)	$\dfrac{\Phi 84 + \Phi 16 - 2 \Phi 50}{2 (\Phi 84 - \Phi 16)} + \dfrac{\Phi 95 + \Phi 5 - 2 \Phi 50}{2 (\Phi 95 - \Phi 5)}$	$\dfrac{\Phi 95 - \Phi 5}{2.44 (\Phi 75 - \Phi 25)}$

coarser. It is calculated by finding the intercept of the 50 percentile in the cumulative curve. It is "the most commonly used, but least accurate, of the measures of average size" (FOLK 1966). The m o d e is defined as the most frequently-occuring grain size, indicated by the peak(s) on the frequency curve. The m e a n (that is arithmetic mean) gives a "weighted" average grain size, and can be calculated in several ways (tab. 3-10).

Sorting is a measure of the total spread of grain-size distribution, the best measure of which is standard deviation (s. below).

Skewness defines the deviation from symmetry of a curve (fig. 3-19).

Kurtosis indicates the peakedness of the distribution curve. Platykurtic curves are broad, whereas leptokurtic distributions have a pronounced peak in the center.

The TRASK sorting (So) is preferred in case of very clayey sands and, according to FRIEDMAN (1962), also for well-sorted sands. For such sands the TRASK sorting has low values, whereas high So-values indicate poor sorting. Distribution curves steeper on the coarse side have a TRASK skewness < 1 (fig. 3-19).

The INMAN sorting yields the standard deviation provided a Gaussian normal distribution is developed. His calculation of mean grain size is unsatisfactory if one is dealing with bimodal and/or skewed distributions (FOLK 1966). His skewness is 0 in symmetrical curves and positive (0—1) with a tail in the fines. His kurtosis is 0.65 in normal curves.

FOLK & WARD's indices enable a more precise characterization of grain size distributions. Their approximation to the mean avoids the shortcomings of INMAN's approximation, because they include 3 points on the cumulative curve (as opposed to 2 for INMAN). FOLK & WARD's formulae approximate the "moments" (see below). Symmetrical curves have a skewness of 0. Normal curves have a Kurtosis of 1, whereas curves with small peaks and broad tails ("leptokurtic") may have values of 1.5—3 (FOLK 1966).

The correct mathematical treatment of grain size analyses uses "moment" measures which, in contrast to the parameters used so far, are not based upon singular, arbitrary points of the cumulative curve, but use all information available, i. e. all grain size fractions (KRUMBEIN 1936). The following moments are used (MARSAL 1967):

1. Arithmetic mean $\bar{x} = (q_1 x_1 + \ldots q_n x_n)/100$

2. Standard deviation $\sigma = \sqrt{\dfrac{q_1(x_1 - \bar{x})^2 + \ldots q_n(x_n - \bar{x})^2}{100}}$

3. Moment coefficient of skewness $\alpha_3 = \{q_1(x_1 - \bar{x})^3 + \ldots q_n(x_n - \bar{x})^3\}/100\,\sigma^3$

4. Moment coefficient of kurtosis $\alpha_4 = \{q_1(x_1 - \bar{x})^4 + \ldots q_n(x_n - \bar{x})^4\}/100\,\sigma^4$.

x is the center point of the size fraction (in Phi units) and q the percentage of this fraction.

Recommended sieve intervals include 1.0 Phi if only the mean diameter is required, and 0.5 Phi if standard deviation, skewness and kurtosis are desired (ISPHORDING 1972).

A linear ordinate is most simple and has the advantage of not attaching any interpretation to the grain size distribution.

In a diagram with an ordinate scaled as a Gauss integral (probability net, fig. 3-15) and an abscissa in Phi grades, the cumulative curve of a logarithmic normal distribution (or "lognormal distribution") is a straight line. Few sediments in nature display lognormal distribution. According to DOEGLAS (1946), TANNER (1959), CURRAY (1960), FULLER (1961), WALGER (1962) and other workers, most sands can be characterized as mixtures of several lognormal distributions, each of which can be separated graphically using the cumulative curve.

The turning points on the cumulative curve define the limits between the various components (for more information see GRY 1938, DAEVES & BECKEL 1948, NEUMANN 1963, WALGER l. c., TANNER l. c., and GRIFFITHS 1967).

Some examples of size distributions are shown in fig. 3-15: The curve on the right is nearly lognormal. The sand on the left is from a desert wadi (a), and is composed of equal parts of eolian sand (b ~ component II) and a coarser lag fraction (III).

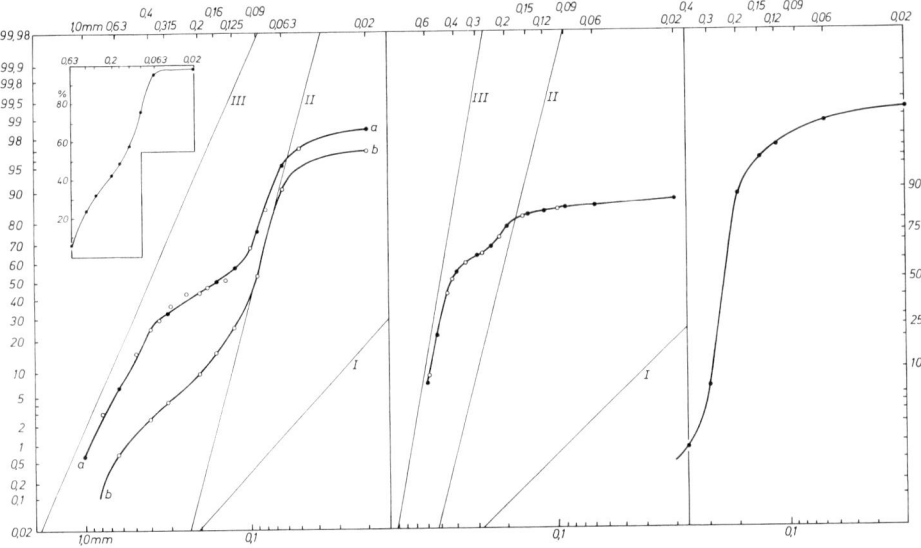

Fig. 3-15. Cumulative curves of various sands, plotted on probability scales. Right: Dogger beta, Lüben near Ülzen. This very well sorted sandstone (So = 1.1) shows a nearly lognormal distribution, with only small admixtures (about 4.5% finer and 0.3% coarser material; the fraction finer than 0.02 mm is mainly composed of diagenetic clay minerals). Center: Valendis sandstone east of Diepholz (average of many grain size analyses), can be separated into three lognormal components I (16%), II (24%), III (60%). Because of its good sorting, III appears to be a lag concentrate. Left: (a) Sand from a Sahara wadi (Libya), (b) eolian sand from the Hamada, close to a, between stones. (a) can be divided into three lognormal components: I (2%), II (50%), and III (48%). (b) is composed of 3% fines < 20 μ, 82% of II and 15% of a component, which corresponds to the upper part of (a). (In the upper left is the cumulative curve of (a) using linear subdivision of the ordinate, for comparison.) The components were determined by graphical fitting. Dots indicate data points, circles represent values determined by synthesis of the assumed components.

This suggests that the components originate from different source areas (see also FOLK & WARD 1957, HAYES 1965). According to WALGER (1962), many distribution curves displaying several peaks are due to sampling errors (i. e. combining material from several individual layers into one sample). By detailed separation of such minute layers in impregnated modern sediments, WALGER (l. c.) showed that the grain size distribution within the individual layers is nearly lognormal.

In cases of real bimodality a combination of the standard deviation and a "dispersion factor" (SHARP & FAN 1963) can be used.

Artificial disintegration of any material yields a straight-lined grain size distribution in a diagram with logarithmic abscissa and an ordinate subdivided according to the ROSIN-RAMMLER-SPERLING relation ("RRS") which is characterized by a finite upper limit of the ordinate (percentage) (MARSAL 1954). Though this has proved useful with volcanic tuffs (KITTLEMAN 1964), it is rarely used in sedimentology.

The TRASK sorting grades indicated in fig. 3-21 are unsatisfactory because they neglect the relationship between sorting and grain size: A plot of the TRASK sorting indices as a function of the median diameters for various sands reveals the best sorting at a median diameter of 0.1—0.2 mm (INMAN 1952). The lower enveloping curve of the dots in such a diagram is relatively distinct and corresponds to thin individual layers about 1 mm in thickness sampled and analysed by WALGER (1962). For this reason SEIBOLD (1963) recommended that the sorting of sandstones be related to this lower enveloping curve, which represents an "elementary sorting" (fig. 3-16). This "relative sorting coefficient", then, is independent of the grain size. Its genetic meaning is not yet clear (see p. 60) but it is preferable to the qualitative sorting terms shown in fig. 3-21 which do not take into account the grain size.

According to SEIBOLD et al. (1961), relative sorting coefficients below 1 originate from the formation of "lag concentrates", i. e. the selective erosion of finer grades leaving coarse grains.

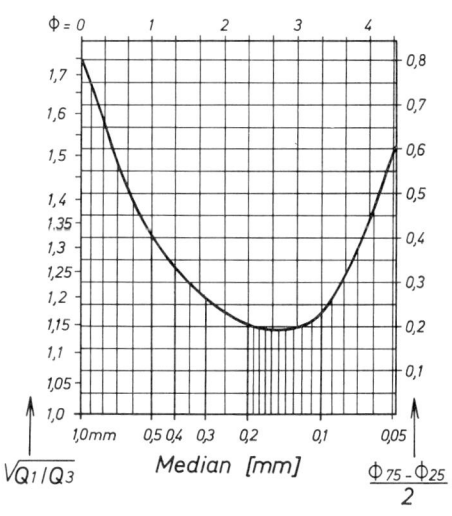

Fig. 3-16. Sorting of individual layers ("elementary sorting", WALGER 1962) of water-deposited sands. By dividing the TRASK sorting of a sandstone by the elementary sorting at the same median diameter (shown by the curve), one derives the "relative sorting coefficient", which may be indicative of the environment. For example a sand with a median diameter of 0.5 mm (Phi = 1) would have an elementary sorting of 0.4. If the sorting coefficient of the sand is 0.6, the relative sorting coefficient therefore is 0.6/0.4 = 1.5.

The combination of various modes of transportation (such as rolling, saltation, suspension; PETTIJOHN 1957, PASSEGA 1957, WALGER 1962) may lead to a sediment with poorer sorting, even within an individual layer (MOSS 1962, VISHER 1967). In contrast, beach sands in which the grains are deposited solely by saltation (FRIEDMAN 1967) generally contain only limited amounts of fine and coarse fractions and therefore are well sorted.

The general absence of grains between 2—4 mm in river and shallow sea sediments (PETTIJOHN 1957, SAXER 1952) justifies the boundary between sand and gravel shown in fig. 2-2 and 3-17. One of the few occurrences of such grains is on beaches (RUSSELL 1968).

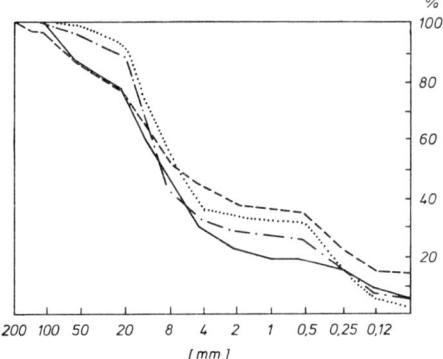

Fig. 3-17. Cumulative grain size curves from sandy conglomerates of the Upper Freshwater Molasse which demonstrate the dearth of grains between 0.5 and 4 mm (from STIEFEL 1957).

Gravel may trap sand: sand which is transported into a porous pebble layer, can be influenced by the stream shadow and thus deposited. For instance, the bimodal distribution of African desert sands can be explained by the protection of selected grains of the saltation load within interparticle voids between the coarse particles of the creep load (WARREN, in press). The observation (ch. 3.213) that most sandstones of lower energy environments (e. g. offshore) have a cumulative lognormal distribution but with a fine grained tail, may be explained in a similar way.

It may be difficult to determine the median of gravels and conglomerates. Instead, the longest diameter of the largest pebble may be measured (PELLETIER 1958), but statistically this requires a minimum number of pebbles for each sample.

3.212 Regional variations in grain size

Grain size generally decreases in the direction of transport (SCHOKLITSCH 1930, FORCHE 1935, KRUMBEIN 1937, PICKEL 1937; SWINEFORD & FRYE 1951: loess; THORARINSSON 1954: Volcanic ashes; POTTER 1955, PETTIJOHN 1957, MCDOWELL 1957, SCHLEE 1957, PELLETIER 1958, CURRAY 1960: marine sediments; HESSE 1965, CONOLLY & EWING 1966: turbidites). This is demonstrated for gravel from different modern and fossil river sediments in fig. 3-11 and for sand from the Mississippi River in fig. 3-18. Such decrease in grain size during fluvial transport may be due to several causes:

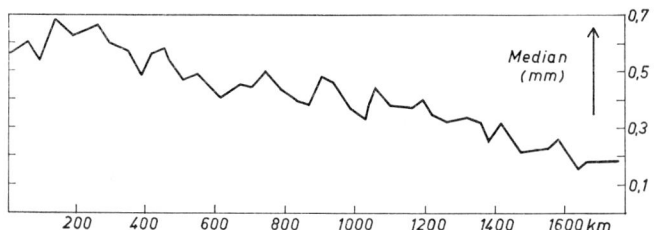

Fig. 3-18. Reduction of sand grain size downstream. Mississippi River below Cairo, Illinois (from Leopold et al. 1964).

(1) Selective sorting by decrease in transport energy (velocity), which depends on the gradient of the river, the volume of water transported, and the geometry of the river channel (approx. = slope × medium river depth), all of which change downstream. Blissenbach (1954) found a positive correlation between pebble size and slope angle in Arizona alluvial fans. Generally the river slope is less in the lower part than in the upper part of the river (Leopold et al. 1964). The width of river channels increases as slope decreases, i. e. downstream. Both influences tend to decrease the transport energy, but the increase in discharge volume from tributary rivers partially compensates for this reduction.

(2) Abrasion, which mainly affects pebbles (Russell 1939, Pettijohn 1957, p. 533—540). Abrasion of sand grains is shown only by an increase of roundness. In gravel-bearing sands this increase in bulk roundness is compensated by new fractures of sand grains. As discussed in ch. 3.12, abrasion may be shown by the decrease in percentage of less resistant pebbles in the direction of transport. Plumley (1948, p. 570) attributed $1/4$ of the reduction in pebble size in a small river to abrasion and $3/4$ to the reduction of transport energy.

Grain size will not decrease in the direction of transport if the river receives only fine-grained material from the source area or if coarser material is contributed by tributary rivers.

Falke (1966) described a small continental basin in the Upper Permian whose sediments were separated into two distinct types (sand and silt), so that the grain size distributions were complementary to one another. By integration of these two types, the provenance could be determined (fig. 3-19).

In beach environments, grain size decreases parallel to the longshore currents. This may be caused by the factors mentioned above for fluvial environments. The higher mean rate of longshore transport of smaller grains (0.61 cm/sec for grains smaller than 0.25 mm compared with 0.41 cm/sec for grains larger than 0.35 mm; Ingle & Schnack 1971) can contribute to the longshore grain size diminution before a steady-state has developed along the coast.

An increase in grain size in the direction of transport was found locally at the shore of the Baltic Sea (Seibold 1963). At the same time, the relative sorting (fig. 3-16) dropped below 1. The coarser sands may therefore represent "lag concentrates" (see p. 55).

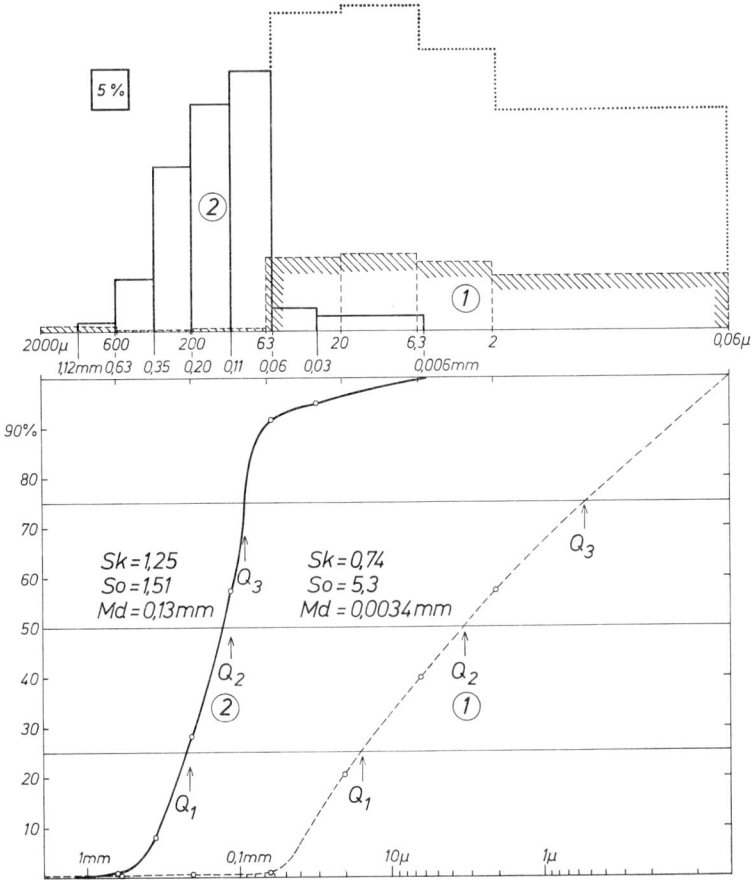

Fig. 3-19. Grain size histograms (above) and cumulative curves (below) of the Rötel shale facies (1, hatched line) and of the Kreuznach sandstone facies (2, solid line), Upper Permian near Mainz, which is supposed to be predominantly fluvial in origin. Representative analyses from FALKE (1966). The histograms are so complementary to each other that sorting (by transport) of a common source material is quite possible (e.g. channel and flood plain deposits). According to FALKE (pers. comm.) the total thickness of the Rötel shale facies is about 4 times that of the Kreuznach sandstone. Assuming an equal regional distribution of both facies it is possible to reconstruct a hypothetical histogram of the source material by taking 4 parts of Rötel shale and 1 part of Kreuznach sandstone. The resulting histogram is shown by the uppermost dotted and solid lines. Median diameter (Md), TRASK sorting (So) and skewness (Sk) are indicated.

Grain size can change also in vertical direction, such as in turbidites, but also in shallow sea sediments as will be discussed later.

3.213 Grain size distribution as an environmental indicator

Many authors have tried to correlate grain size and sedimentary environment in recent sediments in hopes of applying this technique to fossil sediments. With this intention DOEGLAS (1950), VAN ANDEL & POSTMA (1954) and SINDOWSKI (1957) have classified the shape of the cumulative curves. FOLK & WARD (1957), MASON & FOLK (1958), SAHU (1964) and MOIOLA & WEISER (1968) made use of individual parameters and their relationships. FRIEDMAN (1961, 1962), GEES (1965) and BYRNE & MALONEY (1965) have investigated correlations between the various moments.

Many authors have distinguished between fluvial or dune sands and beach sands on the basis of textural analysis (fig. 3-20). Because shearing forces are lower in a river (BRIGGS & MIDDLETON 1965), the percentage of fines in fluvial sands is usually higher than in beach sands. There are exemptions to this rule, however; for example the Mississippi river system which pumps large amounts of silt and clay into the coastal and beach areas (FRIEDMAN 1967). The distinction between fluvial and dune sands is more difficult (and perhaps impossible). Many dune sands are better sorted and often finer than adjacent fluvial sands, but there is considerable overlapping (FRIEDMAN 1961, GEES 1965).

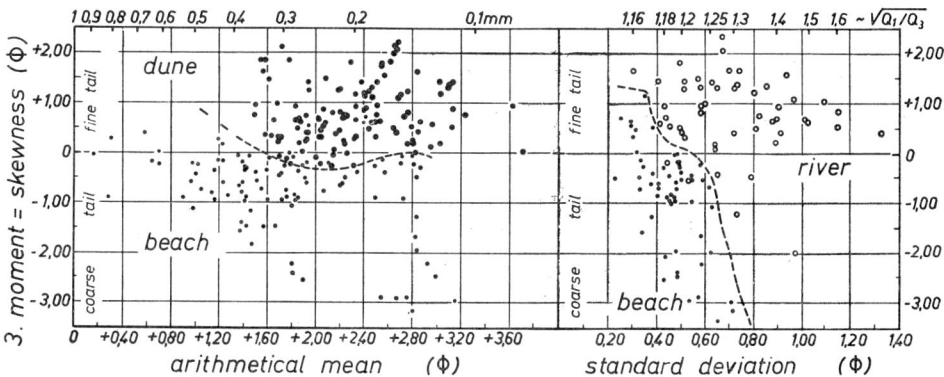

Fig. 3-20. Distinction between dune sands (left) and fluvial sands (right) and beach sands from different localities as shown by arithmetic mean, standard deviation and skewness. In the upper right are the approximate figures of the TRASK sorting. (Modified from FRIEDMAN 1961, 1962.)

Fluvial and shallow marine sands have similar skewnesses, and grain size differences of these environments are missing. However, if one combines the grain size parameters with other sedimentological characteristics (such as shape of grains, type of stratification, trace fossils, and shape of sedimentary bodies), the sedimentary environment often can be inferred. In this regard the "relative sorting coefficient" (SEIBOLD et al. 1961) is recommended (fig. 3-16).

The following review of typical TRASK parameters may serve as a first orientation.

1. fluvial environment
   a) channel and point bar
   sorting — $\sqrt{Q_1/Q_3}$ — generally > 1.2; in braided rivers frequently > 1.3
   skewness predominantly < 1. Marked changes in grain size from one layer to the other are characteristic.

b) flood plain
sorting mostly > 2, skewness always < 1 (tail at the fine end, steeper at the coarse end).

2. eolian environment
   a) dunes
   sorting better than 1 a), skewness similar to 1 a). Median size ranges between 0.15 and 0.35 mm (see also HARRIS 1958). Pebbles are generally absent. Exceptions have been described from the Antarctic, with its high wind velocities and increased air viscosity due to extremely low temperatures (SMITH 1965). Dunes are in general more or less homogeneous with respect to grain size.
   b) wind-driven sand and silt ('loess', in part)
   sorting ≥ 1.4, skewness mostly < 1 (considerable fine material), median diameter frequently < 0.1 mm (e. g. in loess, Md 0.02—0.06 [SINDOWSKI 1957, see ch. 4.222.22], and in fig. 3-15).

3. marine environment
   a) beach
   best sorted sediments generally occur in this environment (mostly between 1.1—1.23). Nearly log-normal distributions are common; skewness often > 1 (3rd moment negative, see fig. 3-20).
   b) shallow marine (tidal and shelf area)
   sorting somewhat poorer, skewness < 1.
   c) deep sea (continental slope and abyssal plains)
   The upper slope generally is silty, the lower slope has clayey silt, whereas on the abyssal plains, (silty) clay is deposited with periodic sand and silt turbidite deposits (see below; BYRNE & MALONEY 1965).

Low and high energy conditions can occur in each of these three major environments. The environments with high energies (1 a, 2 a, 3 a) are characterized by coarser and better sorted sediments, while those with low energy (1 b, 2 b, 3 c in part) have finer, less well-sorted sediments and a skewness below one.

The sizes first eroded by low-velocity water currents are between 0.2 and 0.3 mm; at these threshold velocities, coarser and finer sediments are not eroded. In air, the sizes first eroded lie between 0.05 and 0.12 mm (BAGNOLD 1941). These very low velocities, therefore, should result in better sorted sediments than stronger currents which can erode coarser and finer grains simultaneously. Thus eolian sediments between 0.05 (loess) and 0.12 mm and water-deposited sediments of about 0.2 mm in diameter (fig. 3-16) are especially well sorted. These differences in sorting should enable the differentiation between eolian and subaqueous sediments.

Velocities must be 29.3 times greater in air than in water to obtain the same drag force, in order to move grains of the same size (PETTIJOHN et al. 1972, p. 363).

As shown in fig. 3-21 sorting is not only related to the depositional environment (d, e) but also to the sandstone types (a, b, c) and to "tectonic facies" (d + e versus f).

3.21 Grain size   61

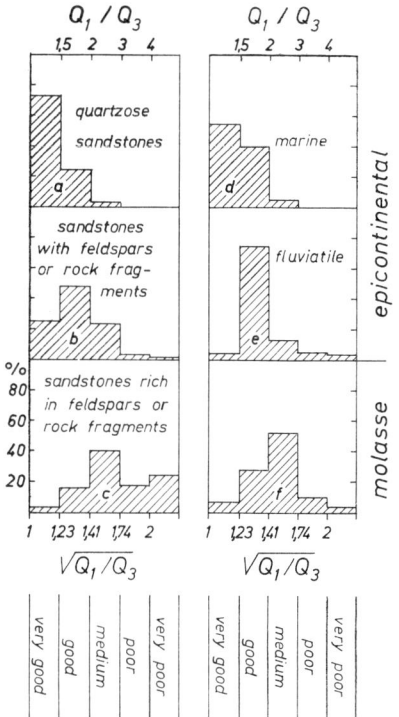

Fig. 3-21. Sorting distributions and grades (from FÜCHTBAUER 1959).

a—c) dependent upon sandstone type
d—f) dependent upon environment and "tectonic facies"

a) 53 sandstones from Northern Germany (predominantly Lower and Middle Jurassic)
b) 178 sandstones from Northern Germany (predominantly Valanginian = Lower Cretaceous)
c) 244 sandstones (Tertiary Molasse, Carboniferous, Buntsandstein)
d) 176 marine sandstones, predominantly from Northern Germany (Liassic-Valanginian)
e) 105 fluvial sandstones from various countries
f) 201 sandstones from the Tertiary Molasse

3.214 The matrix problem

The relative percentage of sand and fines in a sediment depends (1) on the availability in the transport system, (2) on the mechanism of deposition, and (3) on the stability of the depositional area:

(1) A transporting system rich in mud will deposit mud even in a high-energy environment (FRIEDMAN 1967). This may result in mud-supported fabrics of sand-in-mud and probably even pebbles-in-mud ("wackestones", DUNHAM 1962; see ch. 5.2).

(2) A mechanism of transport and deposition such as a turbidity current favors the formation of muddy sands. According to HORN et al. (1971b), most turbidite sands (even coarse sands) in Atlantic abyssal plains contain more than 10% of material < 63 μ. These authors concluded that such sediments were modern equivalents of graywackes.

(3) Rapidly subsiding depositional areas such as molasse troughs and deep basins such as flysch troughs do not favor reworking processes. On the other hand, reworking is frequent in platform sands. Reworking in marine environments, e.g. of deltaic sediments, will reduce the matrix content, because in general the marine currents are low in suspended load. The frequency of reworking increases with the transport distance. The observation e. g. by SEIBOLD (1963, figs. 17, 18) and FÜCHTBAUER

(1964a, p. 242), that sorting of sands improves during transport, can probably be explained by an increasing number of reworking processes.

A difference in sandstone sorting in orogenic and cratonic areas as suggested in (3) is shown in fig. 3-21 (d-f) and in the following example from Germany. Whereas platform sandstones of Middle Jurassic and Lower Cretaceous age and 0.13 mm median diameter contain an average of only 1 percent $< 20\,\mu$, molasse sandstones of Upper Carboniferous and Tertiary age and the same median diameter of 0.13 mm contain about $10\% < 20\,\mu$.

It seems to be a general rule that, if the sandstone sorting is improved, the sorting of adjacent shales is improved at the same time, as shown in table 3-11.

Table 3-11. Synopsis of the overall sorting of three sandstones and adjacent shales.

	cratonic tectofacies		orogenic
occurrence	Lower Jurassic, E of Hannover	Lower Cretaceous, Emsland area	Molasse N of the Alps
environment	marine	marine	marine and fluvial
sandstones color relative silt + clay content	white low	light gray low to medium	medium gray high
shales color relative sand content	gray to black low	dark gray medium (?)	medium gray high
reworking (inferential)	high	medium	low
subsidence of basin	medium	medium	high
distance of transport	high	medium	medium and low
overall sorting	good	fair	poor

The high matrix content of several sandstones, especially the graywacke type, has puzzled many workers. CUMMINS (1962) suggested a **diagenetic origin of matrix** in such rocks. Two lines of evidence have been proposed:

(a) petrographic composition

BRENCHLEY (1969) described shallow water graywackes with 40—60 percent matrix, which apparently formed by diagenetic alteration of volcanic rock fragments. This is supported by the observations (1) that this matrix differs considerably in composition from the adjacent shales, (2) that indistinct grain boundaries are visible in the matrix, (3) that the unstable rock fragments are preserved (or calcitized), where these sandstones are cemented by calcite. — Similar examples of diagenetic matrix forming in sandstones rich in volcanic (unstable) rock fragments are reported by WHETTEN & HAWKINS (1970) and LOVELL (1972). As modern equivalents, sands rich in volcanic rock fragments are frequent in and adjacent to the circumpacific volcanic belt (DUNCAN & KULM 1970) see ch. 3.113.

(b) sedimentation process

On the basis of flume experiments, KUENEN (1966b) concluded that in coarse sandstones containing more than 20 percent fines $< 40\,\mu$, the fines must be at least partly diagenetic.

A primary matrix can also be removed during diagenesis. This was shown by LAJOIE (1968), who compared silica- and carbonate-cemented flysch sandstones from the same occurrence; in the calcareous sandstones, silica cement and matrix were removed prior to carbonate cementation.

3.22 Grain form

The grain form is defined in a descending order of magnitude by three more or less independent properties: (1) shape, i. e. a quantitative answer to the question, whether a pebble or grain is globular or elongated, bladed or platy; (2) roundness, i. e. a quantitative answer to the question, whether a pebble or grain is angular or rounded; (3) surface texture, i. e. (at present) a qualitative description of the minor pits covering a grain.

3.221 Shape

The shape is defined by three parameters:

a (or L) longest diameter ("greatest length")
b (or l) longest diameter perpendicular to a ("greatest width")
c (or E) longest diameter perpendicular to a and b ("greatest thickness")

Three diameters can be measured in pebbles, but in sand grains mounted on slides, only a) and b) can be measured, and in thin sections, usually none can be measured.

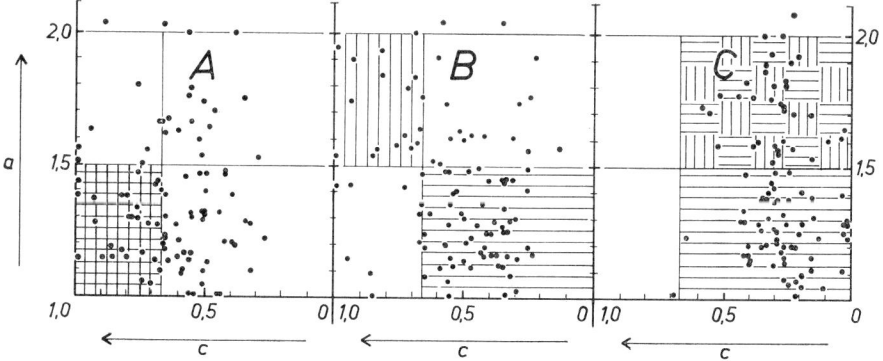

Fig. 3-22. Shape of Pleistocene pebbles of the river Main. a = longest axis, (b = middle axis, taken = 1), c = shortest axis. Chert (A) tends to show equant forms (crosshatched). Graywackes (B) have tabular (horizontally hatched) to rod-shaped (vertically hatched) forms. Shales (C) prefer tabular (horizontally hatched) to bladed (vertical and horizontal hatching) pebbles (from VALETON 1955, limits from ZINGG 1935).

3.221.1 Shape of pebbles

Numerous indices have been proposed to interrelate the three pebble diameters (see KÖSTER 1964). The diagram by VALETON (1955) (fig. 3-22) is particularly clear.

Extremely rod-shaped pebbles ($a > 2$; $c > \cdot 7$) are rare (fig. 3-22). Both the "flatness" $(a + b) / 2c$ (CAILLEUX 1945) and the "sphericity index" $\sqrt{c^2/ab}$ (SNEED & FOLK 1958; recommended by HUMBERT 1968), disregard any differences between a and b.

Pebble shape mainly reflects its structural properties (fig. 3-22): the foliation of shales and graywackes parallel to the bedding planes. Cleavage planes or joint planes always produce flat or rod-like fragments, so there is no chance for round or isometric pebbles to be produced. If no preferred foliation is present, sphericity (see below) increases with increasing distance of transport (PETTIJOHN & LUNDAHL 1943).

The mode of transportation does seem to have some influence on the shape of the pebbles. Generally, beach pebbles are flatter than river pebbles (CAILLEUX 1945, 1961, TRICART 1951, KUENEN 1964a), while the latter are more flat than pebbles from glacial and solifluction material (K. RICHTER 1958). Frequently rodlike shapes are found in river gravel, especially in the lower parts of the river (BLATT 1959, LÜTTIG 1962b, 1964a). In coastal areas pebbles generally undergo more intense abrasion on steeper shores; it is here that spherical pebbles are produced (WENTWORTH 1919, LÜTTIG 1964a). In coastal areas of lower energy flat pebbles collect (KUENEN 1964a). "On sandy, low-wave-energy beaches the smallest pebbles are flattest, while on high-wave-energy gravel beaches the largest pebbles are flattest. The optimum sliding size of pebble is a measure of surf vigor" (DOBKINS & FOLK 1970). For further information see the summaries of CAILLEUX (1945) and LÜTTIG (1962a).

3.221.2 Shape of sand grains

The shape of sand grains is best defined by their elongation $a:b$ (SCHNEIDERHÖHN 1954). The term "sphericity" of WADDELL is defined as

$$\frac{\text{diameter of a circle of the same surface area}}{\text{diameter of the smallest circumference}}$$

though it may be confused with roundness at the upper end of the scale: in nature only well rounded grains can reach values of close to 1 ($=$ sphere).

Sphericity (and roundness) can also change in sands by abrasion during transport, but the degree of change generally increases with increasing grain size, i. e. with stronger bottom friction (PETTIJOHN 1957, fig. 33, 34; J. R. L. ALLEN 1962). The increase of sphericity during transport is suggested by the fact that quartz grains in sandstones are more spherical than those in granites (Moss 1966).

The plane perpendicular to the c axis in quartz has the greatest resistance against grinding abrasion (SCHUMANN 1941); the c-axis and the longitudinal axis of quartz grains, therefore, often are nearly coincident. Nevertheless, in sandstones isometric quartz grains are frequent whereas feldspar (especially plagioclase) grains are more or less rounded cleavage fragments. Eolian transport preferentially selects less spherical but more rounded sand grains; this reduces the sphericity but increases the roundness in dune sands compared with adjacent beach sands (KRUMBEIN & SLOSS 1963, MATTOX 1955, SINDOWSKI 1956, SHEPARD & YOUNG 1961). In the breaker zone, sphericity also diminishes in the direction of transport (PETTIJOHN & LUNDAHL 1943, PETTIJOHN 1957, p. 551).

3.222 Roundness

Roundness is a quantitative measurement of the lack of angularity, independent of shape.

3.222.1 Roundness of pebbles

A number of roundness indices (KÖSTER 1964) have been proposed. According to VALETON (1955), the CAILLEUX (see fig. 3-23) index is not very accurate, and many of the numerical calculations are time-consuming. The classification by KRUMBEIN (1941) of 5 roundness groups, based upon visual comparison (1. angular, 2. subangular, 3. subrounded, 4. rounded, 5. well rounded), is preferred. Because the pebble shape is somewhat different from that of sand grains, roundness scales for one cannot be transferred directly to the other.

In most cases roundness of a pebble is a direct product of its abrasional history during transport. This would suggest that one could relate roundness to distance of transport. However, some rocks, such as basalt, may disintegrate to rounded pieces during weathering. In contrast, easily cleavable rocks may fracture and break apart several times during transport, thus interrupting the continuous increase of roundness. The history of multicycled pebbles during periods of prior transport is also important (LÜTTIG 1964a).

Fluvial transport has been studied by many workers in tumbling barrel experiments (see PETTIJOHN 1957, BERTHOIS & PORTIER 1956, 1957a). KUENEN (1956) better approximated natural conditions by using a round basin in which a rotating current moved the pebbles around the horizontal floor. Using rocks of different resistivity he found that abrasion is highest in sand-free gravels and depends on the current velocity. All experiments show that abrasion is highest during the initial stages of transport; after a few kilometers a considerable roundness is achieved.

Rounding tends to be more intense in beach environment than during fluvial transport. KUENEN (1964a) simulated the breaker zone in a narrow flume 7 m in length. He found that pebbles of 2.5—55 mm in diameter were well rounded after several weeks of continuous small breaker action. The relative loss in weight of the limestone, quartzite and chert pebbles was 10 : 3 : 1. Because pebbles tend to rest on the foreshore and are not continually exposed to breaker action, the time required to round is much longer than experimentally determined. However, a considerable roundness in the coastal area probably is achieved by pebbles only after years or centuries, and in geologic terms, this period is short. Rounded bottle fragments at the beaches are good examples of this.

From these experiments it is not surprising to find that the majority of pebbles are rounded after transport of 15—30 km (POTTER & PETTIJOHN 1963). The most considerable increase in roundness occurs between the first 5—10 km (PETTIJOHN 1957, figs. 130, 131). An example from the Harz Mountains is shown in fig. 3-23. The roundness of the quartzite fragments is not affected appreciably during transport within the moraine (k-m), but subsequent fluvial transport (m-p and a-e) achieves substantial roundness over comparable distances (3—9 km). BLISSENBACH (1954) found a considerable increase in the roundness of gneiss fragments within the first 5 km; over the same interval sphericity did not change. That roundness increases more readily than sphericity also is well established in limestones (KRUMBEIN 1941, experiments; and PLUMLEY 1948, in nature). The most intensive increase of roundness in limestones, metamorphic schists and flint during transport parallel to the coast occurs within the first kilometer (VAN ANDEL et al. 1954). A similar increase in roundness has been shown by PETTIJOHN (1957, fig. 132). As the transport does not run strictly parallel to the coast but zigzags, the true transport distances are considerably longer.

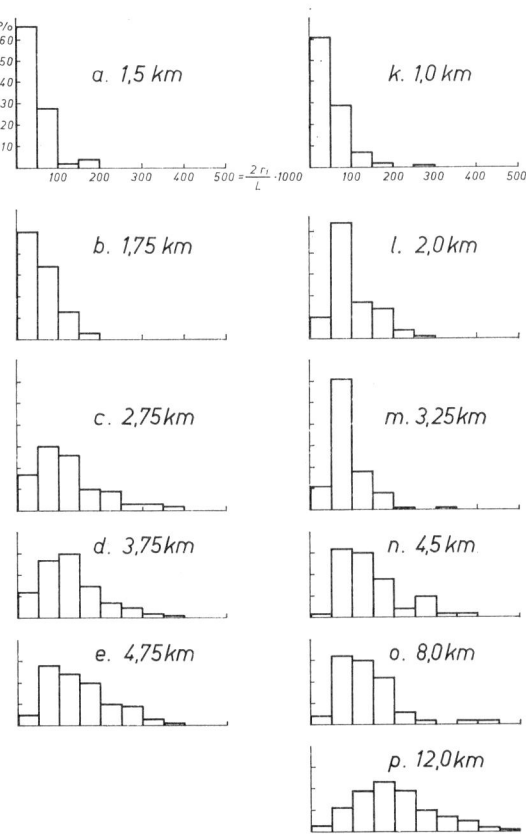

Fig. 3-23. CAILLEUX roundness diagrams. Abscissa = $2r_1 \times 1000/L$; r_1 = smallest radius in the plane of greatest length and width, L = greatest length. Ordinate = percent frequency of occurrence. a—e: change in roundness of quartzitic bed load during fluvial transport of 3.25 km (Mariental, Harz Mts.). k—p: change in roundness during moraine transport (k—m) and subsequent fluvial transport (n—p) of quartzitic gravel (Siebertal, Harz Mts.). a and k are solifluction regoliths; all fragments have weathered crusts 3 mm in thickness, according to KUENEN (1956). From HÖVERMANN & POSER (1951).

Roundness is not only a function of grain size, but also varies between fluvial and marine sediments. In rivers the roundness increases with the grain size. At the coast the situation is more complicated: larger pebbles are moved smaller distances and thus abrasion is less when compared with smaller pebbles and sand. On the other hand, bottom friction decreases with grain size. Thus, there should be an intermediate gravel or sand grain size which shows the highest roundness. This was verified by experiment (KUENEN 1964a), and also in nature: SEIBOLD (1963) found the maximum roundness at 0.5 mm (sand) along the shore of the Baltic Sea. In turbidites, however, the components are not in contact with the bottom during transport, and therefore, roundness is independant of grain size (MIDDLETON 1962).

3.222.2 Roundness of sand grains

As mentioned above, roundness of sands also increases with grain size. The roundness of sand grains is best determined by use of the matching chart of RUSSEL-TAYLOR-PETTIJOHN (MÜLLER 1967c), by which the entire sample or the roundness of each grain can be classified microscopically, and a roundness index calculated. For this purpose RUSSELL & TAYLOR (1937) as well as PETTIJOHN (1957) have based each grade upon the medium number of the "Wadell" roundness (MÜLLER 1967c). It is recommended that these numbers are used only if measurements are done according to the techniques of WADELL (1935). For visual comparison, however, the roundness grades can be numbered as follows (RUSSELL & TAYLOR 1937):

angular	1
subangular	2
subrounded	3
rounded	4
well rounded	5

RUCHIN (1958) used grades of 0—4, but according to PETTIJOHN even newly broken, angular grains have a certain degree of roundness. Thus, the number 1 is used in preference to 0, since the latter infers no rounding.

The number of grains falling within each roundness grade can be multiplied by the corresponding grade number and the sum of these products divided by the number of all identified grains. In this way the mean roundness (and its position on the above table) can be calculated.

HOFMANN (1956) published an experimental method of determining the grain form (shape + roundness) of sands: The specific surface (cm^2 grain surface per g sample) of a sand is calculated from porosity and air-permeability readings and divided by the theoretical specific surface which is calculated from the grain size distribution assuming spherical grains. The quotient is the so-called angularity coefficient. He quotes the following examples:

St. Peter sandstone (Ordovician, Ottawa, Illinois)	1.12
Dune sands	1.18 and 1.27
marine sands	1.30 — 1.37
fluvial sands	1.31 — 1.62

Sometimes "pivotability" (SHEPARD & YOUNG 1961, KUENEN 1964b, G. MÜLLER 1967c) or "rollability" (WINKELMOLEN 1969), properties derived experimentally on both shape and roundness, are more relevant for geological considerations. Rollability increases with increasing grain size in "lag deposits" (e.g. on marine shoals and banks) whereas it decreases with increasing grain size in "receiving deposits" (WINKELMOLEN) e.g. in tidal flats. THIEL (1940) showed experimentally that after 8000 km of transport a diameter of only 62 µ was removed from 1.5 mm quartz grains. We may conclude that fluvial transport will not change the grain size very much. Roundness also increases slowly: Experiments by BERTHOIS & PORTIER (1957b) showed that only 5% of the quartz grains between 1.48—2.18 mm were well rounded after a 2040 km transport. These figures are of qualitative value, however, since they depend upon the experimental conditions. In small rivers of the Black Forest the content of 1 mm grains with abrasion marks increases from 7% at the headwaters to 17% 100 km downstream (ZIMDARS 1958). Experiments by KUENEN (1960a) showed that rounding produced during transport parallel to the shore is slightly more pronounced than during fluvial transport, but the eolian abrasion is 100—1000 times stronger than fluvial abrasion. The kinetic energy of a wind-blown grain is 430 times greater than of a grain moved by water of the same drag force (PETTIJOHN et al. 1972, p. 363). Moreover, cushioning in air prior to impact is minimal compared with water (KUENEN 1960b).

Therefore dune (and beach) sands are better rounded than fluvial sands (VAN ANDEL & POSTMA 1954, BEAL & SHEPARD 1956, WASKOM 1958, ZIMDARS 1958). With predominant onshore winds the roundness in the littoral dunes is better than on the beach. This is explained by the preferred blowing out of rounded grains from the beach (SHEPARD & YOUNG 1961); ZIMDARS (1958) proved that in such littoral dunes there is no real eolian abrasion, since dull grains (which should develop through rounding by wind, see next chapter) are absent.

A decrease in roundness can result from the fracturing of sand grains during transport (MOSS 1966); this occurs mainly in gravel-bearing river beds (BOYER 1949). Increased roundness of quartz overgrowths with distance from the parent rock has been reported from the beach of Bornholm (FÜCHTBAUER & ELROD 1971). In other cases, sorting may play an important role in increased roundness (PETTIJOHN 1957, p. 549).

Feldspars abrade several times more rapidly than quartz grains (KUENEN 1960a, b).

From the foregoing it does not seem possible to determine the sedimentary environment of sandstones simply by measuring the roundness of grains. Roundness, however, can provide a useful distinction of various sand bodies. Occasionally, different sand sources can be established by roundness indices. For instance more than one sand source can be assumed if finer fractions in a sandstone are better rounded than the coarser fractions (KOCH & BLISSENBACH 1960, p. 87, BANERJEE 1964, FÜCHTBAUER & ELROD 1971).

The sand grains in fine sandstones are better rounded than grains of the same diameter in coarser sandstones from the same locality. This is due to the fact that grains in weaker currents (producing fine sand) have more contact with the bottom and therefore, are more intensely worn than in strong currents in which they are kept in suspension (FÜCHTBAUER 1967c).

Diagenesis can obliterate primary grain roundness especially if pressure solution occurs.

3.223 Grain surface texture

The microtexture of grain surfaces, i. e. the shape and depth of the pits covering the surfaces, can be investigated under the binocular (under oblique light) or by electron microscope. The best and easiest grain sizes for oblique light investigation range between 0.3 and 2.0 mm: The grains can be classified as (1) unworn, (2) polished (brilliant; in french: émoussé-luisant), or (3) frosted (dull; in french: rond mat, CAILLEUX 1942, 1943, 1952). Frosted grains generally originate from eolian transport, polished grains from subaqueous transport (v. BRAUN 1953, CAILLEUX & TRICART 1959. SCHNEIDER & CAILLEUX 1959, ZIMDARS 1958).

After extensive exposure to marine conditions more than 50% of all sand grains above 0.5 mm have brilliant surfaces; treatment in fluvial environment results in a considerably lower percentage of polished grains (CAILLEUX 1943, ZIMDARS 1958). An example is shown in table 3-12.

Table 3-12. Surface textures of quartz grains, 0.5—1.0 mm in diameter, from fluvial and marine Tertiary Molasse in Switzerland (v. BRAUN 1953).

	unworn	polished (brilliant)	frosted (dull)	average from
Upper Freshwater Molasse	85%	10%	5%	13 samples
Upper Marine Molasse	40%	55%	5%	46 samples
Lower Freshwater Molasse	80%	10%	10%	8 samples

Transmission electron microscopy has improved grain surface study. Triangular etching figures, about 0.1 μ in width, may occur on the surface of sand grains in aqueous environments. In eolian environments the grains are completely covered with irregular and flat scars, about 0.01—0.1 μ in size (BIEDERMAN 1962). The depth of these scars is similar to the thickness of the amorphous and easily dissolved surface layer of the quartz grains (NAGELSCHMIDT et al. 1952). The main advantage of electron microscopy compared with light microscopy is the ability to observe traces of different environments on the same grain surface.

Scanning electron microscopy is less time-consuming than transmission electron microscopy and produces direct pictures with a great depth of focus. In accordance with the observations by transmission electron microscopy the following environmentally controled features have been found (KRINSLEY & MARGOLIS 1969, GEES 1969):

a) glacial: Very high relief patterns: Conchoidal breakage patterns, largely varying in size; parallel striations of different size (see also COCH & KRINSLEY 1971, BLACKWELDER & PILKEY 1972).

b) littoral: High relief patterns: V-shaped indentations, oriented (chemical etching) and randomly oriented (chemical or mechanical origin; solution possibly enhanced by abrasive forces, SIEVER 1962), straight or slightly curved grooves and scarps.

c) fluvial: Similar to b, but patterns possibly more widely spaced.

d) eolian: Low relief patterns: Flat pitted surfaces; meandering ridges; oriented fractures (see also RICCI LUCCHI & DALLA CASA 1970).

As Schneider (1970) demonstrated, however, the original surface textures of the grains prior to transport should be taken into consideration before making environmental interpretations.

The following mechanism of eolian quartz grain frosting and rounding has been suggested based on experiments by Margolis & Krinsley (1971): Disordering and hydrating of surface layers of quartz grains, which were fractured by continued impacts of other grains, can lead to chemical dissolution and redeposition of silica by the action of "desert dew". Experimentally, supersaturations as high as 300 ppm SiO_2 compared with a solubility of 10 ppm for quartz and 120—140 ppm for amorphous silica have been obtained by grinding sand-sized quartz grains in a bottle.

Other studies in this promising field have been made by Porter (1962), Krinsley & Takahashi (1962 a, b), Krinsley (1965), Bramer (1965), and Margolis (1965).

Valuable as it may be for investigations in many fossil rocks, determination of the sedimentary environment solely on the basis of surface texture is unreliable, for there are numerous interfering factors. The most important are diagenetic changes, e.g. solution and growth seams (Zimdars l. c.) as well as later imprints. For instance, grains frosted by eolian action may quickly be polished by fluvial action.

3.23 Orientation

3.231 Orientation of pebbles

A. Flat pebbles in fluvial gravels commonly show imbrication, i.e. an up-current dip of 15—30° (Cailleux 1945, Ruchin 1958, Doeglas 1962, Johansson 1965). Kürsten (1958) found a dip of 50° in a dry valley of Persia. Near the coast, however, pebbles show a slight dip in a seaward direction (Cailleux 1945, Ruchin 1958). Kalterherberg (1956) found a leeward dip (i.e. in the current direction) in conglomerates with considerable amounts of sand in rocks of various ages. The same was observed in turbidite conglomerates with sandy matrices (Kopstein 1954) whereas Plessmann (1961) and Boccaletti & Micheli (1970) found imbrication in such cases. Imbrication is the most stable position of flat pebbles. It is missing only if the current is too weak, such as at the terminal end of turbidity currents, or if the pebbles have been embedded in sand so rapidly that the current could not rework them.

B. Oblong pebbles roll in a position perpendicular to the current (transverse). When the current velocity drops below the minimum value of rolling, however, the orientation of the pebbles changes to a longitudinal position, i.e. the long axes are parallel to the current. This is how it happens: The pebble touches the ground at several points with its weight distributed unequally. On an ideally flat ground the pebble pivots on its point of contact next to its center of gravity until the longer axis points downstream. If a longer axis is not developed, the obtuse side turns upstream, the pointed side downstream, thus establishing a "streamline" position which is most stable, as the resistance against the flowing medium is at a minimum. If the majority of the obtuse sides points in a similar direction, in a conglomerate longitudinal orientation can be assumed, whereas statistical orientation of the obtuse sides in both directions indicates transverse orientation. This applies also to fossils (A. H. MÜLLER 1957, SEILACHER 1960).

On an irregular bottom orientation becomes poorer. In well sorted conglomerates the main orientation can be obliterated by mutual interference of the pebbles (BRINKMANN 1955). If current velocity drops rapidly, there is no chance for pebbles to become oriented parallel to the river length: large quantities of pebbles accumulate rapidly thus fixing their transverse positions. This is reported from a gravel bed in the Iller River, which formed during a short, but extreme flood (SCHIEMENZ 1960). This happens frequently in fluvial sediments (POSER & HÖVERMANN 1952, SEDIMENTARY PETROLOGY SEMINAR 1964), but mostly a second maximum of pebbles oriented parallel to the current is developed (PETTIJOHN 1957, p. 78—80, SCHIEMENZ 1960, DOEGLAS 1962).

In intermittent rivers a longitudinal orientation of pebbles is generally missing; two nearly transverse orientations are present, probably due to the meandering. In solifluction regoliths and (less clearly) in ground-moraines the orientation of oblong pebbles is predominantly longitudinal. The direction of movement may be shown additionally by scratched pebbles (glacial striations) (K. RICHTER 1932, POSER & HÖVERMANN 1951, LÜTTIG 1954, PETTIJOHN 1957, p. 80, KÖSTER 1964, p. 199).

Also in periglacial blockfields the material tends to be aligned parallel to the local slope direction. As shown by air photo analysis, the degree of this alignment increases with the distance of travel (CAINE 1972).

On beaches the long axes of pebbles frequently are oriented parallel to the coast (KRUMBEIN 1940, RUCHIN 1958, fig. 161). Sedimentation from suspension currents generally is such that after touching the bottom, pebbles are not rolled but only pivoted. Therefore, longitudinal positions prevail (KALTERHEBERG 1956, DOEGLAS 1962).

The type of orientation even of elongated grains may strongly be influenced by the pebble shape. Extreme rod-like pebbles prefer transverse orientation (KALTERHEBERG 1956).

C. The orientation of fossils and pieces of wood varies considerably (see POTTER & PETTIJOHN 1963, WIESENEDER 1962).

3.232 Orientation of sand grains

In sands such investigations are restricted to distinctly oblong grains with length/width ratios > 1.3 or > 1.5. The rules of orientation are the same as with pebbles but longitudinal orientation is more prominent. In fluvial sands imbrication also occurs (RUSNAK 1957, GAURI & KALTERHEBERG 1966). In marine sands, an indistinct transverse maximum nearly always is developed in addition to the longitudinal orientation (SEIBOLD 1963). This transverse orientation is especially pronounced in rapidly deposited sands, as it is in pebbles (SCHWARZACHER 1961). Grains with longitudinal orientation can be imbricated similar to pebbles, or point towards the current with their thick end. Orientation in eolian sands is less clear than in aqueous sands (DAPPLES & ROMINGER 1945). Current lineation (p. 90) and joints may follow the preferred longitudinal direction of the sand grains (SPOTTS & WESER 1964).

Various grain orientations have been reported from turbidites: transverse (BOUMA 1962, p. 84—85, BALLANCE 1964), nearly longitudinal (KOPSTEIN 1954, MC-IVER 1961, MCBRIDE & KIMBERLY 1963, SCOTT 1966, ONIONS & MIDDLETON 1968), and deviating about 40—60° from the "current direction" (SPOTTS 1964). MCBRIDE (1962) reported imbrication. Grain orientation which was about parallel to the directions of sole marks was found only in the middle of the turbidite beds by COLBURN (1968); imbrication in the upper part was opposite to that in the lower part of the bed. BOCCALETTI & MICHELI (1968) reported that grain orientation parallel to the transport direction was better near the top than near the bottom of the graded intervals.

Theoretically grain orientation allows one to distinguish between oblong sand bodies of fluvial and marine origin. The elongation of point bars in rivers parallels the flow direction, which is true also of most long axes of grains. The grain orientation in marine sand bars that parallel the coast, however, coincides with the direction of the wave motion, i. e. about perpendicular to the elongation of the sand bars (CURRAY 1956, SEIBOLD 1963). This relationship may become indistinct offshore; currents running parallel to the coast cause additional grain orientation in this direction (DODGE 1965).

The three-dimensional measuring of sand grains is extremely time-consuming (SCHWARZACHER 1951). Instead, two-dimensional peels covered with grains or thin sections (ZIMMERLE & BONHAM 1962) are used. Contrary to single-grain measurements, aggregate methods cover larger volumes but the determination of grain orientation is only indirect. Such aggregate methods make use of the thermal conductivity, electrical resistance, acoustic anisotropy, elasticity, air and light permeability. Especially effective are those using the anisotropy of the magnetic susceptibility and dielectric anisotropy [MCIVER 1961, POTTER & PETTIJOHN 1963, POTTER & MAST 1963, REES 1965, SHELTON & MACK 1970 (dielectr. anisotr.), v. RAD 1970 (magnet. susc.)].

3.24 Stratification

3.241 Internal structures of sandstone beds

The stratification present in most sedimentary rocks results from vertical change of the type of material or from the orientation of anisotropic particles (e. g. in shales). Organic reef limestones and some massive sandstones, breccias and conglomerates, tillites and some claystones lack apparent stratification. However, laminations can be detected even in some massive sandstones by X-ray "radiographs" (HAMBLIN 1965, G. MÜLLER 1967c). In claystones, stratification becomes visible after weathering. — Monographs and documentations of bedding types have been published by BOTWIN-KINA (1962), PETTIJOHN & POTTER (1964), GUBLER et al. (1966), CONYBEARE & CROOK (1968), RICCI-LUCCHI (1970).

The hydrodynamic interpretation of bed forms and stratification is based on the flow regime concept:

a) Upper flow regime

Shooting flow. "Sand moves in a streaming fashion, lending an appearance to the bottom not unlike that of a sand-covered conveyor belt. Sediment transport rates are relatively high because the grains move almost without stopping rather than intermittently as in ripples or dunes." Water surface is inphase with bed surface (low bed roughness, SIMONS & RICHARDSON 1966). Steep-gradient braided rivers are characteristic (HARMS & FAHNESTOCK 1965).

b) Lower flow regime

Tranquil flow. Resistance to flow ("bed roughness") is large and sediment transport relatively small, compared with the upper flow regime (SIMONS et al. 1965). Low-gradient meandering rivers are characteristic.

With increasing flow velocity, the order of occurrence of the forms of bed roughness in the lower flow regime is: ripples, ripples on dunes, dunes. After a transitional stage, the upper flow regime begins with planar beds followed by antidunes (fig. 4-27, p. 179) and chutes & pools (fig. 4-27). — For bed material coarser than 0.6 mm, dunes form instead of ripples after the beginning of transport at low current velocities (SIMONS et al. 1965). — The transition from the lower to the upper flow regime is accompanied by a transformation of the transversal vortices into longitudinal, helicoidal water or air vortices (J. R. L. ALLEN 1968 b).

The following bedding types are distinguished:

3.241.1 Horizontal bedding

At mm-scale, it occurs in different environments. Examples are:

a) chemical sediments (e. g. interlayering of anhydrite and dolomite)
b) alternations of marlstone layers and limestone beds
c) varved clays
d) alternations of clay and silt layers
e) intercalations of mica in fine sand ("Stromwechselschichtung", BRINKMANN 1926)
f) turbidites and laminites: sediments of turbidity currents with "flysch-type graded bedding" (suggestion for terminology by KUENEN), i. e. fining upward, beginning with a coarse layer containing fines (see ch. 3.242.23 and 6.31)
g) "matrix-free graded sediments", grading from coarse sediment without fine matrix at the base to fine material at the top; different origin, e. g. fluvial conglomerates (SCHIEMENZ 1960, ALLEN 1962, KLEIN 1964, 1965)
h) horizontal lamination in sand: in rivers, in the breaker zone and on sand shoals in the shallow sea, where waves keep sand in suspension (REINECK 1963); generally in the upper flow regime where sand grains predominantly move by saltation (HARMS & FAHNESTOCK 1965; SIMONS et al. 1965)

i) storm-derived sand layers in silt were observed to originate in suspension clouds (Reineck & Singh 1971)

Accordingly, horizontal bedding can form in standing or slightly agitated water, but can also occur in the breaker zone and in torrential rivers. The grain sizes of horizontally bedded deposits therefore range from clay and silt to sand and gravel.

3.241.2 "Flaser" and lenticular bedding

Flasers are wavy silt or clay streaks included in a bed of sand. By increase of the fine-grained material, a transition of flaser bedding into lenticular bedding — cross-bedded sand lenses (= ripples) included in silt or clay — may also occur (Reineck & Wunderlich 1968a).

Flaser bedding, according to Reineck (1962), is formed where thin mud layers are deposited in sand-ripple valleys, and are embedded by another set of sand-ripples. Such mud flasers, which sometimes consist of fecal pellets, are typical of tidal flat deposits (Häntzschel 1936, Reineck 1960a). For other occurrences see ch. 3.252.23, d.

3.241.3 Cross-bedding

3.241.31 Origin and classification of cross-bedding

Cross-bedding consists of beds with oblique layers which in general are tangential to the lower horizontal layers while they meet upper layers at an angle.

In bc-sections (see fig. 3-25) cross-beds often seem to dip in two opposite directions. This was the reason for the term "cross-bedding". "Oblique bedding" (Schrägschichtung) is an alternative term.

Cross-bedding is caused by ripples and dunes which form at moderate wind or water velocities ("lower flow regime") as an equilibrium surface. The reason of the formation of ripples and dunes, however, still remains unexplained. For references of published theories cf. Pettijohn et al. (1972, p. 348).

Ripples and dunes (fig. 3-24) move by erosion of sand on the windward (upcurrent) side and its redeposition (as foreset beds) on the lee side. The lamination visible in cross-bedded sands originates from the steady deposition of the fine grained suspension and saltation load interrupted by small slumps on the lee slope (Reineck 1961, 1963). This mechanism is influenced by an eddy which may develop on the lee side of the ripples. Experimental studies of these phenomena were done by Jopling (1965). Current ripples form in the lower flow regime, between the first and second critical current velocities at which rolling and saltation transport begin. Ripples moving in the direction of current are, therefore, characteristic for the lower flow regime. At higher current velocities, horizontal bedding develops (see above). Antidunes i.e. ripple forms migrating upcurrent, although sediment moves downstream, originate from even higher current velocities by impact of sand on the stoss side of dunes.

Cross-bedding contains grains whose diameters lie within a narrow velocity interval between erosion and deposition. If the erosion and deposition intervals are

Fig. 3-24. Dune consisting of only one set of large-scale cross-laminae dipping towards the left (W). The dune is being eroded by the sea. Beach near Maspalomas, Gran Canaria. Height about 10 m.

large, as is the case with silt and clay (HJULSTRÖM 1939, SUNDBORG 1956), a resuspended grain has little chance to be deposited on the lee side of the same ripple or dune; it remains suspended until it reaches an area of considerably reduced current energy. Theoretically, current ripples should form only in sands between 0.2 and 0.8 mm (v. ENGELHARDT 1973). However, they occur in silts as fine as 0.02 mm and locally in conglomerates (BLUCK 1965; SMITH 1965: eolian, see ch. 5.213; HELM 1971: glacigenic).

The type and size of cross-bedding depends upon:
(a) Mode and energy of water- or wind-movement
(b) Rate of supply
(c) available grain sizes as well as their shape and specific weight (SCHEIDEGGER & POTTER 1967)
(d) water depth and basin subsidence

The size of cross-bedded units increases with the current velocity (CAREY & KELLER 1957, PETTIJOHN et al. 1965, p. 96). This is one of the most important features of cross-bedding.

Individual ripples frequently consist of one set of cross-beds only. The upper part of this unit generally is eroded by the lee-side vortex of the next ripple. The higher the ratio: $\dfrac{\text{migration velocity of the ripples}}{\text{bottom subsidence}}$ the smaller is the portion preserved of each ripple. The migration velocity increases with increasing current velocity and grain size (v. ENGELHARDT 1973).

Larger sand waves and dunes may consist of an individual set (fig. 3-24) or several sets (Reineck 1963). Such a stack of superimposed sets is called a "co-set" (Niehoff 1958). There are two basic types of cross-bedding which sometimes are closely associated (fig. 3-25, upper line, McKee 1965):

a) A steeper ripple with straight foresets that meet the bottom at an angle. The stream lines leave the bottom at the crest of the ripple, forming well-developed lee eddies. On the lee side the grains slide into the current shadow, which is nearly at the angle of repose ($\approx 35°$) ("avalanche ripples"). Even higher angles ($\approx 42°$) have been observed in beach dunes (Land 1964), probably due to some dew or spray moisture. This ripple type forms in the middle part of the lower flow regime. Typical for this type are tabular sets.

b) A relatively flat ripple with high fallout rates of suspended load on the lee-side which is characterized by tangential contact of the foresets with the bottom ("accretion ripple", Imbrie & Buchanan 1965). This ripple dips at angles lower than repose, and forms in the upper- and lowermost part of the lower flow regime. There is less separation of the stream lines from the bottom on lee faces than in type a) though eddies occur, especially at lower velocities (Harms & Fahnestock 1965). According to Jopling (1965b) and Allen (1968a) cross-bedding becomes more tangential with increasing suspension load / bed load ratio, as well as with increasing tractive force. The tangential basal contact is caused by toeset deposits in the zone of backflow owing to eddies (Jopling 1965b). — At higher velocities, the eddies disappear as dunes become low, in the transition to upper flow regime-horizontal bedding. Trough-shaped sets are typical for accretion ripples.

Cross-bedding is described in this book using diagram 3-25 (in part from McKee & Weir 1953). It is also suggested to record the type of cross-bedding in both planes, parallel and perpendicular to the current, if possible.

Trough-shaped cross-bedding and skewed types of ripples predominate in stream and tidal channels as compared with tabular cross-bedding on level areas of the sea floor (J. R. L. Allen 1968a).

"The geometry of the ripple has an important relation to the type of cross-bedding that results from migration of the ripple. Tabular sets result mainly from migration of straight or slightly sinuous ripples. Trough sets result mainly from the migration of less regular ripple forms. Probably most small-scale trough cross-lamination was formed by the migration of linguoid ripples and most large-scale trough cross-bedding was formed by the migration of sinuous or lunate dunes" (Blatt et al. 1972).

The identification of 15 types of cross-bedding (Allen 1963) classified by Greek letters is possible only in very good exposures. A genetic classification is not available yet.

The dip of cross laminae in eolian sands frequently ranges from 30—35° and in subaqueous sands from 25 to 30° (McKee 1957).

Angular planar cross-bedding occurs especially in eolian dunes and in the quiet lower course of rivers. It also is reported from the shallow sea where its dip is considerably lower (Wermund 1965).

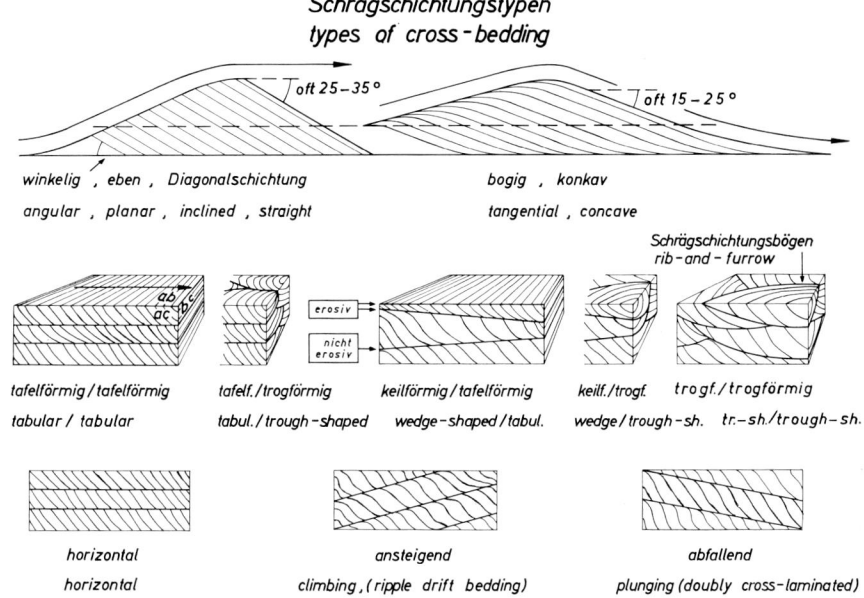

Fig. 3-25. Types of cross-bedding (mainly based on Illies 1949c, McKee & Weir 1953, Niehoff 1958, Reineck 1963, Potter & Pettijohn 1963, Pettijohn & Potter 1964, Wurster 1964, Imbrie & Buchanan 1965). Below each type, the most common terms are noted in German and English. "Tabular/trough shaped" is a type of cross-bedding which appears tabular in the section ac, i.e. perpendicular to the bedding in the direction of current (= longitudinal), and trough-shaped along the bc section, i.e. perpendicular to the bedding and perpendicular to the direction of current (= transverse). In general, only the lower part of each ripple (below the dashed line in the uppermost figures) is preserved. Angular cross-bedding can be steeper than tangential cross-bedding. A complete description according to this diagram consists of at least 4 parts, e.g. "tangential, tabular/trough-shaped, climbing cross-bedding (23°)". In addition, the dimensions of the sets should be indicated.

Smaller units of cross-bedding have been classified by Reineck (1963) into

a) Small-scale ripple bedding: The units are less than 2 cm thick. The curvature radius is relatively small in the bc plane. Horizontal exposures often show trough-shaped cross-bedding with rib-and-furrows (see fig. 3-24). Rib-and-furrows originate as overlapping spoon-like potholes caused by whirling water. In a fluvial environment, small-scale ripples measure about 1 cm in thickness, 5 cm in width and 20 cm in length in the current direction (Wurster 1964). 20 cm wave length separates the small scale ripples from the large-scale ripples (Richter 1926; see also Dillo 1960, Harms & Fahnestock 1965). This ripple type belongs to the lower part of the lower flow regime.

b) Large-scale ripple bedding: The units are more than 2 cm thick (the average in the North Sea is 7.5 cm). The curvature radius in the bc plane is greater than 20 cm. Large-scale cross-stratification is typical in the upper part of the lower flow regime (Harms & Fahnestock 1965).

Wurster (1964) observed a distinct separation of the occurrence of small- and large-scale ripple bedding in fluvial sediments. Similarly Allen (1970a) described two distinct populations of bed waves, the limit between which occurs at a wave length of 60 cm and a height of 4 cm. The smaller ones are called "ripples", the larger ones "dunes" or megaripples (REINECK et al. 1971) (see also ch. 3.242.21 and fig. 3-34). According to PETTIJOHN et al. (1972, p. 346), there is some evidence of a statistical deficiency in abundance of sand waves having heights between 3 cm (ripples) and 7 cm (dunes), and this may reflect a fundamental difference in origin.

The bed form progression from ripples to dunes with increasing stream power in rivers is reported for material of uniform grain size by N. D. SMITH (1971).

According to REINECK (1963), dunes originate from running water of more than 100 cm/s velocity in fine sands, whereas in coarser sands lower velocities are required (see also ALLEN 1970a and DILLO 1960). As a consequence, coarser sands form larger ripples and dunes than finer sands at the same flow velocity. This has been confirmed by HARMS & FAHNESTOCK (1965) in fluvial sediments, by PELLETIER (1965) in marine sediments, and by SCHWARZACHER (1953) and POTTER & BLAKELY (1967).

A classification including size parameters of ripples is suggested by REINECK et al. (1971).

If a sufficiently large sediment supply is available, climbing ripple drift bedding will develop (fig. 3-25; McKEE 1965, 1966a). According to ALLEN (1970b) climbing ripples form when the ratio between the intensity of deposition and the transport rate is high, e. g. in river flood plains and areas of overbank flow (McKEE l. c.). Plunging cross-bedding (fig. 3-25), i. e. forward-dipping sets of cross strata are reported less frequently (McKEE 1962, VAN DER LINDEN 1963, POTTER & PETTIJOHN 1963, fig. 4-6, BIGARELLA et al. 1965). "Scour-and-fill" structures develop where channels are eroded and subsequently filled by sediments concordant or discordant to the bottom. Recumbent-folded crossbedding has been interpreted by ALLEN & BANKS (1972) "as due to the deformation of a liquefied (or perhaps fluidized) sand by current drag following an event in the majority of cases suspected to be an earth-quake shock". The drag force of turbidity currents may be especially effective.

Tangential cross-bedded units of 6 m in thickness and 60 m in length ("dunes") are described from fluvial sands (VAN DER LINDEN 1963). Units from 3 mm up to 33 m in thickness have been reported by POTTER & PETTIJOHN (1963).

In order to compare the dip of cross-bedding in fossil sandstones with that in recent sands, the compaction of the former must be taken into consideration. With an initial porosity of 40%, the original dip α_0 of a cross-bed is calculated from the formula

$$\tan \alpha_0 = \frac{\tan \alpha \cdot 100}{60 + P}$$

α is the dip measured and P the porosity in Vol-% of the rock, including that volume part filled by cement (determined in thin section). The angle of dip may further diminish, if material of the grains has been dissolved (e. g. by pressure solution) and removed from the rock.

3.241.32 Variability of cross-bedding direction

Cross-bedding generally dips in the direction of transport, thus indicating the current direction. In the fluvial environment current direction generally is equivalent to the gross transport direction and reveals the source area of sediment (BRINKMANN 1933). Such correspondence between current directions and gross dispersal directions may occur also in larger areas, as for instance in the Cambro-Ordovician sandstones of the central Sahara (N. Africa) with uniform northward dip of cross-bedding over an area of more than 1000×1000 km (BURROLET & BYRAMJEE 1964, BIJU-DUVAL et al. 1966).

A coincidence of paleo-slope and cross-bedding directions was also found by HATFIELD et al. (1968) in marine sandstones. However, the direction of cross-bedding differed from the direction of the source area by more than 90°.

A standard deviation of less than 78° in eolian and fluvial, and 78—89° in marine sands and sandstones has been calculated for the variance of direction of cross-bedding by POTTER & PETTIJOHN (1963, p. 88/89). On modern coasts, there is usually no coincidence of cross-bedding and paleo-slope: Within sand bars parallel to the coast of the North Sea (REINECK 1963) and Baltic Sea (SEIBOLD 1963) cross-bedding generally dips more or less towards the coast because the surf is generally stronger than the backwash near the coast (VAN STRAATEN 1953). In tidal flat areas (e. g. the Wadden Sea) the variability of cross-bedding dip directions is particularly high, due to the different directions of low and high tides. Bimodal distributions of cross-bedding directions occur (HÜLSEMANN 1955, REINECK 1963, SEDIMENTATION SEMINAR 1966, KLEIN 1970).

"Rip currents are narrow zones along high energy beaches in which water flows seaward through the surf. They are caused by the net onshore transport of wave drift which causes water to accumulate inside the breakers" (COOK 1970). Optimum conditions for their development are high, long period waves, a flat foreshore and a negligible breeze. In the geological record, they are represented by elongate bands of coarse sand oriented normal to the paleo-shoreline and occasionally showing cross-bedding that dips in an offshore direction. An example from the Tertiary Molasse has been published by KÖLBL (1966).

Longitudinal accretions of laminae occur at the slip-face of river beds. The result is cross-bedding that dips perpendicular to the channel direction (WRIGHT 1959 and G. E. WILLIAMS 1966). The same was reported from tidal channels by REINECK (1958 c), see ch. 3.252.23.

3.241.4 Preconsolidation deformations

They originate from the movement of sediment after deposition but before consolidation. Therefore, they develop in the time interval between sedimentation and diagenesis.

3.241.41 Mottled structures and trace fossils (Spurenfossilien)

Whereas mottled structures are due to poorly organized burrowing that blurs the bedding, trace fossils result from distinct borings or excavations that frequently are indicative of the environment (LESSERTISSEUR 1955).

Mottled (churned) structures originate from the activity of endobenthonic fauna that live in the sediment. Such structures are widespread, especially in sandy silt or silty sand (REINECK 1963); they are reported less often in pure silt, clay or coarser sands, possibly because of the difficulty to recognize such structures in homogeneous sediments. Mottled structures are rare though not absent in eolian sediments but common though seldom reported in upper mud of fluvial fining-upward cycles (VAN HOUTEN, pers. commun.). Mottled structures form most frequently in sea floors with slow sedimentation. The details of burrows, e. g. upward and downward movement of some pelecypods living in the bottom can help to distinguish periods of erosion from slow or fast sedimentation (REINECK 1958a).

Gas bubbles, mainly from decomposing organic material, may also produce sediment structures as long as the sediment is sufficiently soft to allow gas bubbles to form (VAN STRAATEN 1954, COLEMAN & CAGLIANO 1965, SCHÄFER 1954). Capillary forces presumably participate in the formation of bubble voids.

Trace fossils in the interior of sandstones are both "Fodinichnia" ("burrows made by hemisessile deposit feeders. They reflect the search for food and at the same time fit the requirements for a permanent shelter", Freßbauten) and "Domichnia" ("permanent shelters dug by vagile or hemisessile animals procuring food from outside the sediment as predators, scavengers, or suspension feeders", Wohnbauten; SEILACHER 1964). The Fodinichnia are constantly shifting whereas the Domichnia are permanent ("deforming" and "constructive" burrows, respectively; SCHÄFER 1956). Fodinichnia belong to SEILACHER's "Zoophycos" facies which may or may not indicate medium water depth; frequently they occur in fine-grained sediments. Domichnia are often well developed in shallow marine sandstones (SCHÄFER 1962, 1963, SEILACHER 1964), e. g. the "Scolithus" sandstone (HALLAM & SWETT 1966) and the "Tigillites" (GUBLER et al. 1966), i. e. sandstones from the Fennoscandian and North American Cambrian and from the North African Cambro-Ordovician, respectively, with vertical burrows about 1 cm in diameter (fig. 3-33).

For a clearly descriptive discussion of these and other sedimentary structures see CONYBEARE & CROOK (1968), for a genetic discussion of Zoophycos see BRADLEY (1973).

3.241.42 Flow structures (Convolute lamination, slumping and others)

Increasing compaction removes a great part of the original interstitial water from a sediment. Therefore, about the same amount of water must return back to sea as was taken from the ocean by the sediment (see p. 6). Whereas this compaction "current" penetrates the deeper strata without great deformation (due to the superimposed pressure and the partial consolidation), in the uppermost sediment, layers may be disturbed by the formation of convolutions and injections.

Due to the different permeability of sand and clay layers, a dewatering of sand-clay alternations is only possible if the lamination is interrupted by burrowing or shrinkage cracking. Otherwise, pore pressure may exceed hydrostatic pressure, and the alternating beds become unstable (RUBEY & HUBBERT 1959). In such a case, several conditions could cause folding ("convolute lamination") or slumping: 1) A slight slope of the strata, little more than 1° (MILNE 1897), associated with earthquakes (CHAMBERLAIN 1964), 2) differential loading (KUENEN 1953, DŻUŁYŃSKI & WALTON 1965, figs. 104—106), or 3) basal friction of currents loaded with sediments (COLEMAN & GAGLIANO 1965, SANDERS 1965, GHENT & HENDERSON 1965, DŻUŁYŃSKI & SMITH 1963, McKEE 1966). Deformation characteristics of sand and clay are different.

a) Convolute lamination (fig. 3-26)

Relatively fine-grained rocks with fine laminations locally show an irregular folding (sharp crests and wide troughs) fading downward (KUENEN 1953, MAYR 1957, POTTER & PETTIJOHN 1963, DAVIES 1965). In many cases these folds are truncated at the top, thus proving their early origin. The thickness of such layers (usually less than 25 cm) generally remains constant for a considerable distance. The individual folded strata are rarely interrupted within the layers. For these reasons, convolute bedding is not attributed to submarine sliding (RABIEN 1956, EINSELE 1963 a). Small lateral differences in pressure probably produce this kind of deformation.

The prerequisite for convolute lamination, the impediment of dewatering, can be expected where sand-clay or sand-silt alternations are deposited quickly and are

Fig. 3-26. Convolute lamination. Upper Devonian (from EINSELE 1963 a). The arrow points toward the top, the scale is in cm.

not disturbed by burrowers. This applies especially to turbidites, in which decimeter-thick layers are deposited within hours. Thus convolute lamination is one of the typical (but not exclusive) features of turbidites. Geometric relations between convolute lamination and current ripples often indicate that the former was triggered by the flow of a suspension current over the sediment (KUENEN 1953).

Dish structures a few cm in diameter are found occasionally in turbidites. Such interrupted layers are characterized by a slightly increased matrix content; a satisfying explanation is missing (CHIPPING 1972), but a secondary liquefaction of sand followed by a differential upward flow of a few centimeters is likely (KRUIT et al. 1972).

b) Slumping

It is triggered by local (SUTTON & LEWIS 1966) or regional bottom disturbances (GRANT-MACKIE & LOWRY 1964). Frequently the direction of slumping corresponds to that of accompanying cross-bedding and ripplemarks (KÜHN-VELTEN 1955, STEWART 1962, POTTER & PETTIJOHN 1963). Generally the individual layer is interrupted by small tears or faults. The thickness of the beds is highly variable. Occasionally torn-off parts of sediment form lenticular bodies ("Phacoids", VOIGT 1962). Recumbent cross-bedding (see MCKEE 1962) indicating initial stages of slumping, may originate from highly concentrated suspension currents (MCKEE, l. c.). Huge unstratified slumping masses of submarine origin, called "olisthostrome" (BENEO 1956, from Greek olisthos = slipperiness, and stroma = nappe), can be very important, e. g. in the Apennine Mountains (Italy). They obviously moved at a very slow speed (GÖRLER 1967, MARCHETTI 1960, GÖRLER & REUTTER 1968). Subaerial slumpings (e. g. quick sand, solifluction) are described frequently: In glacial climates slumps can occur at 2 to 3° whereas in normal climates it takes about 10° (ACKERMANN 1955).

3.241.43 Shrinkage cracks, injection structures and sand dikes

Distinguishing between different cracks can help determine the sedimentary environment. The following classification can be used:

1. Shrinkage cracks

a) mud cracks of subaerial origin (desiccation cracks, sun cracks). Clay laminae generally are bent upward at the edges. Cracks may be associated with pseudomorphs of salt and rain imprints. They open in general from above and can be filled from below or above.

b) syneresis cracks, subaqueous, by osmotic dehydration of the sediments. Clay laminae are not turned up at the cracks. Cracks open from above and are filled from above, generally.

2. Injection structures.

They probably form in areas of relatively rapid deposition of alternating sand and silt (clay) layers, where dehydration was not aided by infaunal burrowing. They need not form only in the uppermost layers but also in deeper sediments, as indicated by the penetration of several clay laminae (fig. 3-27, lower right). They open from below and in most cases are filled probably from below.

3. Sand dikes

Cracks, partly large in size, probably caused by tectonic movement. In Recent sediments they are filled from above or below, in consolidated (older) sediments or crystalline rocks always from above.

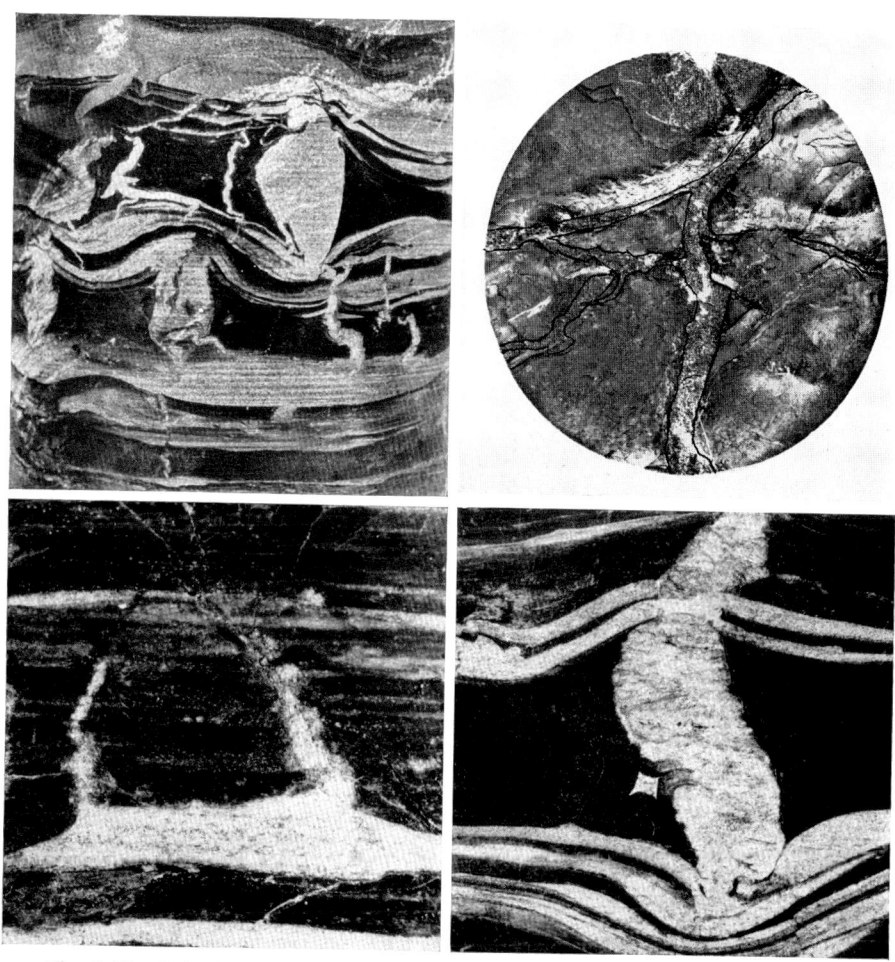

Fig. 3-27. Injection structures. Buntsandstein, Northern Germany. Upper right: Top-view of a bedding plane; cracks in siltstone, filled with sand (diameter of the core: 10 cm). All other figures: wall of vertical cores. Topsides up. Upper left: The wide fillings of sand (light) were rigid and hindered the compaction of the bordering siltstones (dark); the crumbling of small fillings, however, indicate the degree of silt compaction (depth of core recovery: 2500 m, width of the picture 10 cm). Lower left: Cracks in siltstone filled from below with fine grained sand (width of the fig. 2.2 cm). Lower right: Crack filled with sand which widened the crack and intruded the overlying strata (width of the fig. 4.8 cm).

Examples indicating the origin are given below. Whereas most desiccation cracks are filled from above, OOMKENS (1966) observed such cracks which were filled from below: In modern playas and sabkhas, desiccation cracks, 1—10 cm in width, form polygons in the clay surface layer, and underlying water-bearing sands are pushed upwards by their lower density.

Experimental syneresis cracks form below water when the clayey surface layer is dewatered osmotically by a sudden increase in the salt content of the overlying water (BURST 1965). This dehydration called "syneresis" is observed only in clays with expanding clay minerals. The cracks can be filled (by quicksand from below or) by normal sedimentation from above (HARMS 1965, DILLER 1890, LAMING 1964, VITANAGE 1954, VAN HOUTEN 1965a). Filling from above can be seen by the horizontal deposition of shells and micaceous minerals in the cracks (MONROE 1951). Bladed grains pushed up from below are often arranged parallel to the wall. Syneresis was discussed by JÜNGST (1934), VAN STRAATEN (1954), WHITE (1961), DANGEARD et al. (1964) and KUENEN (1965). Such syneresis shrinkage cracks can be oriented more or less parallel to the long axes of ripple marks (DONOVAN & FOSTER 1972).

The instability of newly deposited alternations of sand and clay is not only documented by the flow structures discussed in the last chapter, but also by the fabrics described below. The term "injection structures" is suggested for all cases in which the sediment probably was cracked by the loss of compaction water. In sand and silt alternations of the Buntsandstein (FÜCHTBAUER 1967b) and Lower Permian (OOMKENS 1966) polygonal cracks filled with sand (Netzleisten) are frequent. Whereas the Lower Permian cracks are explained as desiccation cracks, the Bunter cracks frequently are injections as suggested by the frequent fillings from below (fig. 3-27). This seems plausible since the infauna, which might help to dehydrate the sediment equally, are missing. Sand polygons in flysch sediments may have a similar origin (DŻUŁYŃSKI & WALTON 1965). Such structures occur also in limestones (DALEY 1971).

Filled cracks caused by external forces (e. g. by earthquakes or tectonics) are called "sand dikes". Sand dikes filled with sediment from above and oriented parallel to the fold axis are described by JANKOWSKY (1955). Sand dikes up to 7 m in width injected upward and filling two sets of conjugated shear fractures are connected with sand sills which intruded shear planes produced by flexural-slip folding in Tertiary Moreno Shale, California (SMYERS & PETERSON 1971). Sandstone sills also can originate from quicksand entering bedding planes. Larger sand dikes are found occasionally in clay sequences with sporadic sand layers. These occurrences are such that earthquakes can be assumed to have caused these dikes (FAIRBRIDGE 1946). There are sand dikes of as much as 11 m in width and 14,5 km in length (DILLER 1890). Heights of up to 100 m have been calculated (DŻUŁYŃSKI & WALTON 1965, ANDERSON 1951, COLACICCHI 1959, POTTER & PETTIJOHN 1963, STRAUCH 1966).

Though fine-grained sand and coarse grained silt (median diameters between 0.3 and 0.03 mm) are most sensitive to liquefaction and thus are very susceptible to soft sediment deformation (PETTIJOHN et al. 1972, p. 372), clay liquefaction may also occur upon swelling of montmorillonite due to sudden water access. For instance McBRIDE et al. (1968) reported clay dikes in sandstones: Montmorillonite had filled tectonic fissures. Limestone dikes up to 2 m in width and 30 m in length and height were found in lower Carboniferous bioherms (PRAY 1964): Unconsolidated calcareous mud and crinoid arenite was pushed up into the bioherm, which had been cemented previously. Observations like this yield valuable clues as to the time of cementation. FISCHER (1964) reported limestone dikes filled from above in the Calcareous Alps.

Sand volcanoes are small circular eruptions of quicksand (DŻUŁYŃSKI & WALTON 1965, BURNE 1970). Mud volcanoes are larger and mostly associated with the eruption of natural gas (LEVORSEN 1956).

3.242 The bedding plane

3.242.1 Rates of accumulation and formation of sediments

Bedding planes represent either surfaces of discontinuity caused by a change in material or grain size, or parting planes caused by a horizontal orientation of particles as in shales.

Planes of discontinuity indicate interruptions in sedimentation which often begin with a more or less prolonged stratigraphical break (hiatus). Thus the upper and lower surface of beds often represent planes of omission (diastems, see DUNBAR & RODGERS 1957), or even planes of erosion, e. g. in cross-bedded units. The "rate of sediment formation", i. e. the rate at which the thickness of a sequence increases, is a function of (1) the actual "rate of accumulation" which can be observed e. g. if a ripple forms, and (2) the time elapsed during omissions and/or erosion (hiatus).

The total time interval including all periods of non-sedimentation depends upon the supply of material relative to the subsidence of the basin. This latter factor determines how much of the incoming material will be deposited at that site and how much will be transported to other areas of more pronouced subsidence ("contrôle par le fond", LOMBARD 1953). On the shore face with its intensive transportation of sediment the portion of omission is especially high provided the subsidence is low. The rate of sediment formation as calculated from the comparison of thickness and absolute time scale is strongly dependent upon the length of time involved (REINECK 1960 b) (table 3-13).

The rates of sediment formation include only those erosions and omissions which are shorter than the time spans of observation. The decreasing sediment formation

Table 3-13. Rates of sediment formation related to different time intervals of observation (means from REINECK 1960 b).

	cm sediment/1000 yrs with pores	without pores*	interval of observation
A. Tidal flat			
1.	$2.2 \cdot 10^2$	70	2 400 years
2.	$1.2 \cdot 10^3$	400	3 years
3a. Mud	$1.8 \cdot 10^4$	$0.6 \cdot 10^4$	8 days
3b. Sand	$1.4 \cdot 10^6$	$84 \cdot 10^4$	8 days
4a. Mud	$0.9 \cdot 10^6$	$30 \cdot 10^4$	hours
4b. Sand	$1.7 \cdot 10^8$	$10 000 \cdot 10^4$	hours
B. Shallow marine, Texas			
5.	$0.7 \cdot 10^3$	250	11 000 years
6.	$1 \cdot 10^3$	300	65 years
7.	$3 \cdot 10^4$	10 000	80 years**
C. Alluvial plain			
8. Brahmaputra (sand)	$9 \cdot 10^3$	5 500	200 years***

* For calculation of sediment thickness exclusive of pores, mud was reduced to 1/3 of its original thickness (SEIBOLD 1964 a), sand to 6/10. The comparison shows that deposition rates for sand are two orders of magnitude faster than for mud.
** silt in front of one of the outlets of the Mississippi river (SCRUTON 1955).
*** COLEMAN 1969.

rates with increasing time intervals indicate that omissions and erosions become increasingly important the longer the time interval considered.

In table 3-13, the differences in the rates of sediment formation are shown for observation times from hours to thousands of years. In the following two tables the time span is extended from thousands to millions of years.

Table 3-14. Rates of sediment formation in recent oceans (time spans of observation: thousands of years).

Area	cm/1000 years without pores*	Author
1. Trenches (6215—8450 m)	0.02	Yasuo & Sugimura 1961
2. Deep sea sediments	ca. 0.3	from Seibold 1964a
3. Ocean basins W. of California**	4	Emery 1960
4. Interior of the Black Sea	13	from Seibold 1964a
5. Continental slope	17	Rusnak 1964
6. Basins close to the coast, off S. California	19	Emery 1960
7. Arid-semiarid lagoons	33	Rusnak 1964
8. Humid bays and estuaries	67	Rusnak 1964
9. Various tidal flats	70	see table 3-13
10. Gulf of Mexico, neritic	120	from Lombard 1956
11. Shallow marine near Mississippi delta	about 300	see table 3-13 and Rusnak 1964
12. Turbidites, eastern Mediterranean	max. 100	Ryan 1972

* According to Seibold (1964) about 1/3 of the thickness measured; due to prior compaction, the reduction factor might be less, but the differences are not well-known and presumably not very great (Füchtbauer & Reineck 1963).
** without turbidites.

Table 3-15. Rates of sediment formation for older rocks.

Area and period	cm/1000 years without pores*	Author
A. Epicontinental sediments		
1. Eastern Europe, Devonian	3	v. Bubnoff 1950
2. Northern Germany, Mesozoic**	3	Philipp et al. 1963
3. Northern Germany, Maximum (Upper Jurassic)	8.6	Philipp et al. 1963
4. Maximum on cratons	12	Kay 1955
B. Geosynclinal sediments		
5. Ural Mountains, Paleozoic	3	Ronov 1949, from Ruchin 1958
6. Caucasus Mountains, Mesozoic	6	
7. Indonesia, Mesozoic	15	from Seibold 1964a
8. Rhineland, Devonian, shallow marine	35	v. Bubnoff 1950, Reineck 1960b
9. Maximum in geosynclines and marginal deeps	49	Kay 1955
10. Average	2.5	Kuenen 1967a
C. Molasse sediments		
11. Lower Freshwater Molasse, Southern Germany (maximum)	55	(approximately 2000 m in about 3 million years***)

* From Nr. 2, 3, 4, 9 and 11, 18% porosity have been subtracted; Nr. 1, 5, 6 and 8 were not reduced, since they are folded and practically free of pores; Nr. 7 was reduced already and quoted as such.
** Average from subsidence diagrams; (longer) omissions and erosions not included.
*** Length of time according to v. Bubnoff (1947). For more figures see Kuenen (1967a).

A tentative comparison of tables 3-14 and 3-15 suggests that omissions over periods greater than 1000 years account for 50—90% of the total time interval represented in shallow marine sediments (compare no. 8 in table 3-14 with no. 8 in 3-15, and no. 6 and 9 in table 3-14 with no. 2 in table 3-15). As all examples come from areas of deposition, these omissions are probably hiatuses, not full-scaled erosion planes.

In general omissions and small-scale erosion cause long-term rates of sediment formation to be 5 orders of magnitude less than the rate of accumulation of an individual layer on tidal flats (compare no. 4a of table 3-13 with no. 2, table 3-15; see also REINECK 1960b).

As an example of the areal distribution of sedimentation rates, the following figures of the Holocene sedimentation off the Atlantic coast and the Gulf Coast are reported by CURRAY (1965): South Carolina: 90—150 cm/1000 years, SW-Florida: 30—60 cm (carbonates), near Tallahassee (Fla.): 120—180 cm, Mississippi delta: 600—1200 cm, Western Gulf Coast: 120—180 cm, Central Gulf of Mexico and deeper waters of the Atlantic ocean: 2.5 cm of unconsolidated sediment per 1000 years. The maximum sedimentation rates are in the shoreline-nearshore zone with the main concentration in the deltaic complexes. Such an isopach pattern is responsible for accumulations of oil in nearshore Cretaceous sediments between deltaic centers: Arches developed in these positions due to smaller subsidence compared with the adjacent deltas, whereas in the section perpendicular to the coast, the nearshore areas form stratigraphic traps (WEIMER 1970). In the Oligocene of SE-Texas, on the other hand, the interdeltaic sands are barren, compared with the oil & gas-bearing sands deposited near the seaward margin of the delta (GREGORY 1966).

High rates of sediment formation are indicated by the preservation of mud lumps up to 70 cm in vertical diameter, included in Triassic alluvial sandstones (GWINNER 1971).

3.242.2 Bedding surface marks

Sandstone beds alternating with clay layers are particularly rich in bedding surface marks. Frequently bedding marks typical of shallow-water deposits form on the upper surface of the coarse layer, while those of deep sea rocks (turbidites) and of some fining-upward fluvial and shallow-marine beds occur on the lower surface.

3.242.21 On the upper surface (ripple marks, trace fossils)

On the upper surface of beds ("epirelief", SEILACHER 1964) distinction is made between ripples and dunes mainly.

1. Ripples (fig. 3-28). They are classified morphologically (partly according to BUCHER 1919) into
   a) symmetrical ripple marks (oscillation or wave ripples)
   b) asymmetrical ripple marks (current ripples)
       α)   transverse ripples (see "a) Current ripples", below)
       β)   longitudinal ripples
       γ)   linguoid ripples
       δ)   rhomboid ripples
       ε)   complex ripples

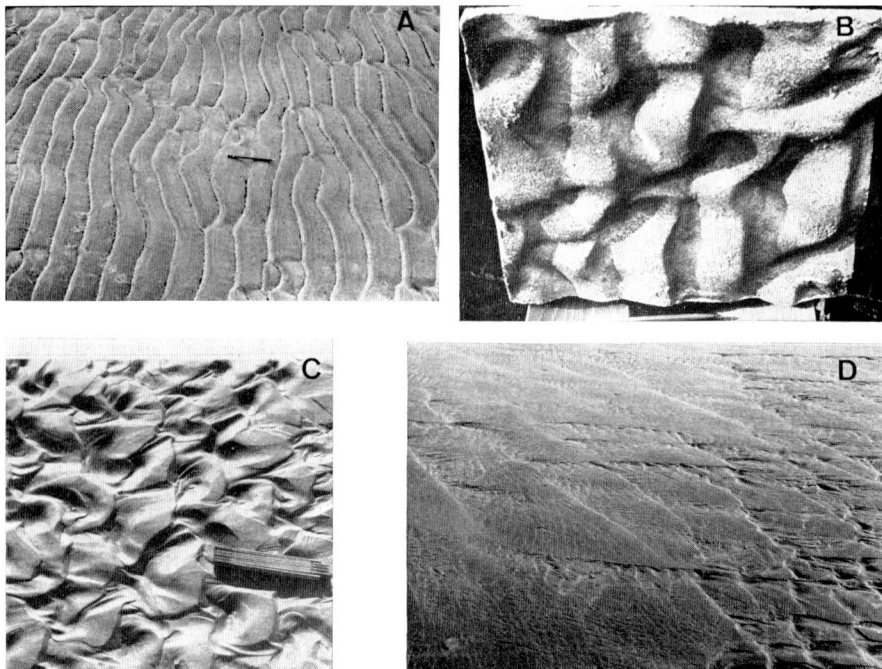

Fig. 3-28. Ripples. A. Oscillation (wave) ripples, asymmetrical because only the swash (from right to left) was able to move sand. Indistinct white lines = intersections of the lee layers with the stoss face; they are visible by slight drying. Pellets and plant remains ("coffee ground") in the ripple valleys. Upper tidal flat, N coast of Mellum Island near Wilhelmshaven, North Sea.
 B. Current ripples with linguoid component (current from left to right); replica of a replica. Courtesy of H.-E. REINECK, Wilhelmshaven.
 C. Linguoid ripples (current from left to right). Mellum Island, North Sea (Plate 84 B of PETTIJOHN & POTTER 1964).
 D. Rhomboid ripples at the landward (= left) side of a beach ridge. Northern beach of the island of Norderney (North Sea). Length of the small rhomboids on the right hand side is approx. 12 cm; length of the larger rhomboids is 30—50 cm. Courtesy of H.-E. REINECK.

A genetical classification has been proposed by HARMS (1969):

a) Current ripples (indicating water velocity or energy): Crests asymmetric and angular; steep faces slope at static angle of repose. Crest spacing 12 ± 1,5 cm.

Current ripples also vary in size and lateral extension with varying current velocity (HARMS, l. c.): "Ripples formed by currents just capable of moving sand grains have relatively continuous but sinuous crests, uniform height and spacing, and asymmetric, angular profiles with downstream faces that slope at the angle of static repose. As current energy increases, crests become shorter and more curved and height and spacing become more variable. At still higher current energy, ripple-covered small dunes with height exceeding 3 cm form." For quantitative interpretation see HARMS (l. c.) and TANNER (1967). For subdivision into avalanche and accretion ripples see p. 76.

b) Combined-flow ripples (indicating the relative importance of currents and waves): Crests asymmetric and rounded; steep faces slope less than static angle of repose (excluding ripple-drift cases).

c) Wave ripples, formed by alternating wave motion (ripples indicating wave dimensions and periods): Crests symmetrical and rounded. Profile sinusoidal, maximum slopes $\approx 25°$. Crests long and very well oriented (straight). Very uniform height and spacing. "Oscillation ripples" is an older term (ILLIES 1949, NIEHOFF 1958, REINECK 1958a). Most wave ripples are $\pm$ asymmetrical because of the higher energy of the swash compared with the backwash of the surf (see fig. 3-28 A) (NEWTON 1968). The ripple index (wavelength divided by height) is smaller in wave than in current ripples (REINECK & WUNDERLICH 1968b).

Linguoid and rhomboid ripples form if the morphology of ripples influences the currents at very shallow depths (less than 3 times the height of the ripple) (VAN STRAATEN 1953, REINECK 1963). Linguoid ripples can be interpreted as miniature barchans welding laterally to form little tongues pointing downwind. The rhomboid ripples have two lee sides, with an acute angle between their crests (the tip pointing in the direction of transport).

The wave-length of wind-ripples corresponds to the mean saltation distance of the grains, which increases with wind velocity (BAGNOLD 1965, v. ENGELHARDT 1973). On the other hand, the "ripple index" (ratio of wave length to height) may range from 30 to 70 in well sorted fine sands but may be as low as 10 to 15 in coarser and less well sorted ones (HARMS 1969). "A distinctive feature that distinguishes wind from water ripples is their concentration of coarse grains at their crests whereas water ripples tend to have coarser grains in their troughs" (PETTIJOHN et al. 1972, p. 365).

Longitudinal ripples in the Dutch tidal flats occur where wave movement and currents are perpendicular to each other. In such cases the ripple crests are oriented parallel to the currents (see ch. 3.241.32).

Complex ripples occur when two systems of oscillation ripples cross each other (interfering ripples, BAUSCH VAN BERTSBERGH 1940), or when flow conditions in a ripple field change frequently and within short periods of time (VAN STRAATEN 1953).

Adhesion ripples (Haftrippeln, Haftwarzen, REINECK 1955) occur, "when wind-blown sand adheres to either continuously or temporarily moist, planar surfaces. Accumulation of sand in this way is common in low-lying coastal and inland areas of deserts (coastal and inland sabkhas)" (NAGTEGAAL 1973).

Ripple marks, mostly current ripples, also occur in the deep sea, and can be formed by contour currents (p. 123). In the upper part of turbidites, small-scale ripple bedding is frequent.

2. Dunes (sometimes the term "sand waves" is used instead) are large ripples (mega-ripples) which form for instance in large tidal channels (REINECK 1963). In the Mississippi River sand waves reach 10 m in height and 250 m in length (POT-

TER & PETTIJOHN 1963). Their size increases with the current velocity. The dunes at the edge of the Bahama platform are composed of several cross-bedded units of oolitic sands (SEIBOLD 1964b). Numerous photographs of sand waves have been collated by PETTIJOHN & POTTER (1964); see also p. 100, ch. 352.11 (eolian dunes), and REINECK et al. (1971) for classification.

3. **Antidunes** are low-amplitude, sinusoidal bed waves broadly in phase with similar but rather steeper water waves on the surface. The waves may move upstream, or downstream, or not at all. Upstream movement occurs mainly if the material transport is high (HARMS & FAHNESTOCK 1965). The wavelength of subaquatic antidunes is about 6.3 times the water depth (ALLEN 1968b). The wavelength L of antidunes is also related to the mean flow velocity v by the formula $v = \sqrt{gL/2\pi}$ (with g = acceleration due to gravity) (ALLEN 1966). Antidunes were observed in coarse sandstones in the basal division of thick turbidites (SKIPPER 1971). They are common in phreatomagmatic tuffs deposited by high energy base surges (FISHER & WATERS 1970, SCHMINCKE et al. 1973) (fig. 4-27).

4. **Rain prints** and desiccation cracks are found predominantly on top of clay- and marlstone beds, but also on top of calcilutite beds (FISCHER 1965a).

5. **Current crescents** are semicircular marks elongated in the direction of current, that form in front of obstacles which frequently are not preserved (RÜCKLIN 1938, GRUMBT 1966).

6. **Wrinkle marks** ("Runzelmarken") are microwaves on surfaces of laminated sediments; most of which are formed by eolian action on sediments that are exposed to air or are covered by not more than 1 cm of water (HÄNTZSCHEL & REINECK 1968, REINECK 1969, TEICHERT 1970).

7. **Trace fossils** on the upper surface of sand layers include "Repichnia" ("trails or burrows left by vagile benthos during directed locomotion", Kriechspuren) and "Cubichnia" (= "shallow resting tracks left by vagile animals hiding temporarily in the sediment, usually sand, and obtaining their food as scavengers or suspension feeders", Ruhespuren; SEILACHER 1964). Trails are made by such animals as crustaceans, molluscs, worms and echinoderms (CONYBEARE & CROOK 1968). Cubichnia are found only in shallow-marine and probably in nonmarine sandstones; in deep water hiding would be worthless because of the absence of light. Cubichnia belong to the shallow water "Cruziana"-facies (SEILACHER 1964) (fig. 3-33).

3.242.22 On planes within the beds (current lineation, fig. 3-29)

Fine steps running parallel to the direction of current are developed on the bedding planes of many horizontally bedded sandstones and become visible by the parting of sandstone layers (SORBY 1856, CLOOS 1938, POTTER & PETTIJOHN 1963). This "current" or "parting lineation" is caused presumably by longitudinal grain orientation (MCBRIDE & YEAKEL 1963, ALLEN 1964a,b, 1968b). It is typical of the upper flow regime (with high velocity currents), in which transport by saltation of the grains is predominant but ripples are missing. Sometimes lineation is caused by mica enriched in longitudinal direction (GRUMBT 1966).

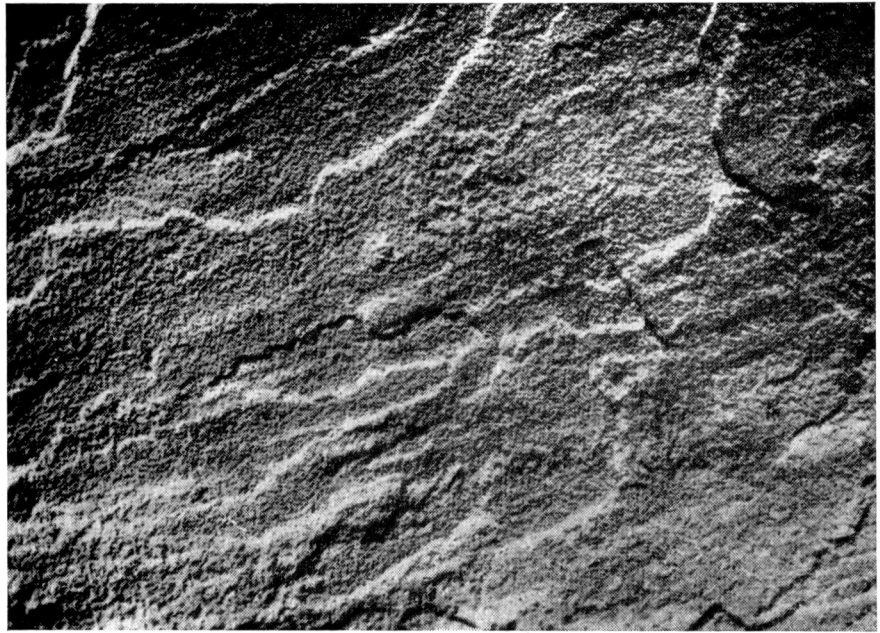

Fig. 3-29. Current lineation (from Cloos 1938). No scale. (On bedding planes of Devonian sandstones).

Another type of current lineation is "streaming lineation" (Conybeare & Crook 1968) which consists of narrow elevations and depressions, rounded in cross section, with their long axes in the current direction. They are, however, developed on the upper surface of beds.

3.242.23 On the lower surface (load casts, current-induced casts, trace fossils)

On the bottom side of sandstone layers the following sole marks can be developed ("hyporelief"):

3.242.231 Load casts (fig. 3-30)

Pocket-like protrusions of sandstone layers, originating from sinking of heavy sand into lighter (more porous) mud (Einsele 1963a). Post-depositional movement may cause orientation of the load casts in one direction. Sporadically they form huge sandstone pipes 50 m in diameter and more than 80 m in height (Schlee 1963). Ball structures (Pettijohn & Potter 1964) and pseudo-nodules (Kuenen 1965) are similar.

3.242.232 Current-induced casts; turbidity currents

On the top of certain shale layers scours are developed which are preserved by sand casts which have been filled during the superposition of a sandstone bed. They are relatively rare in shallow water or fluvial sandstones (de Raaf 1964, Panin 1965,

Fig. 3-30. Load casts on the lower surface of a laminated sandstone, Mississippian, Illinois (Plate 24 A of POTTER & PETTIJOHN 1963, with permission).

GRUMBT 1966, BEAUDOIN & GIGOT 1971), but are abundant in "turbidites", i. e. deposits of suspension (turbidity) currents in deep water, with a few exceptions (KUENEN 1964 c).

The extensive literature on turbidites and sole marks has been summarized by KUENEN & HUMBERT (1964); reviews and collections of plates are published by DŻUŁYŃSKI & WALTON (1965), POTTER & PETTIJOHN (1963), PETTIJOHN & POTTER (1964), GUBLER et al. (1966). For turbidite cycles see ch. 6.31.

The present model of turbidite formation may be summarized as follows: Sediments accumulate at the margin of deep troughs or abyssal plains until there is a rapid mass movement down slope. The triggering mechanism of such a movement can be the exceeding of a critical shear value or initiation by an earthquake (CHAMBERLAIN 1964). Cable breaks starting at the shelf edge and continuing downslope connected with the Grand Banks earthquake (Canada) in 1929 drew attention to the possibility of scouring turbidity currents and their possible role in forming submarine canyons (HEEZEN & DRAKE 1964). Such a coincidence of earthquakes and stepwise cable breaks has been observed also in the western Mediterranean (HEEZEN 1956).

Experiments have shown that earthquakes cause a sudden compaction of sands which leads to a sharp increase of the pressure of interstitial water (VAN DER KNAAP

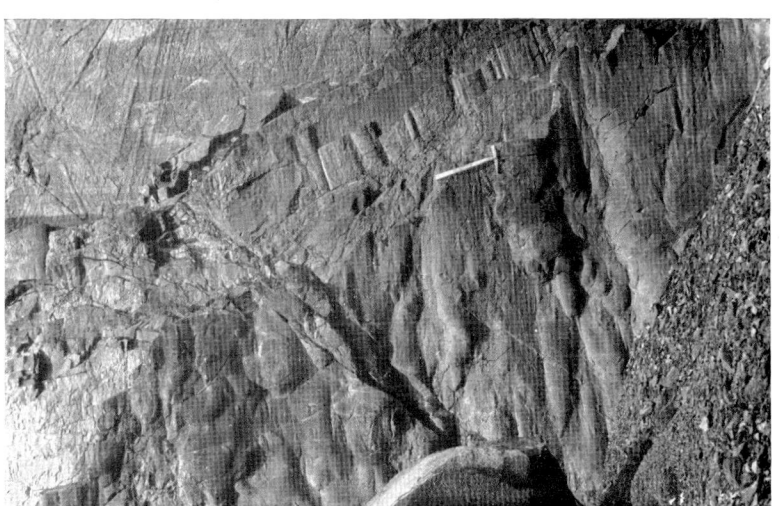

Fig. 3-31. Flute casts, Lower Carboniferous, quarry NW Altes Zechenhaus at the lower Innerste river, Upper Hartz Mountains. L e f t picture: General view on slightly overturned bottom sides; left: flute casts, right: groove casts, which at the extreme right deviate considerably from the general direction. Current was from the left (south). R i g h t photograph is enlarged section of left picture (photographs by PLESSMANN).

& EIJPE 1968). This pressure, if exceeding the hydrostatic pressure, may trigger downsliding of sediment combined with a considerable uptake of water especially in sediments with alternating permeabilities (e. g. in sand-shale alternations): An extremely hyperpycnical (e. g. denser than water) suspension current is formed, directed downslope towards the basin, where it slows in speed, thereby loosing its load while still in motion.

There is probably a large velocity differential between the front and back ends of the current (WALTON 1967). The coarsest components can be transported only in the front part. Therefore they are deposited first; at the same time, however, silt and clay are deposited from suspension. The subsequent parts of the current deposit an increasingly finer grained material; thus resulting in a "flysch-type graded bedding", which is characterized by an upward decrease in grain size but with nearly a constant clay and silt content.

Turbidites may form more or less wide-spread deep-sea fans. As the most prominent recent example, CURRAY et al. (1971) described the complex fan-valley system of the Bengal fan off Bangla Desh: Active channels, thousands of kilometers long and tens of kilometers wide, show meandering and braiding. According to MOORE et al. (1971), more than $3 \cdot 10^6$ km³ of sediment were contributed to this fan by the Himalayan Mountains area only during Quaternary time.

a) Flute casts (fig. 3-31)

Flute casts are club-shaped to conical depressions, beginning sharply and flattening slowly in the direction of current. Their length varies from less than 1 cm to 1 m (mostly 5—8 cm), and their depth from 1 mm to 30 cm (PLESSMANN 1961, POTTER & PETTIJOHN 1963). They originate from erosion of the clay surface by small vortices with horizontal or inclined-to-vertical axes. The faster the current, the larger the vortex. The size of a flute cast increases with the grain size of the sand deposited in it. This observation suggests two conclusions:

(1) these sole marks generally are connected causally to the subsequent sedimentation; (2) a positive correlation between current energy and grain size is valid also for turbidites. Size and shape of flute casts may vary from one layer to the other, but they are rather uniform within any one layer.

The sharp end of the flute casts points upcurrent. This has been shown by comparison with cross bedding or ripple orientation in the upper parts of the same layer.

Longitudinal ridges, a combination of flute casts and load casts, are similar in size and orientation (DŻUŁYŃSKI & WALTON). They can be confused with squamiform load casts with down-current rounded ends, which originate from flow-deformation of the sand casts during or after deposition (POTTER & PETTIJOHN 1963, PETTIJOHN & POTTER 1964). Cone-shaped rows of flute casts may originate from a sequence of vortices (vortex path).

Whereas flute marks are due to scouring vortices ("scour marks") the following marks (b-c) are produced by scraping objects and thus may be referred to as "tool marks":

3.24 Stratification

Fig. 3-32. Groove casts, probably modified by load casts. Left: Detail from fig. 3-31. Right: Eocene flysch, San Remo (Photographs by PLESSMANN).

b) Groove casts (fig. 3-32)

This term was introduced by SHROCK (1948) for casts of mostly sharp-profiled, parallel grooves which cover bedding planes and are similar to plant remains, while others form fine striations. In most cases the grooves are about one millimeter deep, rarely 1—2 cm. The width of the individual grooves is about 1 cm, 30 cm at the most. Occasionally their flanks are curled, as if some sliding object had been dragged along the cohesive sediment (chevron marks). Sometimes the groove cast is missing between these marks: In this case apparently the suction of an object sliding closely above the soft sediment surface was involved (DŻUŁYŃSKI & WALTON 1965). In most cases though the curling of the flanks is missing, suggesting that the sediment was already too stiff to curl.

The mode of formation of these casts is not known. Some smooth grooves are formed by clay fragments (RICCI LUCCHI 1970); channeled grooves can originate from fossil shells or small pieces of wood which occasionally are found at the end of the grooves (DŻUŁYŃSKI & RADOMSKI 1955, KULICK 1960b, TEICHMÜLLER 1960).

Soft animal bodies not preserved in the sediment were suggested as "tools" by TEN HAAF (1959). If, however, the entire bedding plane is covered by deep grooves without any plausible source, another explanation is needed (PLESSMANN 1961). It is possible to imagine a suspension in which larger fragments of highly porous sediment are dragged along the bottom until they fall apart. RICCI LUCCHI (1969) observed armoured balls composed of a mixture of sand, clay and pieces of wood at the end of groove casts.

The direction of the groove casts coincides with the direction of current. However, whereas flute casts and cross-bedding provide a vector direction of transport, this is not the case with groove casts, except with chevron marks. Individual groove casts are not exactly parallel to each other. Large differences of orientation within one bedding plane (fig. 3-31; PLESSMANN 1961) could be explained by several turbidity currents the first of which either did not deposit or was subsequently eroded by the next current (ANDERSON 1965). If the second did not create any grooves, but deposited, a considerable angle between its grain orientation or crossbedding and sole marks can be found (SPOTTS 1964, JIPA 1968). An alternative model of the formation of diverging groove casts is discussed under (d).

c) Impact casts

Objects hitting the ground for one or several short intervals can leave prod casts, bounce casts, brush casts or skip casts. The prod casts usually are small and elongated in the direction of current, with the impact on the downstream end being the deepest. These characteristics together with their sharp borders differentiate prod casts from flute casts.

Bounce casts are very short and deep groove casts, which often are wider than long.

Brush casts are soft imprints which occasionally have a little mound at the lower end.

In prod casts the impact object probably was inclined downstream and hit the sediment after a rather steep fall, whereas in case of the brush casts it was inclined upstream and hit at a relatively low angle (DŻUŁYŃSKI & WALTON 1965). These authors have reported additional marks, e. g. roll marks (by fish vertebra) and crescent marks (wash-outs [in front of obstacles] shaped like horseshoes opened downstream). All these marks (casts) occur mainly in turbidites, though they are found occasionally in shallow-marine sediments (KLEIN 1965).

d) Relation between different marks; discussion of alternative theories

Generally flute and groove casts do not appear on the same bedding plane: if they do, their frequency of occurrence is different. Moreover, the relative frequencies of beds with flute casts and groove casts differ from one formation to the other. Apparently both marks require different types of flow and/or sediment. DŻUŁYŃSKI & SANDERS (1959) and HSU (1959) think that groove casts are caused by a laminar flow of a dense suspension with large components at its base; flute casts are caused by turbulent, less concentrated suspensions. The present writer has found that groove casts frequently are filled with coarser sand than flute casts. This would be in harmony with HSU's opinion (see above), because dense suspensions are capable to transport coarser debris than less concentrated suspensions. According to SANDERS (1965, 207), grooves are attributed to the effects of a "flowing grain bed". The grains and larger objects "moving at the base of the flow may prevent the formation of a stable eddy system at the boundary with the cohesive mud, which is probably the condition required to produce scour marks", e. g. flute casts.

On the other hand, the quality of the clay substratum may play a role (DŻUŁYŃSKI & WALTON 1965). Clay layers have a strong resistance to erosion and thus groove casts are particularly frequent in fine-grained pelagic clay layers between turbidites. A lower resistance to erosion favors flute casts which occasionally are found in coarser shallow-water deposits. If the resistance to erosion is too low, e. g. if one suspension current immediately follows the other, silt and clay deposits of the first current may be removed by the second. Medium resistance to erosion may result in both groove and flute casts. The formation of groove casts also depends upon suitable objects (e. g. pieces of wood or armoured mud balls in the suspension).

Flute and groove marks are found predominantly in rocks deposited in long and (probably) deep troughs. In most cases their orientation is parallel (but sometimes perpendicular) to the axis of the trough; variance of direction is surprisingly low. In some basins there is an alternation of lateral and longitudinal turbidity currents. Due to the greater transport distances longitudinal currents generally produce finer grained turbidites (KUENEN 1966a), although the opposite is also documented (JIPA 1966). The elongated Lake Thun (Switzerland) may serve as an example of "normal" turbidite basins (STURM & MATTER 1971): "The sedimentation in the lake is dominated by density currents from the largest river, Kander, which distributes its sedimentary load by longitudinal transport over the main part of the basin. The smaller tributaries deposit their sediments within 200 m - 300 m from the shoreline."

Some lateral currents have been found to alter their course in a direction parallel to the axis of the basin (DŻUŁYŃSKI & WALTON 1965, BRIGGS & CLINE 1967).

During the past few years several localities have been described in which different marks and structures indicate different directions. SCOTT (1966) investigated a Cretaceous N—S trough in Chile. At its western flank he found graded bedding and slumping towards the east, but no sole marks. — Farther east there are thinner, finer-grained, partly graded sandstones with cross-bedding inclined southward. Also the flute casts point to the South, and the orientation of the grains is North-South. SCOTT assumed that the material was transported laterally into the trough by turbidity currents and then reworked by bottom currents running parallel to the trough. An alternative explanation was brought forth by KUENEN (1967b), who suggested that the turbidity currents ran south parallel to the axis of the trough. The thicker deposits in the western parts of the basin were due to greater rates of subsidence. The slump structures pointing east originate from post-sedimentary movements beginning with folding or North-South faulting. According to KUENEN (l. c.) the currents at the deep sea floor are too weak for major reworking; besides, reworking by normal (traction) currents would not cause flysch-type graded bedding characteristic of turbidites. Similar discrepancies of mark direction are reported by VOORT (1963) and KLEIN (1966, 1967b).

HEEZEN et al. (1966) discovered that the orientation of deep sea currents is mostly parallel to the slope contours ("contour currents", see ch. 3.252.32) rather than down the dip of the slope. This may explain different orientations of ripple marks and slump folds.

The orientation of pelecypod valves may enable one to distinguish suspension currents from sea floor traction currents: concave side up indicates suspension currents, convex up means normal currents (MIDDLETON 1967, but compare p. 207). According to MIDDLETON (1970), "flute and other scour marks ... are difficult to explain in terms of such ocean currents as are known to exist at the present day. Similar features can, however, be formed by currents dynamically similar to turbidity currents, for example, decelerating surges of concentrated dispersions produced in the fluvial environment by crevassing during floods."

The absence of ripple or dune cross-bedding as well as the occasional observation of antidune bedding (SKIPPER 1971) at the base of turbidites point to current velocities above the rippling limit of 100—125 cm/sec. These currents are supposed to have continued for months in order to move all the sand over the bottom for hundreds of kilometers. Actually, such current velocities have never been measured in contact with the bottom of abyssal plains (KUENEN & HUMBERT 1969). According to the same authors, the climbing ripples common in the upper sections of turbidites testify absence of reworking and a high rate of deposition, which is uncommon in normal ocean currents.

3.242.233 Trace fossils

The "Nereites" facies trace fossils are predominant in turbidites. They include mainly "Pascichnia" ("Winding trails or burrows of vagile mud eaters which reflect a "grazing" search for food by covering a given surface more or less efficiently and avoiding double coverage"; Weidespuren; SEILACHER 1964). *Nereites* and *Helminthoida* feed on the pelitic layers. Subsequent turbidity currents may erode the uppermost part of the pelitic layers including some of the trails whereas the trails in the lower part of this covering layer are preserved and are filled with silt or sand. They are "predepositional" with respect to the turbidite above (SEILACHER 1962, EINSELE 1963b). A few burrows, e. g. the twig-shaped fillings of *Granularia*, are "postdepositional". They may penetrate 4 m thick turbidite sands, cut sharply across flute casts and are found on the lower surface (SEILACHER 1962). Repichnia on the lower surface of shallow marine sand layers can be sand casts of trails on the mud surface (fig. 3-33).

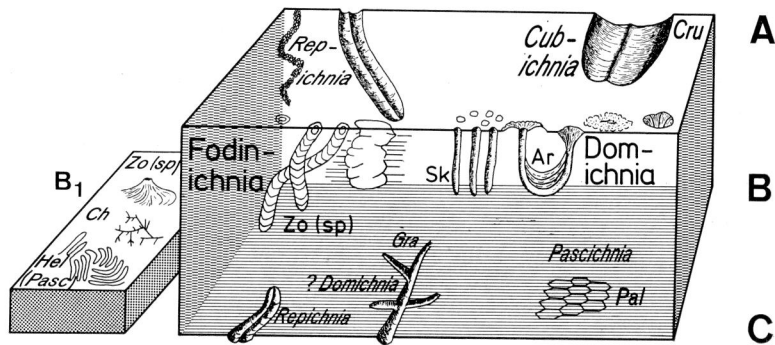

Fig. 3-33. Trace fossils, though bearing mostly Latin names [Ar = Arenicola, Ch = Chondrites, Cru = Cruziana, Gra = Granularia, Hel = Helminthoidea, Pal = Paleodictyon, Sk = Skolithos, (Sp = Spreite), Zo = Zoophycos] can rarely be assigned to known species or genera of animals. But they can be classified according to their main function (resting: Cubichnia; simple crawling: Repichnia; forageing: Pascichnia; mining or trapping food: Fodinichnia; shelters: Domichnia), or to their mode of preservation and topology [epichnia (A); endichnia (B); hypichnia (C)] with reference to the coarser beds, in which they are most commonly found. B 1 refers to parting surfaces of siltstones, marls, and shales. All aspects combined help to distinguish between certain types of "ichnofacies" indicating different environments, particularly those of shallow marine and deep sea sediments. Classification from SEILACHER (1964).

3.25 Sand bodies

3.251 Geometric characteristics

Most sand bodies belong to one of the following types:

1. Sheet sands (blanket sands). Expansive thin sheets of fine grained sand (Md < 0.1 mm), generally forming in the shallow-marine environment. Coarser sands of this type frequently originate from "smeared" marine bars. In this case they usually are connected with unconformities.

2. Extended, tongue-like, wedge-shaped or tabular sand bodies with various grain sizes. They occur mainly in shallow-marine environments, but also in arid fluvial and eolian deposits (KRYNINE 1948, BAARS 1961).

3. Channel sands. Narrow, elongate, predominantly medium to coarse grained sand bodies (frequently called "shoestring sands") which may be divided into two groups of different origin:

a) Fluvial and deltaic channel sands (barfinger sands) filling up a previously eroded channel. Their lower surface, therefore, is convex, and theoretically the upper surface is horizontal. In most cases these sand bodies form meanders and occur in rivers, deltas and tidal flats (POTTER 1963).

b) Marine bar sands originating from longshore sand transport, forming barrier bars (emergent) or sand bars (subaqueous). Their base is horizontal, and theoretically their upper surface is slightly convex. Such bars are frequently meandering or en echelon (RITTENHOUSE 1961). They are restricted to coasts, in which the effects of the surf surpasses that of the tides (tidal range < 1.5 m; HAYES 1967).

In most cases it is difficult to distinguish these two types of channel sands on the basis of shape alone, for either the outcrop is insufficient or the modifying influence of the compaction on the curvature of upper and lower surfaces cannot be eliminated. Additional criteria include the orientation of a sand body with respect to paleoslope: whereas river sand bodies (3a) run about parallel to the dip of the paleoslope, i.e. the slope at the time of deposition (PRYOR 1961), the marine sand bodies (3b) run about parallel to the depositional strike, at right angles to the dip of the paleoslope.

The following sections discuss and compare the characteristics of sandstones from different environments. In most environments, 10 categories of properties are included, partly following PETTIJOHN et al. (1965) (see also POTTER 1967, POTTER & BLAKELY 1967, SHELTON 1967 and SELLEY 1970).

3.252 Sand facies and environment

According to PETTIJOHN et al. (1972, p. 451), alluvial-deltaic sands supposedly are the most abundant sand bodies in the geologic column, forming perhaps 50%, whereas turbidite sands are probably second (perhaps 30%). Sands of all other environments cover the remaining 20%. In this chapter, however, not only sands are discussed. Silt and clay are included in many environments.

3.252.1 Continental environments

3.252.11 Dunes

(partly based on BAARS 1961, STOKES 1961, and McKEE 1966)

Eolian deposits, even those composed of carbonate grains, are called eolianites (PETTIJOHN et al. 1965). The chance of dunes being preserved in the geologic record is considerably less than for marine or fluvial sands. If they are not very thick, they generally are reworked by rivers or marine transgressions and redeposited as fluvial or marine sandstones. Small eolian sand bodies can be preserved e.g. by lacustrine shales as described by WALKER & HARMS (1972).

1. Horizontal association. The following types of dunes are distinguished:

a) Transverse dunes. They form in the lower flow regime and are connected with cylindrical, transverse vortices. Subtypes are "barchans" (i.e. lunate dunes with unimodal direction of cross-bedding, with ends bent forward like a crescent, see also SIMONS 1956, GAY 1962, LONG & SHARP 1964). If the ends are fixed (e. g. stabilization by vegetation) the middle portion may overtake the ends forming "parabolic dunes". Both types originate in areas exposed to one predominant wind direction. Long transverse dune chains may form at higher supplies of sand.

b) Longitudinal dunes. They form at higher velocities, in the upper flow regime. At this velocity the circulation takes the form of longitudinal rollers parallel to the wind direction, in which helical movement occurs (BAGNOLD 1953, ALLEN 1968b). These dunes are often bifurcating upwind as the axes of the helical wind rollers rise slightly downwind (FOLK 1971) (fig. 3-34 bottom). Current lineation replaces the ripples covering transverse dunes. — They are termed "seifs", from the Sahara. Seifs form the world's largest dune fields (NASA 1967) and are more frequent than other dune types. In Algeria they reach altitudes of 250 m. Seifs were found in the Sahara with their crests running EW. In this case, McKEE & TIBBITS (1964) found differing wind directions: Morning winds come from the NE and evening winds from the SE, thus causing bimodal directions of bed inclination.

c) Irregular pyramidal dunes ("star dunes"). These dunes form in areas with shifting wind directions or by interference (fig. 3-34 top).

The percentage of these types in recent Sahara dunes is as follows: a) 5%, b) 72%, c) 23% (JORDAN 1964). In the Colorado Plateau type a) predominates. Whereas barchan dunes are migrating types, seifs and star dunes are nearly stationary. Dunes often are found in or close to morphologic depressions where sand is available from the source rock or from rivers (wadis).

2. Vertical association. In dune deposits, pebbles, silt, and clay generally are missing (loess, an eolian silt, is not found adjacent to or alternating with dunes; it will be discussed in ch. 4).

3. Internal structure. Dunes often are cross-bedded with high-angle foresets (30—35°). They can include many sets by one dune migrating up the backs of other dunes. Stoss side layers also may be preserved; they dip at low angles opposite to the lee side foresets (McKEE 1966b, McKEE & DOUGLASS 1971). The direction of the lee side changes in seifs and star dunes, and with it the direction of dip of the cross-bedding. Planar, tabular or wedge-like cross-bedding is predominant, trough-shaped is rare because small-scale scours are missing. In contrast to fluvial sands, the directions of cross-bedding in eolian sands frequently do not coincide with the direction of paleoslope. Intercalations of adhesion ripples are characteristic (NAGTEGAAL 1973).

4. Thickness of stratified layers. The sets of cross-bedding are especially thick in the lower part of the dunes (10—20 m; fig. 3-24).

5. Bedding planes. Occasionally there are low altitude current ripples parallel to the dip on the steep accretion faces of the sets, caused by wind eddies that sweep laterally along the lee slopes. Moreover, avalanche structures occur (WALKER & HARMS 1972).

6. Mineralogical composition. Eolian sandstones can be recognized by a peculiar grain-size relation of light and heavy minerals (v. ENGELHARDT 1973). Mica generally is rare. In some arid climates, e. g. the western Texas Gulf Coast, dunes composed

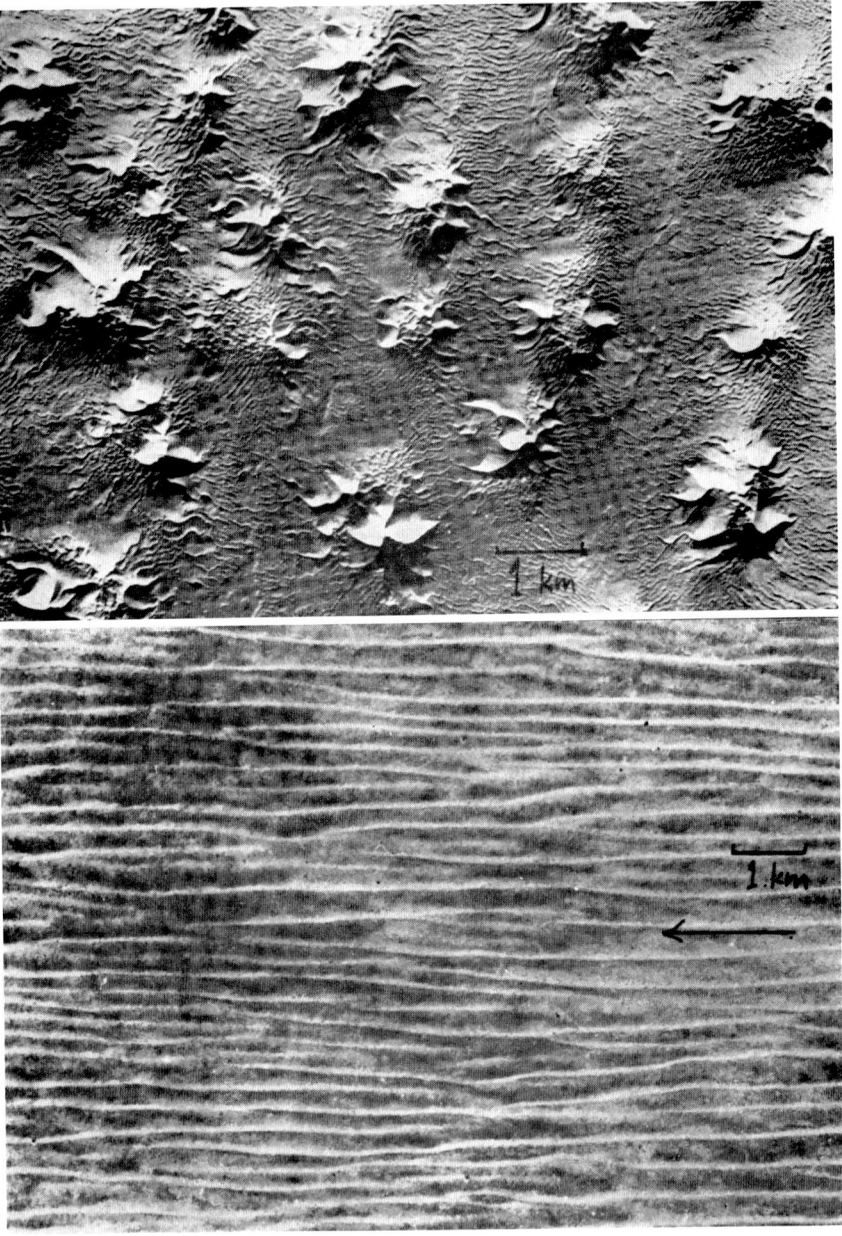

Fig. 3-34. Top: Fully developed "draa" pattern with prominent stellate rosettes (star dunes, about 150 m high): There are two interfering dune trends which in places join to form a fishscale pattern. I. G. WILSON (1972) distinguished by wavelength three major universally discontinuous groups of bedforms: draas (0.5—5 km wavelength), dunes (10—500 m), ripples (0.01—10 m), each subdivided into transverse and longitudinal elements. The gap between the

of clay aggregates occur (HUFFMAN & PRICE 1949); on the Bahama Islands oöid dunes are frequent.

7. Grain size. While the median of eolian drift sand in deserts is generally below 0.1 mm, the medians of dune sands are between 0.14—0.35 mm. They are quite well sorted and have a positive skewness (tail in the fines). Transverse dunes are reported to be better sorted than longitudinal dunes because of the conveyor-belt transport mechanism of the transverse dune types (FOLK 1971). Pebbles are missing in most cases; exceptions are mentioned by ADAM (1950) and MARSAL (1950). Grain-size lamination is not very distinct. Sorting characteristics of eolian silts and fine sands have been discussed in ch. 3.213. Coarse sand and small pebbles can form thin lag deposits on deflation surfaces (WALKER & HARMS 1972).

8. Grain shape. In coastal dunes, the sphericity is lower and the roundness higher than in the adjacent beach sands. The differences do not originate from eolian abrasion but from sorting mechanisms. Eolian abrasion is shown by dull surfaces on sand grains and by ventifact sculptures on pebbles (WHITNEY & DIETRICH 1973).

9. Grain orientation. A uniform grain orientation is not common, at least in larger eolian sand bodies; deposition takes place beyond the influence of the transporting agent, and grains tend to slide down the lee-side slope. A preferential orientation of elongated sand grains parallel to the dip direction, however, has been reported by LAND (1964).

10. Fossils and other properties. Most eolian sandstones are free of fossils, although some animal tracks have been found (GILMORE 1926, 1927, 1928). According to MCKEE (1944) only such tracks are preserved which "show uphill directions, because they were formed on surfaces sufficiently steep (28—30 degrees) to cause slumping with downhill motion and a consequent eradication of downhill tracks". Raindrop impressions have been described frequently (e. g. WALKER & HARMS 1972).

3.252.12 Lakes

Lake sediments mostly are rich in silt and clay (see part II/2). The shore lines are not very clearly developed (VISHER 1965). A spacing of wave-type ripples can be correlative with the size of lakes, and is generally larger than 4—5 centimeter for large lakes and seas, and less for ponds, small lakes and other restricted bodies of water (TANNER 1971). Chemical sediments commonly occur in lakes that lack an outlet (VAN HOUTEN 1965 a) (see ch. 6.34).

3.252.13 Rivers

(ALLEN 1965 b; COLEMAN et al. 1964).

This environment can be subdivided in the "alluvial fan" and the "alluvial plain" (or fluvial proper) subenvironments, though the first is generally

◄
groups, however, exists at different wavelengths for each grain size (NEWTON & WERNER 1972; see also ch. 3.241.31).
Bottom: Elemental longitudinal dunes; dune field with incomplete sand-cover (mean depth < 2 m). Simpson Desert, Australia. Both photographs from I. G. WILSON (1972). Arrow = wind direction.

associated with a semiarid climate and the latter with a humid climate. The fans include the deposits of debris flows, sheet wash, and braided streams. The deposits are characterized by great thickness (exceeding 4500 m in the Connecticut and New Jersey Triassic), rapid lateral facies changes, and preponderance of conglomerates and sandstones (BLATT et al. 1972). In the plain environment, two river types are distinguished:

Braided rivers are formed of anastomosing channels which mostly develop on relatively steep alluvial plains. The channels are labile because they choke in bed load (mainly gravel and coarse sand). This makes the rivers change their course frequently. Rivers with brief but heavy floods tend to be braided. Generally, braided rivers are governed by the upper flow regime with horizontal stratification and even antidunes (HARMS & FAHNESTOCK 1965).

Meandering rivers, on the other hand, are governed by the lower flow regime and therefore contain ripples and dunes. They develop at lower slopes and therefore are poorer in bed load and richer in suspension load ("mixed-load streams", McGOWEN & GARNER 1970). The high content of fines which, once deposited, are less erodible than coarser sediments, may stabilize the meanders which are flanked by point bars, muddy natural levees and overbank deposits. Long, indistinct floods are typical (LEOPOLD et al. 1964, McGOWEN & GARNER 1970). The meander length depends on the water discharge (fig. 3-35) as shown also by smaller meanders developing during dry periods in larger river beds ("arroyos"). More details are included in the following paragraphs (see 2).

1. Horizontal association (fig. 3-36). Distinction is made between

a) Channel sediments. In active channels the sediments consist of sand and pebbles (channel sands); in abandoned channels silt and clay also occur (channel fill in "oxbow lakes").

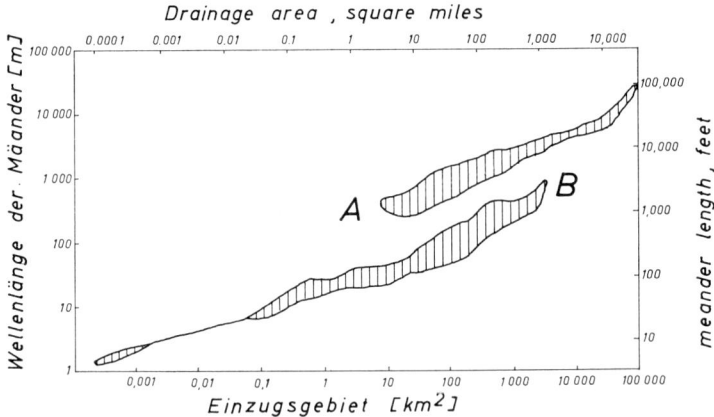

Fig. 3-35. Relation between meander length and size of drainage area, modified from LEOPOLD et al. (1964, Fig. 7-48). A = valley meanders (i. e. = cut into the rocks). Rivers include English rivers, Dnjestr, Don, Maas, Mosel, and tributaries of the Mississippi River; B = River meanders, which are allowed to move freely on an alluvial plain (Maas, Mosel, English rivers and, at extreme left, a small creek).

b) Point bars. They form on the convex side of meanders. Where meanders have not dug into solid rock to become fixed they slowly shift down the valley by erosion and redeposition. In such cases, the inner part of the meander loop consists of groups of point bars.

c) Natural levees. They accompany mainly meander channels and consist of silt and fine-grained sands dipping gradually (often for more than 1 km) towards the flood plain.

d) Flood plain deposits (by overbank flooding). These occur in the marginal parts of a river valley which are flooded only occasionally. They are finest suspension sediment of the river system.

Fig. 3-36. Thickness of the fluvial Trivoli sandstone (Pennsylvanian, Illinois). The thick phase (dotted) is the portion of the sandstone more than 12 m thick (Md 0.29 mm), corresponding to the channel sands. The thin phase (hatched) is the portion of the sandstone less than 12 m thick (Md 0.13 mm), corresponding to non-channel sands. An underlying limestone layer made it possible to calculate the gradient of the main Trivoli valley: 1 foot/mile (0.2 m/km). Arrow = current direction. Modified from ANDRESEN (1961).

a) and b) may be summarized as river channel deposits, c) and d) as flood deposits. A small lateral extension of the beds (1 m — 100 m) which therefore die out soon is typical for continental sediments.

2. Vertical association

Channel sands often have an erosional contact at their base; flood plain deposits are well-bedded. Fining-upwards fluvial cycles are common (see ch. 6.35).

McGOWEN & GARNER (1970) listed the following differences between the vertical sequences in bed-load and mixed-load streams:

Bed-load streams (coarse-grained load and braided rivers prevailing; no grain-size fining)	mixed-load streams (fine-grained load and meandering rivers prevailing; grain-size fining upward)
4. (top): horizontal bedding and thin foresets	4. (top): cross-bedding climbing towards the banks
3. thick foresets	3. horizontal bedding
2. thin foresets and thin trough-shaped cross-beds.	2. horizontal bedding, trough-shaped sets and thin foresets
1. thick trough-shaped sets or homogeneous sediment	1. thick trough-shaped sets

3. Internal structure. Trough cross-bedding, the dip of which is predominantly down the valley. The thickness of the sets increases with the current velocity. The variability of cross-bedding dip azimuths is lower for braided than for meandering rivers (CASSHYAP & QIDWAI 1971). Horizontal bedding is predominant in pebble beds, but also appears in sand bars at high current velocities.

4. Thickness of stratified layers. In river channel deposits the sets are mostly more than 100 cm thick, decreasing towards the top. Flood deposits generally are less than 5 cm thick.

5. Bedding planes. Transverse, asymmetric current ripples are common. At high current velocity flat sandwaves (antidunes) or horizontal bedding with parting lineations occur. Some flute casts are found.

6. Mineralogical composition. Low maturity can be found provided the source rocks are not high-maturity sandstones and chemical weathering of the source rocks is low or erosion is rapid. The distance of fluvial transport is not a very critical factor. Overgrowths of potassium feldspars (FÜCHTBAUER 1967b) occur; red biotites are instable in the fluvial environment (FÜCHTBAUER 1963).

7. Grain size. In broad channels the grain size usually is coarser than in small channels (STOKES 1961). Often the grain size decreases towards the top of a bed. Lamination within the cross-bedded units is caused by a marked change in grain size. Frequently the grain distribution is skewed towards the fines, even in channel sands. Sorting is poorer in flood plain compared to channel sediments because of the breakdown of the transporting energy in this environment.

8. Grain shape. Roundness will be low if little or no previous abrasion took place.

9. Grain orientation. Mostly in the direction of flow; sometimes perpendicular to the flow direction, especially for pebbles.

10. Fossils and other properties. Plant remains (such as roots), vertebrate remains and tracks, and burrows are found occasionally especially in flood deposits. Pelecypods and gastropods are rare in the fluvial environment. Colors: Reddish and greenish ("red beds"); reddish commonly in mud facies, drab frequently in sands (VAN HOUTEN, pers. commun.).

3.252.14 Glaciofluvial and glacial sediments

Intercalations of glacial or glaciofluvial processes and sediments in older formations are significant

(a) by disconformities due to eustatic regression of the sea and subsequent glacial erosion including grooved and striated rock pavements and pebbles as well as crescentic fractures.

(b) by forming characteristic morphologies in the bedrock including U-shaped valleys

(c) by distorted or dislocated bedrock due to ice thrusting (glacial tectonics)

(d) by paleovalley fill consisting of "till" (= moraine) or "tillite", if indurated, i. e. sediment transported below, within, or above a glacier and consisting of poorly sorted boulder clays, matrix-rich breccias and conglomerates, e. g. mixtures containing $15\% < 20\,\mu$ and $15\% > 1$ cm (GILLBERG 1965).

(e) by forming numerous anastomosing channeled sandstone bodies often scoured into the underlying rock. Such glaciofluvial channel sandstones may have formed in a network of proglacial or periglacial braided rivers with very homogeneous main directions over large areas, comparable to but more extensive than certain "sandur" deposits in southern Iceland. Also meltwater lake deposits consisting of varved clays or silts can occur.

(f) by a slow (postglacial) eustatic transgression finishing the glacial event.

Such intercalated glacial formations have been described recently from the Ordovician of North Africa (BEUF et al. 1971, BENNACEF et al. 1971).

CROWELL & FRAKES discussed glacial deposits in different countries, for instance in the Karroo Basin, South Africa (1972).

"Sediments deposited directly by glaciers or indirectly in glacial streams, lakes and the sea together constitute glacial drift" (LONGWELL et al. 1969). The different morphologies connected with glacial deposits, e. g.

drumlins (streamline till accumulations),

ground and end moraines,

eskers (sand- and gravel-filled channels on top of or under stagnant ice),

kames (irregular glaciofluvial sediment accumulations on top of stagnant ice), and

pingos (donut-shaped soil accumulations caused by centrifugal solifluction above ice lenses in permafrost soils) will not be discussed in detail.

3.252.15 Red beds

Red beds are continental deposits composed of predominantly alluvial sand-, silt- and claystones and have a reddish-brown color (as a result of hematite); red beds may or may not belong to molasse units. Until recently it was assumed that the hematite has been supplied by the reworking of lateritic soils (Terra Rossa) and thus formed under tropical to subtropical climates with changing humidity (KRYNINE 1949). However, according to T. R. WALKER (1967 a, b) some modern-day red beds form not in the savannah, but in dry areas, and the process is dominated by in-situ oxidation of ferruginous silicates, such as biotite and hornblende. Such a process, of course, takes place in constantly humid tropical soils as well (T. R. WALKER 1968), but it is favored by frequent drying and wetting at high temperature (see p. 144).

In the Alpine Molasse, which is not a red bed sequence proper, only marine and brackish sandstones contain red-brown biotites, whereas these biotites are weathered into brown-green biotites with simultaneous growth of hematite in fluvial sandstones of the same sequence. Apparently, iron was mobilized in the oxygen-rich environment (FÜCHTBAUER 1963). — Fluvial sandstones and siltstones of the uppermost Carboniferous (Westfalian D and Stefanian) in NW-Germany are red compared with the fluvial sandstones beneath. The flora indicates the onset of an arid climate, simultaneously with the reddening of the strata. Slight evaporation is indicated by an increase in boron in Stefanian siltstones (TEICHMÜLLER 1964, with discussions by JOSTEN). Also the subsequent formations, the Permian and the Buntsandstein (Lower Triassic) accumulated during periods of aridness or semiaridness (as shown by simultaneous evaporites), at least in middle Europe.

SCHLEGELMILCH (1968) and others found that the differences in the iron content between red and gray (or green) shales (or graywackes) of the same sequence can be very small or even absent. This is in harmony with a post-sedimentary origin of the red color.

Red beds are characterized by typical trace fossil associations (SEILACHER 1963).

Under special conditions (oxygen-rich bottom water and slightly ferruginous sediment), red sediments also can form in marine environments as shown by numerous reddish claystones, marl- and limestones (EINSELE 1963 b, HALLAM 1967 b). Even in turbidites, red claystones and siltstones were reported (HUBERT 1965). If the iron content of the sediment and potential oxygen in the bottom water are constant, the red color depends upon the amount of reducing organic substances and the time of exposure, which in turn is dependent upon the sedimentation rate:

	inorganic sedimentation slow ————→ fast
organic sedimentation slow ↓ fast	red / gray

Slow sedimentation was suggested for red Liassic limestones of the Alps (FABRICIUS 1966).

HINZE & MEISCHNER (1968) observed that in the Adriatic Sea the marine sediments reworked from laterites are gray. The sedimentary surface only shows a secondary brown color due to oxidation at the sediment surface. This observation also suggests that red beds must not necessarily be attributed to reworking of red soils (T. R. WALKER 1967b).

3.252.2 Transitional environments

Transitional environments are intermediate in time or space between marine and nonmarine environments. This is the main criterion of such environments to be recognized in older rocks. They can be subdivided in environments connected with a river mouth (deltas, estuaries) low-energy environments (tidal flat, lagoons), larger low- and high-energy transitional environments (marginal seas), and high-energy environments (beaches).

3.252.21 Deltas

[BATES 1965, FISK 1955, 1960, DUNBAR & RODGERS 1957, VAN STRAATEN 1960, MOORE 1966; "Deltas" (M. L. SHIRLEY, ed.), Houston Geol. Soc. Spec. Pub. 1966; "Deltaic sedimentation, modern and ancient" (J. P. MORGAN, ed.), Soc. Econ. Pal. Min. Spec. Pub. 15, 1970; and "Delta systems in the exploration for oil and gas" (FISHER, W. L. et al., eds.) Bur. Econ. Geol. Univ. Texas, 1969.]

In the following paragraphs, the main properties are arranged according to PETTIJOHN et al. (1965).

1.-3. Horizontal and vertical associations and internal structure (fig. 3-37).

A delta is an accumulation of sediments where a river enters a larger body of water giving the coast line a convex configuration. The name was first applied to the Nile delta by HERODOTUS, because the subaerial shape looked like the Greek character Delta. The shape of a subaerial delta is a function of the rate of supply of sediments, the energy conditions, and the subsidence. Beginning with the highest value of the ratio of supply/energy (or subsidence), the following shapes are found: Birdfoot (Mississippi only), lobate, cuspate or v-shaped, arcuate or rounded shape, estuarine delta (filling an embayment or estuary) (CURRAY 1969). During the quick Holocene transgression, most rivers presumably formed estuaries. During the adjustment to the present standstill of the eustatic sea-level rise, many estuaries changed into deltas while most deltas have changed their shape. The longitudinal cross-section of all deltas displays the following units (fig. 3-37, C-D, 3-38):

a) The topset beds ("delta plain") are generally the coarsest sediments, but consist of various environments including marsh and swamp (silty clay), channel point bars (sands), lunate river mouth bars, natural levees (root-disrupted, laminated silty clay), barriers, destructive phase reworked sediments, and a shallow delta front platform (CURRAY 1969).

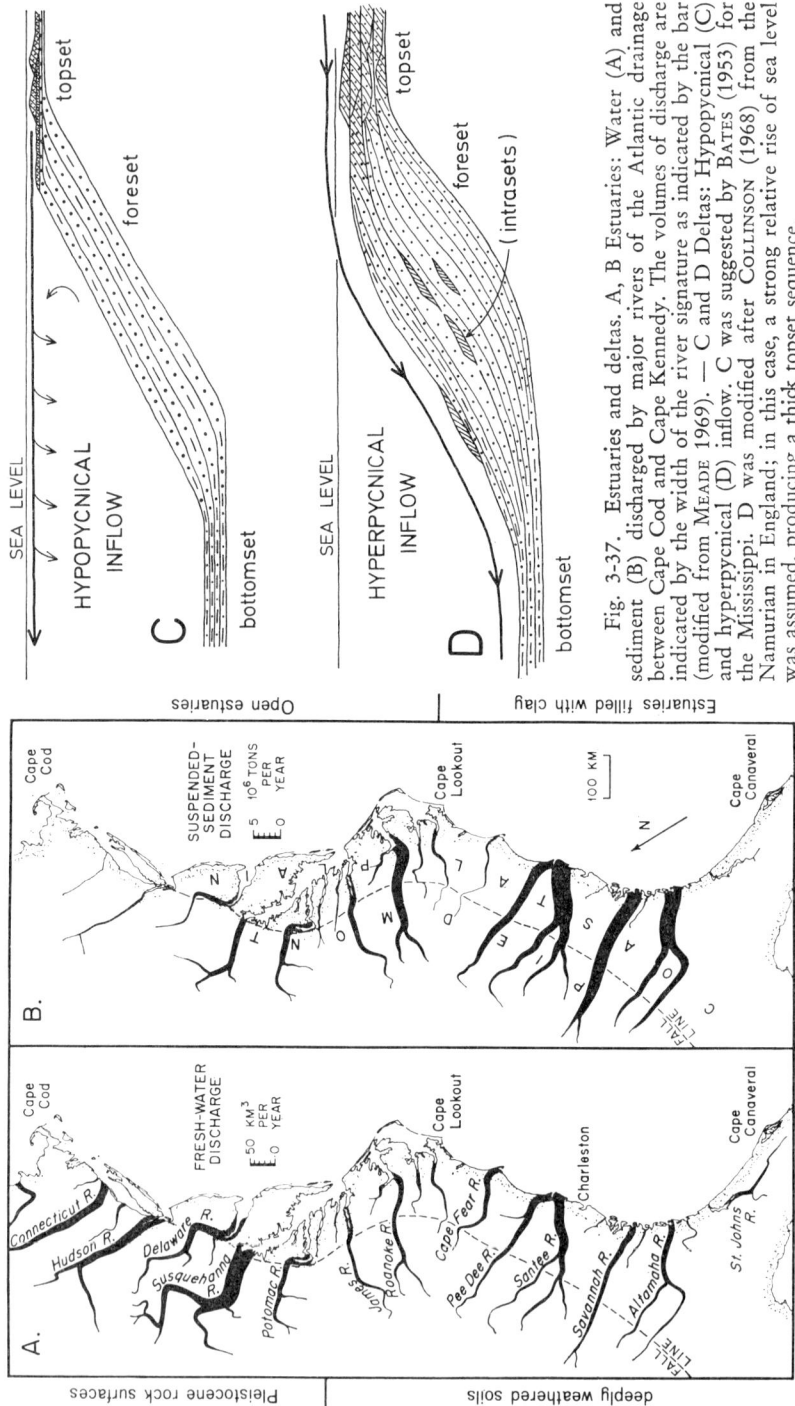

Fig. 3-37. Estuaries and deltas. A, B Estuaries: Water (A) and sediment (B) discharged by major rivers of the Atlantic drainage between Cape Cod and Cape Kennedy. The volumes of discharge are indicated by the width of the river signature as indicated by the bar (modified from MEADE 1969). — C and D Deltas: Hypopycnical (C) and hyperpycnical (D) inflow. C was suggested by BATES (1953) for the Mississippi. D was modified after COLLINSON (1968) from the Namurian in England; in this case, a strong relative rise of sea level was assumed, producing a thick topset sequence.

b) The **foreset beds** ("delta-front") are generally the most rapidly deposited sediments and are of medium to fine grain size. They are thinly interbedded silty sands or sandy silts and silty clays. Small-scale ripple structures are common; burrow-mottled structures occur, provided sedimentation is not too rapid. The quantity of silty clay increases laterally away from the distributary mouth (DONALDSON et al. 1970). The slopes are from less than 1° (Mississippi) to 7—10° (CURRAY 1969).

c) The **bottomset beds** ("pro-delta") are the finest sediments and frequently are laminated silty clays.

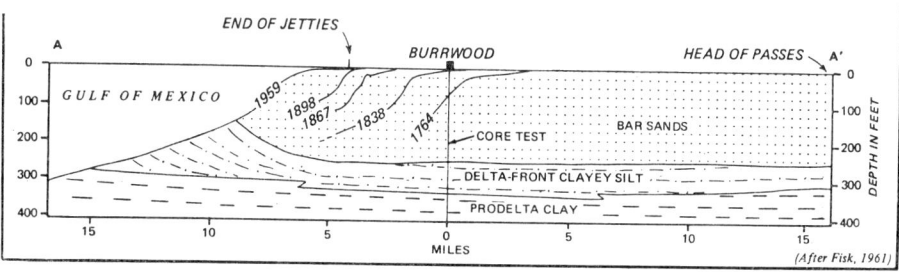

Fig. 3-38. Mississippi delta sedimente; longitudinal section. Southwest Pass bar-finger (from GOULD 1970, with permission) (see also fig. 6-6).

In fine-grained delta complexes like the Orinoco (VAN ANDEL 1967), the topset deposits surrounding the delta shore to a depth of approximately 5 to 10 m consist of dark silty clays interbedded with thin layers of fine, well-sorted sand. They grade abruptly into black sulfidic foreset clays with silt laminae extending to a depth of 35—50 m, on the Atlantic shelf.

Different delta types have been described, in terms of the relative density of river and basin waters (fig. 3-37):

i) GILBERT (1883, 1890) described the type delta (named after him) in a small fossil delta at glacial Lake Bonneville, whose modern remnant is the Great Salt Lake (Utah, USA): Flat bottomsets in the lake are covered by large planar foreset sands, dipping relatively steep towards the lake. With the advancing delta the foresets are covered by topset river sediments. There are two conditions critical to the formation of the classic "Gilbert Delta": 1. The inflowing water has about the same density as the water in the basin (homopycnic; BATES 1953). This results in an easy mixing, and the sedimentary load is deposited close to the river mouth (OOMKENS 1967). 2. Heavy surf which would prevent the buildup of the delta is missing. — Lakes meet both conditions, but most nearshore marine environments do not.

ii) Fresh water with a heavy suspension load is denser than the water in the basin (hyperpycnic), and thus flows down the foresets, smoothing the transition between foresets and bottomsets (AXELSSON 1967).

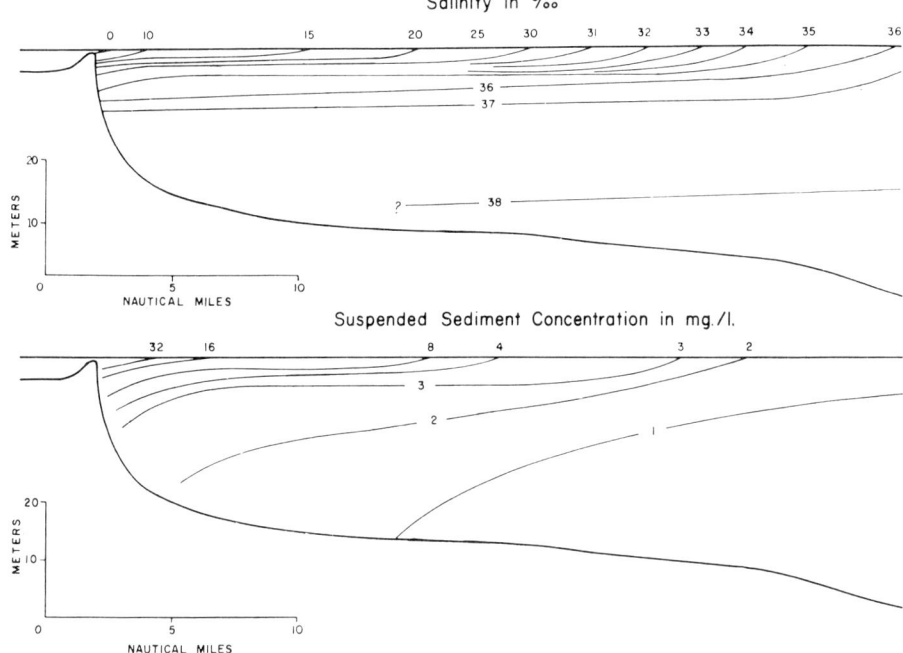

Fig. 3-39. Hypopycnical inflow ("plane jet"). Vertical salinity and suspended sediment sections from the Po River delta towards the Adriatic Sea. The well-stratified salt-wedge system is due to low energy and low tidal range (60 cm) (from NELSON 1970).

COLLINSON (1968) reports from the Namurian (Upper Carboniferous) a delta with cross-bedded unit 40 m in thickness, dipping at 27° and attaching tangentially to the bottom. On top of the foresets smaller cross-bedded units (intrasets) were deposited below the overflowing suspension. This suggests that a hyperpycnic environment prevailed for at least some time (fig. 3—37 D).

iii) In marine deltas the inflow generally is hypopycnic. The fresh water is lighter than the ocean water and remains on top of latter (fig. 3-39). Therefore, the mixing takes place slowly, and the sediments are disseminated farther towards the open sea, with the result that the delta slope is rather flat. Coagulation of clay-size grains due to electrolytic action of sea salts counteracts this influence to a certain degree (pt. II/2). In the Mississippi delta the prodelta slope is less than 1° (SHEPARD 1964). This is due partly to surf and tidal action, which truncate the uppermost parts of the delta and rework them into laminated topsets (SHEPARD, l. c.) or well-sorted sand bodies migrating parallel to the coast. Lamination of sand and silt or clay is typical for deltaic sediments indicating fast deposition which keeps burrowing animals from settling (REINECK 1967a): generally, it indicates an environment hostile to life.

Trough-shaped and planar cross-bedding with considerable thickness of the sets (40 m) was described by VÖLK (1966). COLEMAN et al. (1964) have described the minor sedimentary structures in a prograding distributary.

4.-9. Thickness of stratified layers, bedding planes, mineralogic composition, grain size, grain shape and grain orientation in a delta do not differ characteristically from fluvial sediments that feed the delta. The major exception is the more frequent occurrence of silt in delta deposits. In regressive sequences delta sediments become coarser towards the top, whereas channel-fill sequences are fining upward into silty clay indicating the change from active to abandoned conditions (DONALDSON et al. 1970, OOMKENS 1970). Sedimentary bodies in the Rhone delta are 50—70 m in thickness, and often cover hundreds of square kilometers (OOMKENS 1967). Fossil silt or sand bodies that become coarser towards the top have been described by MELLON (1967). Upward coarsening of sandstones has also been observed in transgressive sequences, due to grading of lagoonal into coastal barrier sediments, but transgressive sequences tend to be thinner and are overlain by marine sediments (OOMKENS 1967). For additional sedimentary structures see table 3-16.

Depending on the hinterland and its tectonic activity, delta deposits are made up either predominantly of clay and silt (Mississippi, Atlantic coast, e. g. Orinoco) or sand (Rhine, Rhone). Sand deltas are more frequent in front of a rising hinterland. Such deltas have been termed "tectonic deltas" by FRIEDMAN & JOHNSON 1966; the Devonian Catskill sediments NW of New York are a typical example. Most deltas in the Tethyan Belt and in the western Pacific are tectonic.

10. Fossils. Plant remains in marine sediments are characteristic of delta regions. The ratio of ostracods to foraminifera is often very high in delta sediments (SHEPARD & MOORE 1955). Burrowing is relatively rare. Consequently the content of organic material is high.

In ancient sediments, mainly according to WEIMER (1970), the recognition of a delta may be based on two or more of the following criteria: (1) an arcuate lithofacies pattern of nonmarine coastal plain strata protruding toward the marine basin; (2) for a designated time-stratigraphic unit, thickest deposits in the general shoreline zone are associated with the deltaic complex; (3) a complex intertonguing of marine (bottomset and foreset) and nonmarine (topset) strata; (4) rapidly changing shoreline-sandstone trends from one time-stratigraphic unit to another; (5) greater abundance of stream deposits than of deposits of other environments of coastal plain; (6) biological criteria in marine strata: benthonic arenaceous foraminifera dominate, whereas pelagic calcareous forms are absent; (7) repetitive lithologic associations (e. g. fine laminations of clay and sand-silt flasers); (8) unusual concentrations of detrital carbonaceous material. (9) persistence of criteria listed above in vertical stratigraphic sequence indicating semipermanency of drainage systems responsible for deltaic deposits, and (10) linear sand bodies branching off in the direction of transport (BUSCH 1971).

The subsidence in deltas may be exceedingly high as shown by the Po river delta (Italy), with a downsinking of up to 13 cm / year (NELSON 1970).

Generally, it has to be taken into account that the modern deltas prograde mostly toward oceanic basins whereas ancient deltas were built frequently in shallow, narrow embayments on a relatively stable continental plate (FERM 1970). Whereas delta systems in rapidly subsiding basins do not prograde appreciably and thus have

Table 3-16. Sedimentary structures of Mississippi deltaic and other depositional environments (from COLEMAN & GAGLIANO 1965, with permission). text. = textural, col. = color variations. D = dominant, A = abundant, C = common, R = rare.

Sedimentary Structures	Shelf — Clastic deficient (carbonate rich)	Shelf — Fine clays	Prodelta	Delta front — Distal Bar	Delta front — Distributary mouth bar	Delta front — Distributary channel	Delta front — Subaqueous levee	Delta front — Subaerial levee	Delta front — Interdistributary bay	Mudflat	Lacustrine	Paludal — Marsh	Paludal — Swamp
Bedding thick >20 cm	D	D	D	D	C-D	D	C-D	C-D	C-D	R	D	D	D
medium 5–20 cm					C-D		C-D	C-D	C-D	C			
thin <5 cm									C-D				
Morphology parallel laminations, text.		R-C	C	A	C	A	C	A	C	R-C	R-C	C-A	R-C
parallel laminations, col.		R	C-A	A	R		A	A	A	C	C	C	C
lenticular laminations		R	C-R						R	C			
wavy laminations							R-C	R-C					
Cross-laminations simple			R	C	A	C	C		C	R-C	R-C	C-A	C
planar			R	C	A	C	C		A				
trough						A	C	R-C	R				
Ripple laminations current		R		C-A	R-C	R	A	C		R-C	R		
ripple-drift		R		C	R-C		A-C	R		C-A	R-C		
wave				C	R-C		R-C	R		R-C			
Scour and fill				C	C	C-A	C	C	C				
Erosional truncations				R	C	C-A	A	C-R	R				
Plant remains distinct particles				R	C	R	R	R					
finely divided					C	C	R	R	C	C	R	R-C	R-C
bedded	R	R	R-C	R	C	C	R	R-C	R-C	C	C	C-A	C-A
Shell fragments						C-A	R-C	R-C	C-A	R-C	R	A	C
Clay inclusions	A	C-A	R	R	R-C	C	R	C	C-A			R-C	R
Load casts						R	C-R	R					
Distorted laminations gas heave						R	R	R					
slump							C-R	R-C					
recumbent folds					R-C	R-C	C-R						
convolute laminations						R	C-A	R	C				
Burrows, small	C-A	A	C-A	C-A	C-A	C-A	C-A	C-A	C-A	C-R	C	C-A	C-A
Burrows, large	C	A	R	R	R	R	R	R	C-A	C-A	C	R-C	R-C
Nodules							R-C	R-C					

facies that tend to persist vertically, deltas on stable cratons prograde extensively so that vertical sections display many different lithologic units. Good examples of vertically variable deltas are the Tertiary deltas of the Gulf Coastal plain in Texas and present-day rivers that occupy the same positions (FISHER & MCGOWEN 1967), and the Lower Triassic (Buntsandstein) delta precursors of the present-day Rhine and Weser rivers (WURSTER 1964, FÜCHTBAUER 1967b).

More information on delta sedimentation is included in ch. 6.33 (p. 377) and fig. 6-7, as well as in the Delta Symposium (Amer. Assoc. Petrol. Geol. 55/8, 1971) dealing especially with well-developed subsurface patterns and their bearing on oil exploration. For delta morphology see WRIGHT & COLEMAN (1973).

3.252.22 Estuaries

[s. "Estuaries" (LAUFF, ed.), Amer. Assoc. Adv. Sci. Bull., 83, 757 p., 1967; VAN STRAATEN 1954, REINECK 1963.]

An estuary is the wide mouth of a river where the tide meets the river currents. It may be formed by the river (drowned river valley) or may be preformed (e. g. tectonically) and then be used as outlet of a river. The general outline of an estuarine deposit should be crudely triangular, and it should grade outward into typical marine sediments and landward into those of fluvial origin (TWENHOFEL 1961, vol. 2).

Estuaries of rivers discharging into salt-water bodies are characterized by a saline layer underlying the fresh water. This saline layer frequently forms a wedge thinning out landward. The interface may be well-defined and move with the tides, but turbulence by wave action, unusually high tides, low fresh-water flows, and wind-induced currents may produce a more mixed condition (IPPEN et al. 1966). Also Coriolis force may deflect flood and ebb flows in the estuary (CURRAY 1969), especially in circumpolar areas. The Cook Inlet (Alaska) with a tidal range up to 12 m is an example: The incoming waters are forced to the eastern margin of the estuary which therefore contains a higher percentage of coarser pebbles than the western side (SHARMA & BURRELL 1970).

Sand is an uncommon sediment in many estuaries, compared with clay and silt. The sediments accumulate especially near the upstream limit of landward bottom flow, i. e. of the salt-water wedge (MEADE 1969). Since the estuaries are low-energy environments, they c o l l e c t suspended and bed-load sediments which cannot settle in the adjacent high-energy beach and offshore environments.

The smaller (southern) estuaries of the Atlantic Coast of the United Staates are in equilibrium with the present sea level; that is, the net long-term accumulation of sediment is zero. The larger (northern) estuaries, however, are still trapping sediment (MEADE 1969). These regional differences are shown on fig. 3-37 A, B: The northern estuaries are open due to small suspension load of the rivers draining an area covered by Pleistocene rock surfaces, and due to seaward dilution of the sediment load concentrated near the fresh water/salt water interface (MEADE 1968). The southern estuaries are filled with suspended load of the small rivers which drain a deeply weathered country.

During the H o l o c e n e t r a n s g r e s s i o n many estuaries developed. With continued stability of sea level in its present position, these estuaries are being filled and many of the rivers are beginning to build deltas (CURRAY 1969). Whereas some of them are still transitional (e. g. the Orinoco delta, VAN ANDEL 1967), others built up already a protruding delta (e. g. the Mississippi River).

On the other hand, estuaries will transform into lagoons if the climate becomes arid and the rivers become weak.

Data from modern estuaries compiled by KLEIN (1967a) showed that many river estuaries are characterized by vertical sequences consisting of a basal lag concentrate of clay pebbles, marine shells, and peat fragments, overlain by cross-stratified sands and capped by flat-bedded, rippled sands, and clay laminae. They record the predominance of upper flow regime during bank erosion and lower flow regime during sediment deposition. A vertical decrease in grain size and bedding thickness is common.

3.252.23 Tidal flats

These are the muddy flats bordering many shallow coasts which are covered with water during high tide (flood) and fall dry during low tide (ebb). Due to the short tidal intervals, however, real drying of the surface layer of sediments hardly occurs. The flats are furrowed by smooth and more or less straight flood channels and by steeper-walled, meandering ebb channels. Because of the configuration of the channels and flats, the average water depth is greater at low tide than at high tide, if only submerged areas are considered. Thus, more suspended sediment reaches the bottom at high tide than at low tide; this results in net deposition within the tidal flat. Also the period of low current velocities is of longer duration at high tide than at low tide because of the above-mentioned differences in water depth. Moreover, the velocities decrease landward in the tidal flat (CURRAY 1969). These phenomena explain the following observations:

(1) A net influx of sediment into the tidal flats from the sea.
(2) An increase of grain size towards the open sea.

This grain size distribution is opposite to the normal offshore zonation. Muddy flats grade seaward into fine-sand plates; both are interrupted by tidal channels which move laterally through the tidal flats by erosion of the sandy mud and redeposition of the sand fractions: "Longitudinal cross-bedding" is frequent in the (ebb) channels (VAN STRAATEN 1954, REINECK 1958c), i. e. layers of sand or interbedded sand and silt are deposited on point bars dipping towards the channel at low angles; the included sand layers generally show fine ripple stratification dipping parallel to the channel's axis (REINECK 1970).

DE RAAF & BOERSMA (1971) noted the following criteria of tidal environment in older rocks, in descending order of importance (see also KLEIN 1971a, b, 1972):

a) directional bimodality of cross-bedding occurs in different dimensions (herring-bone structures)

b) large-scale cross-bedded sands are coupled in super- or juxtaposition with small-scale cross-bedded sand-silt units. Horizontal lamination-units may be included. This indicates highly variable hydrodynamic conditions.

c) small-scale ripples ascend the mega-foreset slopes in opposite directions. Silt and clay may be deposited on these ripples.

d) flaser (= silt lenses) bedding of sand and silt is wide-spread and quantitatively more important than in other environments, in which they occur, e. g. in coastal lakes (Rhone delta plain), proximal fluviomarine areas (Rhone prodelta), lagoons (Zuiderzee), marshes, interdistributary bays (Mississippi delta), and some low-energy fluvial environments

e) larger ripples with wave-reworked tops point to emersion or at least to significant differences in the water level

f) Intercalation of strongly bioturbated layers in several types of sandy and/or muddy units is frequently reported especially from older tidal flat deposits

Other important criteria include

g) laterally moving tidal channels, sometimes with accumulations of shell beds, clay intraclasts or fine-grained sediments at their bottom

h) air bubbles which, however, generally will not survive the first stages of diagenesis

i) diverging directions of cross-bedding sets and the ripples on their top

j) algal mats which are interrupted by rippled areas about one meter in diameter, in which the mats have been removed mechanically before the development of ripples (D. B. Mackenzie 1972)

k) lateral interruption of an oscillation ripple-field by small channel-like areas with rhomboid or linguoid ripples typical of very low water depth

l) deep dinosaur footprints indicating that most of the anima's weight was borne by the bottom, hence the water must have been quite shallow.

m) mud cracks intercalated in a marine sequence (cf. Klein 1970).

n) According to Seilacher (1967b), suspension-feeding burrowers with vertical burrows and "spreite" lamellae are typical to the agitated shallowest-water environment including tidal flats. Burrowing is missing in the channels.

o) Generally, the tidal flat fauna is poor in species but rich in individuals, as it is typical also of other restricted environments. The restriction of the life conditions is mainly due to the sudden decrease of salinity caused by strong rainfalls during low-tide.

Such lines of evidence should be supplemented by data on the position of the shore-line and the paleoslope.

Though the probability of preservation of complete intertidal sequences is low, such successions may occasionally be important indications of intertidal sediments. Grohne (1956) observed the following vertical sequences in the Jade bay (Germany)

(Top) salt marsh ("Grodenschichten") with roots of halophytes
       (see van Straaten 1954)
       muddy tidal flat deposits
       mixed tidal flat deposits   regressive sequence
       sandy tidal flat deposits

sandy tidal flat deposits
mixed tidal flat deposits
muddy tidal flat deposits   transgressive sequence
(salt marsh with roots)
brackish or freshwater-clays
(Bottom) peat

The internal fabrics of these tidal flat sediments are described by REINECK (1962) from the German Wadden Sea: Salt marshes are irregularly interbedded sands and silts which are perforated by roots. — Muddy tidal flat deposits consist of silt and clay with sand streaks; frequently these strata are intensively burrowed and obliterated. — Mixed tidal flat deposits contain flaser and lenticular bedding as well as "tidal bedding" (irregular alternations of sand and silt/clay layers); burrowing occurs. — Sandy tidal flat deposits are horizontally bedded or cross-laminated; flasers occur. The preservation of the above sequences and textures, however, depends on the rate of subsidence of the area.

Many details on the intertidal environment of the German Wadden Sea are collected in the publication "Das Watt" (REINECK, ed. 1971).

3.252.24 Lagoons

A lagoon is defined as a shallow lake connected with the sea, often elongate parallel to the coast line. Therefore it represents a transitional environment, normally of small scale.

Lagoons as for instance the Laguna Madre (Texas) are filled mainly by material from the open sea or from the barrier island (fine sand) transported by tidal currents, washover storm waves, and wind transport. An average sedimentation rate in the northern Laguna Madre amounts to 120 cm / 1000 years, which is 2—4 times slower than in the more humid Texas bay to the north of Laguna Madre (RUSNAK 1960b).

Sedimentation rate, however, is not critical to lagoons nor to any other environment but depends on the rate of subsidence or post-glacial sea-level rise.

Other attributes of lagoons are oyster bioherms, the frequency of fecal pellets, and the vertical burrows mentioned for the tidal flats.

3.252.25 Marginal seas and Gulfs

("Nebenmeere", for references see SEIBOLD 1970).

Marginal seas are comparable to, but larger and deeper than lagoons, and receive water from one or more rivers. Present examples of humid marginal seas include the Baltic Sea, the Hudson Bay, the Black Sea, and the Gulf of Siam; examples of arid marginal seas include the Mediterranean Sea, the Red Sea, the Persian Gulf, and the Gulf of California (VAN ANDEL 1967).

In humid marginal seas, a stable salinity stratification (high salinity near the bottom, as found in estuaries) is developed due to salt water influx from the ocean. This results in oxygen deficiency near the bottom, as well as in distinct lithofacies zonations (e. g. sand, silt). Laminated sediments are frequent because bioturbation is low. Near the outlet, erosion and formation of lag deposits occur: Transport of cobbles 10 cm in size was observed in 14 m of water depth. Humid marginal seas may be bordered by peat swamps and brackish lagoons.

In arid marginal seas, carbonate sediments are frequent. Due to higher convection providing the bottom water with oxygen, bioturbation frequently destroys the stratification. The circulation of surface water from the ocean into the marginal sea introduces many planktonic skeletons into the marginal sea. Arid marginal seas may be bordered by sabkhas and evaporitic lagoons.

3.252.26 Beaches
(VAN STRAATEN 1959, REINECK 1963, CURRAY 1969)

"Sand may be supplied to the beach by rivers, by wave erosion of unconsolidated sediments along the shore, by eolian transport from the interior and, in some cases, by shoreward transport reworked from the shelf. Sand is exported from the beach either landward by dunes, or by longshore drift or to the deeper part of the shelf. If supply locally exceeds export, the beach progrades seaward; if supply is less than export, the beach regrades." (PETTIJOHN et al. 1972, p. 481.)

In this chapter only sandy and pebbly beaches are included. Cliffed coasts, which are rimmed mainly by cobbles if sediments are formed at all, are not considered. Sand eroded on such cliffs will not settle in place.

The shore zone may be subdivided into the following sand bodies, in offshore direction (from CURRAY 1969, in part modified):

a) Cheniers

These "are isolate elongate bodies of sand lying in and on coastal mud flat and marsh. They are formed during alternate periods of local transgression and regression" due to fluctuation in the rate of supply of sediment to coastline. "The coastline progrades by depositional regression and building of mud flats seaward when rate of sediment influx is high. Subsequently, when rate of supply is reduced, the previously unsorted material is winnowed by wave action and the available coarse material is piled up as a chenier" (erosional transgression). "The type example lies in southwestern Louisiana downcurrent for discharge of much of the sediment from the Mississippi River" (FISK 1955).

b) Beach ridges composing barriers or barrier islands as well as the shore face.

They "are formed by successive accretion to the coastline of new beach foreshores during continuous depositional regression". This may result in a thick deposit of barrier sand (e. g. the sand bars on the Dutch west coast, VAN STRAATEN 1965), in contrast to the thinner deposits of the transgressive type of coastline (fig. 3-40). "Beach ridges are rather common components of barriers throughout the world. A typical spacing between ridges appears to be approximately 50 m." "A Barrier Complex may commonly consist of a shore face, a beach, dunes, sand flat or barrier flat, and a lagoon beach. A barrier may form a barrier island, barrier spit, or bay barrier, depending on whether it is in the form of an island, is attached to the mainland at one end, or is attached on both ends."

DAVIES et al. (1971) investigated modern and ancient barrier environments and found the following upward gradation:

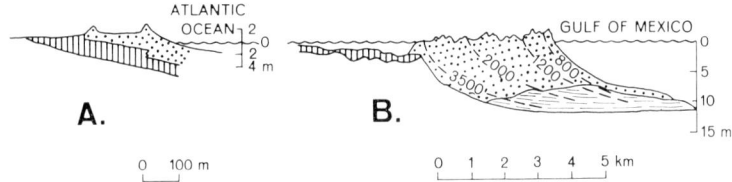

Fig. 3-40. Simplified sections through barriers. A = Transgressive barrier, Sapelo Island, Georgia; B = Regressive barrier, Galveston Bay and Island, Texas. Dotted lines are isochronous surfaces of earlier stages as indicated by radiocarbon dates of shells from drill holes (years B. P.). (Signatures: dots = barrier sand; vertical hatching = lagoonal, marsh, lake, tidal flat sediments; horizontal hatching = marine (interbedded sand and clay); lower surface = Pleistocene. In transgressive sequences, the overlapping units are increasingly marine; other examples (Nayarit) would show offshore marine layers overriding the beach sands. In regressive sequences, the overlapping units are increasingly nonmarine (from CURRAY 1969 and SWIFT 1969 a).

(1) irregular interlaminations of silt and clay (= lower shoreface)
(2) burrowed and generally structureless sand (= middle shore face)
(3) low-angle and microtrough cross-laminated sand (= upper shoreface-beach)
(4) structureless and rooted sand (eolian environments) (= top)

c) Longshore bars

These are "sedimentary deposits, generally of sand (or gravel), lying submerged under all but the lowest of tide levels, seaward of and parallel to a beach". "They may exist in the surf zone at the plunge point of the breakers", and "can only approach water level to about one-half the original water depth" (all citations from CURRAY 1969).

From a more general point of view, the shore zone may be subdivided into emerged sand bodies including the beach ridges and the (less important) cheniers, and submerged sand bodies: the longshore bars.

The material is supplied by longshore currents which may be fed by rivers or by adjacent segments of shore zone, or it is pushed toward shore from the continental shelf by waves, or supplied by tidal currents (SWIFT 1969a). According to CURRAY (1969), approximately half of the sand of the Nayarit beach ridge barrier (Mexico) was derived from the inner continental shelf. PIERCE (1968), in an analysis of the sediment budget of some of the barrier chains of the east coast of the United States, concluded that "almost half of the annual input of sand to these barriers comes from reworking the relict basal transgressive sands of the adjacent continental shelf" (CURRAY 1969). On the other hand, sand can migrate seaward, e. g. aided by rip currents, which are seaward return flows of water through the surf zone to deeper water, and form if two longshore currents of contrary direction meet in the shore zone.

The following paragraphs are also arranged according to PETTIJOHN et al. (1965).

1.—2. Horizontal and vertical association

Elongate sand bodies predominate which run oblique or perpendicular to the shore channels at tidal coasts, but parallel to shore channels at coasts with wave action (sand bars) (HAYES 1967). The "Nearshore Modern Sand Prism" reaches twenty meters in thickness, 0.1 to 3.0 km in width and up to 100 km in length. The maximum width must be greater in areas with less relief. Regressions and transgressions also may simulate a wider sand belt. Diachronous sand bodies can form (= sand bodies crossing the time correlations, KRUMBEIN & SLOSS 1963) (fig. 3-40, fig. 6-4). If the coast subsides continuously over a long period of time, as was the case with the Oligo-Miocene of the Gulf Coast (Frio formation), these sand bars may reach considerable thickness (more than 400 m). The sands in these important oil-bearing strata were transported along shore from the Rio Grande delta (BOYD & DYER 1966).

3. Internal structure. Planar cross-bedding (in megaripples) and festoon (= trough-shaped) cross-bedding (with small-scale ripples), unimodal (or bimodal) dip perpendicular to the longitudinal axis of the sand bar.

4. Thickness of stratified layers. Mostly less than 5 cm, maximum 20 cm, frequently decreasing towards the top of the section.

5. Bedding planes. Many types of ripples (VAN STRAATEN 1959). In shallowest water linguoid and rhomboid current ripples and micro-wave ripples.

6. Mineralogic composition. High maturity is possible if circumstances (source et al.) are favorable. Concentrations of heavy minerals (placers) are typical.

7. Grain size. Coarsest on the fore shore (= between mean high-water and mean low-water line; North sea: median 0.2—0.3 mm), finer on the shore face (= from mean low water line to 7 m depth; North Sea: median 0.14—0.18 mm, REINECK 1963). Best sorting is found on the shore face. The distribution curves generally are skewed towards coarse sizes (i. e. fine sizes missing). The grain size may increase towards the top in regressive sequences; it may decrease in transgressive ones (see also CHAPEL 1967).

8. Grain shape. In the coastal area, the environment influence on the shape is seen mainly in pebbles. Sand grains are smoothed and polished only.

9. Grain orientation. Ideally, elongated grains are perpendicular to the longitudinal axis of the sand bars.

10. Fossils and other properties. Burrow structures are found in quiet water. They are missing on the fore shore. They also are rare on the shore face. — Coastal sediments generally are grayish.

3.252.3 Marine environments

3.252.31 Shallow sea (about 7 m to shelf edge)

The sediments on the present continental shelves in general are modified Pleistocene sediments which were deposited in different marine and nonmarine environments when the shore line was near the edge of the continental shelf. This Quaternary blanket is up to 100 meters thick and even more, especially to the north (EMERY &

MILLIMAN 1970). The modification of these sediments during the rapid transgressions and regressions connected with the alternating warm and cold stages formed a complex pattern of morphologies and sediment distributions (HOPKINS 1958).

The hydraulic regime of the inner shelf fashions this material into the "Nearshore Modern Sand Prism" (SWIFT 1969a). The outer shelf likewise cannibalizes its substrate to generate the "Shelf Relict Sand Blanket" (SWIFT 1969b).

Modern sands extend to depths of but 10 to 20 meters and, in response to diminishing wave and tidal currents, show a seaward diminution in grain size (PETTIJOHN et al. 1972, p. 490).

The information gained on the present continental shelf, however, obscures more than clarifies the conceptional models of ancient marine sediments. The main difference is that at present the shallow sea is mainly bordering the continents and extending only a few tens to hundreds of kilometers from the strand line to the shelf edge, while by contrast, the Ordovician epeiric sea of North America was over 4000 kilometers wide, inundating the lower portions of the continent. Also the slope of modern shelves is steeper (0.1 to 1 m/km) compared with ancient shelves (about 0.1 m/km) due to Pleistocene erosion. The epeiric seas may have been 30 meters or less deep over thousands of square kilometers; gradients of 5 centimeters or less per kilometer have been estimated by SHAW (1964).

Climatically-controlled chemical sedimentation was favored in such epeiric seas, because of their shoal nature and resultant restriction of circulation, and because of their proximity to peneplained land areas shedding little sediment (SWIFT 1969c).

1.—2. Horizontal and vertical association. Widespread sheet sands are possible in this environment. They may be lag deposits if intercalated at the base of a transgression.

3. Internal structure. Horizontal lamination; close to the shore trough-shaped cross-bedding.

4. Thickness of stratified layers: 5 cm, but there is a positive correlation to the grain size.

5. Bedding planes. Transverse wave ripples close to the coast, farther offshore frequent horizontal bedding.

6. Mineralogic composition: Depending on source, distance of transport, climate etc., the maturity may be high.

7. Grain size: Fine sand, well sorted, with fine tail, as well as silt and clay deposits. Close to the coast on flats and in channels coarser sand. In modern shallow seas considerable modifications occur due to Pleistocene rearrangements.

8. Grain shape: like coastal sands.

9. Grain orientation: untypical.

10. Fossils and other properties: Surface trails, burrows, permanent shelters and resting tracks are typical ("Cruziana" facies, SEILACHER 1964), but only in ancient rocks according to MILLIMAN (pers. commun.). Glauconite is common in the shallow sea mainly, but is not universally present.

3.252.32 Deep sea

Deep sea sands can be formed by different agents:
a) turbidity currents (KUENEN 1938), see ch. 3.242.232.
b) contour currents (HEEZEN et al. 1966).
c) storms (e. g. west of North Africa).

"Contour currents" or "geostrophic currents" move parallel to the slope. They can develop as countercurrents underlying surface currents such as the Gulf Stream. The undercurrent moving southward along the east coast of North America is pressed towards the continental slope by the Coriolis force. It moves by velocities of 2—20 cm/sec. compared with up to 2500 cm/sec in turbidity currents. The contour currents produce short-crested ripples and probably do not carry much sediment with them but rework and redistribute the sediments of the continental slope (HEEZEN et al. 1966). Rippled deep sea sands have been described by HEEZEN & JOHNSON (1964), and by KOLPACK (1964).

The following charateristics apply to turbidites (for mechanisms, see also ch. 3.242.232 and 6.31).

1. Horizontal association. Far extending, linguoid sand sheets (SULLWOLD 1961, STANLEY 1967) with decreasing grain size in direction of movement. The thickness of sandstones, and, under certain circumstances, claystones decrease in the same direction (VENZLAFF 1965). Correlation of individual layers for more than 100 km has been reported from ancient turbidites (HESSE 1965).

2.—3. Vertical association and internal structure. A sandstone with graded bedding (a) is overlain by (b) horizontal sand layers, (c) an alternation of sand and silt with small-scale festoon cross-bedding and convolute bedding, and (d) horizontally bedded silt and clay laminae (see ch. 6.31).

Layers of inverse graded bedding are included in many turbidites, especially near the bottom. The thickness of such layers may be related to the total thickness of the turbidite involved (PASSERINI 1966). No satisfactory explanation exists for this phenomenon. Possibly it results before the buildup of the viscosity enables the turbidity current to move larger pebbles (H. FLÜGEL, pers. commun.). SANDERS (1965) believes that larger particles normally move toward the top of grain flows, i. e. concentrated suspensions of cohesionless grains, whereas FISHER (1971), more convincingly, claims the same process for debris flows.

Coarse-grained turbidites with low clay content, that seldom have graded bedding or well-developed sole marks (with the exception of load casts), have been termed "fluxoturbidites" (DŻUŁYŃSKI et al. 1959, see also STAUFFER 1967). They form in the basal parts of proximal regular turbidites. Their thickness mostly is greater and more irregular than that of turbidites. WALKER (1967) defines them as proximal turbidites. Channels from 3 to 20 m depth are typical for areas proximal to the source (WALKER 1966), as are concave slump planes (LAIRD 1968). Other criteria of "proximity" are reported by WALKER (1970).

4. Thickness of stratified layers : cm to m (one turbidite).
5. Bedding planes. Typical sole marks (see ch. 3.242.23).

6. Mineralogic composition. Sandstones of low maturity (graywackes) are frequent, depending on source.

7. Grain size. Flysch-type graded bedding is typical: decrease in grain size upward yet with a fine matrix distributed throughout the layer (see fig. 6-2). Generally, the thicker the layer, the coarser the grain size at the base. In the northeast Pacific, distal turbidites are perfectly graded whereas proximal turbidites are frequently not (Horn et al. 1971a).

8. Grain shape. Roundness is independant of grain size as long as there is not intensive reworking in shallow marine environment, or the material is inherited from older sediments.

9. Grain orientation. It is variable as discussed above in detail (ch. 3.232.).

10. Fossils. Fossils of shallow marine environment are predominant in turbidites. In the clay layer separating the turbidites pelagic fossils are present. Moreover the trace fossils are typical for these environments: tightly spaced grazing patterns are developed ("Nereites facies", Seilacher 1964). They form in the mud ("pre-depositional" compared with the turbidite) and are partially eroded and filled by the turbidite. "Post-depositional", elongated and branching trails from sand-burrowers cross the flow marks (Seilacher 1959, 1964). Primitive agglutinating foraminifera are found in the clayey part of ancient turbidites (Pflaumann 1967).

3.26 Sandstone facies and tectonics

The relations between tectonics and petrography are discussed mainly for clastic sediments. If "tectofacies" is defined as the combination of rock characteristics which are typical for a certain sedimentary-tectonic environment (Sloss et al. 1949), then the following tectofacies can be distinguished schematically (partly from Pettijohn 1957):

A. Cratonic tectofacies (on continental lithosphere)
B. Orogenic tectofacies (at plate margins), in chronological order:
   1. Geosynclinal facies
   2. Leptogeosynclinal facies
   3. Flysch facies
   4. Molasse facies

3.261 Cratonic tectofacies

The deposits form on a stable (or unstable) shelf. The sedimentary basins tend to be irregular or equidimensional in shape, in contrast to the elongated troughs of orogenic tectofacies.

Sandstones are poor in clay, and are predominantly quartzose. Frequently such clean sandstones alternate with dark claystones or siltstones (tab. 3-11, p. 62), but also occur with shallow-marine limestones and dolomites as well as evaporites.

Conglomerates are rare, and when present are thin. Most fluvial series also are thin, and may consist of arkoses i. e. feldspar-bearing mostly kaolinitic sandstones, which derived from weathered granites or gneisses.

Synsedimentary faulting and tilting of blocks may interrupt the uniform thickness of these deposits producing wedge-like rock bodies with the thick end of the wedge pointing towards the source area (KRUMBEIN & SLOSS 1963). This may happen on a local scale (denudation of the deposits on the "horst" and redeposition in the adjacent parts of the "graben". SCHAD 1964; "yoked basins" KRUMBEIN & SLOSS l. c.), or on a regional scale (marginal elongate troughs as a transition to orogenic tectofacies) (VOIGT 1963; "marginal basins", KRUMBEIN & SLOSS l. c.).

3.262 Orogenic tectofacies

3.262.1 Geosynclinal facies

This tectofacies sometimes differs only slightly in petrographic character from cratonic facies: Quartzose sandstones and carbonate rocks are found in the USA and shallow marine limestones in the Triassic of the Eastern and Southern Alps (ZANKL 1967). The typical distinguishing features are the great thicknesses and the elongated basins of geosynclines (TOLLMANN 1967). This is the reason why the Gulf Coast with its thick sequence of shale and sandstone has been referred to as a geosyncline. In view of plate tectonics, however, only the downwarping arc-shaped troughs at consuming plate margins should be termed geosynclines.

Other typical "orogenic" sediments include certain marlstones and sand- and clay-bearing limestones deposited in moderately deep water, like the alpine "Bündner" slates ("schistes lustrés") (TRÜMPY 1960). It is difficult to keep geosynclinal facies and flysch facies apart. In the Alps, the latter is more uniform.

3.262.2 Leptogeosynclinal facies (TRÜMPY 1960)

This facies accumulates during periods when local movements cause a rather fast subsidence of individual troughs, with which sedimentation cannot keep pace. Sediments with low rates of accumulation form mainly in deeper water ("starved sedimentation", PETTIJOHN 1957).

As to the sediment types, PETTIJOHN (1957) reported black shales, cherts, siliceous pelagic limestones, and phosphorites; red limestones also occur in such starved basins or on submarine ridges. In the "Rheinisches Schiefergebirge" (Germany) radiolarian cherts ("Lydite") were deposited in the Lower Carboniferous, whereas dark and reddish carbonates formed in the Upper Devonian (EINSELE 1963b). "Starved" sediments in the Jurassic of the Alps include red pelagic limestones with omissions and submarine solution ("subsolution", HOLLMANN 1964), radiolarites (= radiolarian cherts), and barren shales (TRÜMPY 1960, AUBOUIN 1964, 1965).

Deep basins between and in front of middle Devonian reefs received turbidity currents loaded with reef detritus ("allodapic limestones", MEISCHNER 1964). A turbidite origin was suggested for Jurassic and Cretaceous breccias at the flanks of steep

3.2 Texture, fabric and environment

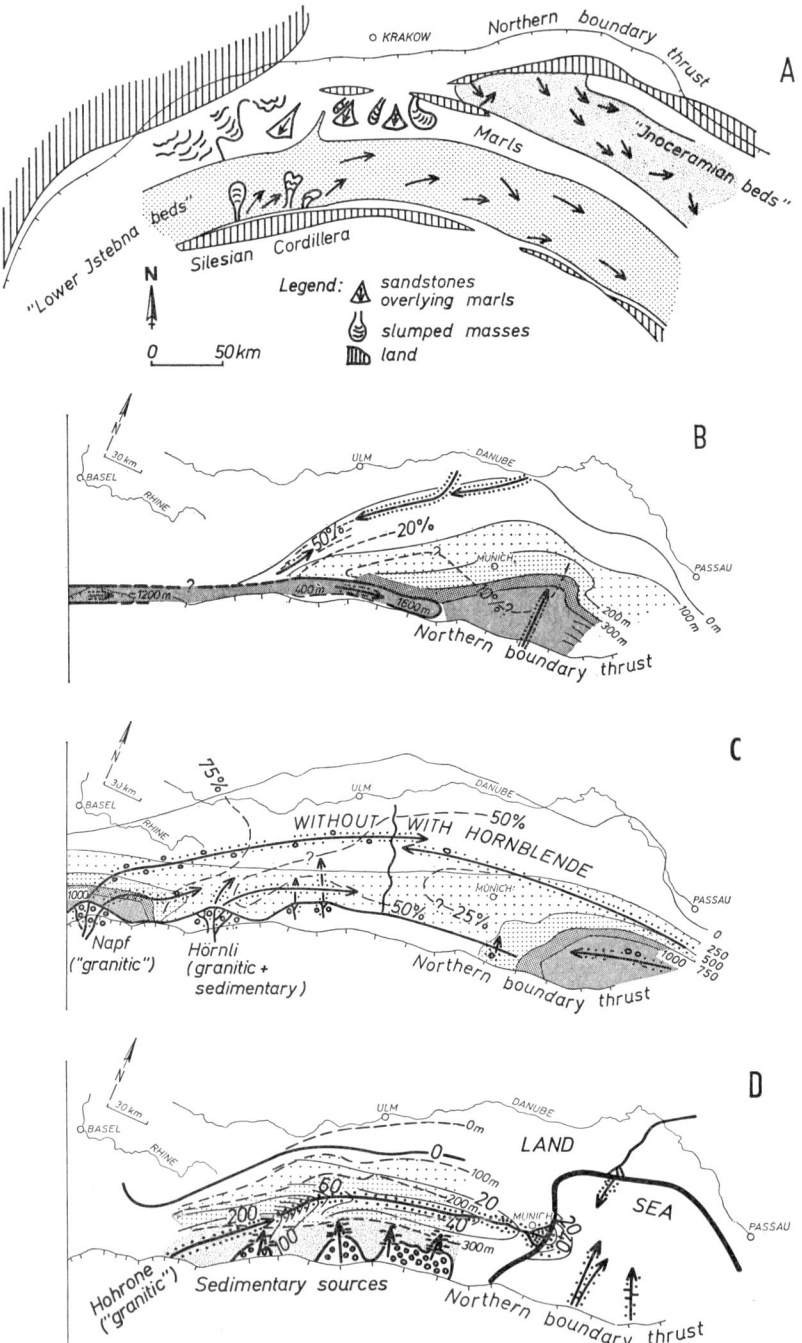

troughs in the Alps (AUBOUIN 1965); "sedimentary micaschists" (a micaschist breccia with a marly sedimentary matrix) formed in large slump blocks (LEMOINE 1967). These rocks can be considered transitional to flysch tectofacies.

3.262.3 Flysch facies

The next stage is called "flysch facies" (see SEILACHER 1967a). The term "flysch" originally meant marls, calcarenitic sandstones and sandy limestones from the Swiss Alps. Mainly because of KUENEN's work, however, this term has been modified to include the world-wide thick sequences of turbidites which filled up elongate basins within growing mountain chains (fig. 3-41 A). An important aspect of deposition was the syngenetic development of orogenesis: small ridges rose above sea level and concomitant erosion resulted in supply of detritus. Eventually this detritus slid down into deep waters, in the process forming a concentrated, poorly-sorted suspension. The sandstones resulting from this process are called "graywackes" by PETTIJOHN (1957).

Fig. 3-41. Basin patterns of flysch and molasse basins.

A. Upper Senonian flysch in the Polish Carpathians (from DŽUŁYŃSKI & WALTON 1965, based on KSIAZKIEWICZ 1962 and UNRUG 1963). The Lower Istebna beds consist of polymict conglomerates ("fluxoturbidites") interbedded with coarse-grained, feldspathic sandstones and occasional partings of dark shales or siltstones. Intercalated fine-grained calcareous sandstones (not included in the figure) derived from the opposite shore (N) show an opposite current direction. The slumps partly consist of pebbly mudstones.

B. Middle Oligocene marine Molasse, north of the Alps (Deutenhausen + Tonmergel fm.). Example of a transition of flysch into molasse (for legend see C.). At left is the retreating E—W flysch trough of the Deutenhausen fm. (calcareous sandstones and marls). To the east, where the trough was not developed, marly sediments were dispersed into the northern foreland. The western trough was filled completely during "Tonmergel" time, whereby flysch-type sedimentation was finished; the trough was then overridden by the dispersal systems of the Molasse (N.E.-arrow in center). Quartz sand, free of carbonate, was supplied from the North. (Capital letters were only used for the Alpine Molasse).

C. Middle Miocene marine Molasse N. of the Alps (B. and C. modified from FÜCHTBAUER 1967a, B in the western part from Hsü 1960). Legend for B and C: petrography shown by conventional signatures along the arrows which indicate the mean directions of dispersal based on heavy and light mineral analyses; lines = thickness of the series under consideration; hatched lines = percentage of sandstone (and conglomerate) beds. Petrographic informations of fig. B-D are based on core analyses of nearly 100 borings in the foreland Molasse and on samples from about 400 outcrops in the subalpine Molasse; isopach curves are based on electric logs and on outcrops.

Two main transport directions, one opposite to the other, probably were caused by alternating (tidal?) currents. (Similar features were reported form the present-day shelf around England by KENYON & STRIDE 1970). Minor supplies came from the South.

D. Upper Oligocene (Upper Chattian) fluvial Molasse N of the Alps (modified from FÜCHTBAUER 1964a; the eastern part also is based on PAULUS 1964 and KÖWING et al. 1968). Legend: lines = bulk thickness of sandstones; thin, hatched lines = bulk thickness of marlstones. This embryonic stage of the river system developing parallel to the Alps is characterized by a rapid decrease in sandstone thickness towards the East. Two general points are worth of mentioning, (a) whereas the bulk thickness is similar in the alluvial and the marine area, the sandstone thickness in the delta area E of Munich is higher, compared with the immediately adjacent alluvial area; (b) whereas the sandstones increase towards the granitic source area in the W., the marls increase in a southerly direction, towards the large conglomerate fans in the subalpine molasse. Because the source rocks of these fans are limestones, marls and flysch (sandy limestones), only gravel and mud but no sand was supplied; the gravel formed giant "Nagelfluh" fans when entering the Molasse plain, the mud was dispersed all over the basin.

He considers them to be the prime characteristic of the flysch tectofacies (see also ch. 3.213). The definition of graywacke used in this book (impure sandstone rich in rock fragments) frequently coincides with the flysch sandstones; depending on the source rock, however, quartzose and feldspathic sandstones and calcarenites also can form (WIESENEDER 1967, ANGELUCCI et al. 1967, p. 396).

The impurity of the graywackes can be explained in part (see ch. 3.213) by the absence of major synsedimentary reworking and sorting in these quickly subsiding basins, resulting in a high content in matrix and unstable rock fragments which diagenetically transformed into matrix (CUMMINS 1962, WHETTEN & HAWKINS 1970). Such unstable rock fragments are derived from basic volcanic and tuffaceous rocks eroded in andesitic mountain chains or andesitic and basaltic island arcs, in "eugeosynclines".

Flysch sediments formed in the Appalachians during the Ordovician and the Devonian (PETTIJOHN 1957, Fig. 165), in California during Plio- and Pleistocene times (Ventura basin, NATLAND & KUENEN 1951), in Great Britain during the Cambrian (KOPSTEIN 1954), in the Variscian mountains during late Devonian (EINSELE 1963b) and early Carboniferous time (see ch. 3.113). In the Alpine mountain belts (figs. 3-41, 42) flysch developed in Cretaceous and Eocene time (CROWELL 1955, DŻUŁYŃSKI et al. 1959, BOUMA 1962, HESSE 1965, and others). "Wildflysch" consists of very poorly sorted slumps and turbidites with large boulders. Within one orogen different tectonic facies may occur at the same time, as shown by fig. 3-42 for the Hellenides (Greece). Here the leptogeosynclinal facies ("Pre-flysch", "P") and flysch ("F") as well as flysch and molasse ("M") developed at the same time.

3.262.4 Molasse facies

The term "molasse" also comes from Switzerland. It was used to describe unconsolidated Tertiary marlstones, sandstones and conglomerates, which accumulated as detritus in the foreland of the uplifted, folded Alpine mountain chains (figs. 3-41 C, D, 42 c-e). In contrast to the flysch troughs, molasse troughs were not preformed but subsided simultaneously with deposition, perhaps aided by increasing sediment load. Thus fluvial and shallow marine sediments accumulated, not deep-sea sediments. The term "molasse" has been used to describe similar sediments of corresponding tectonic position in other areas.

The petrography of molasse sandstones varies considerably depending upon the source area. In the foreland of the Alps they are dolomite-arenaceous sandstones, usually cemented with calcite; the feldspar and rock fragment content varies (FÜCHTBAUER 1964a). According to PETTIJOHN (1957), they may be classified as "subgraywackes". The molasse sandstones of the coal-bearing Upper Carboniferous north of the Variscian mountains and west of the Appalachians are subgraywackes poor in carbonates. Many accumulations of coals are supported by the same factors which are also active during molasse sedimentation: alternations of marine and nonmarine environments. This explains the common occurrence of coal bearing sediments in molasse basins.

Whereas dark-gray colors prevail in the Upper Carboniferous, the colors in the Tertiary molasse of the Alpine foredeep are highly differentiated: marine sandstones are greenish, partly from glauconite; fluvial sandstones are light gray, but

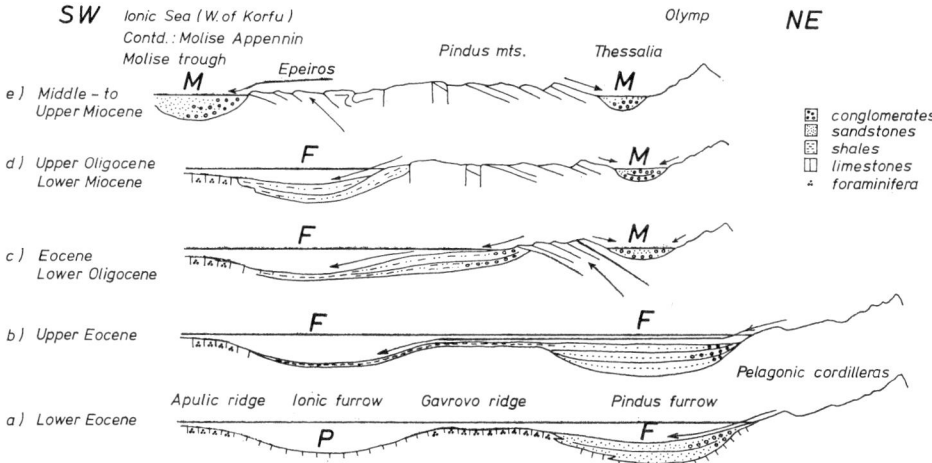

Fig. 3-42. Sequence of orogenic tectofacies (P = Pre-flysch, F = flysch, M = molasse) in the "Hellenides" (from AUBOUIN 1965). Horizontal line = sea level. Only those sediments deposited during the time indicated at left are marked by signatures.

a) While the eugeosynclinal Pindus "furrow" (= basin) is filled with flysch from the Pelagonian Cordillera, there is a leptogeosyncline with slow deposition of lime in the miogeosynclinal Ionian furrow due to lack of clastic source areas.

b) The filling of the Pindus furrow is completed; fine detritus crossing the former Pindus furrow and the Gavrovo ridge is deposited in the Ionian flysch basin.

c) The Pindus basin is folded, and the derived detritus fills the Ionian flysch basin. At the same time a small molasse trough forms in Thessalia ("Mesohellenic furrow").

d) The Ionian furrow still is a sand trap ("barrière en creux"), so that the detritus from NE cannot reach the Apulian ridge area.

e) The Ionian basin is folded before being filled up. The Apulian ridge subsides and receives up to 3000 m of molasse sediments (Molise trough of southern Italy).

A similar model can be applied to the Alps, by changing directions, SW to N, and NE to S (TRÜMPY 1960, 1965).

yellow in outcrops. Marine marlstones are gray, fluvial ones are multicolored greenish, violet, brownish and reddish (LEMCKE 1955). Other molasse rocks, e. g. the Upper Devonian Catskill formation in the Appalachians and the Tertiary "Capas Rojas" of Eastern Peru (KOCH & BLISSENBACH 1960) are predominantly reddish.

3.263 Facies maps (fig. 3-41 B-D)

Facies maps give a more detailed picture of regional tectonic movement, especially in cratonic basins (KRUMBEIN & SLOSS 1963). Such maps also can indicate probable sources for sediment supply and thus can be used to construct paleogeographic and paleogeologic maps. For this purpose the facies distribution, grouped according to the classes of a sand-clay-carbonate diagram (fig. 2-1), is superimposed on an isopach map of the stratigraphic section in question. — In addition to the sand-shale ratio, the clastic ratio, the cumulative sand thickness and the number of sand layers are useful parameters. More suggestions can be found in KRUMBEIN & SLOSS (1963), and SLOSS et al. (1960).

3.3 Diagenesis

3.31 Diagenetic processes in sandstones

3.311 Definitions

The term "diagenesis" as used in this chapter includes all changes of a sediment between deposition and metamorphism but excludes all weathering processes (CORRENS 1939a), which are discussed on pp. 1—2 and in Part III. The chemical changes of the components in marine and brackish seas during and immediately after the deposition ("halmyrolysis") are negligible in sands (some exceptions are mentioned in ch. 3.326). Diagenesis can be recognized geometrically (changes of fabric, especially of porosity) and chemically (dissolution and precipitation). The following discussion of the interrelation between chemical and geometric processes refers only to samples homogeneous in composition and fabric.

Precipitation in the free pore space is referred to as "neoformation" ("cementation" in part; see below). Precipitation immediately replacing dissolved primary components is called "replacement". If it follows the contours of the replaced mineral, it is termed "pseudomorph". Kaolinite (tab. 3-17, line 3a) for instance can be "pseudomorphous" after feldspar. Often it is not possible to discover whether a mineral was replaced step-by-step or was dissolved, and the pore space was filled with a new mineral (cementation): "Replacement" (German: Verdrängung) may be reserved for the gradual substitution of one mineral by another whereas "substitution" (Ersatz) may be used in a broader sense including later cementation. "Authigenesis" includes all terms discussed above. "Alteration" occurs if parts of the parent lattice are used to build up the new mineral, for instance if illite reorganizes into muscovite.

The term "cement" should be applied only to neoformations which result in a reduction of porosity within an area of homogeneity. This occurs mainly during allochthonous neoformation, i.e. when material is supplied from outside the area of consideration. Since autochthonous neoformation requires dissolution within the area of consideration, porosity is reduced only if the supporting rock framework, e.g. quartz, is dissolved. (Line 2a of table 3-17). Otherwise porosity will increase, since not all of the dissolved material is reprecipitated within the same area. For instance, line 2b shows that porosity increases when feldspar alters to kaolinite — provided the supporting framework is not attacked —, since kaolinite can occupy only about half of the feldspar's volume, because of its lower silica content.

The relations between the various aspects of diagenesis are listed in table 3-17, which is based on a detailed classification (with examples) of chemical processes indicating whether the porosity is influenced in a positive (increase) or negative (decrease) way. Moreover, the degree to which porosity and compaction are influenced is also noted.

3.312 Mechanical compaction

Increasing overburden leads to "compaction", and thus a reduction in rock thickness with time at the expense of porosity.

3.31 Diagenetic processes in sandstones

Table 3-17. Chemical processes in sandstones and their influence on porosity and compaction.

Chemical process	Example	Influence on porosity	Influence on compaction
1. Dissolution			
a) pressure solution	Pressolved quartzite (see p. 134)	— strong	strong
b) intrastratal solution	Dissolved heavy minerals	+ full	none
2. Neoformation (cementation in part)			
a) autochthonous, homogeneous*	Quartz grains grown at the expense of matrix-SiO_2	+ weak	none
	Quartz grains dissolved at pressure points, enlarged in the pressure shadow	— double	strong
b) autochthonous, heterogeneous*	Growth of kaolinite in the pore space at the expense of feldspar which dissolved at other places in the sandstone	+ weak	none
c) allochthonous, homogeneous*	Silicification by supply from outside (cementation)	— full	none
d) allochthonous, heterogeneous*	Cementation by anhydrite or Fe_2O_3 or $CaCO_3$	— full	none
3. Replacement			
a) autochthonous*	Kaolinitization of feldspar	+ weak	none
b) allochthonous*	Dolomitization of orthoclase	none	none

* autochthonous: derived from material within the sandstone layer in question
allochthonous: supplied from outside
homogeneous: formation of a mineral already present in the sandstone
heterogeneous: formation of a mineral not yet present in the sandstone
(all terms from v. ENGELHARDT 1960).

Porosity reduction without chemical reactions including dissolution-reprecipitation processes is restricted to shallow burial, because of the short time and low temperature involved. At greater depth, chemical processes increase in importance. Both mechanical and chemical processes contribute to the porosity decrease with increasing depth as shown in fig. 3-43.

The following relations are shown in fig. 3-43:

3.3 Diagenesis

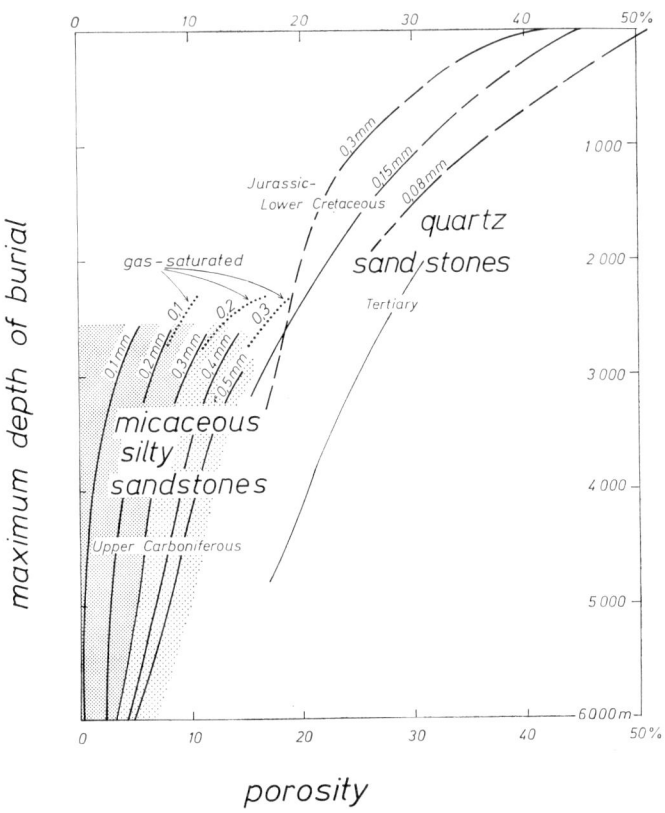

Fig. 3-43. Decrease of sandstone porosity with increasing maximum burial depth for sands with different median diameters. The solid lines of Jurassic-Lower Cretaceous and Upper Carboniferous sandstones are based on about thousand porosity and grain size analyses conducted by Elwerath Oil Co., Hannover (FÜCHTBAUER 1967c); Tertiary = Frio formation (Gulf Coast), average figures from MAXWELL (1964, fig. 7 and 8); values between 30 and 270 m from KALTERHERBERG (1968); uppermost values from experimental trickling below water (FÜCHTBAUER & REINECK 1963). The stippled lines on the left side are porosities observed in gas-saturated Upper Carboniferous sandstones. The realm of pressure solution is stippled.

1. The porosity of fine-grained quartzose sandstones low in clay and carbonates decreases slower at shallow depths but faster at greater depths than the porosity in coarse-grained sandstones.

The same difference is obtained experimentally by shaking sand of different grain sizes but uniform sorting both dry and in water: coarse sand becomes more densely packed than fine sand (FÜCHTBAUER & REINECK 1963). This points to a mechanical type of rearrangement as the main agent of porosity decrease within the topmost 1000—1500 m. Presumably the coarser sand grains which are better rounded slide more easily into denser packing (FÜCHTBAUER 1967c, Tab. 1).

Below a depth of 1000 or 1500 m, in the realm of increasing chemical activity, fine-grained sandstones make up for the porosity loss of the coarse sandstones. This again can be explained by the different roundness of coarse and fine sands: Whereas for well-rounded grains (coarse sands) 1.5% of the grain volume must be dissolved to yield a compaction of 10% (as a minimum; FÜCHTBAUER 1967c), much less than 1.5% dissolution is needed for angular grains (fine sand) to yield the same amount of compaction. — Beginning with a depth of 500 m, quartz overgrowths develop in Jurassic sandstones of northern Germany (fig. 3-50) though stronger quartz cementation begins only between 1000 and 1500 m. As discussed in ch. 3.321, pressure solution and clay mineral alterations in the adjacent siltstones presumably provided the silica for the overgrowths.

2. The duration of burial is also important: Tertiary sandstones are more porous than are Jurassic and Cretaceous sandstones at the same depth.

3. The porosity of fine-grained Upper Carboniferous micaceous sandstones is lower than that of coarse-grained. This results from a primary increase in silt and clay with decreasing median diameter of the sample:

median 0.1 mm: about 21% < 0.02 mm   0.4 mm: about 12% < 0.02 mm
0.2 mm: about 16% < 0.02 mm   0.5 mm: about 9% < 0.02 mm
0.3 mm: about 14% < 0.02 mm

Moreover, the phyllosilicates enforce pressure solution, as discussed in the next paragraphs.

3.313 Dissolution

Dissolution of minerals is not restricted only to the zone of weathering, but extends to far greater depths.

Within the supporting rock framework preferential solution takes place at the points to which the overburden pressure is transmitted (pressure solution). The pressure dependence of quartz solubility is low, according to the experiments of KENNEDY (1950), but solubility increases through elastic deformation from unequal pressure (VERHOOGEN 1951, SIEVER 1962) or tectonic stress (SIEVER 1959). One example is the points of contact of superimposed sand grains, where the transmitted lithostatic pressure is 2—2.5 times higher than normal hydrostatic pressure in interstitial fluids. Therefore, pressure solution is found at quartz — quartz contacts between superimposed sand grains, where parts of adjacent grains have been dissolved (THOMSON 1959). Such pressure solution seams may form small stylolites which can enlarge horizontally at a later stage; the solution may even attack quartz overgrowths which formed earlier (TAYLOR 1964). The amount of dissolution sometimes can be estimated by comparing the stable heavy mineral concentration at the seams with that of the neighboring rocks (HEALD 1955, 1959).

Besides pressure-induced dissolution of the supporting rock framework, dissolution of isolated minerals, e. g. feldspars and heavy minerals ("intrastratal solution", ch. 3.116.33) also occurs. Whereas intrastratal solution generally increases porosity, pressure solution leads to a decrease in porosity. The volume to be dissolved for any decrease in porosity increases in later stages of compaction since the grain contacts that transmit the pressure become more and more horizontal, causing the sand grains to flatten: at a burial depth of 1000 to 1500 m pressure solution becomes an important factor of compaction. The final stage is a "pressolved quartzite" nearly free from pores (SKOLNICK 1965). A complete cementation by autochthonous quartz would require a pressure solution of 21% of the quartz grains (FÜCHTBAUER 1967c).

Pressure solution in a broader sense is increased by the presence of clay or silt seams (HEALD 1955, 1956a, THOMSON 1959, FOLK 1960, LERBEKMO & PLATT 1962) or mica sheets (CAROZZI 1960, LOWRY & DE RUDDER 1966) between the quartz grains (fig. 3-44a). Thus pressure solution in silty and micaceous sandstones of the Buntsandstein and the Upper Carboniferous is stronger than in matrix-free sandstones of the Buntsandstein, Rotliegendes, and Dogger beta of northern Germany.

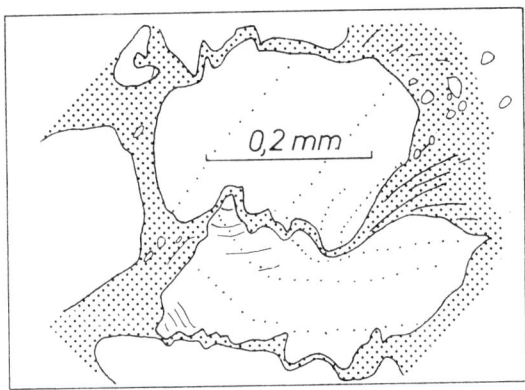

Fig. 3-44a. Pressure solution at a clay seam between superimposed sand grains. Buntsandstein (FÜCHTBAUER 1967b).

There are two possible explanations of the increased pressure solution in the presence of clay or silt seams:
(1) release of K^+ from the clay films produces an alkaline environment which increases the solubility of silica (THOMSON 1959),
(2) diffusion of dissolved matter is enhanced by the clay films (WEYL 1959).

Transport of dissolved matter is supposed to be the active mechanism in the stylolite formation in limestones where the stylolites are the main pathways of solutions (p. 296). In sandstones, however, this principle cannot work, because the main pathways are the interparticle pores rather than the clay seams between sand grains. (2) may apply to clay seams, but not to quartz-mica contacts; these do not differ much from quartz-quartz contacts. (1) is a more probable mechanism. Above 100°C, the solubility of silica is increased already between pH 7.5 and 8.5, i.e. in the normal pH range of diagenesis (OKAMOTO et al. 1957).

Basically pressure solution is the same process as quartz dissolution along cleavage planes (BREDDIN 1930, HOEPPENER 1956, PLESSMANN 1964). It is expected that selective pressure solution of quartz and feldspar grains also takes place in more intensely compressed siltstones. This may modify the rock considerably and provide silica for the cementation of the neighboring rocks as well as filling available fissures. Pressure solution of feldspars became evident by comparing the mineralogy of Devonian shales with embedded concretions, in folded rocks (KNOKE 1966).

The magnitude of pressure solution as estimated from the hydrostatic pressure effect on solubility is greater for carbonate than for quartz (PETTIJOHN et al. 1972, p. 423).

3.314 Neoformation (cementation, in part)

The following important minerals are found to precipitate in sandstones in order of decreasing frequency: quartz, calcite, dolomite, siderite, anhydrite, muscovite, kaolinite, chlorite, orthoclase, albite, hematite, halite, barite, celestite, and zeolites (such as analcime).

3.314.1 Criteria for determining the sequence of precipitation

Sometimes the sequence of neoformations can be deduced from fabric relations. Even then it should be kept in mind that diagenetic mineral precipitation generally covers a long time span, therefore resulting in an overlap of precipitation of individual minerals. Moreover, fabric relations allow alternative interpretations in many cases. For these reasons it is highly desirable to develop quantitative criteria which enable a more or less direct determination of the succession of neoformations.

1. Quantitative criteria fig. 3-44 b

a) "Minus-cement-porosity" is the porosity which would occur if the cements were dissolved; this indicates the porosity of the sand prior to cementation (ROSENFELD 1949). Cements filling secondary solution pores should not be counted,

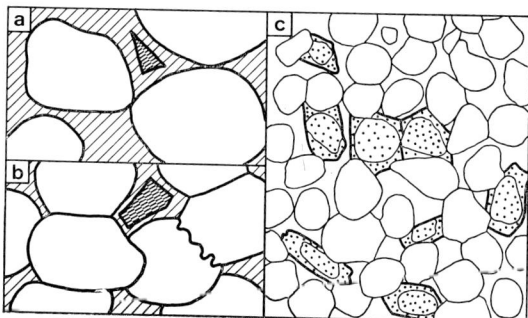

Fig. 3-44 b. Minus-cement-porosity (a, b) and contact strength (c).

The minus-cement-porosity including cement (hatched) and porosity (stippled) is 33 Vol-% in (a) and 16 Vol-% in (b), indicating an earlier cementation of (a) compared with (b).

The contact strength of grains separated by albite cement (dotted) is 1.2 [10 point contacts, 3 long contacts; $(10 + 2 \times 3) / (10 + 3) = 1.2$], whereas the contact strength of grains separated by quartz cement (white) is 1.6. Detrital albite and quartz grains are dotted and white, respectively. Since the contacts become closer with increasing burial depth, before the main cementation occurs, albite cementation occurred earlier than quartz cementation in this example from the Buntsandstein.

however. According to fig. 3-43, a minus-cement-porosity of about 40% (e. g. ALLEN 1962) would indicate a very early cementation, as this is about the porosity of newly deposited sands. This was confirmed in the Dogger Beta by the preservation of unstable heavy minerals in calcareous sandstones with minus-cement-porosities of about 40%; in calcareous sandstones of lower minus-cement-porosity values these heavy minerals were removed by intrastratal solution (DRONG 1965). It may be difficult to use this method — e. g. by point counting — if detrital grains are replaced marginally by cement minerals, thus causing an over-estimate of volume. Moreover, it should be kept in mind that only the beginning of cementation is documented by both the minus-cement-porosity and the contact strength (see below): In cases such as the heavy mineral preservation quoted above, early cementation during shallow burial can lead to a nearly impermeable rock, but in other cases it may not.

b) "Contact strength". With the number of tangential ($=a$), long ($=b$), concave-convex ($=c$) and sutured ($=d$) grain contacts (TAYLOR 1950) within a homogeneous area of a thin section the formula

$$\frac{1a + 2b + 3c + 4d}{a + b + c + d}$$

renders a measure of contact strength. Since, however, the distance to adjacent grains is similar for b, c and d, a shorter formula $(a + 2e) / (a + e)$, with $e = b + c + d$ may be used instead, which makes no difference between b, c and d. The closer the grain contacts, the greater the indices. With tangential contacts only, the index is 1, with closer contacts (e) only, it becomes 2. According to TAYLOR (l. c.) the contacts become closer with increasing burial depth. Therefore it should be possible to estimate the relative age of two cement minerals A and B by comparing the contact strength of sand grains in an area cemented by A with that of an area cemented by B. This will be demonstrated by an example from the Buntsandstein in northern Germany (tab. 3-18, using the longer formula).

Table 3-18. Contact strength of sand grains with different cementation (FÜCHTBAUER 1967 b, c).

Cement Mineral	Number of Samples	$\bar{x}$	s	Student-t distribution
Albite	7	1.20	± 0.075	± 0.11
Quartz	19	1.59	± 0.145	± 0.1
Anhydrite	13	2.15	± 0.21	± 0.19
Without cementation	6	2.26	± 0.19	± 0.34

$\bar{x}$ = arithmetic mean; s = standard deviation; in the last column is the variance around $\bar{x}$ according to the Student-t test on the 99% probability level (MARSAL 1967).

The table shows that the contact strength for albite, quartz and anhydrite are significantly different. It also shows that the minerals probably crystallized in this order which is in accordance with the qualitative fabric relations (see below). After the crystallization of anhydrite the entire rock body apparently was "frozen", so that there was no further compaction, even in areas lacking cement.

c) Number of contacts per grain (TAYLOR 1950) may be used in well-sorted sandstones of rather uniform grain size in about the same manner as the contact strength.

2. Qualitative criteria

In most cases qualitative criteria are ambiguous and require additional data. Some of the qualitative criteria which can be used are shown in the following examples from the Buntsandstein.

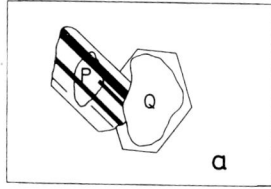 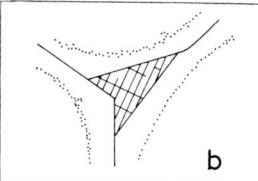 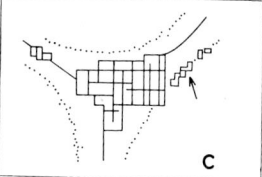

Fig. 3-45. Some fabric relations in cements. Examples from the Buntsandstein (FÜCHTBAUER 1967 b).

a) The overgrowth of the plagioclase grain (P) is older than the one of the quartz grain (Q).

b) Three quartz grains with overgrowths. The remaining pore space was filled with anhydrite.

c) Like b), but the anhydrite marginally displaced the quartz and entered between the quartz grain and its overgrowth (see arrow) as described by HEALD (1956 a) for calcite and quartz grains.

a) The secondary enlargement of quartz grains was interrupted at different times by the neoformations of calcite and anhydrite. Because the secondary quartz seams associated with calcite cement are smaller than those associated with anhydrite, the calcite cementation must have been earlier. The interpretation that calcite and anhydrite precipitated previously and were displaced by quartz at different rates is highly improbable, for there are no indications of calcite displacement by quartz, not even in calcitic ooids.

b) Younger growth seams are interrupted by the older ones. In fig. 3-45 a quartz seams are interrupted by those of feldspar. The other possible interpretation is that the growing plagioclase displaced the quartz. This, however, is unlikely, since the plagioclase replaces only the quartz overgrowth, not the detrital quartz grain. Besides, displacements of quartz by feldspar have not been observed in the Buntsandstein by other criteria.

c) In large pores, the youngest cement occurs generally in the middle of the pore (fig. 3-47). On the other hand, it is conceivable that quartz-overgrowths replaced an earlier pore-filling cement (fig. 3-45 b), although the amount of replacement involved is unlikely in a cemented sandstone. Another argument can be added: The fabric shown in fig. 3-45 c permits two explanations:

1) after formation of quartz overgrowths, the remaining pores were filled with anhydrite, which displaced the quartz marginally.

2) anhydrite filled the entire pore volume, but quartz replaced it, and in turn anhydrite replaced the quartz marginally.

The last possibility seems highly unlikely in this compact rock.

3.314.2 Quartz neoformation

Secondary quartz rims in many cases form clear homoaxial (i. e. optically continuous) overgrowths on detrital quartz grains. These rims may or may not be separated from the detrital grains by "dust" lines (fig. 3-46), which are in part void space as suggested by later infill (fig. 3-45 c) and according to scanning electron micrographs by PITTMAN (1972): "Secondary quartz probably eventually infills many of these voids, which destroys the 'dust' line unless impurities are present also." This may be the explanation, why in many sandstone formations such "dust" lines are visible, whereas in other formations it is nearly impossible to recognize the overgrowths. In the latter case, the overgrowth may be recognized by luminescence microscopy (SIPPEL 1968). Stronger luminescence is developed in the detrital grains due to chemical impurities compared with the overgrowths. Elongations in the c-axis direction are frequent in the overgrowths.

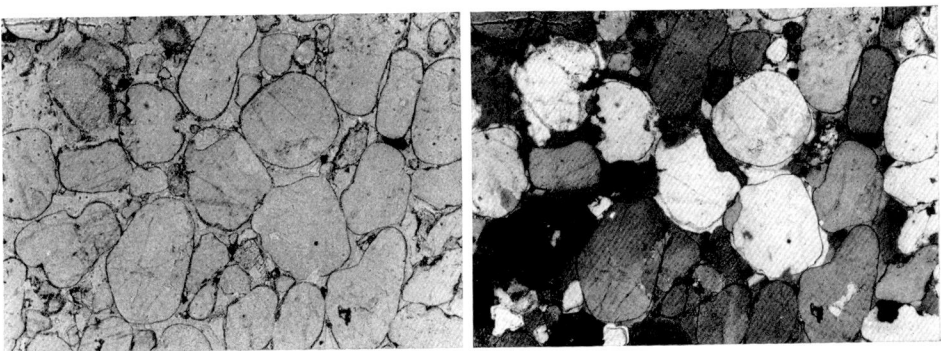

Fig. 3-46. Quartz-cemented sandstone (Penrith sandstone, England), on the right side (with crossed nicols), the optical continuity between core and growth seam is seen (width of the photograph: each 4 mm).

The silica is autochthonous if it originates from inside the sandstone where it may have been mobilized by pressure solution and precipitated in the immediate pressure shadow or at least within the homogeneous domain (WALDSCHMIDT 1941, HEALD 1956 a). Autochthonous silica also originates from the chemical reactions in the sandstone, e. g. by kaolinitization of feldspars. Aluminum oxide which is set free during weathering of feldspars may cause amorphous precipitates rich in silica, even from very dilute silica solutions, according to experimental studies by HARDER & FLEHMIG (1967). Kaolinite and quartz may crystallize from such colloidal precipitates. The general formula of the kaolinitization of potassium feldspar reads

$$4\,KAlSi_3O_8 + 4\,H_2O \rightarrow Al_4(OH)_8Si_4O_{10} + 2\,K_2O + 8\,SiO_2$$
Potassium feldspar   kaolinite   quartz

The conversion of montmorillonite to illite can also supply dissolved silica to pore waters (PETTIJOHN et al. 1972, p. 427).

Reprecipitation as overgrowths of silica set free by pressure solution can be hindered by clay seams (KRYNINE 1948, PETTIJOHN 1957, SIEVER 1959, HORN 1965). On the other hand, pressure solution is intensified in such sandstones (p. 134). The clay content, therefore, is doubly important for the movement of silica in sandstones.

As an example, much silica was pressolved in Upper Carboniferous silty sandstones, but presumably only a small portion of this silica was reprecipitated in the Upper Carboniferous rocks. In contrast, the overlying well-sorted sandstones of the Lower Permian are rich in quartz overgrowths but poor in pressure solution: Silica released from the Upper Carboniferous probably precipitated in the Lower Permian. Such a migration of silica from below is clearly demonstrated by HEALD & ANDEREGG (1960): Zones free from cement in quartz-cemented sandstones were found on top of small pebbles, i. e. in their shadow.

Transport of silica on a smaller scale has been observed in the Buntsandstein. In silty sand layers quartz was removed by pressure solution; in the neighbouring well-sorted sand layers, quartz cement was precipitated. Cementation may even change from one grain to the other as demonstrated by smear slides from homogeneous sandstones: grains lacking any overgrowths occur together with doubly terminated quartz crystals in the same sample. This may be due to changing microenvironment or to differential coating of grain surfaces by organic substance or clay minerals.

Differences in solubility can also cause chemical transport. Increased pressure solution was found in portions of a Lower Cretaceous sandstone enriched with chert, whereas silica was precipitated in those layers poorer in chert (SLOSS & FERRAY 1948).

Frequently cementation increases in the upper- and lowermost 1—2 meters of thick sandstone bodies between beds of shale. These outer zones may have only 5% porosity as compared with 10—15% in the middle of the layer in the Buntsandstein, and with 20—25% in the Dogger Beta (FÜCHTBAUER 1967c). In the Buntsandstein this cementation originates from quartz, anhydrite and carbonate, in the Dogger beta from quartz only. Calcitic and dolomitic cementation is also concentrated at the upper or lower borders of sandstones, but distribution generally is uneven. FOTHERGILL (1955) explained such occurrences by ionic filtration of compaction water moving from sandstones into claystones. This explanation without doubt applies in cases where small cement seams are found at thin clay layers, either within or on the borders of sandstones (WERNER 1961). In most cases, however, increased cementation at the upper and lower borders of sandstones probably originate from ionic diffusion from the claystones into the sandstones, for the following reasons:

a) Filtration would not cause zones 1—2 m thick with only a weak reduction of porosity, but compact cementation rims directly at the boundary.

b) Cementation at the bottom of sandstone beds caused by filtration should be (but is not) an exception, since the compaction current usually is directed upwards.

c) Some cementation has been observed even where an early oil saturation had made filtration impossible (FÜCHTBAUER 1967c).

Such ionic diffusion may be explained as follows. At first, the ions are retained in the clay- and siltstones due to filtering during compaction (v. ENGELHARDT &

GAIDA 1963), but later they diffuse into bordering portions of the sandstones. Part of the silica ions are provided by pressolution of quartz grains coated by clay or mica, especially in siltstones, as described above for silty sandstones. As early as 1920, JOHNSON had assumed that an important part of the quartz cement could be derived from the shales.

Locally fibrous or cryptocrystalline quartz seams are found instead of homoaxial overgrowths (fig. 3-47). These seams indicate a rapid precipitation from concentrated silica solutions as suggested by hydrothermal experiments using a rapid supply of concentrated solutions (HEALD & RENTON 1966). In nature, fibrous or cryptocrystalline quartz cement is found in sandstones with easily soluble silicic components, e. g. needles of sponges (CAYEUX 1929), intercalations of tuff (FRYE & SWINEFORD 1946), in the matrix of graywackes containing volcanic rock fragments, and also in most surface-cementations in arid areas (see also HELLMERS 1949). In all these occurrences, opal is formed first (FRYE & SWINEFORD 1946), then during diagenesis it is dehydrated to chalcedony and finally into fibrous or cryptocrystalline quartz (M. FRIEDMAN 1954, CAROZZI 1960).

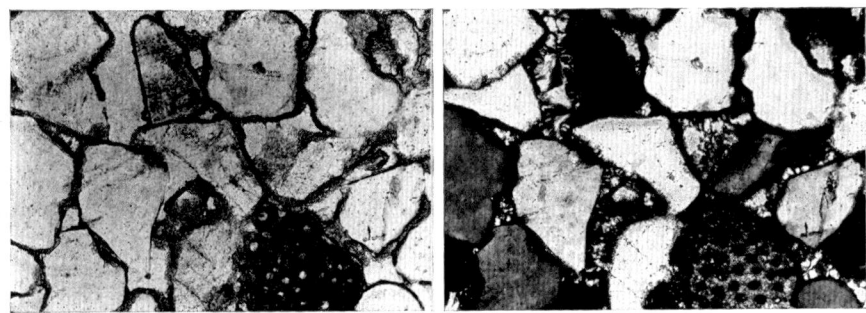

Fig. 3-47. Silica-cemented sandstone (Eocene, outcrop near Heverlee, Belgium). Right: Nicols crossed. Gradual cementation: 1. opal (dark in both photographs), 2. chalcedony (fibrous on the right photograph) (width of the photographs each 1.5 mm).

If pressure solution is found between overgrowths of adjacent quartz grains, it is possible that these grains were deposited with overgrowths from older sandstones.

The exceptional property of secondary quartz is that it is mineralogically identical to the chief sand constituent. This fact is shown in the fabric: In sandstones with a narrow alternation of coarse- and fine-grained layers, the quartz cement prefers the latter (FONDEUR 1964, MELLON 1964, HELING 1965, FÜCHTBAUER 1967b) because the intensity of cement precipitation (assuming constant supersaturation) is dependant upon the number of nuclei. Inasmuch as the quartz grains act as nuclei of the SiO_2-cement, the number of nuclei increases with decreasing grain size. In addition, in the fine-grained sandstones, the porosity (and also permeability) is reduced more by seams of equal thickness than in coarse-grained sandstones. Therefore, subsequent circulation of carbonate or sulphate solutions predominates in the coarser layers, thereby causing their cementation.

The origin of the secondary silica in sandstones can be summarized as follows:

Early diagenetic micro- to cryptocrystalline cement composed of quartz as well as chalcedony and opal (in the desert) requires either evaporating weathering-solutions (in a dry climate) or highly soluble varieties of silica in the source area or volcanic glass shards, silica skeletons, and chert grains in the sediment (STORZ 1931, AHRENS et al. 1960, MILLOT 1960, MCBRIDE et al. 1968, RAPSON-MCGUGAN 1970). As shown by the enrichment of silica in kaolinitic "tonsteins" and in Tertiary quartzites associated with coals, there must be especially favorable conditions in peat bogs. A possible mechanism of silica precipitation is a diagenetical pH-increase beginning with very low pH-values, because the solubility of silica increases not only toward higher but also toward lower pH. Fine quartz dust formed during abrasion of sand grains in eolian dunes has been claimed to be another possible source of silica (BIEDERMAN 1962).

Early diagenetic well-crystallized quartz overgrowths have been reported from the zone of mixing of river and sea water in tropical New Caledonia by BALTZER & LE RIBAULT (1971).

The most important source of late diagenetic homoaxial quartz cement is pressure solution in sand- and siltstones, at grain contacts covered with clay or mica. This only happens with an adequate depth of burial though occasionally homoaxial overgrowths can form even during early diagenesis. Other sources of late-diagenetic quartz cement include kaolinitization of feldspars and replacement of silicates by carbonates, which is a major source of silica in deep burial sandstones (WALKER 1960). The distinction between shallow and deep burial is more valuable for clastic rocks than the distinction between early and late diagenesis; the latter terms are used for carbonate rocks in this book.

"Quartzite" is a field term for strongly lithified sandstones with nothing but quartz cement. Three subtypes are distinguished:

a) Cemented quartzites = silica-cemented sandstones, predominantly poor in clay. Further differentiation can be made by indicating the crystal size of the cement.
b) Pressolved quartzites = sandstones lithified by pressure solution.
c) Metaquartzites = metamorphic recrystallized sandstones. Main characteristic is intensive intergrowth of frequently platy crystals which usually have few inclusions (SKOLNIK 1965).

3.314.3 Feldspar neoformation (fig. 3-48)

Authigenic albites and potassium feldspars are found in sandstones as well as in carbonate rocks.

In sandstones, most authigenic feldspars occur as overgrowths on detrital feldspars. Such overgrowths generally are homoaxial if the seam and core have equal composition. Twin lamellae developed in the detrital grains may or may not continue in the overgrowths (GOLDICH 1934) or gradually die out (HEALD 1956b).

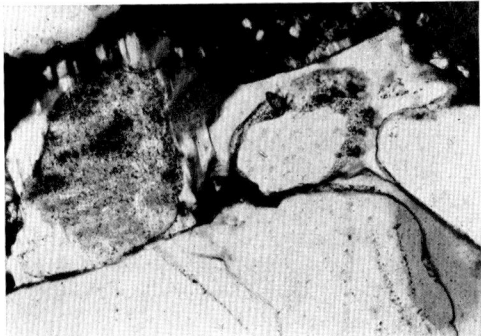 

Fig. 3-48. Two secondary albite seams around plagioclases (above), and quartz overgrowths (at right); Buntsandstein (long side 0.55 mm).

Fig. 3-49. Two dolomite crystals (gray) replacing potassium feldspar grains but sparing quartz grains. Buntsandstein (Lower Triassic) (long side 1.39 mm).

In the Buntsandstein, albite overgrowths on albites preferably developed parallel to (010). This is particularly characteristic of albite seams on potassium feldspars. (For mineralogic details of authigenic feldspars see ch. 5.225.) Feldspar cement often is an early cement (CAROZZI 1960, HELING 1965, see also table 3-18, p. 136), which can even be influenced by the depositional environment (see ch. 3.324). It is widely distributed in Triassic basins. Occasionally it may have been produced by partial diagenetic alteration of early analcime (VAN HOUTEN 1965b). Intensive authigenesis of K- and Na-feldspar in tuffs has been reported (FÜCHTBAUER 1950, WEISS 1954, HAY 1966). Other occurrences of intensive K-feldspar formation are not understood yet (e.g. BERG 1952).

For further references of authigenic feldspar in sandstones see KIRSCH & HALLBAUER (1960), MAUREL (1962), GLOVER (1963) and HELING (1965), and a compilation by HAY (1966).

3.314.4 Carbonate neoformation

In sandstones as well as in calcarenites, the early, fast-growing carbonate cement is often micro- to crypto-crystalline (i.e. less than 100 µ) (KÜHN 1948) or fibrous (GARRISON et al. 1969). The later calcite cement generally is micro- to macrocrystalline (i.e. larger than 10 µ); crystals up to 20 mm in size (HELING 1963) may enclose several sand grains.

Early dolomite and siderite cement also are generally crypto-to microcrystalline, although siderite can be macrocrystalline. The early dolomite cement of the Buntsandstein is found predominantly in siltstones, the late cement is an iron-bearing dolomite in sandstones. Similarly, of the two generations of siderite in the Upper Carboniferous (WEBER 1966), the first generation formed in siltstones and coal during or immediately after sedimentation (coal-ironstones, MÜLLER 1952), whereas the coarser, second generation includes dolomite, ankerite and siderite cements in sandstones. Magnesite cement has also been reported (HELING 1965). Isomorphous admixtures of Mg and Fe in calcite cement have been used as diagenetic thermometer by MURAVYOV (1970).

Because of a high minus-cement-porosity (ch. 3.314.1) calcite cement in some sandstones probably is indicative of the shallow burial stage. Whether large crystal size is primary or is due to later recrystallization is not known, because of the lack of data regarding early sand cementation. GARRISON et al. (1969) reported a fibrous calcite cement generated in deltaic sands, but the mechanism of formation is not known.

Decementation (PETTIJOHN et al. 1972, p. 405), i. e. the reversal from precipitation to dissolution of a cement, becomes visible either by meniscus-like remnants of cement between the grains or by a diagnostic relic texture of corroded detrital grains (e. g. quartz) that had been partly replaced by the former cement (e. g. calcite) which disappeared during decementation. This may occur e. g. "by uplift of a sandstone from a deeply buried position in which the pore water was a slightly supersaturated or saturated brine to a new position close to the surface where it may be invaded by undersaturated surface waters of meteoric water derivation".

3.314.5 Sulphate neoformation

Like deep burial calcite cement, anhydrite cement generally forms large crystals, which enclose several sand grains. Frequently these anhydrite crystals are more extended in the bedding plane than perpendicular to it, indicating solution migration parallel to the bedding plane. Anhydrite is not restricted to evaporite formations, but also may be present in normal marine sandstones and in some marine-fluvial transition deposits (e. g. Upper Carboniferous, Northern Germany).

Small quantities of barite are widespread in sandstones, as shown by its frequency in heavy mineral separations. In the Buntsandstein, long, spearlike crystals of barite and baritocelestite are found. Part of the barium may originate from K-feldspars, which occasionally are replaced by barito-celestite. Celestite itself, however, is rare in sandstones (CAROZZI 1960).

3.314.6 Clay mineral neoformation

Overgrowths of tri- or dioctahedric chlorite (KOSSOVSKAYA & SHUTOV 1958, v. ENGELHARDT 1960, CARRIGY & MELLON 1964) and chamosite (HORN 1965), similar in shape to fibrous calcite cement ("A"), often are early crystallizations. Aggregate rolls of kaolinite mostly (but not always; GREENSMITH 1965) are connected with the shallow-burial stage decomposition of feldspars. Grain coatings and sheaf-like flaky aggregates of illite, mixed layer minerals and montmorillonite also belong to the early phase of clay mineral neoformation.

With increasing depth, some of the early-formed minerals, like montmorillonite, mixed layer minerals and kaolinite, become unstable and disappear in favor of 2 M-muscovite and chlorite (see tab. 3-19, 22). If the connate water is slightly acid and poor in the cations K and Mg, kaolinite cement can be found in considerable quantities even at depths of 4300 meters (see p. 154). Kaolinite can also be prevented from illitization by oil impregnation (MILLOT 1970, p. 246).

The frequent formation of kaolinite in shallow depths of burial, and of muscovite in greater depths is explained by the increase of ion concentration and of pH with increasing depth. Rare neoformations, like talc (GÖRZ 1962), are not considered here.

3.314.7 Other neoformations

Zeolite cementation, mostly analcime and laumontite, are characteristic of vitric tuffs or of sandstones with volcanic components (Kossovskaya & Shutov 1955, Brown & Thayer 1963, Carrigy & Mellon 1964, Enlows & Oles 1966, Hay 1966, Raam 1968). Analcime also forms in the absence of volcanic material, in evaporitic environments (Keller 1952, van Houten 1962, 1965, Hay & Moiola 1963), and as an early diagenetic cement in accompanying sand- and siltstones (Iijima & Utada 1966, Füchtbauer 1967b). Laumontite also is found in coal-bearing beds (Bur'yanova & Bogdanov 1965/1967), in which the coal diagenesis only slightly surpassed the lignite stage.

Halite cement is not rare in sandstones that contain nearly saturated interstitial water, e. g. in the North German Buntsandstein; its distribution is very uneven.

Hematite (α-Fe_2O_3) as autochthonous neoformation is widespread in red beds. According to Schwertmann (1966) and Schellmann (1959) it forms directly by dewatering of amorphous or very poorly crystalline $Fe(OH)_3$ and not, as formerly assumed, via the intermediate stage goethite, α-$FeO(OH)$. Hematite is favored over goethite, even in the presence of water, by increased temperature, low pH-values, higher concentration of the amorphous hydroxide, as well as by ageing due to repeated drying. These conditions are met, at least partly, by arid conditions producing red beds. Under such conditions, an iron-stained montmorillonite is derived from alteration of iron-bearing silicates such as hornblende and biotite (T. R. Walker 1967a).

The presence of magnetite as a neoformation is mentioned only rarely (S. A. Friedman 1954). In nearly all sandstones, with the exception of red beds, pyrite is common.

Authigenic heavy minerals have been mentioned on pages 27—30 and 337. Authigenic anatase is most frequent in sandstones. The ratio of detrital to authigenic titanium minerals as an indicator of the degree of diagenesis has been used by Prozorovich (1970) to investigate the time of oil migration into a reservoir.

The selective cementation of sandstones by uranium minerals, as well as by quartz and calcite, on the Colorado plateau (U. S. A.) has been discussed in numerous papers (e. g. Phoenix 1956).

3.315 Replacement and alteration

"Replacements" are reactions in which one crystal grows at the expense of and in place of another. The ions released during this reaction may (1) react immediately with the ions present in the pore fluid or (2) they may influence the precipitation of another mineral, or (3) the authigenic mineral uses only the place provided by the dissolving unstable mineral. The authigenic and the replaced mineral are connected by a thin film of fluid. Only this explains why both the interior (fissures, inclusions) and exterior fabric are preserved during replacement. If the form is maintained, we deal with a "pseudomorph". If, however, the first mineral is completely dissolved before the secondary mineral grows, we are not dealing with a real replacement, but rather with cementation of a secondary pore (see p. 130).

"Alterations" of a mineral use major parts of the lattice of this mineral for the formation of a new mineral. Frequently this includes incongruent dissolution, by which only part of the components of a mineral are dissolved.

With increasing subsidence porosity decreases and therefore the amount of replacement increases compared with cementation. This is explained also by the fact that the physico-chemical conditions (composition of interstitial water as well as temperature and pressure) differ increasingly from the depositional environment, so that the original minerals become unstable and are replaced by others in order to bring about equilibrium.

An early diagenetic replacement which utilizes fully the substances present in the available minerals is dolomitization (ch. 5.121). Though dolomite is theoretically stable in sea-water, it apparently forms too slowly as to be able to crystallize directly from sea-water. In clastic rocks early diagenetic dolomite is found in crypto- to microcrystalline stringers in siltstones and fine-grained sandstones; presumably it replaced calcite or aragonite. The magnesium comes from the interstitial water.

Also the replacement of feldspar by kaolinite belongs to an early stage of diagenesis (shallow burial stage). In this case, only water is needed in addition to the available ions. From a diagram published by GARRELS & CHRIST (see Vol II, pt. 2), one can predict that, if log a_{K^+}/a_{H^+} values fall much below 5 or 6 and H_4SiO_4 concentrations drop much below 10^{-3} moles/liter, K-feldspar will kaolinize. If log a_{K^+}/a_{H^+} rises above 6.5, K-mica becomes stable instead of kaolinite at H_4SiO_4 concentrations below 10^{-4} moles/liter (a_{K^+} and a_{H^+} being the activities of potassium and hydrogen ions in solution which become equal to the concentrations, in very dilute solutions). In the transition zone of shallow and deep burial stage, crystal enlargement of the illitic matrix begins. The illite crystallinity (see below) may increase at the same time. Later, in the deep burial stage, muscovite is formed, in part as a replacement of feldspar and kaolinite (KISCH 1966), in part as a neoformation (sericite, SCHERP 1963). In deeply buried geosynclinal sediments, KOSSOVSKAYA & SHUTOV (1955, 1958) found a transition from biotites (which were already partly amorphous) to sheaths of chlorite and muscovite. In the early metamorphism, they convert back into biotite.

In accordance with the above findings, feldspars have been replaced by kaolinite and quartz in the upper part, and by hydromica in the lower part of a 500 m thick sequence investigated by VATAN (1962).

The substitution of quartz and feldspar by carbonate, anhydrite or barite is governed by the chemistry of the interstitial water alone. These replacements belong to a late stage of diagenesis. In the Buntsandstein, quartz grains are replaced by anhydrite, whereas K-feldspars are dolomitized (fig. 3-49) presumably owing to increased pH, causing higher solubility of feldspars and quartz and lower solubility of carbonates. Carbonate can also replace siliceous matrix (LAJOIE 1968). Anhydrite as a replacement mineral forms at elevated temperatures because the solubility of anhydrite decreases with increasing temperature, in contrast to most silicates and quartz (v. ENGELHARDT 1973).

In the deepest burial stages carbonates can replace large parts of the rock.

While diagenetic reactions generally take place between only one mineral and the interstitial water, several reactions occur simultaneously in metamorphism (FYFE et al. 1958).

Alterations during diagenesis occur particularly with mica and clay minerals. This is reflected by the increase of crystallinity (see ch. 3.33) and of the intensity of (002) compared with (001) in X-ray diagrams of illite (DUNOYER DE SEGONZAC 1969) and micas and is in part caused by minor alterations in the charge of the different layers (DUNOYER DE SEGONZAC et al. 1969). The disappearance of montmorillonite and mixed layer minerals and the appearance of 2 M-muscovite with increasing depth can also be classified as an alteration.

In acid environments, an alteration of kaolinite into dickite (and nacrite) can be observed with increasing diagenesis. The depth of alteration depends on the chemical conditions (DUNOYER DE SEGONZAC 1969, 1970). In Cambrian sandstones of the Sahara, FERRERO & KUBLER (1964) found dickite cement which apparently formed from detrital kaolinite of the adjacent siltstone (see also table 3-19). Generally, chlorite is more stable in rocks of the deep-burial stage of diagenesis than the kaolinite-type minerals.

The albitization of K-feldspar in graywackes (MIDDLETON 1972) is another example of alteration. Other common alterations include magnetite to hematite and ilmenite to leucoxene.

3.32 Examples of diagenetic sequences

There is no uniform pattern of diagenesis for all types of sandstone. The primary composition of sandstones influences the process of diagenesis to a considerable extent, as shown by KOSSOVSKAYA & SHUTOV (1958, 1970). Their synoptic table is cited. below. The papers by COOMBS et al. (1959) and PACKHAM & CROOK 1960) were consulted for zeolites.

"Epigenesis" in table 3-19 means late diagenesis or deep burial stage, "metagenesis" refers to the transition zone between diagenesis and metamorphism. Line 1 predominantly contains neoformations, lines 2 and 3 also show alterations and replacements whereas in lines 4 and 5 parts of the rock have been replaced combined with destruction of the primary fabric. Some statements of this table shall be discussed in greater detail, in the following chapters.

One important feature of increasing diagenesis is the continuous loss of water of the clay minerals and zeolites (KOSSOVSKAYA & SHUTOV 1965). These authors pointed out the convergency of all impure sandstone types with increasing diagenesis. Moreover, they found the same processes in sandstones, siltstones, and claystones. However, due to higher permeability, diagenesis in sandstones begins earlier and takes less time than in claystones; this may, however, be counteracted by the smaller grain size of claystones compared with sandstones.

Table 3-19. Neoformation and replacement in sandstones in relation to primary facies and degree of diagenesis (modified from KOSSOVSKAYA & SHUTOV 1965).

	quartzose sandstones	sandstones with rock fragments	acid arkoses*	intermediate arkoses**	volcanic graywackes
	Increase of unstable primary components $\longrightarrow$				
Early epigenesis	—	—	Vermiculite		heulandite, analcime
Late epigenesis	quartz; kaolinite into dickite	rock fragments into chlorite + illite	quartz, albite; clay into 1 M-illite	heulandite, plagioclase; glass into laumontite	
Early metagenesis	quartz, pyrophyllite, (muscovite)	chlorite; 1 M- into 2 M- muscovite		laumontite, prehnite, chlorite; illite into muscovite	
Late metagenesis		epidote, muscovite, chlorite, stilpnomelane			
Regional-metamorphism	kyanite?	biotite			

* chief constituents: quartz, K-feldspar, acid plagioclase, muscovite
** chief constituents: intermediate plagioclases, biotite, hornblende

KARPOVA (1969) makes use of the transition from the orthohexagonal chlorite — polytype Ib to the monocline polytype IIb in distinguishing epigenesis from metagenesis (according to X-rays, differential thermal-analysis and infrared spectroscopy) (see also TRÖGER 1967, p. 570).

3.321 Diagenesis of quartz sandstones

The increase in diagenesis with increased depth of burial is shown by the decrease in porosity. In early stages, chemical changes are subordinate. The decrease in porosity from about 40 to 30% in the uppermost 1000 meters predominantly originates from mechanical rearrangement; this is higher in sandstones with greater roundness (see p. 132). Beginning at a depth of about 400 meters the influence of chemical compaction increases, as shown by the facets of quartz overgrowths on the detrital grains. A count of these facets shows that cementation increases continuously with increasing depth (fig. 3-50). This method of quantitative evaluation of quartz cementation ends at about 2300 m depth, when most of the grain surfaces are covered by facets. At the same depth according to PROZOROVICH (1970), the increase of convexo-concave grain contacts with depth becomes more pronounced, in Jurassic and Cretaceous sandstones of Siberia (1000 m : 8% [porosity 27%]; 2300 m : 20% [porosity 22%]; 3000 m : 65% [porosity 8%]); possibly this method can replace the above method, at greater depths. The Siberian sandstones contain a certain amount of clay matrix.

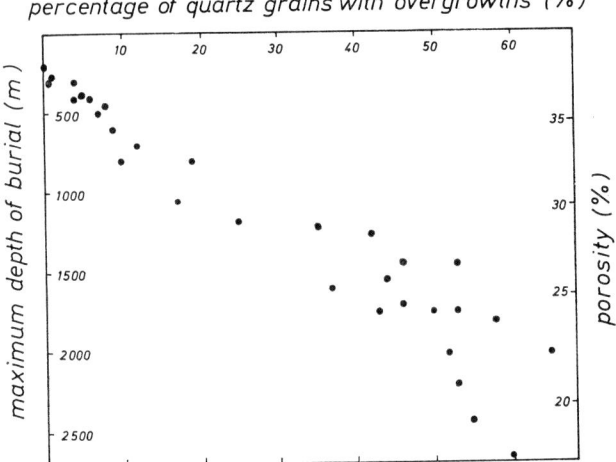

Fig. 3-50. Increase of quartz cementation in Dogger Beta quartz sandstones with increasing depth of burial. The "percentage of quartz grains with overgrowths" was counted on loose grains in transmitted light (magnification 100 ×) and calculated according to the following formula

$$\frac{100 \cdot (0.5 \, b + c)}{a + b + c}$$

a = number of quartz grains with few or no facets (= overgrowths)
b = number of quartz grains with facets covering less than 50% of the surface
c = number of quartz grains with facets covering more than 50% of the surface.

The estimates have proved to be independent of grain size. The dots represent means of larger numbers of samples (modified from PHILIPP et al. 1963).

According to the minus-cement-porosity (3.314.1), many sandstones received a lithifying quartz cement preventing further compaction at a burial depth of approximately 1000 m (FÜCHTBAUER in press).

The fact that diagenesis is strongly reduced if the pores are filled with hydrocarbons can be used for the determination of the approximate burial depth at the time of the oil accumulation (FÜCHTBAUER 1961, PHILIPP et al. 1963). In fig. 3-50 the amount of quartz overgrowth is plotted against maximum depth of burial; this allows an estimation of maximum depth of burial because quartz crystals retain their overgrowth in case of later uplift.

The main factor influencing the increase of quartz overgrowth with increasing depth is presumably temperature, but time of burial may also be an important additional factor. If the volume of particles less than 20 microns is more than 10%, the silica cementation is strongly reduced (PHILIPP et al. 1963), because clay minerals hamper quartz overgrowth.

The origin of the silica has been discussed above. Pressure solution of quartz grains in the sandstones probably was less important than pressure solution of quartz grains in the adjacent silt- and claystones. This is shown by the increased sandstone cementation towards the adjacent shales (fig. 5 and table V in FÜCHTBAUER 1967c).

Kaolinite, mostly replacing feldspar, is an early diagenetic product in quartz sandstones, but later generations of kaolinite have been reported also (ZIMMERLE 1963, FONDEUR 1964).

Calcite cement that forms without the replacement of adjacent grains, generally originates in an early stage of diagenesis. This is testified by the relatively high minus-cement-porosity of such calcareous sandstones (see ch. 3.314.1). In outcrops, calcite cement may be dissolved by weathering (PETTIJOHN 1957, GRAF & LAMAR 1950, ADAMS 1964). Whereas early-diagenetic calcite cement does not replace and even not affect adjacent sand grains, carbonate apparently becomes less soluble than quartz with increasing burial depth, presumably due to increasing pH (see also p. 332): macrocrystalline calcite cement frequently replaces quartz grains in deeply buried sandstones (WALDSCHMIDT 1941, HEALD 1950, FOTHERGILL 1955, DAPPLES 1959, SIEVER 1959, PAGE & CAROZZI 1962 [in dolomites], ZIMMERLE 1963, SHARMA 1964).

Late cements occurring in low quantities are anhydrite, barite and celestite. A characteristic sequence of neoformations and replacements can be put up for quartz sandstones:

1. Kaolinite, partly replacing feldspars
2. quartz
3. calcite, dolomite
4. sulphates
5. calcite replacing quartz and feldspars

3.322 Diagenesis of calcareous sandstones

Calcareous sandstones may be classified into calcite-cemented sandstones and calcarenitic sandstones in which the carbonate is detrital like the silicic components, or biodetrital. While in the calcite-cemented sandstones the diagenesis of the grains is interrupted if the cement precipitated early enough, strong diagenesis normally occurs in calcarenitic sandstones, including transformation and dissolution-reprecipitation processes.

This is exemplified by the Lower Marine Molasse north of the Alps which is composed of dolomite-arenitic sandstones containing a primary matrix of clay minerals and calcite. Part of the calcite was dissolved and reprecipitated chiefly in such layers containing skeletal calcite grains, whereas kaolinite is present in layers devoid of such grains. The decrease in porosity with increasing depth is greater in these calcareous sandstones than in quartzose sandstones. Replacements of quartz and feldspar by calcite are observed at depths greater than 3500 meters (FÜCHTBAUER 1964a, 1967c).

3.323 Diagenesis of feldspathic sandstones (fig. 3-51, left)

Two Upper Triassic sandstones of Southern Germany may serve as examples. They are separated by only 60—70 meters of multicolored marls yet are quite different: The "Schilfsandstein" (sandstone) at the base and the "Stubensandstein" above.

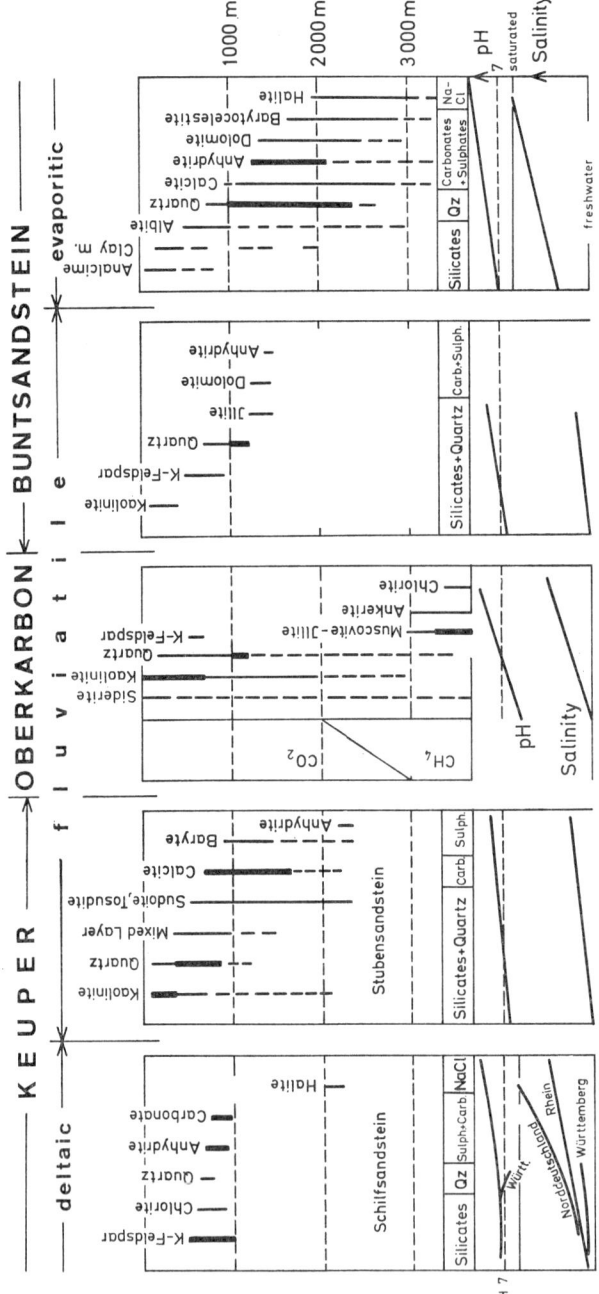

Fig. 3-51. Sequences of cementation in fluvial, deltaic-marine, and evaporitic sandstones of Triassic and Upper Carboniferous age in Germany. Thick lines indicate strong cementation, interrupted lines indicate uncertain cementation. Salinity and pH are indicated schematically (from FÜCHTBAUER, Geol. Rundsch., in press).

Table 3-20. Comparison of diagenetic processes in two Triassic sandstones (from shallow boreholes near Stuttgart).

		Schilfsandstein (HELING 1965)	Stubensandstein (HELING 1963)
Depositional environment		deltaic	fluvial
Source of material		from North	from East
Thickness		about 40 m	80—85 m
medium grain size		0.1 mm	0.2—0.4 mm
Detritus > 0.02 mm	quartz	45%	75%
	K-feldspar	22%	6%
	plagioclase	4%	19%
	rock fragments	14% (sericitic slate)	?
	mica	11%	present
	glauconite	4%	0%
< 0.02 mm		10% (I, Ch*)	about 15% (K*, I, ML*, S*)
Carbonate		trace	0—30% (average 15%) calcite*, dolomite*
Sequence of neoformations and replacements		1. K-feldspar	1. kaolinite replaces feldspars
		2. chlorite replaces biotite	2. quartz
		(3. quartz)	3. mixed-layer muscovite-montmorillonite
		(4. anhydrite replaces quartz and feldspar)	4. sudoite
		(5. carbonate replaces quartz, feldspar, and anhydrite)	5. calcite replaces quartz and feldspar
			6. barite

The percentage figures represent gross averages. The fraction < 0.02 mm contains quartz, feldspar, and the following clay minerals: Ch = chlorite, I = illite, K = kaolinite, ML = mixed layer, S = sudoite.
* predominantly authigenic. ( ) = subordinate

The main difference is the behavior of the feldspars: Whereas the orthoclase and plagioclase grains of the Schilfsandstein have secondary rims of K-feldspar, such overgrowths are missing in the Stubensandstein: In fact, feldspars became unstable and kaolinitization is observed.

Only a tentative explanation can be given: Both sandstones contain coal stringers that release acidic humic solutions into the newly deposited sands. The percolation of these solutions was stronger in the fluvial Stubensandstein than in the deltaic Schilfsandstein, for different possible reasons:

1. The percolation may be stronger in fluvial than in deltaic-marine sediments because of the alternating water flow from the river towards the groundwater and vice versa, depending on the groundwater table.

2. The permeability of the Stubensandstein is higher due to the coarser grain size, thus allowing for more percolation.

3. The average pH of freshwater is 6.5, of seawater, however, 7.5. Therefore, acidic pore-fluids were more probable in the Stubensandstein than in the Schilfsandstein. In the latter, an alkaline environment enabled the precipitation of potassium feldspar from potassium-rich weathering solutions which can be inferred from the high percentage of detrital K-feldspars. Kaolinitization, in general, is typical in nonmarine environments, as shown also by its occurrence in coal-bearing deposits (e. g. Upper Carboniferous).

The silica in the quartz overgrowths in the Stubensandstein was derived from pressure solution and from kaolinitization of feldspars (HELING 1963). Neoformation of mixed-layer minerals and sudoite is suggested by a comparison of sandstones poor in carbonate with sandstones in which an earlier stage of diagenesis has been preserved by carbonate cementation: Whereas sandstones poor in carbonate contain considerable amounts of mixed-layer minerals and sudoite, the cemented layers contain only traces of mixed-layer minerals and no sudoite. In deeper boreholes the kaolinite: sudoite ratio shifts in favor of the latter (KULKE 1969).

In addition to the primary composition of the sandstones the chemistry of the interstitial water, which still partly reflects the depositional environment, is of considerable importance for diagenesis. This is also demonstrated in the following section.

3.324 Diagenesis of feldspathic, slightly evaporitic sandstones (red beds)

The Buntsandstein consists of alternating layers of red siltstones and mostly red sandstones. A river system transported the material predominantly in northerly directions discharging into a basin of variable salinity. In the fluvial area sandstones are dominant, in the brackish to evaporitic areas siltstones predominate. The sorting is poor: layers of fine and coarse sand alternate within vertical intervals of less than one mm. The sequence of events was determined using the criteria of ch. 3.314.1 (tab. 3-21, fig. 3-51, right).

Brackish to evaporitic sandstones of the Buntsandstein (Lower Triassic) are characterized by neoformation of Na-feldspar (overgrowing sodium and potassium feldspar grains), anhydrite and analcime; fluvial sandstones by K-feldspar (forming small overgrowths on potassium feldspar grains). Apparently these neoformations are induced by the depositional environment, which exerted its influence via the interstitial water even at burial depths of 500 to 1000 m. Although lamellibranchs suggest that marine transgressions periodically reached far to the South, these oscillations did not much influence the pore fluids. The partial conformity between detrital and authigenic feldspars in both areas may be explained (1) by the supply of many K-feldspars from the southern source area whereas Na-feldspars came from the East and are present only in the North and East, (2) by the weathering solutions rich in potassium, which percolated the silty overbank sediments and later on caused the formation of authigenic K-feldspar in the fluvial sandstones, whereas the sodium-rich evaporitic environment caused the formation of Na-feldspar. Beginning with a burial depth of about 1000 meters (estimated from the minus-cement-porosity) the neoformation of albite in the North was surpassed by a strong quartz cementation. Sandstones buried at various depths (2000—5000 meters of maximum burial depth) only differed in the degree but not in the kind of mineralogy of cementation.

Within any layer, fine sands have more quartz and feldspar cement, coarse sands more anhydrite cement (see ch. 3.314.2). Moreover, cementation is more intensive in thin sandstone layers than in thick ones; near the walls of thick sandstones porosity also is reduced (ch. 3.314.2).

Table 3-21. Diagenetic processes in the Buntsandstein from different depositional environments (from VALETON 1953 and FÜCHTBAUER 1967 b, c).

		Northern Germany	Southern Germany
Depositional environment		brackish to evaporitic	fluvial mainly
Source of material		from S, E, N	from SW and SE
Detritus > 0.02 mm	quartz	60—70%	73—78%
	feldspar	15—25 (albite, K-feldspar)	10—15 (K-fsp. ≫ albite)
	mica	1	1
	rock fragments	10—15	10—15
	calcitic oöids	varying	—
< 0.02 mm		4% (I, Q, H*, F, Ch*)	
Sequence of neoformations and replacements		1. hematite, analcime, vermiculite, mixed layer, illite 2. albite (3%), chlorite 3. quartz (15%) 4. calcite (4%) 5. anhydrite (10%), occasionally replaces quartz 6. dolomite (5%), often replaces K-feldspar 7. baritocelestite, replaces quartz and feldspar 8. halite	1. K-feldspar (0,5%) 2. quartz (15%) 3. dolomite (1%)

The maximum content (in vol. %) is indicated for the more frequent minerals. Ch = chlorite, F = feldspar, H = hematite, I = illite, Q = quartz, * predominantly authigenic.

3.325 Diagenesis of coal-bearing micaceous sandstones

KNEUPER (1957), ESCH (1962) and especially SCHERP (1963) gave the following description and interpretation of the sandstones of the coal-bearing Upper Carboniferous in Northwestern Germany (tab. 3-22, fig. 3-51).

In such sandstones, the crystallization of siderite and the kaolinitization of feldspar presumably starts shortly after deposition, particularly in the acid environment of the coal swamps (SCHERP 1963, PACKHAM & CROOK 1960). The low pH is retained in the pore fluid during the subsequent burial because of the release of H_2O containing humic acids and CO_2 during the early stages of carbonization. For further discussion, see ch. 3.315. The ordering degree of the kaolin minerals improves with increasing depth ("fireclay" → kaolinite; ECKHARDT & v. GAERTNER 1962). This takes place earlier in sandstones than in silt- and claystones (FÜCHTBAUER 1967c, fig. 12).

In a second stage of carbonization, beginning at about 2500 meters burial depth, the CO_2 formation generally decreases in favor of that of methane (JÜNTGEN & KARWEIL 1962). The interstitial water may remain acid if there is not much carbonate in the sandstones. This may apply to sandstones in the U.S.S.R. des-

Table 3-22. Diagenetic processes in Upper Carboniferous sandstones.

Depositional environment		limnic, with marine layers		
Detritus > 0.02 mm	quartz	75%		
	feldspar	10% (K-feldspar, plagioclase)		
	rock fragments	10%		
	mica	5%		
	plant remains	varying		
< 0.02 mm		10—20% (K*, I, Ch*)		
Carbonates		10% (S*, A*, C*)	depth** (meters)	temperature** (°C)
Sequence of neoformations and replacements		1. siderite (cryptocrystalline)	(1788)	(70)
		2. kaolinite replaces feldspar	(1788)	(70)
		(3. quartz)		
		4. ankerite replaces kaolinite; sericite replaces kaolinite and feldspar	3000 3000	100 100
		5. chlorite replaces biotite	3350	120
		6. sericite (intensive)	5300	190

A = ankerite, C = calcite, Ch = chlorite, I = illite, K = kaolinite, S = siderite.
* Predominantly authigenic.
** The present depth and temperature at which the replacements or neoformations have been found in the borehole Münsterland 1, according to observations by SCHERP (1963) and HEDEMANN (1963). The Upper Carboniferous was reached at 1788 m.

cribed by KOSSOVSKAYA & SHUTOV (1965). Here, with increasing depth, kaolinite changed into dickite. This occurred, depending on the time of burial, at 2500 m depth in unfolded Mesozoic rocks, but at 1000—1500 m depth in unfolded Paleozoic and Precambrian rocks. On the other hand, kaolinite has been found in an Upper Paleozoic sandstone of Northern Germany at a depth of 4300 m (FÜCHTBAUER 1967c).

In carbonate-bearing sandstones, such as the northwest German Upper Carboniferous, a neutral to basic interstitial environment develops in this second phase of carbonization, and the kaolinitization of feldspars gives way to sericitization of kaolinite and feldspar (ESCH 1962, HECHT et al. 1962, SCHERP 1963, KISCH 1966, see also DAPPLES 1959 and KOSSOVSKAYA & SHUTOV 1958). In addition, cements of ankerite and calcite partly replace kaolinite. The lower boundary of the acidic environment in the borehole investigated by SCHERP is presumably at a depth of 3000 m. According to HEDEMANN (1963), the present temperature here is 100° C, and coalification nearly reached the anthracite stage. The present depth just about equals the former maximum burial depth but diagenesis was accelerated due to a geothermal gradient of presumably less than 15 m / 1° C during the Carboniferous (TEICHMÜLLER 1962).

At somewhat greater depth biotite becomes instable and is replaced by chlorite and muscovite (SCHERP, l.c., KOSSOVSKAYA & SHUTOV 1958). The iron released by this alteration either forms ankerite or hematite pigment; the released titanium forms rutile. The ankerite cement in the interstices and joints is richer in magnesium with increasing depth. The filling of the joints corresponds to the cement minerals of the adjoining rocks. Therefore, the joint fillings can be explained by lateral secretion (STADLER 1963) rather than by hydrothermal action.

The reduction in **porosity** with increasing depth for the Upper Carboniferous sandstones of several northwest German boreholes is shown and explained in fig. 3-43. The filling by natural gas decreased diagenesis, and a higher porosity by about 3 vol-% of the rock corresponding to a depth reduction of about 400 m, was preserved.

Comparing the diagenesis of these sandstones with the matrix-poor sandstones described in the previous chapters, it is obvious that neoformations prevail in the sandstones described earlier and replacements are more frequent in the micaceous sandstones described in this chapter. Probably a high portion of primary clay and mica flakes interfered with cementation and enabled further compaction which was enhanced by pressure solution (see also ch. 3.314.2).

3.33 The transition from diagenesis to metamorphism in sandstones

According to WINKLER (1970), the "**very low stage of metamorphism**" begins at about 200° C and is characterized by **laumontite**, in the realm of burial metamorphism with a normal geothermic increase of temperature (30° / 1000 m). At about 300° C, **pumpellyite and prehnite** develop. At about 380° C, the low-stage metamorphism with zoisite — clinozoisite and actinolite is reached. Several workers, e.g. FREY & NIGGLI (1970) inserted the term "**anchimetamorphism**" between diagenesis and metamorphism, including approximately the laumontite and the pumpellyite zone; the present writer is not convinced that this is a valuable term, and prefers WINKLER's "very low stage of metamorphism", instead.

Laumontit is present especially in rocks with volcanic components (PACKHAM & CROOK 1960, BROWN & THAYER 1963, KOSSOVSKAYA & SHUTOV 1965, HAY 1966, IIJIMA & UTADA 1966). But it is also reported from a non-metamorphic Miocene sandstone with a maximum burial depth of 3300 m (KALEY & HANSON 1955), and a Permian sandstone with 900 m maximum burial depth (RAAM 1968). It is doubtful whether temperatures of 200° C were reached in these rocks. Frequently, diagnostically significant facies minerals are not present so that there is a need for other criteria. For the zeolite facies see COOMBS et al. (1959) and ch. 3.326, 4.423.26, 4.6.

Many workers used the **coalification rank** of coals and carbonaceous material dissipated in the sandstones as a criterion of diagenesis and metamorphism, but the limits proposed vary considerably: KOSSOVSKAYA et al. (1957), KISCH (1969) and FREY & NIGGLI (1971) reported the occurrence of laumontite together with flame coal (approx. 39% volatiles) and gas coal (approx. 31% volatiles), prehnite and pumpellyite together with ess coal (15% volatiles), lean coal (14% volatiles), and anthracite. On the other hand, K. WEBER (pers. commun.) found pumpellyite in slates containing meta-anthracitic carbonaceous material (table 3-23, M. WOLF, in print). According to KISCH (1969), the transition of heulandite to laumontite is connected with less coalified measures in younger than in older strata. This suggests a stronger influence of time in coalification than in the adjustment of mineral equilibria.

Probably a more specific criterion is the illite crystallinity as first used by WEAVER (1960) and KUBLER (1964), see ch. 4.6. In order to be independent of the experimental conditions, K. WEBER (1972a) divided the peak width at half height on illite diffractograms (fraction 2—6 μ or polished rock surface) by the peak width at half height of quartz added as an external standard. Even these figures, however, vary considerably, so that the mean of several samples and several measurements of each must be taken (table 3-23).

Table 3-23. Comparison of coalification of carbonaceous inclusions ($R_{max.}$ = maximum reflectance; WOLF, in print) with the illite crystallinity ($Hb_{rel.}$ = peak width at half height, compared to quartz; K. WEBER 1972b) for slates of Middle Devonian age, Sauerland, Germany.

coal rank	$R_{max.}$ % (vitrinite)	Hb rel. (2—6 μ)	other minerals
anthracite	4	1.5	pyrophyllite
meta-anthracite	5	1.3	
meta-anthracite	6	1.15	{ paragonite*, pumpellyite } $Hb_{rel.}$ = 1.0—1.2

* WEBER, pers. comm.

Slates are used in table 3-23, because more carbonaceous material occurs in them. The conditions of diagenesis probably are different in sandstones, because of the primary differences in composition and permeability. As mentioned above, the thermal history also influences the relation between coalification and clay mineral diagenesis. Abnormal relations occur locally in the same area (WEBER 1972a).

The table which covers most of the "very low stage of metamorphism" is based on the maximum reflectance of vitrinite instead of the volatiles, because in the anthracite rank, the reflectance changes more rapidly than the percentage of volatiles. Anthracite has less than 10 percent of volatiles and $R_{max.} > 2.5$, meta-anthracite has less than 4 percent of volatiles and $R_{max.} > 4.5$, according to WOLF (in print).

Pyrophyllite is found in rocks poor in K and Mg (Fe) at a depth corresponding to about 200° C, whereas in other rocks it may appear at considerably higher temperatures (KISCH 1969). On the other hand, it has been found already in about 600 m depth (well Assejmi, DUNOYER 1969).

Fibrous stilpnomelane was found by FREY & HUNZIKER (1973) in glauconitic rocks as the first indication of metamorphism corresponding to about 200—300° C.

Kaolinite is generally found only to a depth of about 3000 m (DUNOYER et al. 1968, KISCH 1969, DUNOYER 1970). In places it overlaps with the laumontite zone

(KISCH l. c.). In the boring Münsterland, however, only 100° C were found near the bottom of the kaolinite zone, at a depth of 3000 m (ch. 3.325). A transformation into dickite was observed at greater depth (DUNOYER 1970) as well as in very old "diagenetic" sandstones (Cambrian of N. Africa, FERRERO & KUBLER 1964), in acid environments. If enough cations were present, mica instead of dickite formed in alkaline environments; with sea water as pore fluid, this may happen at temperatures as low as 100° C (DUNOYER 1970).

Paragonite-quartz assemblages have their lower thermal compatibility limit at 315—335° C (CHATTERJEE 1973).

K-feldspar in general becomes unstable near the diagenesis-metamorphism transition. At the same time, illite regenerates into muscovite by K-uptake.

According to WINKLER (1970) metamorphic conditions should normally begin at a depth of about 6000 m corresponding to about 200° C. The following observations made in such rocks can be used as general criteria of the transition diagenesis-metamorphism.

Increasing reactions between the primary components and a decrease of sedimentary fabric including porosity is characteristic of this depth. Most of the feldspars are sericitized in deeply buried Upper Carboniferous sandstones. Replacement of entire rock portions of quartz, feldspar and phyllosilicates by various carbonates is also observed in these sandstones (FÜCHTBAUER 1967c). Although no changes in texture occur in the zone of "early metagenesis" (deeper than 6000 m) in the Permo-Triassic sediments of the Werchojansk geosyncline (Siberia), intensive intergrowths of grains as well as sericitization and chloritization of the clay matrix are found (KOSSOVSKAYA et al. 1957). Most of the biotites was altered into an "amorphous phase" or into chlorite-sericite bundles. During "late metagenesis" (tab. 3-19) the grains became lenticular with spine-like intergrowths.

In an area near Osnabrück, which underwent weak thermo-metamorphism during early Tertiary time, the Upper Carboniferous sandstones contain nests of macrocrystalline siderite and Fe-rhipidolite ($n_z = 1.64$, $2 V_x = 20°$, abnormal interference colors) surrounded by opaque seams. An equivalent burial depth of 8000 m has been inferred from the porosity of these sandstones (fig. 3-43). The same rhipidolite is also found in phyllites. A similar rhipidolite ($n_z = 1.635$) was found in the Upper Carboniferous below 5100 m filling joints together with albite and apatite (SCHERP 1963). Apparently the chlorite nests in the rocks reported above indicate a higher degree of mobilization than joint-fillings (FÜCHTBAUER 1967c, see also KARPOVA 1969).

Further criteria include the occurrence of large pyrite crystals formed on pyritic worm tracks, and the appearance of especially large authigenic feldspars (in carbonate rocks).

Since equilibria generally are not reached during diagenesis, the diagenetic mineral association strongly depends upon kinetics. For this reason, grain size, permeability and other parameters play a more important role in diagenesis than during metamorphism.

3.4 Porosity, permeability, and oil saturation

Porosity is defined as the portion of pores expressed in parts of the bulk volume of the sample. Frequently it is expressed in percent. The void ratio is defined as the ratio of voids and solids in a sample.

Permeability is defined by Darcy's law:

$$k = \frac{\eta \cdot L \cdot Q}{F \cdot \Delta p \cdot t} \quad [\text{darcies}]$$

with η = air viscosity, L = length [cm] of the rock cylinder used for the permeability test, Q = air volume streaming through the cylinder [cm^3], F = cross-section of the rock cylinder, Δp = pressure difference in front of and behind the cylinder [atm], t = time [sec]. The corrected formula for the pressure in the mid of the rock cylinder (G. Müller 1967c) reads

$$k = \frac{\eta \cdot L \cdot Q \cdot 1000}{F \cdot \Delta p \cdot (1 + \Delta p/2) \cdot t} \quad [\text{millidarcies}].$$

In grainstones including sandstones and calcarenites with interparticle pores, the permeability is related to the porosity by the Kozeny-Carman formula (v. Engelhardt 1960):

$$k = \frac{\varepsilon^3}{5 \, S_0^2 \, (1-\varepsilon)^2}$$

with ε = porosity, S_0 = specific surface (cm^2/cm^3 solid material) which increases with decreasing grain size. If porosity, sorting, and grain shape are held constant, then the approximation $(S_{01})^2 / (S_{02})^2 = M_2^2/M_1^2$ (with M = median diameter) is valid (v. Engelhardt & Pitter 1951), or, by using the above Kozeny-Carman formula:

$k_1/k_2 = M_1^2/M_2^2$, or, for constant specific surface,

$$\frac{k_1}{k_2} = \frac{\varepsilon_1^3 \, (1-\varepsilon_2)^2}{\varepsilon_2^3 \, (1-\varepsilon_1)^2}$$

The thick line in fig. 3-52 is based on this equation, for an assumed porosity-permeability pair, in order to show the general shape of the relationship for constant S_0, i.e. for constant grain size. For lower grain size, the points would deviate from this curve vertically downwards. The diagonal elongation of most cluster areas reflects the fact that in fine sandstones the porosity decreases with decreasing grain size because of silt and clay filling the pores. This does not apply to the coarser, clay-free sandstones, as shown by the upper portion of the shaded cluster area. The decrease in porosity with increasing permeability (and grain size) is probably caused by stronger mechanical compaction of the better rounded coarser sand grains (see fig. 3-43).

The oil saturation of the pore space depends on the inner surface related to the porosity (S/ε; v. Engelhardt 1960). (The inner surface S [cm^2/cm^3 rock] is connected with the specific surface S_0 by the formula $S_0 = S/1-\varepsilon$). The inner surface of rocks is water-wet, as a general rule. Therefore if oil replaces the interstitial water,

3.4 Porosity, permeability, and oil saturation

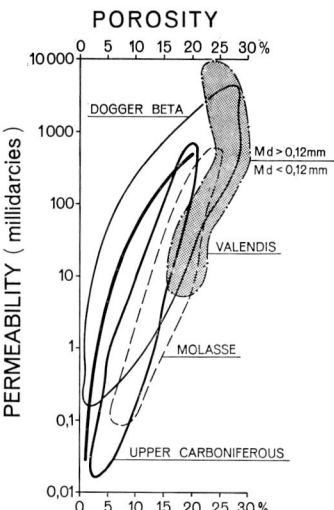

Fig. 3-52. Relationships of porosity and permeability of different sandstones. Thick line is theoretical relationship using an assumed porosity-permeability pair for construction (see text). Encircled areas engulf about 90% of the points in point clusters. Dogger Beta = quartzose sandstones (in part with carbonate cement); Valendis = sandstones, not cemented; Molasse = dolomite-arenitic sandstones with carbonate cement; Upper Carboniferous = pressolved micaceous sandstones (modified from FÜCHTBAUER 1967 c).

a certain amount of water will remain in the interstices (meniscuses) between the grains. It is termed "irreducible water saturation" or briefly "connate water" (assuming that this is preserved sea water). Because its amount depends on the inner surface which is connected with the permeability by the grain size (see above), permeability and irreducible water saturation are also correlated (table 3-24) within a given oil pool in which a certain porosity/permeability relationship exists.

Table 3-24. Irreducible water saturation S_{wc} (in percent of the pore space) related to the permeability, of Lower Cretaceous (fine and medium) sandstones, Emsland area (from fig. I/5, LÜBBEN 1969).

permeability:	3.16	10	31.6	100	316	1000	3160 md
connate water:	77	61	47	34	23	13	7% S_{wc}

As shown in this table, the connate water content becomes very high in silty sandstones and sandy siltstones with low permeability. Even if they were oil-saturated, diagenesis would continue in such rocks.

4. Pyroclastic rocks[1]

The fragmental components of pyroclastic rocks are formed by volcanic processes and are chiefly of volcanic composition. There are all transitions from pyroclastic to epiclastic sedimentary rocks, i. e. those that are derived from erosion of preexisting rocks. Pyroclastic rocks form in many different ways and are therefore difficult to classify (see reviews by WENTWORTH & WILLIAMS 1938 and FISHER 1966).

At present it seems most appropriate to classify pyroclastic rocks by a combination of descriptive characteristics (grain size, composition) and genetic processes (process of fragmentation, mode of transportation, and site of deposition).

In this book major distinctions are made: a) between pyroclastic sediments formed under subaerial and those formed under subaqueous conditions; and b) between fall and flow deposits, the two main modes of transport. There are transitions between both types of subdivision, however, as exemplified by the base surge deposits. This newly recognized type of pyroclastic deposit forms at the ground-(water-)atmosphere interface, and the base surge mode of transport results in pyroclastic sediments showing characteristics of both fall and flow deposits. Underlying all these subdivisions, however, is the grain size classification.

4.1 Grain size classification

Volcanic fragments, regardless of grain size, which are ejected from the vent during a volcanic eruption are called tephra, a term coined by ARISTOTLE (THORARINSSON 1954). Tephra may consist of particles from the magma itself (dense or inflated) (juvenile fragment), solidified volcanic rocks from previous eruptions (accessory fragment) or fragments of broken solid country rock which may be of diverse petrographic composition (accidental fragment), crystals, or mixtures of all these materials. Tephra is classified according to grain size, but grain size limits differ among various authors. According to FISHER (1966), whose classification is here adopted, fragments over 64 mm are blocks (if accidental or accessory) or bombs (if juvenile), those between 2—64 mm lapilli, clasts between 2 and $^1/_{16}$ mm are ash. Particles below $^1/_{16}$ mm are volcanic dust. Most classes are represented in fig. 4-1. Indurated pyroclastic rocks composed predominantly of ash, lapilli, blocks and bombs are, respectively, called tuff, lapillistone, and pyroclastic breccia. Indurated mixtures of ash and lapilli are lapilli tuffs, mixtures of ash and blocks are tuff breccias, and

---

[1] An early draft of this chapter, prepared in 1970, was critically read by R. V. FISHER, R. S. FISKE, P. E. POTTER, and R. L. SMITH.

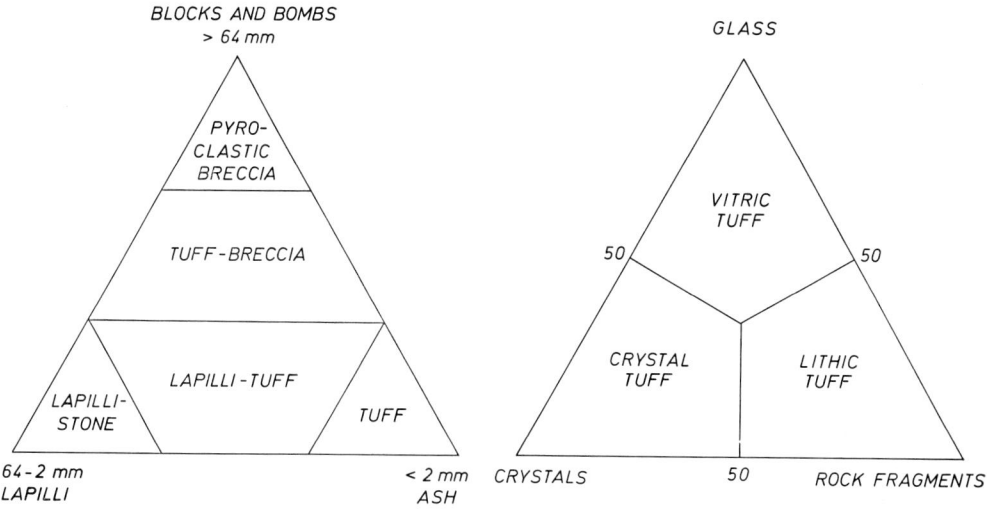

Fig. 4-1.   Fig. 4-2

Fig. 4-1. Classification and nomenclature of indurated tephra according to grain size (modified after FISHER 1966).

Fig. 4-2. Classification and nomenclature of tuffs according to their composition (from PETTIJOHN 1957, fig. 79).

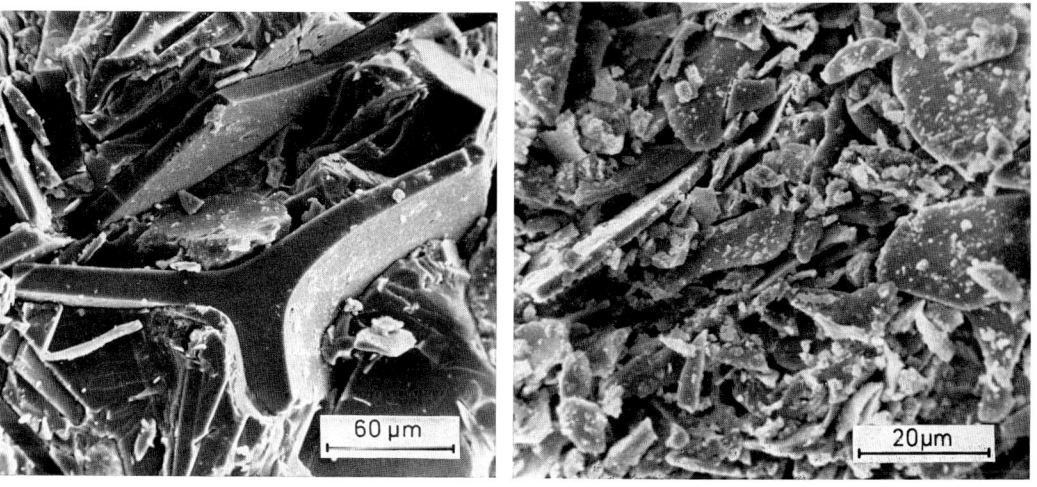

Fig. 4-3.   Fig. 4-4

Fig. 4-3. Scanning electron micrograph of glass shards of rhyolitic composition representing thick bubble-wall junctures showing typical Y-shape. Center of accretionary lapillus. Pliocene Ellensburg Formation, Washington (USA).

Fig. 4-4. Scanning electron micrograph of thin platy glass shards representing broken bubble walls of rhyolitic ash. Rim of accretionary lapillus. Pliocene Ellensburg Formation, Washington (USA).

mixtures of lapilli and blocks may be called lapilli breccias. FISHER (1966, p. 293) did not give exact percentage boundaries for the mixtures, "because they vary from perhaps as much as 40% to as little as 15%, depending upon the prejudices of the individual".

Pyroclastic rocks whose grain size is amenable to microscopic petrographic study (especially tuff) have been classified for some time according to the proportions of rock fragments, crystals or glass shards (figs. 4-2—4-4) (PIRSSON 1915). Chemical composition of the magma can be inferred roughly by determining pyrogenic mineral phases, in particular feldspars, and the refractive index of glass.

4.2 Pyroclastic rocks produced under subaerial conditions

4.21 Pyroclastic fall deposits (tephra)

4.211 Origin of tephra

Volatiles (mostly H_2O, CO_2 and S-compounds) are dissolved in a magma at deep levels. The amount of these volatiles depends mainly on the chemical composition of the melt, e. g. K-poor tholeiitic basaltic magmas contain less than 0.4%, Hawaiian tholeiites 0.3—0.6%, and some alkali-rich basaltic magmas 0.5—1% by weight of H_2O (MOORE 1970b); silicic melts may contain several per cent by weight of H_2O. When approaching and breaking through the earth's surface, the lowered confining pressure allows the gases to exsolve, and expansion of these gases causes vesiculation and rupture of the melt resulting in volcanic explosions, the energy of an explosion generally increasing with the water content of the magma and its viscosity. MCBIRNEY & MURASE (1970) distinguish 2 main factors resisting the disruptive force of gas pressure in bubbles of volcanic liquids: (1) The tensile strength of the liquid when the proportion of bubbles in a vesiculated liquid is small, and (2) the surface tension of walls of bubbles in a highly vesiculated liquid. The surface tension is strongly dependant on composition, temperature, and water content of the liquid and thus controls coalescence of bubbles.

Higher water content and viscosity are the reason why tephra of intermediate and silicic composition is much more abundant than that of mafic composition. Highly inflated glassy blebs of magma are pumice (of silicic composition) and scoria (of mafic composition). Ash and lapilli (also blocks) commonly contain or consist of pumice or scoria. Shapes of vesiculating liquid ejecta of low viscosity (basaltic magmas in general) are determined to some degree by surface tension and deformation by acceleration and air resistance. With increasing viscosity, the morphology of ash particles is determined more by the shape of the vesicles, tubular vesicles leading to elongate grain shapes, equant, undeformed vesicles to equant grain shapes (HEIKEN 1972). Individual pumice fragments, however, may be rounded in dense eruptive clouds by abrasion. Completely disintegrated froth, or finely ground pumice, can be recognized by the characteristic Y-shaped glass shards (fig. 4-3) representing bubble

Fig. 4-5. Layers of accretionary lapilli in vitric tuff of Pliocene Ellensburg Formation, Washington (USA). Cores of lapilli are dark and coarse-grained shards (fig. 4-3), white rims are fine-grained shards (fig. 4-4).

wall junctures or by crescentic or flat platy shards (fig. 4-4) representing broken bubble walls.

Accretionary lapilli (or pisolites) are ellipsoidal to spheroidal lapilli generally less than 1 cm in ϕ consisting of generally one core made of large glass shards, crystals or rock fragments surrounded by concentric envelopes of finer grained ash (fig. 4-5). Accretionary lapilli are usually formed in wet turbulent ash clouds by the accretion of small ash particles around a moist nucleus (MOORE & PECK 1962). They are characteristic of base surge deposits (see section 4.23, table 4-1).

4.212 Transport and deposition of tephra

Liquid or solid materials which are ejected from volcanic vents and transported in the air are deposited at varying distances away from the eruptive center.

Larger clasts (blocks and bombs) may be ejected along ballistic trajectories during an explosion. The distance to which they are thrown is a function of: (1) ejection velocity; (2) ejection angle; (3) mass of the projectile; (4) projected cross-sectional area of the projectile; and (5) the drag coefficient (which is in turn dependant upon the density of the atmosphere, the shape and surface roughness of the projectile and its velocity) (FUDALI & MELSON 1972). Using the most efficient ejection angle of 45°, FUDALI & MELSON (1972, Fig. 3) produced a graph showing the maximum range of a projectile as a function of size, mass, and initial velocity (fig. 4-6). Velocities of the moving gas stream in the vent and in the expanding cloud are not taken into account.

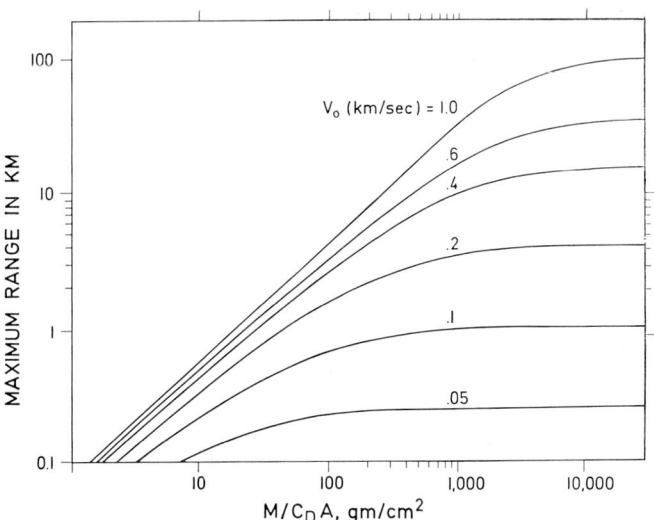

Fig. 4-6. Maximum range a spherical ejected block is thrown as a function of its mass (M), drag coefficient (C_D), projected cross-sectional area (A), and initial velocity (V_0). From FUDALI & MELSON 1972, fig. 3.

More or less vertical eruption columns, however, transport lapilli and ash to great heights. Most ash, carried to heights of up to 15 km, is deposited in the earth's gravity field or is washed back to earth by rain or snow within days after an eruption. Volcanic dust, however, projected to heights of over 50 km may form world-wide dust veils (LAMB 1971). For example, volcanic dust of the 1883 eruption of Krakatau circled the earth many times before it settled. The contribution of small volcanic particles on global atmospheric turbidity and its effect on weather and climate has recently been reviewed in detail (DEIRMENDJIAN 1973).

Widespread ash layers resulting from large eruptions make excellent stratigraphic marker horizons (figs. 4-7—4-11). THORARINSSON (1954) has initiated the modern studies of tephrochronology and there are numerous accounts on the stratigraphic distribution of Recent and fossil ash falls on land and in marine sediments, for example: WILCOX (1965), LIRER et al. (1973), KITTLEMAN (1973), NINKOVICH & HEEZEN (1965), KELLER & NINKOVICH (1972), HORN et al. (1969), HUANG et al. (1973).

Pyroclastic particles (shards, pumice, crystals, lithic fragments) have different size, density, and shape which may result in quite different terminal velocities. WALKER et al. (1971) determined experimentally and theoretically terminal velocities for different types of particles. They found that pyroclastic particles should be treated as cylinders rather than spheres in their aerodynamic behavior and produced a graph showing terminal fall velocity versus diameter, density, and shape of particles.

The basic unit in pyroclastic air fall deposits is the fall unit (NAKAMURA 1964). Many parameters of individual fall units show systematic changes as a function of

Fig. 4-7. Air fall tephra of phonolitic composition showing mantle bedding. 50 cm thick massive layer just above road is thoroughly zeolitized (fig. 4-34). Adeje, Tenerife (Canary Islands). Person for scale.

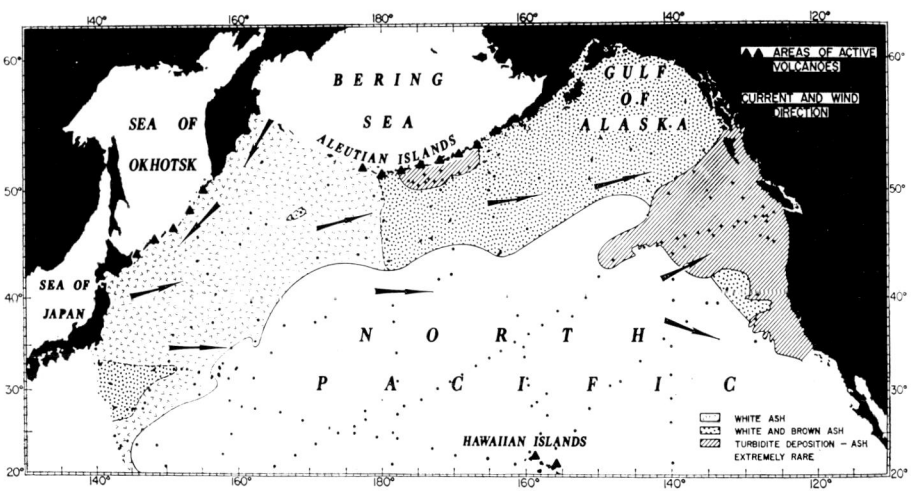

Fig. 4-8. Distribution of white and brown ash in sediments in the North Pacific. From HORN et al. 1969, fig. 1.

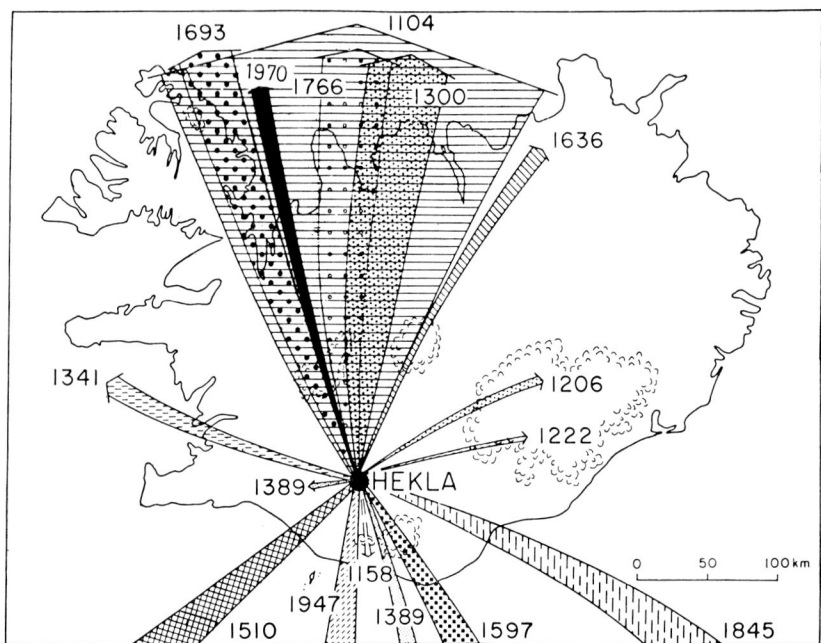

Fig. 4-9. Map of Iceland showing date, direction (arrows) and relative volume (width of arrows) of the 15 historic tephra eruptions of volcano Hekla. From THORARINSSON & SIGVALDASON 1972, fig. 1.

the distance to the eruptive center as revealed by theoretical considerations (SCHEIDEGGER & POTTER 1968) and actual measurements (see summaries by FISHER 1964 and WALKER 1971):

Thickness of units (as shown by isopach maps, figs. 4-10, 4-11), median grain size (figs. 4-12, 4-13) and diameter of the largest clasts (pumice or lithic fragments, fig. 4-14) decrease exponentially away from the vent.

The sorting coefficient (INMAN) also decreases, i. e. the sorting becomes better (FISHER 1964, WALKER 1971; fig. 4-13); and the composition may change because of eolian differentiation separating crystals, or different types of crystals, from pumice, lithic fragments, and glass shards. Isopach maps of air fall deposits commonly show a general asymmetric, elliptical distribution of ash layers with respect to the eruptive center (fig. 4-10, 4-11) because of prevailing wind direction or, less commonly, from direction of the blast. The median may not decrease when blasts are directed or winds blow in different directions at different altitudes. Most Recent ash deposits are elongate east or west of the source volcano because upper air winds at altitudes between about 5 000 — 15 000 m have pronounced E—W flow directions (EATON 1964).

Good sorting in many tephra fall deposits results in well defined bedding structures (fig. 4-7, 4-15, 4-16, 4-17) which contrasts with the massive appearance of flow deposits (figs. 4-15, 4-18, 4-20). However, tephra deposited close to the vent and

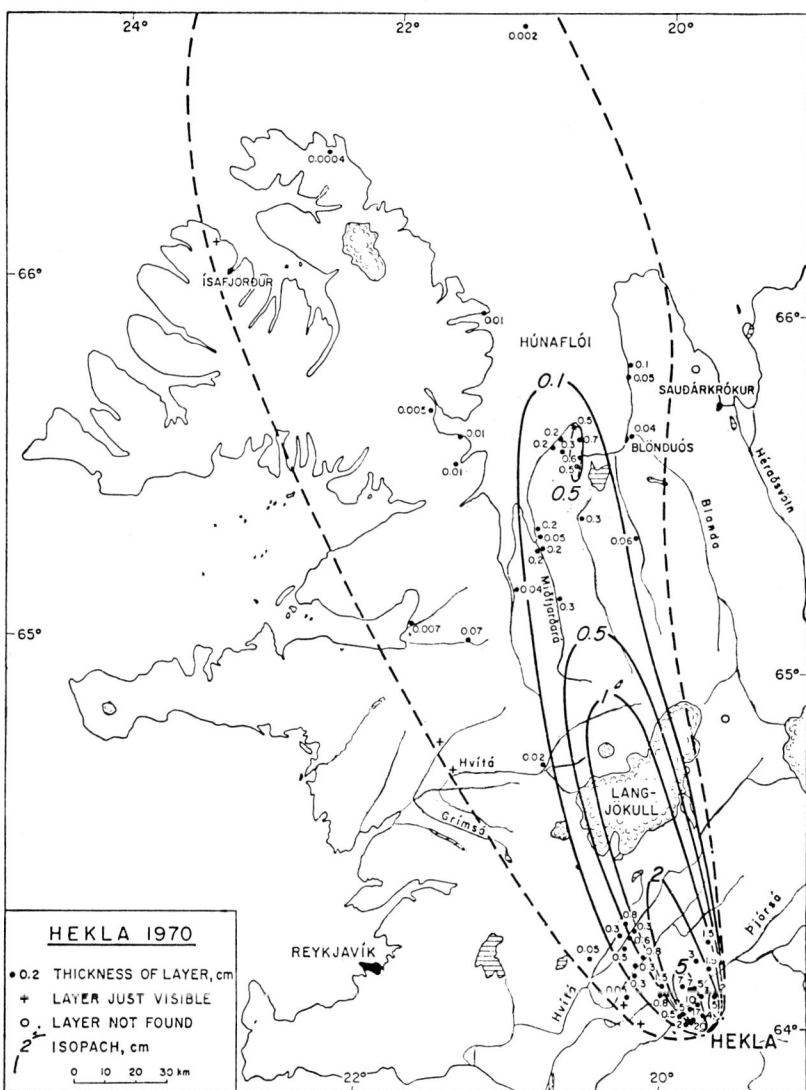

Fig. 4-10. Isopach map of tephra layer deposited during the 1970 Hekla eruption (Iceland). From THORARINSSON & SIGVALDASON 1972, fig. 5.

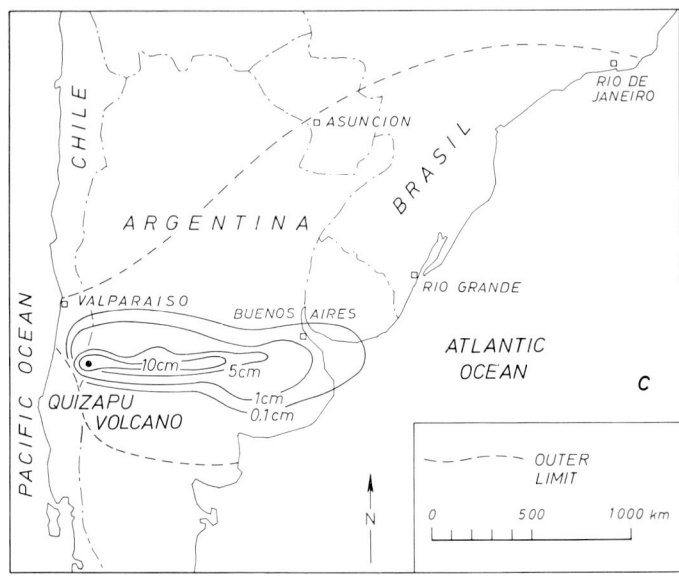

Fig. 4-11. Isopach map of tephra deposited during the eruption of Quizapu Volcano, April 1932 (from LARSSON 1937).

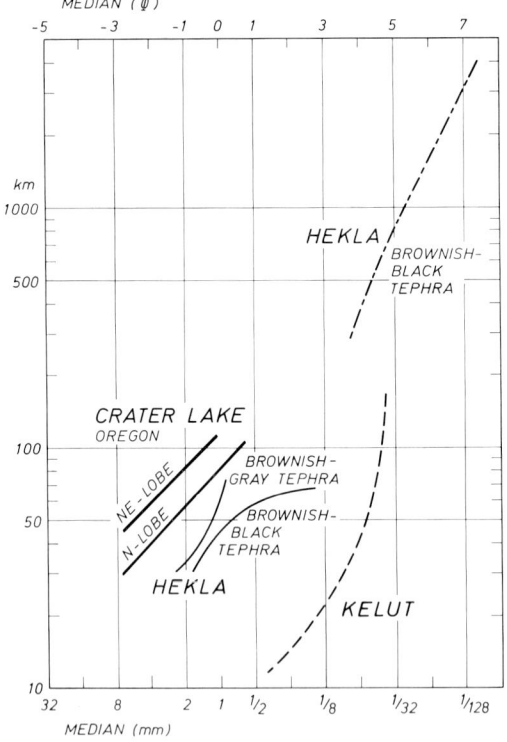

Fig. 4-12. Median diameter versus log distance for tephra eruptions from Crater Lake (Oregon, USA), Hekla (Iceland) and Kelut (Java). From FISHER 1964, fig. 7.

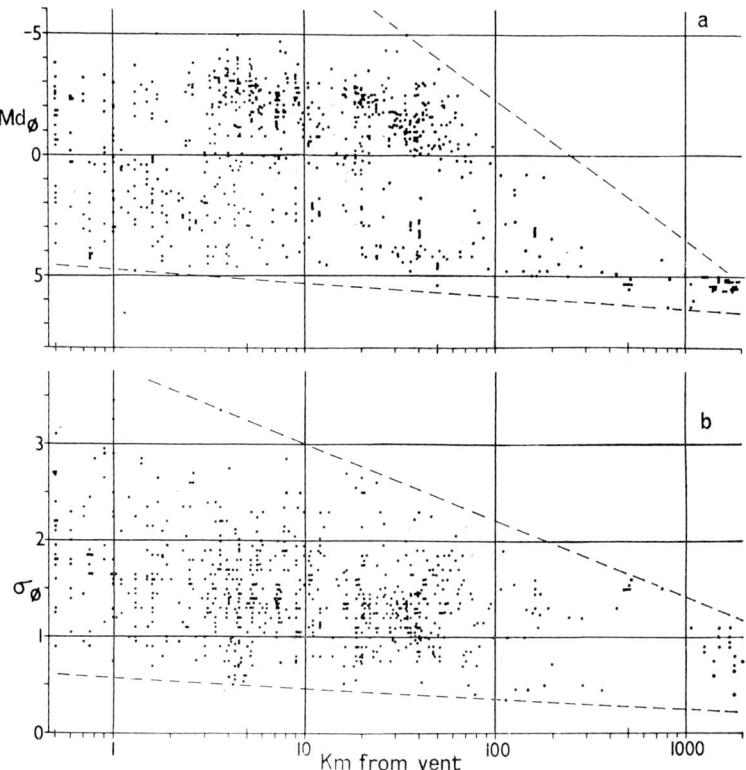

Fig. 4-13. Decrease of median (Mdϕ) and sorting ($\sigma\phi$) with distance from vent for pyroclastic fall deposits. From WALKER 1971, fig. 7.

tephra deposited from blasts continuing for some times result in rather massive beds (e. g. Plinian pumice deposits; figs. 4-15, 4-16, 4-17). Moreover, soil-mixing agents such as frost-heaving, plant-wedging and soil-living animals may completely destroy bedding structures in slowly accumulating wind-deposited volcanic ash and dust (FISHER 1966).

Basaltic subaerial pyroclastics generally do not form sheets but are restricted to cinder cones where they are produced by fire-fountaining (Strombolian or Hawaiian type of activity). The characteristic features of these deposits as contrasted with the common basaltic pyroclastics formed by steam explosions (Surtseyan) are listed in table 4-1.

The settling velocity may strongly contrast between particles within the same flow or fall unit, e. g. if pumice and crystals are deposited together. Small and dense clasts (crystals) may have the same settling velocity as large and light-weight clasts (pumice) and they will be sedimented together resulting in a poorly sorted deposit. Density differentiation is particularly marked when tephra is deposited in water: the porous pumice can float for some time until it becomes waterlogged, while the heavy

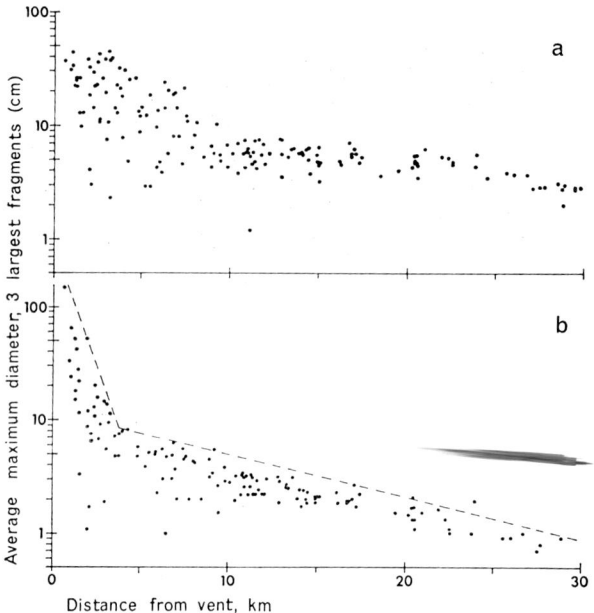

Fig. 4-14. Plot of the diameter of the three largest pumice fragments (a) and the three largest lithic fragments (b) against distance from vent for Plinian-type air fall pumice deposits of Fogo 1563 eruption (São Miguel, Azores). From WALKER & CROASDALE 1971, fig. 25).

Table 4-1. Comparison of characteristic properties of two types of basaltic pyroclastic deposits (from WALKER & CROASDALE 1972, table 1). The term achnelith is proposed by WALKER & CROASDALE for fragments whose external form is partly controlled by surface tension.

	Strombolian/Hawaiian	Surtseyan
Median diameter on or near cone	Usually 1/2 to more than 16 mm ($\varphi + 1$ to more than -4)	Usually 2 to less than 1/8 mm ($\varphi - 1$ to more than $+3$)
Deviation, σ_φ, on or near cone	Usually less than 5	Usually more than 1.5
Spatter	Common	Absent
Particle shape	Achneliths common in all grades. Bombs also common	Achneliths absent
	Often markedly elongated	Equant or nearly so
Accretionary lapilli	Absent	Common
Impact structures ('bomb sags')	Rare or absent	Common
Thickness of individual beds on or near cone	> 1 cm, commonly > 5 cm	Commonly < 1 cm, to as little as 1 mm
Alteration	Always reddened by steam oxidation near the vent	Never reddened
	Sometimes palagonitised	Palagonitised in all but the youngest examples

4.21 Pyroclastic fall deposits (tephra)

Fig. 4-15. Pyroclastic sequence at Mendig (Eifel, Germany) showing, from base to top: bedded lapilli-tuffs (LT), overlain along erosional unconformity (E) by fine-grained (dust to fine ash) pyroclastic flow (PF) deposit showing extreme thinning outside paleovalley, air fall pumice lapilli layers (AP) showing mantle bedding (for detail see fig. 4-16), deposit from wet "pyroclastic cloud" (PC) thinning to layer of even thickness away from paleovalley, and dark pyroclastic flow (lahar?) deposits (PF). About 3 km from source (Laacher See). Scale marked in 10 cm intervals.

Fig. 4-16. Plinian air fall pumice lapilli. Note absence of bedding and ash size fraction and angular shape of pumice lapilli. Detail of fig. 4-15.

Fig. 4-17. Bomb sag and coarse pumice lapilli breccias rich in dark accidental and accessory lithic fragments alternating with thin, dark deposits of "mud clouds". East of Laacher See (Eifel, Germany), about 0.5 km from vent. Scale marked in 10 cm intervals.

components (crystals, rock fragments) sink immediately. The pumice clasts may sink later together with the very fine ash and thus form the upper part of a reversely graded pyroclastic bed (fig. 4-32). FISKE (1969) has given criteria for recognizing pumice in ancient marine pyroclastic rocks. Reverse grading, e. g. in pumice deposits, can also result from increasing intensity during continuous gas blast eruptions. When pyroclastic material is deposited in water or is reworked soon after deposition it may become admixed with epiclastic debris in all proportions. The result is mixed pyroclastic-epiclastic rocks which are called tuffites in German and Russian and tuffaceous sediments in Anglo-American usage. The pyroclastic components in such mixed rocks can be recognized by the presence of glass (as shards, pumice, glassy coatings around minerals, or glassy rock fragments), euhedral outlines of crystals, or the presence of high-T minerals (e. g. sanidine) or minerals that are very unstable in sedimentary rocks (e. g. olivine).

4.22 Pyroclastic flows

Pyroclastic flows may be defined as "all fragmental flows or avalanches composed of pyroclastic material irrespective of temperature of emplacement" (SMITH in ARAMAKI & YAMASAKI 1963, p. 90). There is no generally accepted classification of pyroclastic flows (for some recent discussions see MACGREGOR 1955, SMITH 1960a,

4.22 Pyroclastic flows

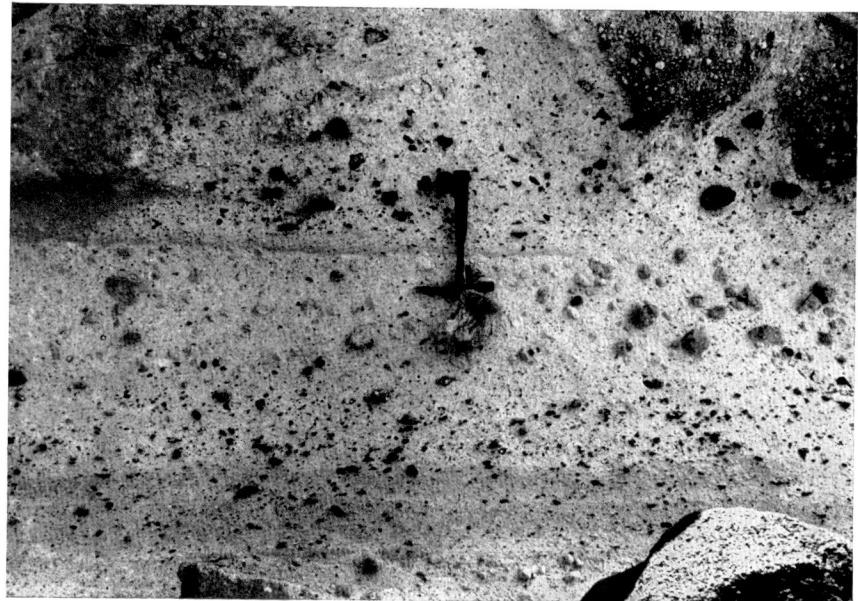

Fig. 4-18. Two flow units in Quaternary unwelded low energy pumice flow deposits, Tenerife (Canary Islands) showing pronounced normal density and reversed size grading. Dark platy dense phonolite rock fragments (partly aligned) are concentrated at the base and light large phonolitic pumice lapilli and blocks at the top of each flow unit.

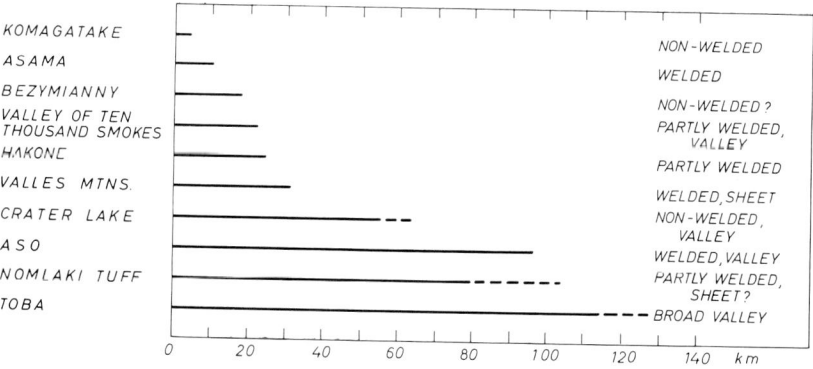

Fig. 4-19. Distance travelled by pyroclastic flows from their source (after SMITH 1960a, fig. 2).

4.2 Pyroclastic rocks produced under subaerial conditions

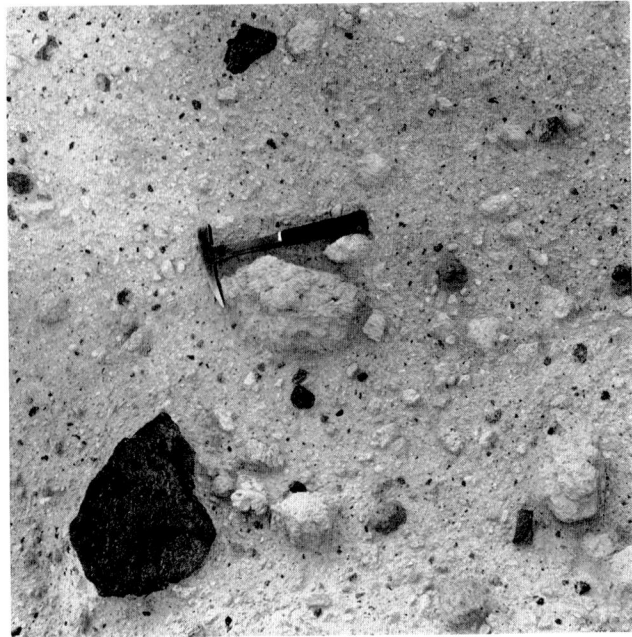

Fig. 4-20. Poorly sorted, unwelded, unbedded, high energy pumice flow deposit made up mainly of dacitic pumice and glassy dacitic rock fragments. Caldera margin (Santorin, Greece). This flow was erupted during the climax of the great prehistoric Minoan eruption that produced a wide-spread ash layer in the Eastern Mediterranean, 1500 BC (NINKOVICH & HEEZEN 1965).

Fig. 4-21. Lower part of rhyolitic welded ash flow tuff (ignimbrite). Miocene Mogan Formation (Gran Canaria, Canary Islands). The lensoid fragments are collapsed pumice lapilli and blocks. Hammer for scale.

ARAMAKI & YAMASAKI 1963, MURAI 1963). Excluding the volcanic mudflows or lahars discussed in section 4.24, three main types of pyroclastic flows (with many transitions) can be distinguished, based on different modes of eruptions and resulting differences in the deposits, the common denominator being turbulent ground transport resulting in poor sorting (fig. 4-24) and no or vague bedding of the deposit (figs. 4-18, 4-20, 4-21).

The Peléan type flows originate from the explosion or collapse of volcanic domes, commonly of andesitic to dacitic composition. The bulk of the resulting deposits is coarse-grained, monolithologic with abundant lithic fragments, unwelded, restricted to smaller topographic depressions (river valleys) and generally does not exceed 1 km^3 in volume.

The pyroclastic flow at Pelée (1902) — like many other pyroclastic flows — consisted of two quite distinct parts: the basal hot high-density avalanche, confined to valleys and transporting the bulk of the solid material and the spectacular billowing glowing low density ash cloud (nuée ardente) advancing across the terrain but carrying little ash (PERRET 1937, ROSS & SMITH 1961).

The St. Vincent type flows result from the backfall of higher eruption columns and may therefore be crystal-or pumice-enriched avalanches which may be, however, like the Peléan flows, several hundred degrees C hot, although they are not welded. They may be more siliceous in composition and therefore contain more highly vesiculated pumice. Their volume rarely exceeds 10 km^3.

By far the most voluminuous deposits (which may exceed 100 km^3) result from fissure type eruptions in contrast to the other two types which commonly originate

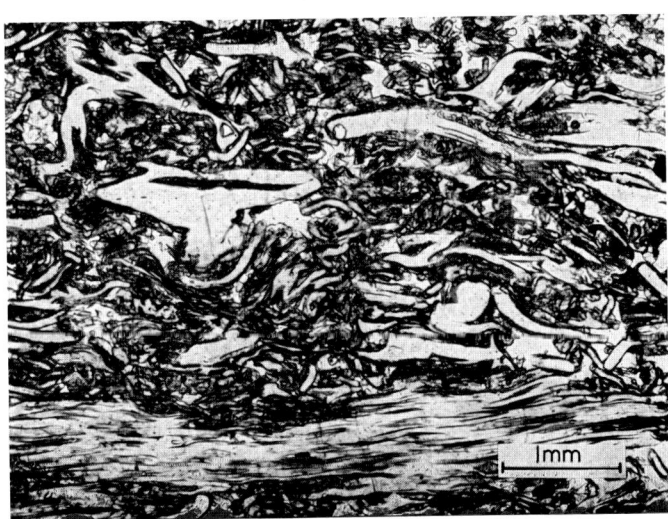

Fig. 4-22. Photomicrograph of rhyolitic (pantelleritic) welded tuff from Calla del Alca (Pantellaria, Italy) showing partly welded glass shards and collapsed pumice lapilli (base of photograph) with traces of collapsed long-tube vesicles.

on volcanic cones (fig. 4-19). SMITH (1960 a, b) and Ross & SMITH (1961) review the most pertinent details of these eruptions, their deposits, and the rocks (see also PETERSON 1970 for a review of the more recent literature). Large volumes of highly inflated and comminuted generally rhyolitic ash (fig. 4-23) or lapilli fragments are generated in low pressure effervescence or low eruption columns and travel as hot, highly mobile, inflated ash (or pyroclastic) flows. The high mobility of these flows is probably due to several factors (which are indirectly inferred because no historic fissure type eruption has been observed): (1) Expansion of gases exsolved from the magmatic components and probably also heated, entrapped air result in high kinetic energy and reduction of friction between the particles; (2) Low permeability resulting from high concentration of fine-grained (silt-size or less) particles will retard rapid loss of gases and further enable pyroclastic flows to travel for 100 km or more even on nearly horizontal ground. During the final stages of flow, the pyroclastic flow deflates and turbulent motion (fig. 4-20) may be replaced by laminar movement (fig. 4-18). Depending upon the particular combination of temperature, gas content, chemical composition and viscosity, glass shards and pumice fragments may start to become sintered or welded together during flowage but most welding occurs after the flow comes to rest (figs. 4-21, 4-22), the minimum welding temperature being above 500° C (SMITH 1960a, BOYD 1961). Pyroclastic flow deposits are commonly composed of several flow units (fig. 4-18) which are deposited rapidly one after the other and may form a simple or compound cooling unit (SMITH 1960b). Cooling units show systematic succession of zones of different degrees of welding and crystallization (SMITH 1960b). The top of a cooling unit is commonly unwelded because of rapid heat loss and little overburden and the rapidly cooled base may also show no or little welding while a zone of dense welding in which the porosity of the rock may be

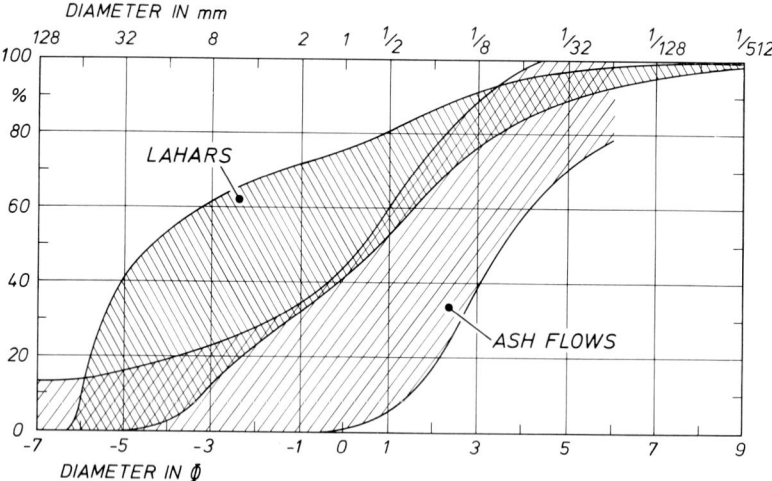

Fig. 4-23. Grain size distribution in ash flows (after SMITH 1960a, fig. 4) and Recent lahars from the Cascade Mountains (USA) (after MULLINAUX & CRANDELL 1962, fig. 2).

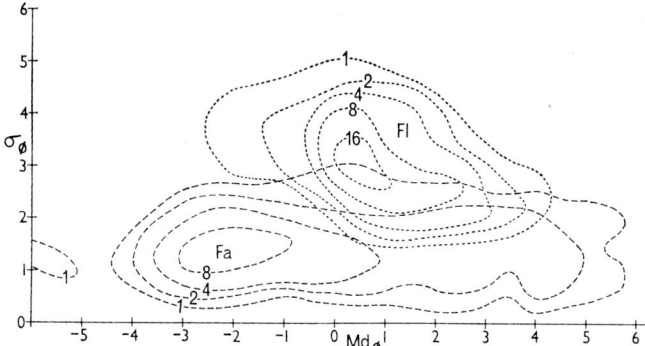

Fig. 4-24. Contoured plot of median (Mdϕ) versus sorting ($\sigma\phi$) for 300 samples of pyroclastic flow (Fl) and 1300 samples of pyroclastic fall deposits (Fa) showing poorer sorting and smaller median for the flow deposits. Figures are percentages. From WALKER 1971, fig. 2.

reduced to nil generally occurs in the lower third of a cooling unit. The welded rocks are commonly called welded tuffs and are recognized in thin section by welded glass shards and collapsed pumice lapilli (fig. 4-22).

Several cooling units form pyroclastic (or ash) flow fields which may cover areas exceeding 200 000 km^2. The term **ignimbrite** is in general use in Europe and New Zealand and is, in its most restricted sense, a synonym for the rock type welded tuff. However, because confusion arises from its loose usage to cover the deposit, flow and eruptive mechanism as well, there is a tendency to call deposits of all types of pyroclastic flows ignimbrite, whether the rocks are welded or not (e. g. WALKER 1971).

4.23 Base surges

While phreatic or phreatomagmatic — i. e. steam explosions due to rising magma coming in contact whith water — have been known for some time, the characteristics of deposits resulting from such explosions were recognized only recently (MOORE et al. 1966, MOORE 1967, FISHER & WATERS 1970, WATERS & FISHER 1971, SCHMINCKE et al. 1973). During some phreatomagmatic blasts, high-velocity steam, ash, and lapilli-laden surges spread radially away from the eruptive center. Three stages in radial dispersal have been recognized:

"(1) Surges of white steam with only a few solid particles spread immediately from the base of the emerging column as water vapor, concentrated along its periphery, escapes and condenses;

(2) black plumes of solid particles shoot radially on ballistic trajectories from the walls of the rising eruption column as it is torn apart by internal steam bursts; and

(3) the turbulent mixture of solid particles, water, and air tumbles en masse to the ground, and much of it flows away on the heels of the steam surge and incorporates it" (WATERS & FISHER 1971, p. 5596).

Fig. 4-25. Typical rim sequence of base surge and air fall basaltic pyroclastics produced by phreatomagmatic explosions during formation of maars. Juvenile sideromelane is present chiefly in the ash fraction, all larger lapilli and blocks being accidental lithic fragments. Caldera de los Marteles, Gran Canaria (Canary Islands). White measuring rod just behind hammer is 2 m long.

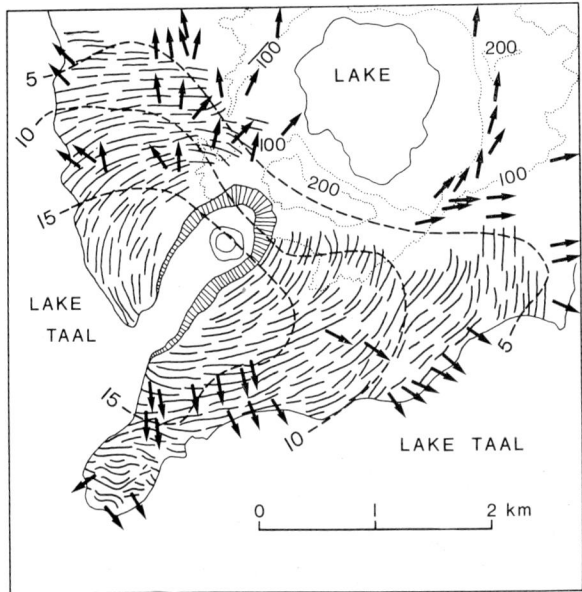

Fig. 4-26. Map of the southwest part of Volcano Island, Taal Volcano (Philippines), showing the 1965 explosion crater and the trend of dune crests (short, discontinuous lines) as mapped from aerial photographs. Heavy broken lines are wavelengths of dunes in meters. Dotted lines are topographic contours in meters. Arrows show direction of major base surge movement as measured in the field by the direction of sandblasting, tilting, and mud-coating of trees and houses. (Modified from MOORE 1967, fig. 12.)

4.23 Base surges

Fig. 4-27. Quaternary Gray Laacher See Tuffs at Laacher See, Eifel (Germany), largely of base surge origin, overlying air fall pumice lapilli (at scale) and mud cloud deposits, shown in more detail in fig. 4-15. The medium-grained pyroclastics show dune structures resembling experimentally procuced antidunes and chute and pool structures (white arrows) with steep stoss side and gentle lee slope bedding. Crests of largest dune bodies migrate upstream in successively higher beds (left side of photograph). Note thinning of coarse-grained beds at dune crests. This material is probably deposited as air fall modified by lateral base surge transport. Transport direction from left to right, slightly uphill. About 2 sm from eruptive center. Scale marked in 10 cm intervals. For more detailed discussion see SCHMINCKE et al. 1973.

Low-angle cross-bedding, antidunes, chute and pool structures, large scale undulations, poor sorting, and a mixture of juvenile debris (commonly sideromelane) — the larger fragments commonly having a dense, cracked and cauliflower-shaped outer crust —, abundant accidental rock fragments, absence of bedding sags beneath large blocks — evidence of horizontal transport — in some beds and round "agglutinate" only slightly vesicular large lapilli, many with rock fragments as cores, are characteristic features (figs. 4-25, 4-26, 4-27).

Bedding sags, vesiculated tuffs (LORENZ 1974), accretionary lapilli, and convolute laminations are evidence of the cohesive and therefore moist nature of the material during deposition. Most maars in the world are now thought to be due to steam explosions because structures characteristic of base surge deposits have been recognized in many horizontal rim sequences surrounding maars (fig. 4-25).

Bedding structures and grain size characteristics of phreatomagmatic base surge deposits were also found in deposits from hot low-density high velocity nuée ardentes and the term "ground surge deposit" was proposed as a general name with base surge being a type of ground surge (SPARKS & WALKER 1973). At present, it seems preferable to extend the well-established term base surge to non-phreatomagmatic deposits in which magmatic gases and/or entrapped air — and not external-water vapor — are the main fluidizing phase. Many more "base surge type"-deposits will probably be found in the near future and revision of classification and nomenclature is likely when more sedimentological parameters have been determined and transport mechanisms more precisely analyzed.

There is yet another type of deposit, particularly associated with phreatomagmatic eruptions but also with Plinian pumice deposits which shows mantle bedding over mountains but thickens in valleys (figs. 4-15, 4-17). These are fine-grained massive rocks showing evidence of having been moist during deposition. They could have been deposited from (wet) ash clouds that were inflated enough to cover high elevations with a thin veil but close to the ground to converge toward the valleys where most load is dropped. They are low-energy deposits in contrast to those of base surges or many types of pyroclastic flows.

4.24 Lahars

Pyroclastic flows with water as transporting agent are called volcanic mudflows or lahars (figs. 4-28, 4-29, 4-30). Lahars may form by a number of processes (ANDERSON 1933): ash mantling the slopes of a volcano may become mobilized by heavy rains and flow downslope as mudflows (cold lahars); during eruption through crater lakes, tephra may become mixed with water and spill over the crater lip as hot lahars; when hot pyroclastic flows enter rivers or lakes they may continue as lahars.

Lahar deposits show extremely poor sorting (figs. 4-28, 4-29) although the rounding of rock fragments (several rock types may be present; pumice is rare) is usually better than in pyroclastic flow deposits. Lahar deposits are of large areal extent and their distal parts are commonly interbedded with fluvial deposits. Lahars are known from Indonesian (VAN BEMMELEN 1949) as well as from numerous other Recent and fossil volcanic areas, in particular from andesitic volcanoes (e. g. Western USA: ANDERSON 1933, MULLINEAUX & CRANDELL 1962, SCHMINCKE 1967a).

4.25 Volcanic breccias

Fig. 4-28. Lahar of andesitic-dacitic composition overlying conglomerate. Ellensburg Formation, Washington (USA). Note poor sorting, absence of bedding and fine-grained basal layer. Hammer for scale (white arrow). Compare with fig. 4-29.

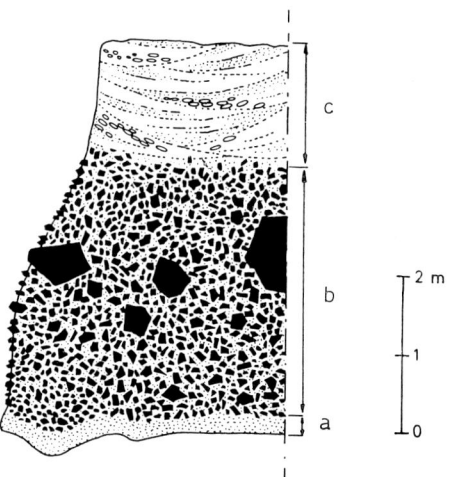

Fig. 4-29. Schematic cross-section through typical Ellensburg lahar, showing three-fold subdivision into a) fine-grained base, b) poorly sorted main part, and c) cross-bedded top containing well rounded pumice pebbles. From SCHMINCKE 1967a, fig. 9.

4.25 Volcanic breccias

Volcanic breccias are rocks composed of angular volcanic fragments greater than 2 mm in size. The brecciation and/or emplacement is the result of volcanic, tectonic or erosional processes (WRIGHT & BOWES 1963, p. 83, FISHER 1966, PARSONS 1969). Volcanic breccias can be further subdivided into autoclastic breccias (including those formed by friction such as flow breccias), alloclastic breccias (including intrusion-

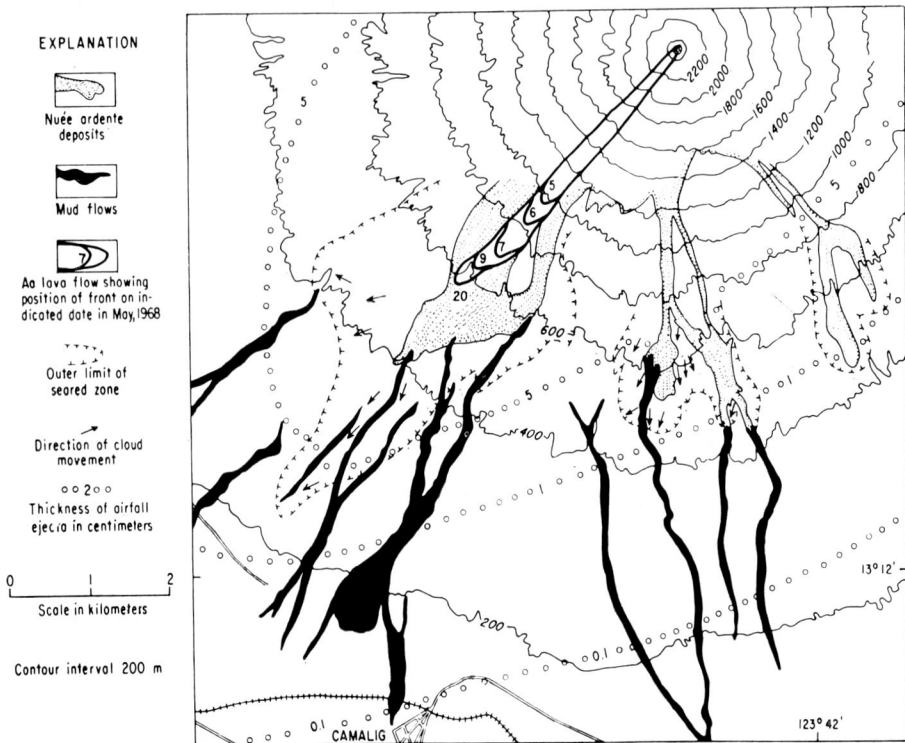

Fig. 4-30. Lahars (mudflows) formed at the termini of nuée ardente deposits by torrential rains washing down loose air fall ash and unconsolidated material from the top of nuée ardente deposits of the 1968 eruption of Mayon Volcano, Philippines. From MOORE & MELSON 1969, fig. 2.

breccias, explosion breccias, and intrusive breccias), and pyroclastic breccias. The alloclastic breccias, which include the complex vent breccias, will not be discussed in this chapter (see CLOOS 1941, GATES 1959, JOHNSTON & LOWELL 1961, WRIGHT & BOWES 1963, PAPENFUSS 1963, FISHER 1966).

Most lava flows, with the exception of basaltic pahoehoe lava, become brecciated during flow to some degree as a function of their viscosity. Because viscosity is mainly dependent on temperature, gas content, and chemical composition, brecciation is most pronounced in highly silicic flows, especially in the parts of the flow which have travelled the longest distance. The cooler, more viscous upper layer of a silicic lava flow breaks during movement. Individual blocks degas and may become pumiceous. The pumiceous or scoriaceous top breccia thus formed may tumble down the front of the flow and be overridden by its still molten central portion. Thus, even flows which are incompletely brecciated may have top and basal breccias. Many autobrecciated lava flows may be recognized by the presence of an unbrecciated central part. Completely brecciated flows consist of dense to vesicular, angular to subrounded blocks, and may have a matrix made of finely comminuted material. Monolithologic composition, poor sorting, absence of bedding, and paucity of fine ash are generally characteristic. The feeder dikes themselves may be highly brecciated and may have erupted breccia flows. Brecciated flows and dikes are particularly common on andesite volcanoes (e. g. CURTIS 1953, FISHER 1960, FISKE et al. 1963, SWANSON 1966, PARSONS 1969).

4.26 Peperites (lava-sediment mixtures)

When lava intrudes (from below) or flows (from above) into soft unconsolidated pyroclastic or epiclastic sediments it may shatter and form intimate mixtures of — commonly chilled — lava fragments and sediment (commonly baked). Such mixtures are called peperites (LACROIX & BLONDEL 1927, SCHMINCKE 1967b). The origin of rocks at the type locality of peperite at Gergovia in France, however, is still a matter of debate. GLANGEAUD (1957) and JONES (1969b) cite abundant evidence that the peperites are not intrusive (as formerly thought: MICHEL 1953) but an alternation of waterlaid pyroclastics and marly sediments. Peperites are very common on the sea floor where basaltic lava is erupted or intruded into soft, water-rich sediments. Peperites, hyaloclastites, and pillow lavas may form a close association in this environment. Peperites formed at the soft sediment — basaltic basement interface make it commonly difficult to decide if true basement is reached or a sill intruded much later entirely into sediments.

4.3 Pyroclastic rocks produced under subaqueous conditions

Pyroclastic rocks produced in underwater eruptions are widespread, particularly in submarine deposits, although they have been little studied as yet.

4.31 Hyaloclastites and subaqueous tuffs

Hyaloclastites (RITTMANN 1962) or aquagene tuffs (CARLISLE 1963) are volcanic clastic rocks — generally of basaltic composition — formed by the quenching and fragmentation of lava coming into contact with water. The term hyaloclastite is used here in a broad sense to include pillow rind breccias, granulated lava, and the voluminous pyroclastic rocks produced at shallow water depth or under surface conditions by steam explosions and accompanying expansion of magmatic gases. Hyaloclastites may form when lava flows into rivers or lakes (FULLER 1931) or the sea (FISHER 1968, MOORE et al. 1973), when lava is erupted subglacially (SAEMUNDSSON 1967, JONES 1969a, 1970) or by eruptions in sea water (CARLISLE 1963, McBIRNEY 1963, PICCOLI 1966, MOORE & FISKE 1969).

Theoretically, processes of fragmentation can be subdivided into those occurring below and above the critical confining pressures at which the melt becomes saturated in magmatic gases. At higher pressures (greater water depths, 500 m for K-poor tholeiites, 1800 m for alkali basalt, MOORE 1970b) fragmentation is due to rapid cooling of the melt that causes stresses which are released by shattering of the glass (thermal shock) resulting in essentially non-vesicular fragments. In addition, rolling of pillows down steep inclines, slumping of pillow deposits, and spalling off of pillow rinds lead to the formation of coarse-grained pillow breccias. These processes may also occur at shallow water depths but at these low confining pressures a combination of expansion of magmatic gases and repeated steam explosions leads to a comminution of melt into hyaloclastites composed of more or less vesicular small fragments (figs. 4-31, 4-35). Thus, oceanic — and similarly subglacial — volcanoes commonly show a characteristic vertical succession of different rock types as a function of water depth (confining pressure): Pillows and pillow fragments in the lower and central part, overlain and marginally grading into vaguely bedded tuffs, and topped at water level by well stratified and crossbedded, and in part fine-grained hyaloclastite tuffs.

Steam explosions are generally called phreatic (resp. phreatomagmatic) when the evaporated water is ground-(phreatic)water. The term phreatic eruption is increasingly used for steam explosions in sea water. WALKER & CROASDALE (1972, p. 304) propose the name "Surtseyan" for the type of basaltic pyroclastic activity exemplified by the explosive opening stages of the Surtsey eruption, a basaltic volcanic island formed in the Atlantic Ocean south of Iceland in 1964. Characteristic features of Surtseyan are compared with Strombolian deposits in table 4-1 and fig. 4-31.

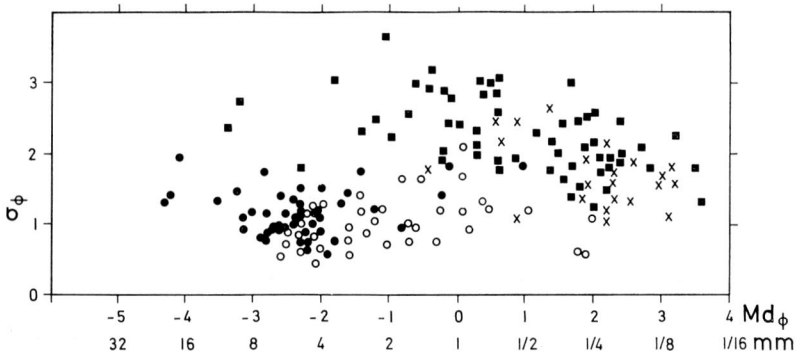

Fig. 4-31. Sorting (INMAN) versus median of Surtseyan (squares and crosses) and Strombolian (circles) basaltic ashes. Dots and sqares represent near vent samples from ash ring or cinder cone. Crosses and circles represent samples taken from further from the vent. For discussion see table 1. Modified after WALKER & CROASDALE 1972, fig. 2.

4.32 Subaqueous pyroclastic flows

Subaqueous pyroclastic flows consist ideally of freshly erupted and quenched debris that travel downslope into deeper water in turbulent, low-temperature flows. Numerous deposits formed by flows of this type have been found in Japan and the USA in recent years (FISKE 1963, FISKE & MATSUDA 1964).

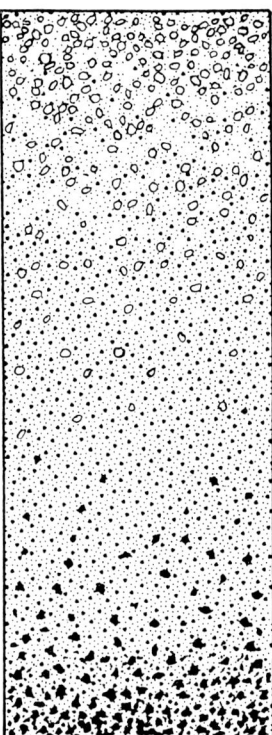

Fig. 4-32. Schematic section of graded submarine pumice flow deposit. The grading is defined by upward increase in pumice clasts. From FISKE 1969, fig. 1.

FISKE & MATSUDA (1964, p. 95) describe the formation of dacitic pumice flows of the Tokiwo Formation (Micocene) in Japan as follows (fig. 4-33):

"A. Beginning of eruption. Vesiculating dacite magma is erupted into cold sea water. Submarine eruption column starts to boil up above the vent.

B. Climax of eruption. Large volumes of dacite magma fountain above the vent, and the submarine eruption column carries much debris high into suspension. The eruption column may have burst through the sea surface into the air. Intense sorting splits the debris into various fractions. Buoyant pumice floats; dense dacite fragments, large crystals, and compact pumice lapilli settle around the vent and slough laterally in a subaqueous pyroclastic flow; most ash remains in suspension.

C. End of eruption. Amount of debris falling from submarine eruption column gradually decreases and is insufficient to maintain a steady pyroclastic flow. The flow is therefore replaced by repeated turbidity currents (shown here following close behind each other). Because the volume of material raining downward around the vent gradually decreases and becomes finer-grained and less dense, the later turbidity currents become more infrequent, and they carry finer and less dense ash. Much silt-sized ash remains in suspension and is dispersed by slowly moving currents."

Deposits of such subaqueous pyroclastic flows are commonly better sorted than subaerial pyroclastic flows. Graded bedding is very characteristic (fig. 4-32).

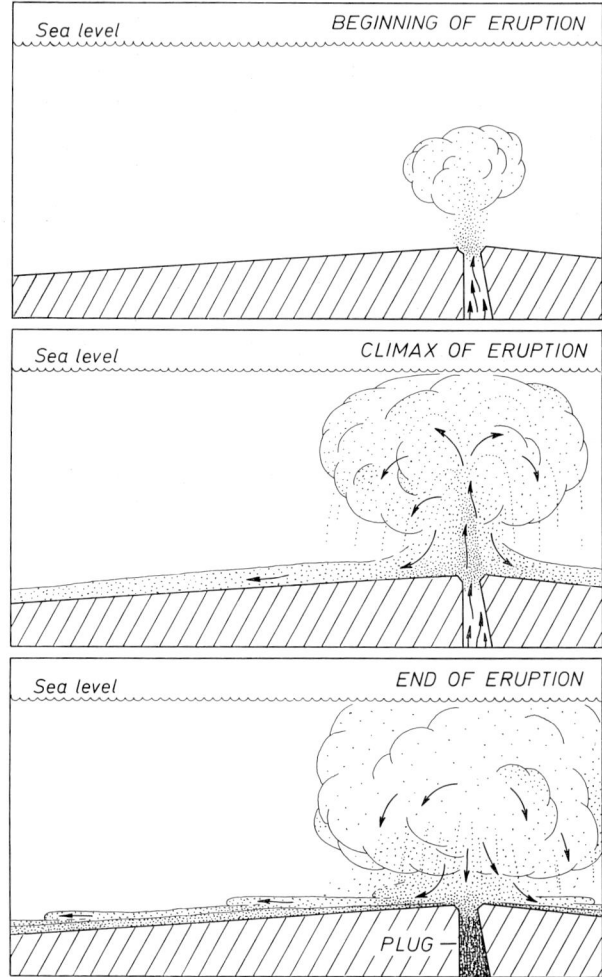

Fig. 4-33. Schematic representation of 3 stages of submarine eruption of Wadeira Tuff O, Japan. From FISKE & MATSUDA 1964, fig. 6.

4.4 Alteration of volcanic glass

Volcanic glass is unstable in nearly all conditions on and within the upper crust. It is altered during weathering and diagenesis to a number of minerals, commonly zeolites (fig. 4-34), clay minerals, particularly montmorillonite which may form entire deposits — bentonites — and silica polymorphs. Rhyolitic volcanic glass (obsidian) which contains a few tenths of a percent of H_2O after emplacement may take up several percent by weight of H_2O from its environment (ROSS & SMITH 1955)

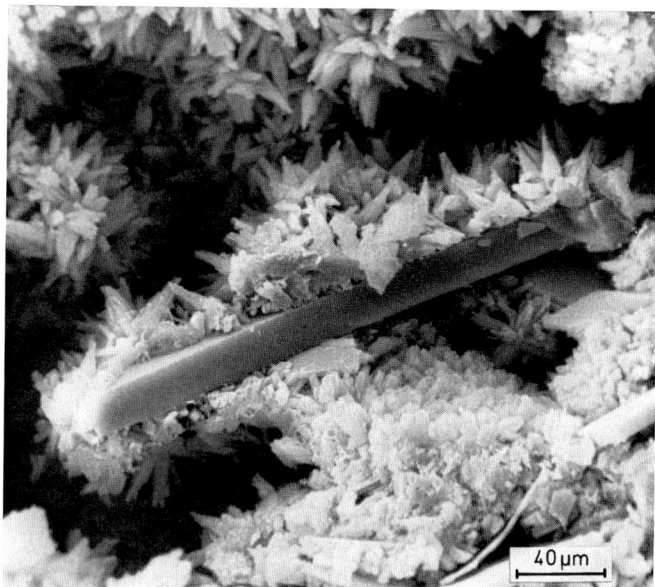

Fig. 4-34. Scanning electron micrograph of strongly zeolitized vitric accretionary lapilli tuff of phonolitic composition (philippsite crystals on feldspar fragment). Adeje, Tenerife (Canary Islands).

and becomes hydrated to perlite. A number of volatiles can be expelled from the hydrated glass and appreciable amounts of certain elements (particularly Na) may be lost and gained during leaching of the glass by ion exchange with the groundwater (Ross & Smith 1955, Lipman 1965, Noble 1967). Friedman & Smith (1960) and Friedman et al. (1966) demonstrated that hydration is a time and temperature dependent process:

$$x^2 = k\,t$$

where

x = depth of penetration of water in microns
k = constant for a given temperature
t = time in years since the beginning of the diffusion process.

They used this relationship to date obsidian artifacts, the hydration rates ranging from $0.4\,\mu^2/10^3\,a$ at $1°C$ to $11\,\mu^2/10^3\,a$ at $30°C$. The thickness of the hydrated layer is mainly dependent upon the temperature, humidity, chemical composition of the obsidian and erosion of the hydrated layer.

Hay (1966), Sheppard (1971) and Utada (1971) review the occurrence and origin of zeolites in pyroclastic rocks. The six most common zeolites are analcime, chabazite, clinoptilolite, erionite, mordenite, and philippsite. Silicic vitric material reacts with pore waters by a solution-precipitation mechanism, the formation of zeolites being favored by high pH and high activities of alkali ions in the interstitial water. Analcime, however, may form during later diagenesis from alkalic, silicic zeolite precursors. Volcanic glass of basaltic composition (sideromelane) is altered rapidly during weathering and diagenesis, its common alteration product being an amorphous yellow substance called palagonite (fig. 4-35). The palagonitization during submarine weathering seems to follow the above equation and seems to be strongly dependent on temperature. MOORE (1966) found that palagonite rinds on pillow basalts at water depths larger than 1000 m (mean temperature less than 5° C) formed at rates of 480—2000 μ^2 while at 30 m water depth (mean T 25° C) the rate of palagonitization is 2400—5400 μ^2 / 1000 years (MOORE 1970a). MOORE (1966) and HAY & IIJIMA (1968) also demonstrated large chemical differences between sideromelane and palagonite. Microprobe studies by HAY & IIJIMA (1968) of subaerial palagonite tuffs showed that during the alteration of sideromelane into an equal volume of palagonite, one third of the SiO_2, half of the Al_2O_3 and 3/4 or more of CaO, Na_2O and K_2O were lost and deposited, from first to last, as: phillipsite, chabazite, thomsonite, gonnardite, natrolite, analcime, and montmorillonite, together with opal and calcite cement. Palagonite forms more rapidly in the zone of percolating water than below the water table, and forms in a particular chemical environment over a relatively short period of time. Temperature, rainfall, grain size, permeability, and original composition are important factors in determining the degree to which sideromelane is altered to palagonite and the composition of the authigenic mineral assemblage which is formed in the pore spaces between the pyroclastic fragments. MORGEN-

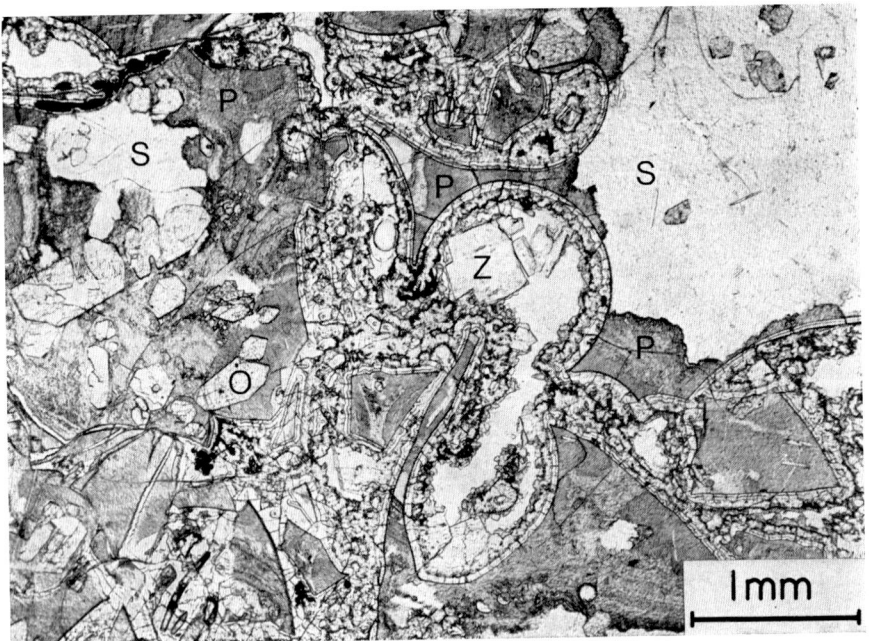

Fig. 4-35. Palagonitized sideromelane tuff (Labrador). S = sideromelane, P = palagonite, O = olivine, Z = zeolite (chabazite). Specimen courtesy B. G. J. UPTON.

STEIN (1972) showed that alteration of sideromelane in the marine environment starts with hydration along syngenetic cracks in the sideromelane. This hydration leads to volume expansion resulting in stresses and diagenetic fracturing and a geometric increase in glass failure with time. During the alteration of sideromelane to palagonite, Na, Mg, Si, Mn, and Ca are lost while K, Ti, Fe, and O are added. According to MORGENSTEIN (1972), potassium behaves quite differently in terrigeneous and marine palagonites. In the latter, potassium temporarily enters the palagonite from sea water during alteration but is released during subsequent solution of palagonite and enters authigenic clays and zeolites, e. g. K-rich philippsite. HONNOREZ (1972) discusses the products of palagonitization in rocks at Palagonia (Sicily) in detail.

5. Carbonate Rocks[1]

5.1 The primary sediments and their origin

Calcareous sediments consist primarily of calcite and aragonite. The mineral species formed in inorganic sediments depends primarily upon the water chemistry whereas in biogenic sediments it also depends on the type of lime-secreting organisms.

5.11 Calcilutites

Definition

GRABAU (1913) defined calcilutites as limestones whose fabric cannot be resolved either by the naked eye or by touching it. Such rocks presumably were deposited as aragonite or calcite muds and underwent no major crystal enlargement. This definition therefore mingles primary with diagenetic characteristics. Perhaps one should use "calcilutite" and related terms primarily to define constituent particles rather than crystal size:

calcirudite = primary particle diameter: larger than   2000 μ,
calcarenite = primary particle diameter: 2000—63 μ,
calcisiltite = primary particle diameter: 63— 2 μ,
calcilutite = primary particle diameter: less than   2 μ.

Many workers combine the terms calcisiltite and calcilutite, and refer to calcilutite as being finer than 63 μ, because of the difficulty to determine primary particle diameters.

Since present crystal size also reflects the diagenetic history, it is not considered as an important criterion in the above nomenclature of limestones. Limestones with crystal sizes of 10 to 20 microns (microsparites or pseudosparites, FOLK 1965b) may be termed calcilutites as well as those with crystals less than 4 microns (micrites, FOLK 1959, 1965b; calcite mudstones of BATHURST 1959a), if they contain no microscopic structures other than crystal outlines. It must be remembered, however, that primary sedimentary textures, such as shell fragments, may have been obliterated by diagenesis.

---

[1] For a critical review of the German edition, my thanks go to Prof. G. M. FRIEDMAN (Troy, N. Y.), for the improvement of ch. 5.15 I am indebted to Prof. E. FLÜGEL (Erlangen). Dr. J. D. MILLIMAN (Woods Hole) improved the English edition considerably with many valuable suggestions.

For convenience, the following descriptive terms are suggested for homogenous limestones:

macrocrystalline limestone (crystal size $> 100\,\mu$)
microcrystalline limestone (crystal size $10-100\,\mu$)
cryptocrystalline limestone (crystal size $< 10\,\mu$);

or, if inspected macroscopically,

coarse crystalline, if crystals are recognized with the naked eye,
fine crystalline, if crystals are not recognizable with the naked eye.

FRIEDMAN (1965b) suggested the following terms:

micron-sized (crystals smaller than 10 microns in diameter)
decimicron-sized (crystals 10—100 microns)
centimicron-sized (crystals 100—1000 microns)
millimeter-sized (crystals 1—10 mm)
centimeter-sized (crystals 1—10 cm)

A frequent mode of the crystal size of calcilutites is $2\,\mu$ (BATHURST 1971b).

Alternative terms which are frequently used have been introduced by FOLK (1962, 1965b):

micrite = cryptocrystalline calcilutite $< 4\,\mu$,

microspar = a micrite which underwent crystal enlargement to form a limestone with a crystal size of 4-10(-50) microns,

pseudospar = as microspar, but composed of crystals larger than 10(-50) microns.

Whereas the use of micrite is recommended, microspar and pseudospar are somewhat confusing, because "spar" is generally used in connections such as "biosparite" corresponding to a limestone composed of skeletal grains and cement.

Origin

Three modes of formation of calcilutites can be distinguished, of which in many cases more than one may be operative.

(1) precipitation from supersaturated solutions.
(2) disintegration of skeletons (biocalcilutite and -siltite).
(3) supply of detrital-terrigeneous lime mud in suspension.

1. Precipitation from supersaturated solutions

Precipitation of lime may be caused by warming of the water. Precipitation is also facilitated by carbon dioxide uptake through photosynthesis and by microbial activities in the reduction of sulfur and nitrogen-bearing compounds (DREW 1914, CORRENS 1939, LALOU 1957, CLOUD 1962b). In addition, an abundant supply of suspended nuclei is required. Many of these conditions are realized in warm shallow seas which frequently are supersaturated with $CaCO_3$. MÜLLER, IRION & FÖRSTNER (1972) suggested the following modes of formation of primary inorganic carbonates in lakes; (1) loss or extraction of CO_2 as a result of change in p-t-conditions (Plitvickih lakes, Yugoslavia) or plant assimilation (Ob-i-Istada, Afghanistan; Lake Balaton, Hungary), (2) evaporation concentration (Tuz lake, Turkey), (3) mixing of different water bodies (Van Lake, Turkey). The physicochemical conditions of the precipitation of $CaCO_3$ in general will be discussed in Part III of this work.

Whether calcite, magnesian calcite, or aragonite precipitates, depends partly on the molal proportion Mg/Ca within the solution, as shown by numerous experiments (LEITMEIER 1910, MURRAY 1954, LIPPMANN 1960): If Mg/Ca $<$ 1, calcite is formed. This is valid for most fresh water lakes and rivers (a mean average of Mg/Ca is 0.275, according to CLARKE 1924, RANKAMA & SAHAMA 1950, and PEARSE & GUNTER 1957). At Mg/Ca $>$ 1 (in sea water: 5.26), aragonite is precipitated if the solubility product is surpassed. This condition is common in most shallow marine environments. The following explanation is suggested by LIPPMANN (1960): Magnesium ions are preferentially incorporated in the calcite lattice. They are, however, protected by hydrate envelopes, which tend to hinder the precipitation of calcite in Mg-rich environments (LIPPMANN 1960). On the other hand, the crystallization of aragonite can be suppressed by adsorbed organic anions such as the citrate, the pyruvate and the malate of sodium (KITANO & KANAMORI 1966, see also KITANO & HOOD 1965).

If calcite precipitates from a solution rich in Mg, magnesian calcite should form, although it is more soluble than aragonite. Thus, aragonite forms preferably (WEYL, in BATHURST 1971b). On the other hand, Mg-calcite requires lower supersaturation to grow than aragonite, whereas the growth rate of Mg-calcite is much smaller than of aragonite (LIPPMANN 1973). This discussion shows that the question of precipitation of Mg-calcite or aragonite is complicated and far from being resolved.

Natural occurrences in which either aragonite or calcite are precipitated are described in the following paragraphs.

A. Aragonite

One of the most striking examples of inorganic precipitation in highly saline environments is the Dead Sea (Israel): turbid clouds of aragonite suspended in the water (commonly called "whitings") form every year during the hottest periods. Aragonite deposits form light layers alternating with dark layers of calcite, which are interpreted as products of bacterial decomposition of gypsum (NEEV 1963, FRIEDMAN 1965c). In the Red Sea, which is characterized by warm bottom water (22° C) with elevated salinity (40.5—41‰; MILLIMAN & MÜLLER 1973), alternations of inorganically precipitated layers of aragonite and Mg-calcite mud occur (GEVIRTZ & FRIEDMAN 1966, MILLIMAN et al. 1969).

During evaporation of Tuz Gölü (Turkey), aragonite and magnesian calcite are precipitated (IRION 1970). In the Van Lake (Turkey) whitings due to aragonite precipitation form in front of the river entrances; the pH in this alkaline lake ranges from 9.5—10 (IRION, pers. commun.).

Similar "whitings" are reported from the Coorong of Australia (VON DER BORCH 1965) and from the Persian Gulf (KINSMAN 1969). They are composed of suspended aragonite and generally are confined to surface layers of water (WELLS & ILLING 1964). Precipitation has been suggested to be inorganic in this area since the strontium content of the sediments is significantly higher than in the skeletons of aragonite producing organisms living in the area (KINSMAN 1964, 1969). The origin of whitings, however, is still under discussion.

Bacteria also may precipitate carbonate; aragonite-encrusted bacteria which are found frequently in Florida Bay sediments enrich their cell membranes considerably in calcium compared with sea water (GREENFIELD 1963).

Aragonite also can form in hot springs (e. g. Karlovy Vary, ČSSR), depending on the chemistry of the solutions.

In Lake Balaton (Hungary), aragonite encrusts the leaves of *Potamogeton* whereas at the same time magnesian calcite is precipitated by large phytoplankton blooms (G. MÜLLER 1970 a, and in press).

In the shallow sea, photosynthesis e. g. by seaweed may play an important role in the precipitation of $CaCO_3$, but in general no inorganic precipitation occurs. Instead, epibiont (e. g. coralline algae, foraminifera) growth on the leaves of *Thalassia* may use the supersaturation caused by the CO_2-uptake of the leaves. A carbonate mud production (Mg-calcite and ? aragonite) by epibionts of 2800 g/m^2-year was estimated by PATRIQUIN (1972) at Barbados.

Moreover, algae and sea grasses are lime traps (fig. 5-1; GINSBURG & LOWENSTAM 1958, LYNTS 1966).

Fig. 5-1. Lime-trapping sea grass. *Thalassia* banks adjacent to a tidal delta, Florida Keys. Aerial photograph, view toward the North.

B. Calcite and magnesian calcite

Whereas in lakes the Mg/Ca mole ratio determines the occurrence of calcite ($<$ 2), Mg-calcite (2-12), and aragonite ($>$ 12) (MÜLLER, IRION & FÖRSTNER 1972), the question is still open for marine sediments, though Mg-calcite is preferably found in less concentrated water than aragonite.

(Low-magnesium) calcite lutites smelling of H_2S have been found in late Pleistocene bottom sediments of the Red Sea by FRIEDMAN (1972). He suggested a bacterial origin by reduction of solid or diluted sulfates for these lutites as well as for many "evaporitic" gray-colored calcilutites in the geologic record. The vast majority of modern calcarous deep sea sediments, however, are biogeneous, mostly planktonic foraminifera (LISITSYN & PETELIN 1967).

Calcite and magnesian calcite (10—12 mole percent $MgCO_3$) are the main carbonates in eastern Mediterranean deep-sea sediments (J. MÜLLER & MILLIMAN 1971). "The calcite is derived mainly from coccoliths and planktonic foraminifera, while the magnesian calcite is represented by discrete crystals and crystal clusters. Most magnesian calcite is concentrated in the lutite fraction, but also can serve as the cement and main component in lithic layers and fragments." Layers rich in Mg-calcite alternate with low-magnesian calcite layers in the eastern Mediterranean, which is intermediate with respect to the temperature (13.7° C) and salinity of the bottom water (38.4—38.6 ‰) between the Red Sea [see above (A)] and the normal ocean water (0.5—2.5° C; 34—36 ‰), in which only low-magnesium calcite is present. With increasing sediment depth, however, the Mg-content of the Mg-calcite decreases whereas Mg in the pore water increases, due to diagenetical exsolution of Mg from the Mg-calcite (MILLIMAN & MÜLLER 1973).

In marine environments, magnesian calcite seems to be restricted mostly to skeletal material and to early diagenetic cements (MILLIMAN 1971).

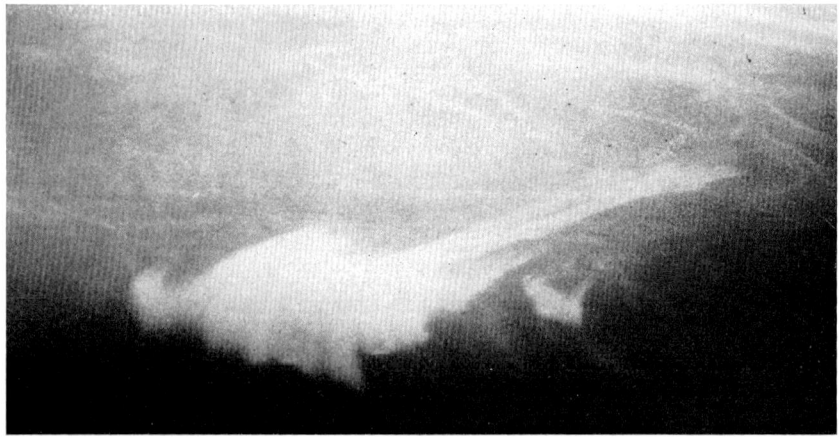

Fig. 5-2. Whiting (suspended aragonite) in the shallow sea, western part of Florida Bay; about $1/2$ kilometer long (Aerial photograph). A direct precipitation is not proven.

Calcite precipitation in fresh water is reported from Lake Constance (north of Reichenau, MÜLLER 1966a), from Lake Neuchâtel (Switzerland, KÜBLER 1962, pp. 282, 286), and from a freshwater lake on Andros Island (Bahamas, NEWELL & RIGBY 1957). Calcite is formed also as a terrestrial surface crust in dry climates (caliche), in springs (tufa) and in caves (dripstones).

Magnesian calcite is precipitated in the Lake Balaton (Hungary; MÜLLER 1970a) and in the coastal lakes of South Australia (ALDERMAN & SKINNER 1957) (see ch. 5.331). Mg_{20}-Calcite (20 mole percent $MgCO_3$) has been found by KÜBLER (1958) in an Upper Tertiary cryptocrystalline lacustrine limestone. Since it is improbable that this unstable mineral formed diagenetically, inorganic precipitation is assumed.

2. Disintegration of skeletons

The most important calcilutite-forming process at present is the disintegration of calcareous skeletons (STIEGLITZ 1972). One possible reason is discussed at the end of ch. 5.151, p. 211. In older periods, however, the diversity of carbonate-secreting organisms which extracted $CaCO_3$ from sea-water was lower, and thus the probability of inorganic carbonate precipitation was higher (STRAKHOV 1969). For instance in formations such as the Triassic "Hauptdolomit" (Calcareous Alps) which were deposited in a restricted environment and therefore are nearly barren of fossils, inorganic or algal-supported precipitation of lime mud may have occurred.

Disintegration is accomplished by mechanical abrasion, such as by surf action, and by biological activities, predominantly by boring algae (Cyanophyceae), echinoids *(Diadema)*, mollusks, crabs, holothuria, fish, and sponges (see 5.155) (GINSBURG 1957, HAMBLETON 1962, SWINCHATT 1965, MATTHEWS 1966, BATHURST 1971b). Currents and organisms may transport this material a considerable distance from the locality where it was formed (GINSBURG 1956, MATTHEWS 1966).

The calcareous skeletons of organisms in shallow seas consist mostly of aragonite and high-magnesium calcite, whereas they consist of low-magnesium calcite in the deep sea (planktonic foraminifera, coccoliths) (excepted pteropods). Corresponding compositions are found in the sediments of these different environments (STEHLI & HOWER 1961, TAFT & HARBAUCH 1964, FRIEDMAN 1965a, G. MÜLLER 1966b).

Fine-grained sediments of the Florida Bay contain 34% Mg-calcite (12—16 mol-percent $MgCO_3$), 60% aragonite and 6% calcite (MÜLLER & MÜLLER 1967). The Mg-calcite originates from disintegration of skeletal material, mostly foraminifera (CHAVE 1952, GINSBURG 1956). Presumably the aragonite fraction, which dominates the fine sediments, is formed also by disintegration of skeletons, since the coarse fraction investigated by MÜLLER & MÜLLER (l. c.) contains even more aragonite than the fine fraction. Moreover, aragonite skeletons, e. g. most of the gastropods and pelecypods, are far more frequent than calcitic and Mg-calcitic ones: In the Florida Bay, mollusks compose 70 to 95 percent of the fraction > 0.125 mm (GINSBURG 1956). According to tumbling barrel experiments by FORCE (1969), mollusks tend to break down mechanically to particles 500—250 µ (layers), 32—4 µ (sublayers), and 0.5—0.125 µ (single crystals) in size. Grain sizes < 500 µ which are produced mainly in the wave zone tend to accumulate on the outer shelf and in lagoons, off West Florida. This means that the fine material accumulates far from the places where it forms.

The influence of aragonite secreted by codiacean algae, notably *Penicillus* and *Halimeda*, is probably more important. STOCKMAN et al. (1967) calculated a deposition rate of 1—2 cm / 1000 years of fine carbonate mud produced by the green alga *Penicillus*, based on the mean frequency of this alga in shallow marine environments. These algae are very productive, thus enabling them to be extremely efficient sediment-forming organisms inasmuch as the turnover time is about 1—2 months (STOCKMAN et al. 1967), see also LEWIN (1962) and HOSKIN (1963).

The fine-grained sediments on the Great Bahama Bank also consist of aragonite and 10—20 percent Mg-Calcite (CLOUD 1962, FRIEDMAN 1964). It seems quite possible that organically-derived aragonite-rich muds, when stirred up by fishes and other organism, may account for the "whitings" seen in many areas, e. g. in the Florida Bay (fig. 5-2) (GINSBURG 1956, CLOUD 1962, BROECKER & TAKAHASHI 1966; based on ^{14}C). Theoretically, this material may provide nuclei for inorganic precipitation (CORRENS 1950, PURDY 1963), but adsorption of organic compounds onto carbonate minerals probably reduces reaction rates of inorganic carbonate equilibria (SUESS 1970). The aragonite needles in suspension are 0.1×1 to 0.6×6 micron in size (CLOUD 1962b) and have been interpreted by LOWENSTAM & EPSTEIN (1957) as fragments of algae and other organisms, based on stable isotope investigations. CLOUD (l. c.), on the other hand, prefers inorganic precipitation.

Fig. 5-3. Coccolith in a calcilutite. Paleocene flysch of Zumaya, Spain (Electron micrograph by S. HONJO; width of photograph: 9 micron).

According to scanning electron micrographs, the bulk of the fine-grained material in the Bimini lagoon is the result of the disintegration of organic skeletons (STIEGLITZ 1972). For further discussion see BATHURST (1971b).

The conspicious role of skeletal material in deep-sea calcilutite is also evident. Nannoplankton mud rich in coccoliths and foraminifera *(Globigerina)* is the most frequent sediment in the present oceans (see also Part II/2 and 5.54).

MÜLLER & BLASCHKE (1969a) found that the white calcite layers in the Black Sea sediments consist of coccoliths too, and are not precipitated inorganically as previously assumed (RUCHIN 1958). Similar bandings of coccolith-rich calcite layers have been reported in the Lower Jurassic Posidonia shales of Northern Germany, in the Oligocene "Heller Mergelkalk" of Bavaria, and in other bituminous shales of Tertiary age (fig. 5-26b; MÜLLER & BLASCHKE 1969b and 1971). Mesozoic and Cenozoic massive calcilutites which have been deposited in deep water also contain many coccoliths (fig. 5-3; FARINACCI 1964, HONJO & FISCHER 1964, MINATO et al. 1967, E. FLÜGEL 1967, GARRISON 1967). Perhaps the best known example of coccolith limestone is chalk. Coccolithophorids evolved during the early Mesozoic. Therefore, no carbonate nannofossils have been detected in deep-water Paleozoic limestones (FISCHER et al. 1967, p. 25).

The presence of calcareous mud is by no means a reliable criterion of low-energy environments: Skeletal remains can act as efficient mud-trapping baffles; the matrix content is governed by the relation between production and removal of lime mud (PURDY & IMBRIE 1964).

LEES & BULLER (1971) observed, that carbonate sediments in shallow, temperate waters are dominantly skeletal; carbonate muds are rare. This has to be taken in mind in climatic interpretations of carbonate rocks.

3. Supply of detrital-terrigeneous lime mud in suspension

Lime mud originating by mechanical decomposition and redistribution of older limestones is only a minor constituent of calcilutites (at least in those containing little insoluble residue). The reason for this is that proper limestones disintegrate by erosion mainly into conglomerates; fine grained debris is a relatively minor component. On the other hand, marls can disintegrate directly into mud or into soft pebbles. Because of the rapid production of the readily disintegrated marl mud, only a small amount will be dissolved. Consequently, lime and clay tend to be redeposited together as marls. Examples of such secondary marlstones are those in the Tertiary molasse north of the Alps (FÜCHTBAUER 1964a), the modern sediments in the Lake Constance (G. MÜLLER 1966a) and in the northern Persian Gulf (SARNTHEIN & WALGER 1973).

5.12 Clotted limestones

Definition

Frequently encountered variants of calcilutites are limestones composed of grains 0.02—0.1 mm in diameter which have indistinct borders and which differ only by their lower crystal size (mostly cryptocrystalline) from the surrounding matrix ("structure grumeleuse", "Krümelkalk" = lumpy, grumelous texture; CAYEUX 1935) (fig. 5-4). There are no sharp limits between clotted, pelletal, and intraclastic limestones (ch. 5.13 and 5.14).

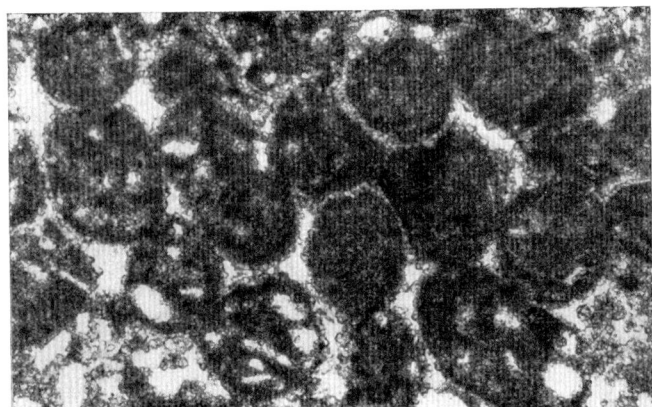

Fig. 5-6. Pelletal dolomite. Pellets, about 0.5 mm in size, were partly compacted soon after deposition. White areas are pores. Paleocene, Libya (width of photograph 4.4 mm).

Origin

Most grains corresponding to the above description are probably fecal pellets. These pellets play an important role in modern sediments (WETZEL 1923, HECHT 1935). In older rocks, however, they are preserved only under favorable conditions. Because fecal pellets are originally soft, they often disintegrate into a structureless lime mud soon after burial (GINSBURG 1957). Later diagenesis may also obliterate the pellets and give them the appearance of clots (ch. 5.12; SHINN 1969, BATHURST 1970). Pellets composed of clay and silt (for instance), are prominent constituents of many shallow marine sediments, such as the intertidal sediments of the North Sea. However, these pellets generally are not preserved beyond the point of initial sediment con-

Table 5-1. Size (in mm) and origin of fecal pellets of different recent occurrences.

length	width	originator	author
0.08	0.04	snails on seaweed	D. MOORE, W. D. BOCK (pers. comm.)
0.075—0.15 [1]		snails, 20—25 mm in size	D. MOORE, W. D. BOCK (pers. comm.)
0.9	0.16	snails	KORNICKER & PURDY 1957
0.6	0.3	worms	ILLING 1954
0.7		worms	GINSBURG 1957
ca. 1.0	ca. 0.5	worms	MAYER 1956
0.5	0.1 [2]	crustacea	EARDLEY 1938
0.2—2	0.2—1 [3]	crustacea	BRÖNNIMANN & NORTON 1960
1.0	0.4	echinoderms	VOIGT 1929

[1] Some sediments in eastern Florida Bay and the Bahamas (PURDY 1963) are made up predominately of these aragonitic pellets.

[2] Pellets consisting of clayey, cryptocrystalline aragonite, form about 30 percent of the Great Salt Lake (Utah) sediments.

[3] Sculptured by delicate channels parallel to the long axis; deposited only by anomural decapods, e. g. Favreïna (upper Jurassic and lower Cretaceous; MOORE 1933, BRÖNNIMANN & NORTON 1960, ELLIOT 1962).

According to scanning electron micrographs, the bulk of the fine-grained material in the Bimini lagoon is the result of the disintegration of organic skeletons (STIEGLITZ 1972). For further discussion see BATHURST (1971b).

The conspicious role of skeletal material in deep-sea calcilutite is also evident. Nannoplankton mud rich in coccoliths and foraminifera *(Globigerina)* is the most frequent sediment in the present oceans (see also Part II/2 and 5.54).

MÜLLER & BLASCHKE (1969a) found that the white calcite layers in the Black Sea sediments consist of coccoliths too, and are not precipitated inorganically as previously assumed (RUCHIN 1958). Similar bandings of coccolith-rich calcite layers have been reported in the Lower Jurassic Posidonia shales of Northern Germany, in the Oligocene "Heller Mergelkalk" of Bavaria, and in other bituminous shales of Tertiary age (fig. 5-26b; MÜLLER & BLASCHKE 1969b and 1971). Mesozoic and Cenozoic massive calcilutites which have been deposited in deep water also contain many coccoliths (fig. 5-3; FARINACCI 1964, HONJO & FISCHER 1964, MINATO et al. 1967, E. FLÜGEL 1967, GARRISON 1967). Perhaps the best known example of coccolith limestone is chalk. Coccolithophorids evolved during the early Mesozoic. Therefore, no carbonate nannofossils have been detected in deep-water Paleozoic limestones (FISCHER et al. 1967, p. 25).

The presence of calcareous mud is by no means a reliable criterion of low-energy environments: Skeletal remains can act as efficient mud-trapping baffles; the matrix content is governed by the relation between production and removal of lime mud (PURDY & IMBRIE 1964).

LEES & BULLER (1971) observed, that carbonate sediments in shallow, temperate waters are dominantly skeletal; carbonate muds are rare. This has to be taken in mind in climatic interpretations of carbonate rocks.

3. Supply of detrital-terrigeneous lime mud in suspension

Lime mud originating by mechanical decomposition and redistribution of older limestones is only a minor constituent of calcilutites (at least in those containing little insoluble residue). The reason for this is that proper limestones disintegrate by erosion mainly into conglomerates; fine grained debris is a relatively minor component. On the other hand, marls can disintegrate directly into mud or into soft pebbles. Because of the rapid production of the readily disintegrated marl mud, only a small amount will be dissolved. Consequently, lime and clay tend to be redeposited together as marls. Examples of such secondary marlstones are those in the Tertiary molasse north of the Alps (FÜCHTBAUER 1964a), the modern sediments in the Lake Constance (G. MÜLLER 1966a) and in the northern Persian Gulf (SARNTHEIN & WALGER 1973).

5.12 Clotted limestones

Definition

Frequently encountered variants of calcilutites are limestones composed of grains 0.02—0.1 mm in diameter which have indistinct borders and which differ only by their lower crystal size (mostly cryptocrystalline) from the surrounding matrix ("structure grumeleuse", "Krümelkalk" = lumpy, grumelous texture; CAYEUX 1935) (fig. 5-4). There are no sharp limits between clotted, pelletal, and intraclastic limestones (ch. 5.13 and 5.14).

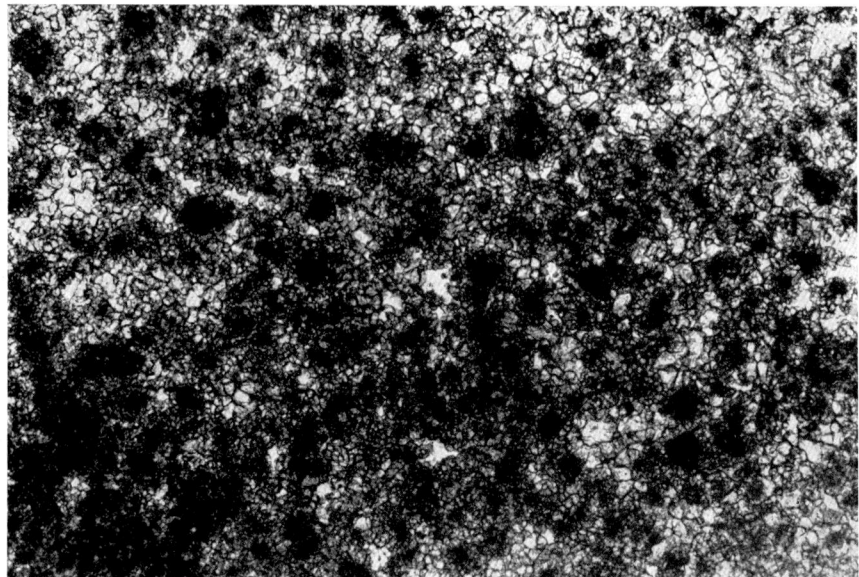

Fig. 5-4. Clotted limestone. The grains, 0.1 mm in diameter, have been interpreted as diagenetic in origin. Upper Jurassic, Flügelstein/Altmühl (Germany) (width of photograph 3.7 mm).

Origin

Sedimentary formation of these blurred clots has been observed in both modern freshwater and marine sediments. Near the shore line in the Lake Constance, oncoids (see 5.152) and precipitated lime mud are clotted (G. Müller 1966a; fig. 5-5). Bahama Bank sediments contain clots and cryptocrystalline grains composed of more than 80% aragonite (Illing 1954). Their origin has not been investigated in detail. An observation by Monty (1967), however, may lead to a better understanding of these and the Lake Constance occurrences: Blue-green algal mats frequently contain calcitized monocellular algae which form cryptocrystalline grains of a few tens of microns to 2 mm in size (ch. 5.152, C 2 a). The smaller grains closely fit the clots under consideration.

Clotted textures can be preserved in sheltered areas, e. g. below pelecypod valves, whereas they merge outside. This has been observed in rocks of Triassic age by Bachmann (1973). Most clotted limestones seem to be indicative of low energy environments. Beales (1958), in this sense, combined them with the pelletal limestones into the "bahamites".

Different authors found globular aggregates, 0.02—0.1 mm in size, to be composed of aragonite (Shinn 1969, Taylor & Illing 1969) or magnesian calcite (Amiel et al. 1971, Schulze & v. Rad 1971) in modern sediments. Presumably these are formed in the pore fluid of calcarenitic sediments by nucleation in an highly supersaturated shallow marine or intertidal environment (Taylor & Illing l.c.). This then would represent a transition between "precipitation" and "cementa-

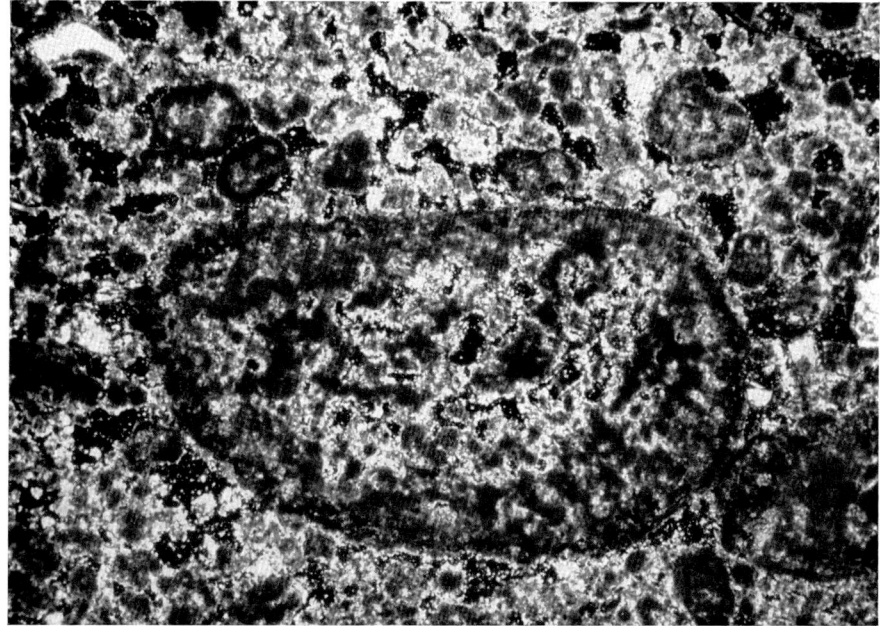

Fig. 5-5. Recent clotted oncolith. Gnadensee, part of Lake Constance (width 3.6 mm, + nicols; from G. MÜLLER 1966 a).

tion". Biochemical influences were suggested by SCHULZE & V. RAD as an explanation of this "internal pelleting" (SHINN, l. c.).

Not all clotted textures, however, are necessarily sedimentary in origin. Some may have been formed diagenetically by sporadic crystal enlargement in cryptocrystalline limestones. Such microcrystalline centers may finally merge and leave isolated cryptocrystalline clots (fig. 5-4; CAYEUX 1935, MAXWELL 1962). Large intervals between the clots tentatively can be used as a criterion of diagenetic origin (fig. 5-4), since sedimentary grains are supposed to form mostly packed aggregates. Another diagenetic explanation was suggested by BATHURST (1970), who believes that pelletal limestones may transform into clotted limestones by crystal enlargement which begins in the matrix and extends into the pellets, leaving only central areas of unaltered pellets. Similarly SHINN (1969) described a clotted texture caused by vague preservation of fecal pellets replaced marginally by magnesian calcite.

5.13 Pelletal limestones

Definition

Pelletal limestones are composed of sharply bordered grains, 0.1—1 mm in size, which are generally well rounded and oval, sometimes spherical. Typically they are uniform in size (at least within the same deposit) and patchy in distribution (fig. 5-6).

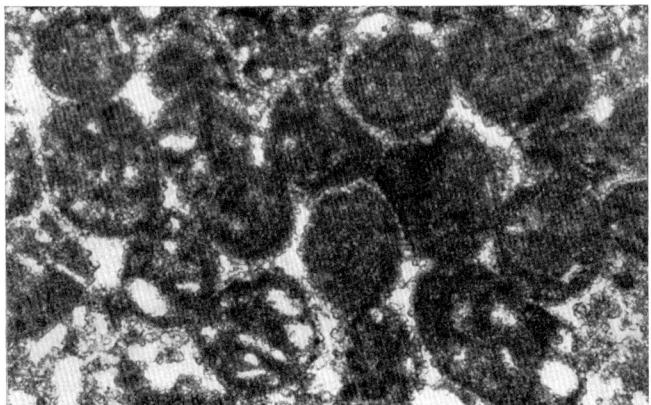

Fig. 5-6. Pelletal dolomite. Pellets, about 0.5 mm in size, were partly compacted soon after deposition. White areas are pores. Paleocene, Libya (width of photograph 4.4 mm).

Origin

Most grains corresponding to the above description are probably fecal pellets. These pellets play an important role in modern sediments (Wetzel 1923, Hecht 1935). In older rocks, however, they are preserved only under favorable conditions. Because fecal pellets are originally soft, they often disintegrate into a structureless lime mud soon after burial (Ginsburg 1957). Later diagenesis may also obliterate the pellets and give them the appearance of clots (ch. 5.12; Shinn 1969, Bathurst 1970). Pellets composed of clay and silt (for instance), are prominent constituents of many shallow marine sediments, such as the intertidal sediments of the North Sea. However, these pellets generally are not preserved beyond the point of initial sediment con-

Table 5-1. Size (in mm) and origin of fecal pellets of different recent occurrences.

length	width	originator	author
0.08	0.04	snails on seaweed	D. Moore, W. D. Bock (pers. comm.)
0.075—0.15 [1]		snails, 20—25 mm in size	D. Moore, W. D. Bock (pers. comm.)
0.9	0.16	snails	Kornicker & Purdy 1957
0.6	0.3	worms	Illing 1954
0.7		worms	Ginsburg 1957
ca. 1.0	ca. 0.5	worms	Mayer 1956
0.5	0.1 [2]	crustacea	Eardley 1938
0.2—2	0.2—1 [3]	crustacea	Brönnimann & Norton 1960
1.0	0.4	echinoderms	Voigt 1929

[1] Some sediments in eastern Florida Bay and the Bahamas (Purdy 1963) are made up predominately of these aragonitic pellets.

[2] Pellets consisting of clayey, cryptocrystalline aragonite, form about 30 percent of the Great Salt Lake (Utah) sediments.

[3] Sculptured by delicate channels parallel to the long axis; deposited only by anomural decapods, e. g. Favreïna (upper Jurassic and lower Cretaceous; Moore 1933, Brönnimann & Norton 1960, Elliot 1962).

solidation. In some modern environments, however, pellets can be preserved by hardening soon after formation, sometimes presumably by drying (ILLING 1954, KORNICKER & PURDY 1957).

Durable, well-formed fecal pellets are secreted by many marine invertebrates, especially worms and snails (D. MOORE and W. D. BOCK, Miami, Marine Laboratory, personal communication); generally, they are recognized by their cylindrical shape and greyish color (MORET 1940). Some pellets made by lamellibranchs and gastropods have sculptured surfaces (MOORE 1939). Similar fossilized sculptured pellets 0.4 mm in diameter, have been found in the gut tracts of fossil molluscs (COX 1960, CASEY 1960).

Size ranges of other pellets can be found in a species key for fecal pellets of marine organisms in the southern Florida region (MANNING & KUMPF 1959).

In addition to carbonate fecal pellets phosphatic pellets also should be mentioned. According to CAROZZI (1960), phosphatic pellets can reach lengths as great as 20 mm (crustacea pellets in phosphatic cretaceous sediments). With such dimensions, they are better named "coprolites" (FOLK 1965a), the lower limit of which is 10 mm (PETTIJOHN 1957). An annotated bibliography of coprolites has been published by HÄNTZSCHEL et al. (1968). Phosphatic fecal pellets can be recognized by their luminescence (ZUMPE 1964).

In this book, the term "pellet" is restricted to these grains of presumed fecal origin. If a fecal origin is not suggested, the term "intraclast" may be applied (ch. 5.14 B). Difficulties arise, where transitions between intraclasts and assumed pellets exist such as in many Paleozoic limestones in the Appalachians (FRIEDMAN, pers. comm.). The following synopsis may be helpful for the identification of non-skeletal grains.

	homogeneous grains	laminated grains
regular shape and size	$>$ 0.1 mm, elongated, distinct: pellets (or micritized oöids) $<$ 0.1 mm, globular, blurred: clots	oöids
irregular shape and size	intraclasts or detrital grains or micritized skeletons	oncoids

5.14 Detrital limestones

Definition

Detrital limestones are composed of carbonate rock fragments which were more or less indurated when eroded. According to shape and size of the components, the following subtypes may be distinguished:

Carbonate conglomerates	(rounded pebbles, $>$ 2 mm in size)	⎫	
Carbonate breccias	(angular pebbles, $>$ 2 mm in size)	⎬	(A)
Calcarenites	(detrital calcareous grains, 0.063—2 mm)	⎭	
Intraclastic limestones	(resedimented fragments of different size)		(B)

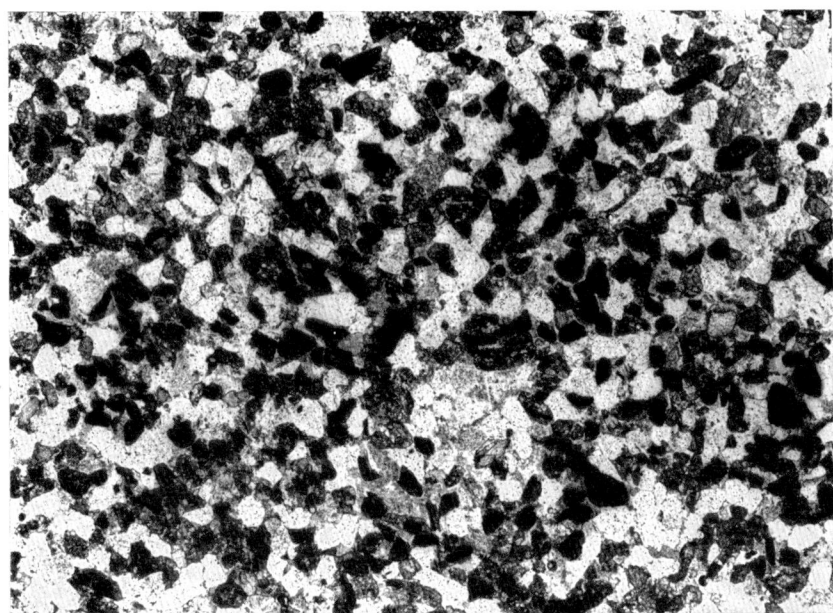

Fig. 5-7. Strongly dolomite-arenitic sandstone (43 percent dolomite grains). The dolomite grains (black) correspond in size to the sand grains (white). Bausteinschichten (Oligocene Molasse, well Schwabmünchen 1 near Augsburg; thin section, width is 8 mm).

(Calcareous sandstones, however, are [quartz] sandstones with calcite cement or calcite matrix.)

Origin

These rocks may have originated by erosion of older limestones (A) or by the reworking of penecontemporaneous sediments (B). The demarcations between (A) and (B) and also between them and pelletal limestones are sometimes difficult. Calcarenites and calcareous conglomerates, which contain quartz grains and pebbles of the same size as the carbonate fragments (as in figs. 5-7 and 5-85), belong to group A. Rocks of group B may contain curved or lobate lumps of very different size (fig. 5-8).

A. Calcarenites are less common than calcareous conglomerates and breccias (BEALES 1958, CAROZZI 1960). This is in contrast to quartz sandstones which are more common than quartz conglomerates. The reason for this difference is that granites and gneisses, the major source rocks of coarse-clastic silicate rocks, disintegrate mainly into material of sand grain size, whereas most carbonate rocks, which are not dissolved chemically during weathering, disintegrate into pebbles.

A good example of this relation is found in the fluviatile molasse sediments north of the Alps (FÜCHTBAUER 1964a). The fans originating in granitic source rocks consist mainly of sandy material and extend as tongues 100—2000 m in thickness, as far

5.14 Detrital limestones

Fig. 5-8. Dolomitic oncolite with oncolite intraclasts. Upper Permian Zechstein 3, well Norddeutschland 5 near Bentheim. Polished rock surface; core diameter 8.2 cm.

as 350 km into the foreland. On the other hand, fans originating in the Calcareous Alps consist of conglomerates ("Nagelfluh") which drop abruptly basinwards and continue as marls (SCHIEMENZ 1960).

Instructive photographs of Cambrian dolomite-arenites have been published by ZADNIK & CAROZZI (1963). ZIEGENHARDT (1962) described calcarenites which originated from fluvial reworking of tufa.

Marine breccias which are common in Mesozoic rocks in the Alps and in the Dinarids (fig. 5-9), indicate steep slopes (especially in the subaeric section) and the absence of a shore face. Transport may have been initiated by exceeding the angle of repose or by earthquakes. Therefore such deposits are frequently encountered in orogenic zones. If sufficient lime mud is present, turbidity currents can form and transport the breccias far into the basin (TRÜMPY 1960, LEMOINE 1967, MEISCHNER 1964, SANDERS & FRIEDMAN 1967).

B. Marine breccias and calcarenites may be "resediments" (SANDER 1936), that is reworked fragments of compacted or partially lithified sediments of the same sequence. These rocks are also termed "intraclastic limestones"; individual particles are called "intraclasts" (FOLK 1959). Such sediments occur frequently in shallow marine carbonate rocks. A tectonically triggered early diagenetic brecciation has been reported by J. WENDT (1965).

Normally, lithification is accomplished by (a) temporary emergence or (b) submarine diagenesis:

Fig. 5-9. Limestone breccia. Island of Krk, Yugoslavia (width of section about 1 m).

a) Desiccation cracks (see ch. 3.241.43) can be formed in sediments which periodically are exposed above mean high water level. These cracks may be enlarged and form a breccia composed of petrographically uniform fragments which maintain their identity even upon later transgressions of the sea (FREYTET 1965, HÖTZL 1966), and thus will survive reworking (BRAUN & FRIEDMAN 1969). In this way, early diagenetic dolomites may form brecciated dolomites (fig. 5-10) and intraclastic dolomites. A subtype is the "inhomogeneity breccias" (SANDER 1936), which supposedly form by early diagenetic fracturing of harder layers between softer ones; sometimes the fragments even may still fit together (see also FISCHER 1965a).

b) Oölites frequently contain intraclasts composed of eroded pieces of oölite with the margins cutting through oöids (fig. 5-11). Such fragments point to an early lithification, which makes the distinction from older reworked rocks difficult (ZIEGENHARDT 1966). Very flat pebbles of massive calcilutite generally are thought to be intraclasts.

One example of intraclasts of submarine origin are the "grapestones" (lumps) of the Great Bahama Bank (ILLING 1954). These aggregates of different particle types are up to 1 mm in size (PURDY 1963, FRIEDMAN 1964, KENDALL & SKIPWITH 1969). They are bound by cryptocrystalline aragonite rodlets (MILLIMAN 1967, KINSMAN 1969, FABRICIUS 1972) and can merge into larger crusts (ILLING 1954). An influence of blue-green algae (WILSON 1967) and organic mucilage (BATHURST 1967b) in the formation of such crusts has been established. According to FABRICIUS (1972), the grapestones are the first stage in the formation of composite oöids.

SHINN (1969) discovered crusts formed by submarine cementation in the Persian Gulf; the Recent age of these crusts was indicated by the presence of pottery frag-

5.14 Detrital limestones

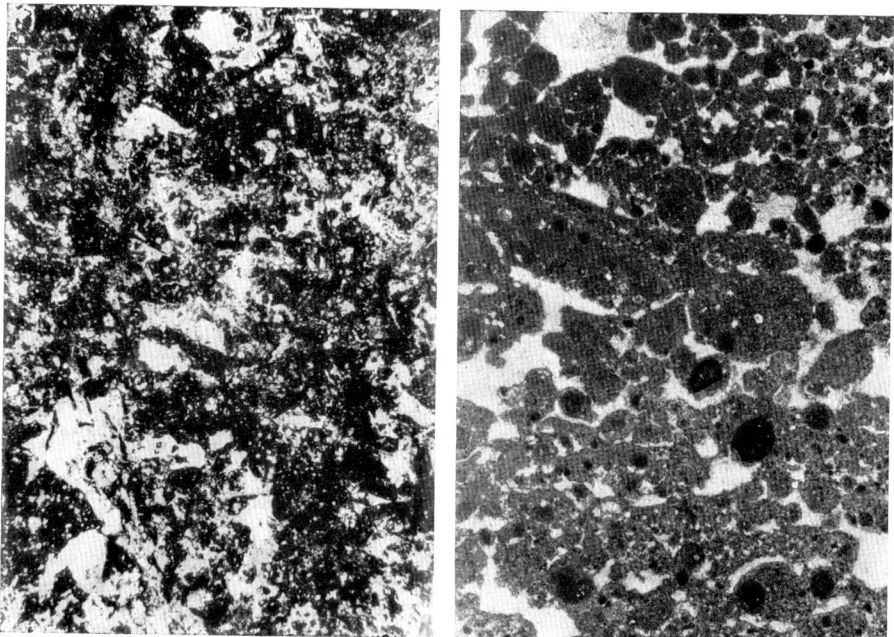

Fig. 5-10. Brecciated early-diagenetic dolomites. Paleocene, Libya. Left: partly porous, partly with celestite cement (width 18 mm). Right: cemented by calcite (width 9 mm).

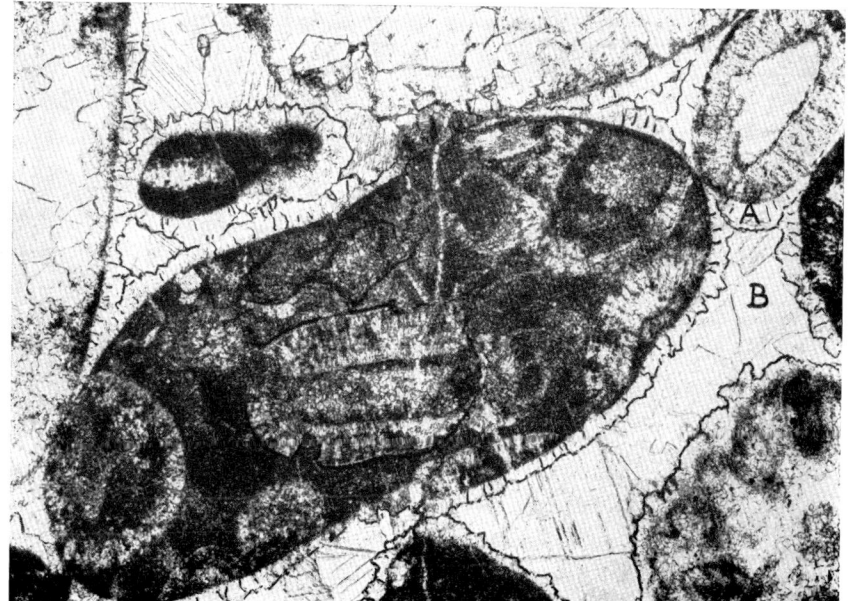

Fig. 5-11. Intraclast of oölithic calcilutite, surrounded by cements A (radial) and B (isometric). Lower Jurassic, Northern Alps (thin section; courtesy F. FABRICIUS; photograph slightly retouched, width 2.5 mm).

ments incorporated into the crusts. These crusts are buckled, overthrusted and brecciated in many places. Crystallization pressure during cementation or the increase in volume connected with the transformation of aragonite into calcite (which has been observed in the crusts) are suggested as possible agents of this brecciation. A similar early diagenetic buckling has been reported by LINDSTRÖM (1963) from the Ordovician.

Rocks which are speckled by crystal enlargement may contain a granular (ch. 5.12) or even a brecciated texture. Such late diagenetic breccias differ from early diagenetic ones by having fragments with blurred margins (ORME & BROWN 1963).

Structureless grains of more or less spherical shape are frequent in carbonate rocks. They may form from oöids, onkoids or skeletal grains by micritization (see p. 285), a process by which the grains become cryptocrystalline. They were termed "pelletoids" by BLATT et al. (1972), whereas PURDY (1963) and BATHURST (1966) described such grains from low-energy environments of the Bahama Bank as "matrix grains". Similar grains have been reported from the Gulf of Aqaba (Red Sea) by FRIEDMAN (1968a).

5.15 Skeletal limestones

5.151 Ecologic and mineralogic considerations, and identification key

Definition

Limestones containing more than 50 percent (by volume) of preserved or fractured calcareous skeletons of plants or animals are termed "skeletal limestones". Various limestones can be distinguished by their dominant components, for example, foraminiferal limestones, coral limestones, crinoid limestones, and so forth.

"Biocalcarenites" are limestones composed of well-rounded fragments 0.063—2 mm in size (fig. 5-12). In "biocalcirudites" the fragments are greater than 2 mm, in "biocalcisiltites" they are 0.002—0.063 mm in diameter. Biocalcarenites (and biocalcirudites) often display cross bedding similar to that found in sandstones.

Origin and ecology

Skeletal limestones can be biocoenoses (life assemblages), thanatocoenoses (death assemblages), or taphocoenoses (grave communities, residue assemblages) (H. SCHMIDT 1935, W. H. BERGER 1971). The skeletons can be dissolved or broken down by mechanical abrasion (currents and surge), as well as by "bioerosion" (NEUMANN 1966) which occurs through the activities of boring organisms. If the skeletal particles become rounded and sorted, biocalcarenites result. The original organisms can generally be determined only if the grains are larger than 0.1 mm (FERAY et al. 1962). By scanning electron microscopy, however, virtually all material coarser than 15 microns and much of the smaller material can be identified (STIEGLITZ 1972).

High energy environments modify the biocoenoses not only by sorting processes but also by the abrasion and subsequent elimination of delicate skeletons, such as

Fig. 5-12. Biocalcarenite with oöids and intraclasts, cemented by calcite. Grains with micritic envelopes. Jurassic, Ethiopia (width 5.5 mm).

Fig. 5-13. Biocalcirudite, cemented by calcite. Micritic envelopes. Lower Jurassic. Northern Alps (thin section by F. Fabricius, width 20 mm).

echinoderms, bryozoa and green algae (Chave 1964). For this reason, well sorted biocalcarenites give a less reliable impression of the life community than do poorly sorted skeletal limestones.

The influence of paleo-currents can be recognized in skeletal limestones in which shells are embedded in a convex-up position. This position presents a minimum restistance against the current. The convex-up texture is well-developed only if the shells do not impede one another, and if they rest on a smooth bottom, for example, single shell-layers in clays. Furthermore, the valves should not sink into a soft substratum preserving a concave-upwards position (R. Richter 1942), nor should they be incorporated in a rapidly deposited turbidite (Middleton 1967). In small ripples, a predominant concave-up position also can be observed (Clifton & Boggs 1970). Predators and scavengers may turn the shells as shown by experiments (Clifton 1971). They are effective in producing a dominantly concave-up orientation.

In biocoenoses, the remains of benthonic organisms are embedded on their place of life, therefore reflecting the environmental conditions: The substratum is highly important. Many organisms need a firm substrate. Thus submarine rock terraces and coral reefs often have dense populations of epifauna while soft mud bottoms are dominated by infaunal burrowers. Quiet sand bottoms are more easily populated than agitated environments, in which only rounded fragments are found (e. g. submarine oölite dunes). The benthic population in muddy bottoms decreases as the sedimentation rate increases. In addition, the few shells tend to be diluted by the inorganic material. On the other hand, the shells in such an environment rarely exhibit the extreme degree of bioerosion seen in higher energy environments. Oxygen deficiency at the bottom suppresses benthonic life, and in such environments only taphocoenoses can be formed.

Another example of taphocoenoses is the sinking of planktonic and nektonic skeletons to the bottom.

The concentration of Ca and CO_3 in the water and in the sediments may influence the absolute shell thickness. KLÜPFEL (1916) observed that in sedimentary sequences the walls of shells are thicker in limestones than in interbedded shales. This difference, however, may be due to diagenetic alteration rather than depositional environment. With increasing water depth the number of benthonic organisms decreases considerably. This is caused in part by the decreasing light intensity, and thus the decreasing number of herbivores.

Other environmental factors influencing the composition of benthonic communities are water agitation, temperature and salinity, as well as the interactions of these factors.

In order to infer the environment of a biocoenosis, the mutual influences of the organisms must be considered. These influences include the food-chain as well as all types of symbioses (communities of mostly mutual benefit), commensalisms (communities with normally one-sided benefit but without detriment for the partner) and parasitisms (communities with one-sided benefit at the expense of the partner; s. Geol. Soc. Amer. Memoir 67, I. 1957).

In order to reconstruct quantitatively the composition of a former biocoenosis from the composition of a skeletal limestone one must consider the mean life span of the major organisms. This can be computed from the formula

$$P_B = \frac{P_F \cdot 1 \cdot 100}{\Sigma P_{F_i} l_i}$$

with P_B = percentage of the organism under consideration in the biocoenosis
P_F = percentage of this organisms in the skeletal limestone
1 = mean life span of this organism.
$P_{F_i} l_i$ = sum of the products of percentage and life span of all kinds of organisms in the limestone.

Important are the findings by HERTWECK (1972) in the Georgia coastal region (U.S.A.): Lebensspuren and in-situ skeletal remains are preserved only from 7.1% of the 268 living species.

Biostromes and bioherms are special types of biocoenoses. They will be treated in ch. 5.18.

Mineralogic composition of calcareous skeletons

The composition of recent calcareous skeletons has been investigated and discussed by CLARKE & WHEELER (1922), BØGGILD (1930), MAYER & WEINECK (1932), LOWENSTAM (1954), CHAVE (1954, and references therein) and DODD (1967). Excellent summaries have been published recently by BATHURST (1971b) and HOROWITZ & POTTER (1971).

Aragonite seldom survives in fossil skeletons; generally it is dissolved or replaced by calcite. A direct determination of the original mineralogy therefore is possible only if both calcitic and aragonitic skeletons are protected from diagenetic alteration. This occurs for instance in communities embedded in oil or oil shales, since the transformation of aragonite into calcite requires aqueous solutions. On the other hand, aragonite transformation is indicated by the loss of the original skeletal texture. Aragonite is replaced by equant granular calcite (fig. 5-43); the original shell structure can become hazy (BØGGILD, p. 241), or will dissolve and the subsequent void will be filled by a drusy mosaic (cement). Skeletons, however, which are composed of fibrous, curved, undulose calcite crystals, suggest a primary calcitic composition (fig. 5-41). This has been confirmed by observations in oil-embedded fossil communities from the Carboniferous to Cretaceous (LOWENSTAM 1963, STEHLI 1956, HALLAM & O'HARA 1962, HUDSON 1962, FÜCHTBAUER & GOLDSCHMIDT 1964) (figs. 5-40, 42).

Other possibilities of determining the original mineralogy are provided by chemical composition: The calcite lattice is capable of containing more magnesium than the aragonite lattice (CLARKE & WHEELER 1922, ZELLER & WRAY 1956). Therefore the Mg-contents of primary calcitic skeletons are higher than those which have been transformed from aragonite (LOWENSTAM 1954, 1963).

The mineralogic composition of skeletons is more dependent upon phylogeny than upon the physicochemical environment. No phylum of animals or plants has carbonate skeletons that are composed completely of aragonite. On the other hand, brachiopods, arthropods, and echinoderms use only calcite as carbonate material. Most phyla, however, use calcite or aragonite (as described more fully in the following chapters), and several phyla can contain mixtures of the two polymorphs.

The evolution of the different phyla with time has resulted in a variable production and sedimentation of calcite and aragonite in the geologic past (LOWENSTAM 1963): In the Paleozoic calcite organisms predominated, in the Mesozoic era aragonite increased, whereas foraminifera and coccoliths caused an increase of calcite in the Cenozoic era.

An influence of the physicochemical conditions on the mineralogy of skeletons is observed in coelenterates, bryozoa, annelids, and mollusks (LOWENSTAM 1954, 1963): Aragonite is more prevalent in warm-water species (figs. 5-14, 15): thus aragonitic green and red algae live in warm seas, whereas calcitic red algae occur in temperate and polar seas as well (REVELLE & FAIRBRIDGE 1957).

The magnesium content in calcitic skeletons is directly dependent upon water temperature (CLARKE & WHEELER 1922, CHAVE 1954b), increasing with elevated temperatures during the main growing phase. This temperature influence appears to be more distinct in the more primitive organisms, such as coralline algae. The latter are especially rich in Mg, e.g. the red alga *Lithothamnion* produces calcite with up to 25 weight percent $MgCO_3$. Systematical differences of the Mg-content in different parts of the body are observed, e. g. the Mg-content in the spines of sea-urchins is

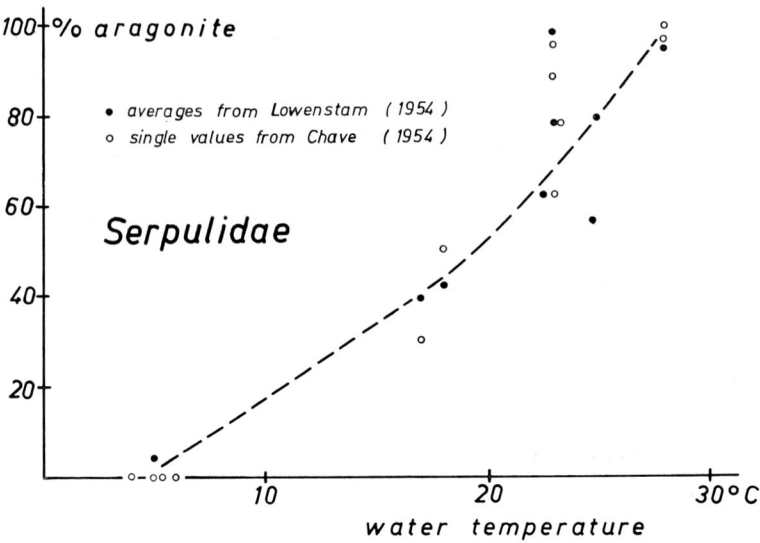

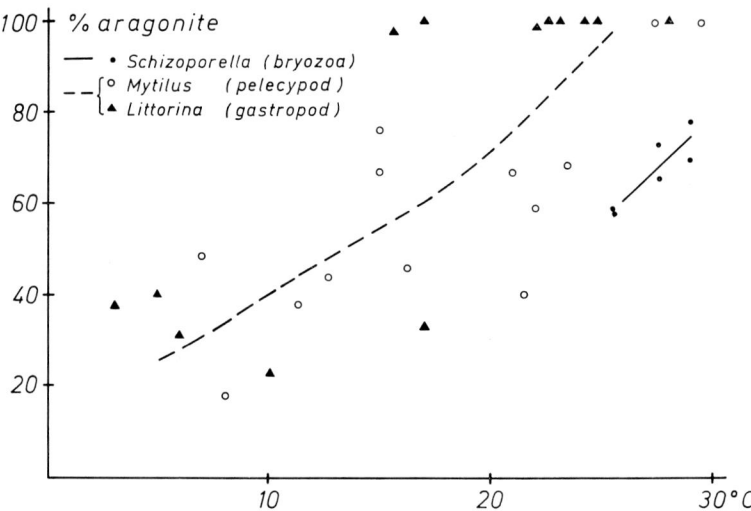

Figs. 5-14 and 5-15. Relationship between aragonite and water temperature in different skeletons (fig. 5-15 from LOWENSTAM 1954). As abscissa, the mean annual temperature is used for the serpulids and for the gastropod *Littorina,* the warmest monthly mean for *Mytilus,* and the temperature at the time of sampling for the bryozoan *Schizoporella.*

lower than in the body plates (fig. 5-16). Their teeth are particularly rich in Mg (up to 43.5 mole-% $MgCO_3$; SCHROEDER et al. 1969).

How do animals and plants precipitate calcium carbonate? One possible answer is given by the following observations (MITTERER 1972): Amino acid analyses show that organic matrices from calcitic and aragonitic skeletons of many species of calcareous algae and invertebrates (also from oöids) are characterized by a predominance of acidic amino acids, particularly aspartic acid which generally comprises from 30% to 70% of the total protein. On the other hand, amino acids extracted from non-calcified organic tissues, e. g. from a molluskan periostracum, from sabellariid worm tubes, and from blue-green algae, are poor in aspartic acid. Such acidic amino acids dissociate in water by releasing the H^+-ions from both COOH-side chains, thus becoming negative ions, which attract e. g. Ca^{++} from the solution. This Ca-concentration around the aspartic acid ions causes them to act as crystallization nuclei by epitaxis (see also TRICHET 1971).

Such a mechanism would explain why inorganic precipitation of calcium carbonate is weak or even missing in the presence of organisms producing acidic amino acids.

In the following chapters, those groups of plants and animals which contribute significantly to carbonate rocks are discussed. Emphasis is placed on properties which

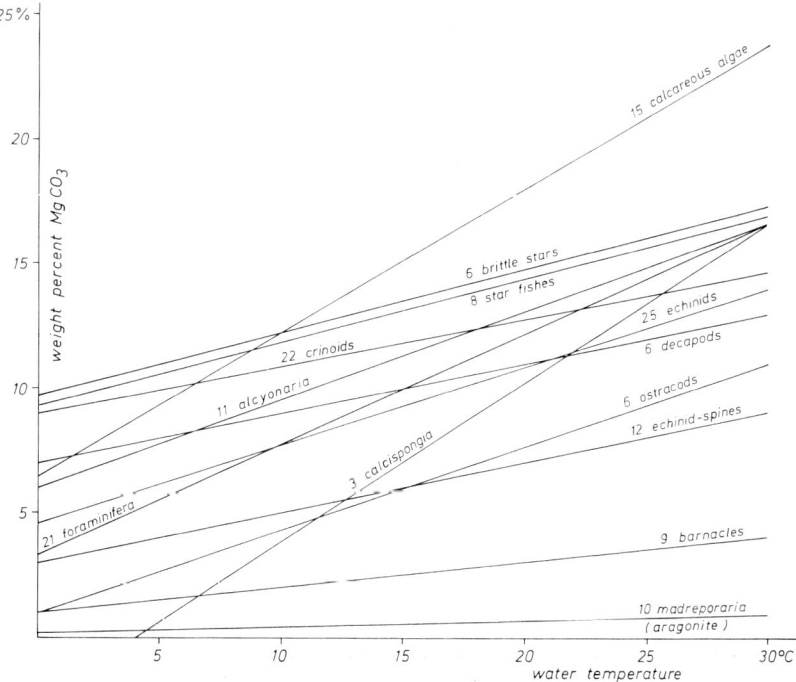

Fig. 5-16. $MgCO_3$-contents of calcitic skeletons, in relation to water temperature (regression lines of least squares, CHAVE 1954a).

are of major importance to sedimentologists, and which often are mentioned only marginally or not at all in paleontologic textbooks: composition, structure and size of skeletons, as well as ecologic data which enable an interpretation of the sedimentary environment (see also EKMAN 1953).

Key to identification of skeletal and related grains in thin section

The following key is in part based on HOROWITZ & POTTER 1971 and MILLIMAN 1974; figures are approximate smaller diameters in millimeters; C and A indicate original mineralogy (calcite and aragonite).

1.	Cryptocrystalline — disordered	
1.1	Swiss-cheese appearance	
1.1.1	Irreg. grains with tubes (up to 0.05 mm)	— Halimeda (green alga)
1.1.2	Irreg. grains (0.4) with several regul. distrib. tubes	— *Favreïna* (fecal pellet)
1.2	Irreg. network (0.2) with open areas (canals, 0.2) (C)	— Calcisponges
1.3	Laminations (0.1—0.4) crossed by pilae; sometimes fibrous (A)	— Stromatopores
1.4	Globul. or elong. skel. with chambers of diff. shape and arrangement (0.05—0.2) (C)	— Porcellan. or agglut. foraminifera
1.5	Branch., nodul. or layered skel. with circ. or rectang. netw. of small quadr. or elong. cells (0.01) (C)	— Coralline = Red. algae
1.6	Closely packed sinuous tubes (0.01—0.02), "spaghetti" aspect	— Blue-green algae
1.7	Structureless grains or cryptocryst. surface and coarser interior (due to dissolution of arag. shell and subsequ. cementation)	— Micritized grains
2.	Fibrous	
2.1	Fibers (= optical axes) in part tangentially oriented	
2.1.1	Reticulate fabric, rounded or angular chambers (zooecial openings) (0.1—0.4) (C, A)	— Bryozoans
2.1.2	Circular and tubul. sect. (1 mm diam., wall 0.1) (C, A)	— Serpulid worms
2.1.3	Globular and egg-shaped grains or crusts, diam. 0.1—1, regular lamination (A)	— Recent oöids
2.1.4	Swiss-cheese appear.; fibers only adjac. to tubes (A)	— Halimeda (s. 1.1.1)
2.2	Fibers (= optical axes) approx. radially oriented	
2.2.1	Shells (wall diam. 0.1—1); laminae with tangential or oblique orientation	
2.2.1.1	Sinuous or straight shells with diff. types of laminat. (Moll: C, A; Bra.: C)	— Mollusks, brachiopods
2.2.1.2	Outer shell: prisms, inner shell: isometr. calcite (previously A)	— Mollusks

5.15 Skeletal limestones

2.2.1.3	In tangent. sect. somet. honeycombs (= prisms, 0.05—0.1) (A, C)	— Pelecypods
2.2.1.4	Very regular laminae or nets (0.1—1) (C)	— Rudistids
2.2.1.5	Typical large chambers: shells with cross-lamination (A)	— Gastropods
2.2.1.6	With holes (= puncta, diam. 0.03, distance 0.1) (C)	— Punctate brachiopods
2.2.2	Shells without distinct laminations	
2.2.2.1	Charact. shape, Wall diam. 0.1—0.2 (C)	— Trilobites
2.2.2.2	Wall diameter 0.01—0.05, length 0.5—4 (C)	— Ostracodes
2.2.2.3	Vase-shaped. Wall diam. 0.01, length 0.06 (calcified)	— Tintinnines
2.2.3	Chambers	
2.2.3.1	Diam. 0.2—1. Walls with bundle extinction (A, C)	— Corals
2.2.3.2	Diam. 0.2 — very large, globul.; Walls with cross-lamin. (A)	— Gastropods
2.2.3.3	Diam. 0.05—0.2. Walls sometimes perforated (C)	— Hyaline foraminifera
2.2.4	Circular shape	
2.2.4.1	Cylinders (3—10) with or without growth bands (C)	— Belemnites
2.2.4.2	Cones (1, walls 0.1) with ribbing in longit. sect. (C)	— Tentaculites
2.2.4.3	As 2.2.4.2, but without ribbing (C)	— Styliolines
2.2.4.4	Cylinders (0.2—0.6), in part hollow, laminated (C)	— Brachiopod spines
2.2.4.5	Hollow spheres (0.5) with regul. protuberances (C, A)	— Charophyte oögonia
2.2.4.6	Well-rounded (elongate) grains, regul. lamin.	— Pre-Pleistocene oöids
2.2.4.7	Spheres (0.05), filled with cement (C)	— Oligostegina
3.	Micro- to macrocrystalline	
3.1	Single crystals	
3.1.1	Irregul. fragm. with or without screen patterns (C)	— Echinoderms
3.1.2	Cylinders (0.2—2), hollow in part, with radial patterns (C)	— Echinoid spines
3.1.3	Bisymmetric, tender shells (0.02), 1—2 single crystals (C)	— Lombardia (Saccocoma ?)
3.1.4	Small spines (0.05—0.1), one or a few crystals (C)	— Sponge spicules*
3.1.5	Spheres (0.03—0.2); cryptocr. wall, cement in center often 1 cryst. (C)	— Calcispheres, e. g. Oligostegina
3.2	Branched shape	
3.2.1	Segmented longit. sect.; circul. cross sect. (0.2—1) with radial lines; punctate stems in tangent. sect. (A)	— Dasycladacean algae
3.2.2	Tufts of filaments (0.03), or tubes (0.5—1) with vesicular walls (A)	— Codiacean algae
3.3	Shell fragments, transformed to calcite (Lamellae preserved) or dissolved and voids cemented; sometimes with micritized surfaces (A)	— Mollusks (Pelecypods, Gastropods, Cephalopods)

4. Non-carbonate skeletons
 4.1 Silica (opal, transformed to chalcedony and fibrous quartz during diagenesis)
  4.1.1 Small spines (0.05) — Sponge spicules*
  4.1.2 Spheres (0.1—0.2) with punctate walls — Radiolarians
  4.1.3 Elongated valves (0.01) with fine screens — Diatoms
 4.2 Calcium phosphate (brownish, very low birefringence)
  4.2.1 Lamellar plates, 0.1—0.2 in diameter — Inarticulate brachiopods
  4.2.2 Vesicular fragments — Bones
  4.2.3 Dense teeth and blades — Fish teeth & scales
  4.2.4 Lamellar toothlike blades and cones (0.5—3) — Conodonts

* Silica spicules may be calcified, calcite spicules silicified.

Index

Belemnites	2.2.4.1	Gastropods	2.2.1.3, 3.3
Blue-green algae	1.6	Green algae	1.1.1, 2.1.4, 3.2
Bones	4.2.2	Halimeda (green alga)	1.1.1, 2.1.4
Brachiopods, inarticulate	4.2.1	Lombardia	3.1.3
Brachiopods, punctate	2.2.1.4	Micritized grains	1.7, 3.3
Brachiopods, spines	2.2.1.4	Mollusks	2.2.1.1—2.2.1.3, 3.3
Bryozoans	2.1.1		
Calcispheres	3.1.5	Oligostegina	3.1.5
Cephalopods	3.3	Oöids, Pre-Pleistocene	2.2.4.6
Charophyte oögonia	2.2.4.5	Oöids, Recent	2.1.3
Codiacean algae	3.2.2	Ostracodes	2.2.2.2
Conodonts	4.2.4	Pelcypods	2.2.1.1, 3.3
Coralline algae	1.5	Radiolarians	4.1.2
Corals	2.2.3.1	Red algae	1.5
Dasycladacean algae	3.2.1	Rudistids	2.2.1.2
Diatoms	4.1.3	Saccocoma (Echinod.)	3.1.3
Echinoderms	3.1.1—3.1.3	Serpulid worms	2.1.2
Echinoid spines	3.1.2	Sponges, calcareous	1.2
Favreïna (fecal pellet)	1.1.2	Sponges, siliceous	4.1.1
Fecal pellets	1.1.2	Sponge spicules	3.1.4, 4.1.1
Fish scales	4.2.3	Stromatopores	1.3
Fish teeth	4.2.3	Styliolines	2.2.4.3
Foraminifera, agglutinating	1.4	Tentaculites	2.2.4.2
Foraminifera, hyaline	2.2.3.2	Tintinnines	2.2.2.3
Foraminifera, porcellaneous	1.4	Trilobites	2.2.2.1

5.152 Calcareous algae and Schizophyta (JOHNSON 1961, MASLOV 1956, 1962)

According to their living color, algae are subdivided into the phyla of green, red, brown and blue-green algae.

The colors are important to the occurrence of algae: Green algae mostly absorb in the red band. The red light, however, is almost completely absorbed in the uppermost layers of the sea; this is an important limiting factor in the depth distribution of green algae. The red algae and the blue-green algae contain red and blue pigments absorbing green, yellow (and also ultra-violet) light which is available in deeper water. Moreover, the blue-green algae developed a special adaptation mechanism to absorb most strongly the incident radiation which enables them to thrive even in turbid or deep water. For those at depths greater than 1000 m, in the Indian Ocean, it is still unknown whether they feed on iron bacteria or even became autotrophic chemosynthetic organisms using the Fe/Mn redox system as a source of energy. Both mechanisms would explain the Precambrian layered iron ore stromatolites (MONTY 1971).

Several algae are able to extract lime from sea water and to incorporate it in or deposit it around their tissue. The former applies to the red algae only, so that their cell walls are preserved. Some green and perhaps some blue-green algae precipitate calcium carbonate crusts around their cell membranes (POBEGUIN 1954, MONTY 1967); most blue-green algae, however, act merely as mud traps (sediment binders).

In addition to their role as sediment-forming organisms, the algae are important as environmental indicators; for example luxuriant growths of green algae is observed only in warm waters, in depths less than 10 meters (JOHNSON 1962, p. 251, VOIGT 1956). Some algae that do not deposit a hard skeleton occasionally can be preserved in outline by animals or plants that live on them (e. g. stalky cavities in bryozoan colonies, VOIGT, l. c.).

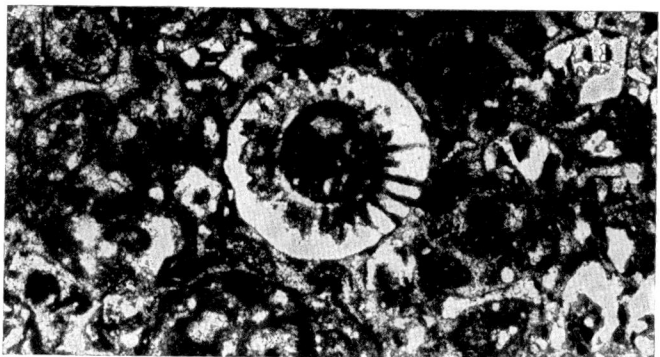

Fig. 5-17. Cross section of dasycladacean alga with vugs. Paleocene limestone, Libya (width of photograph 3.3 mm).

A. Green algae (Chlorophycophyta)

1. Dasycladaceae (figs. 5-17 to 19)

Appearing in Ordovician time at the latest, this family culminated in the mediterranean Triassic of the Alps (Diplopora, v. PIA 1942), and in the upper Jurassic (e.g. Arab formation, POWERS 1962). However, it was still abundant in the Mediterranean Tertiary, and at present is found in many tropical lagoon environments (see JOHNSON 1951, 1963). The Dasycladaceae have preferred warm seas since at least Triassic times (FRITSCH 1959, following v. PIA).

Tiny branches are arranged in whorls at regular distances around a central stem, which is wrapped by radial aragonite needles (POBEGUIN 1954). The aragonite

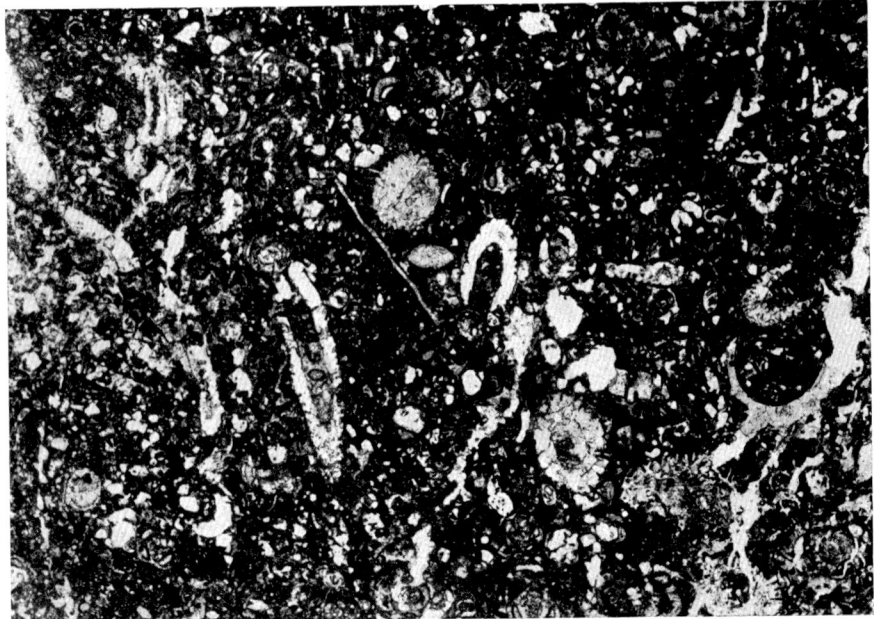

Fig. 5-18. Dasycladaceae and foraminifera in cryptocrystalline limestone. Paleocene. Libya (width of photograph 13 mm).

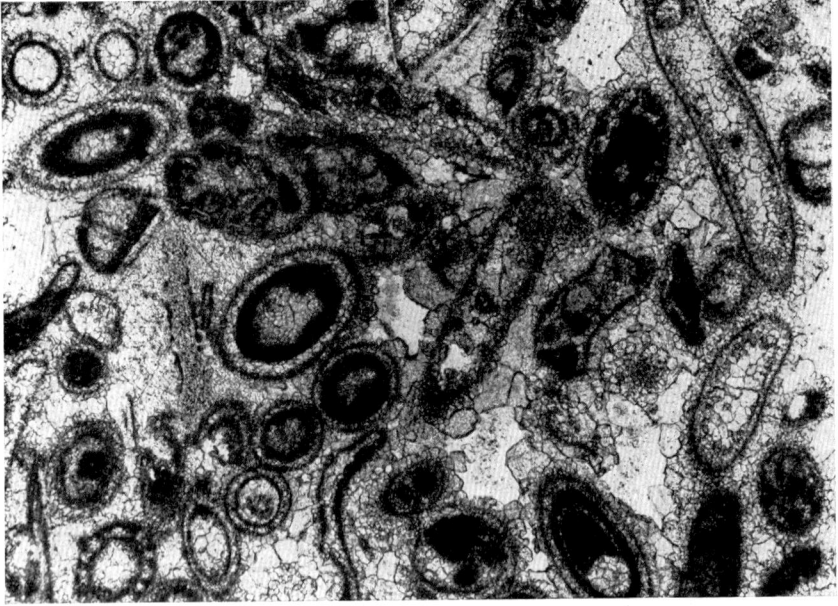

Fig. 5-19. Dasycladacean limestone, cemented. Upper Cretaceous, Libya (width of photograph 4 mm).

generally is transformed in limestones, whereas the stem may be filled by mud matrix (figs. 5-18, 19). In other cases the aragonite envelopes are dissolved and preserved as voids penetrated by small branches and gametangia (figs. 5-17, 18 in part). Stem diameters differ widely from species to species, ranging between 0.1 and 5 mm. Dasyclad s p o r e s are frequent in the alpine Permian (E. FLÜGEL 1966). They form spheres of 0.16 mm median diameter with structureless cryptocrystalline walls and are filled with fibrous cement.

2. Codiaceae

These green algae first appeared in the Cambrian (KONISHI 1961). Examples of these algae in the fossil record include *Hedstroemia, Garwoodia,* and the rock-forming *Microcodium* in the Cenozoic (JOHNSON 1961).

In modern seas, *Penicillus* and *Halimeda* are common. Like the dasyclads, they are mainly found in protected areas (GINSBURG 1956), although *Halimeda* occurs also in the leeward parts of reefs.

The clublike, platy or cylindrical segments of *Halimeda* consist of a tissue of loosely connected, partially branching medullary filaments about 0.05 mm in diameter and finer cortical utricles embedded in a cryptocrystalline fibrous aragonite matrix (fig. 5-20). For lime-producing activity see ch. 5.11 (2).

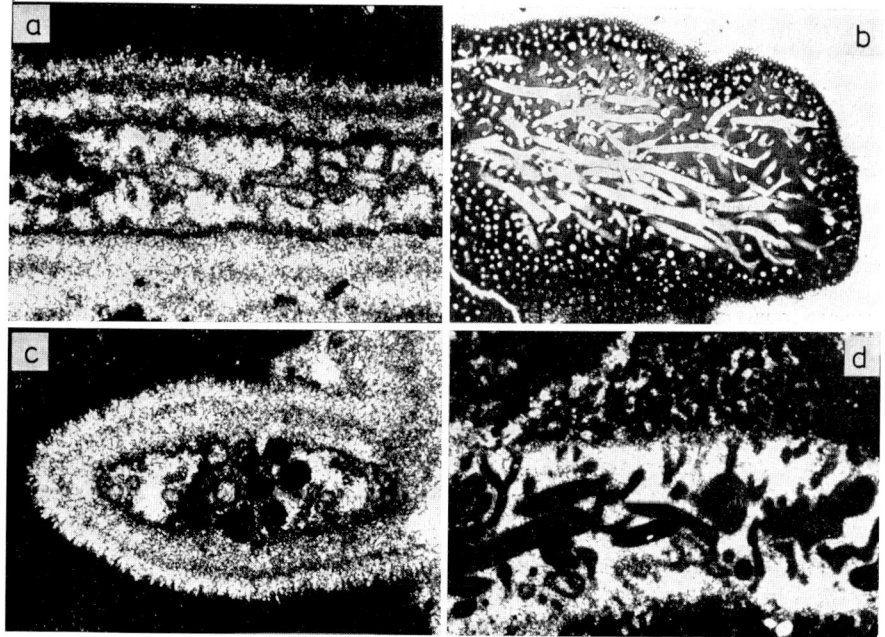

Fig. 5-20. Green alga *Halimeda*. a, c, d = Paleocene, Marsica (Abruzzo Mts., Italy; thin section, courtesy E. OTT); b = recent, Bermudas. a, b, d longitudinal sections, c cross section. a, b, utricles (tubes) cemented or empty, c, d, utricles filled with micrite. a, c, show "aggrading recrystallization" (FOLK 1965 b) around the alga prograding towards the micrite. (Width of each photograph 2.1 mm.)

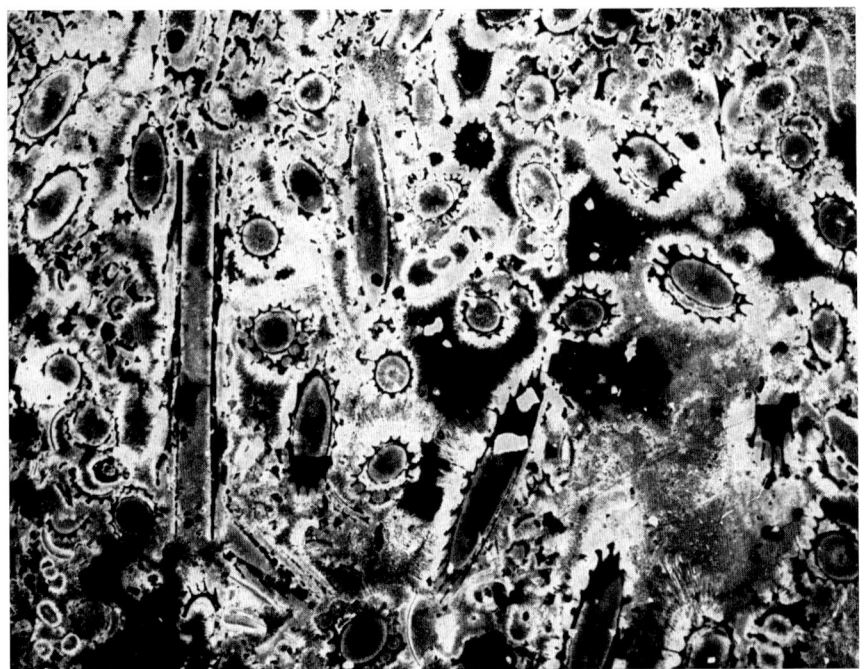

Fig. 5-21 a. Characea stalks, Miocene limnic limestones, Steinheimer Becken, Württemberg. Cortex of the stalks (primary aragonite, due to evaporitic tendency) is dissolved, but fibrous aragonite cement A is preserved. In the lower center is a central tube surrounded by cortical tubes. Negative print, width 12 mm (courtesy M. WOLFF).

3. Charophyta

Charophytes appear in the lower Devonian and are rock-forming in formations such as the Jurassic limestones of Colorado (JOHNSON 1961, p. 274) and Upper Cretaceous limestones of Eastern Peru (KOCH & BLISSENBACH 1960). At present these algae are found only in brackish and fresh water.

Of these bushy plants, only the calcified stalks and the oögonia, 0.5 to 1 mm thick capsules with delicately spiraled etchings, are preserved. Unlike the marine green algae, these skeletal parts consist of calcite (for exception s. fig. 5-21 a).

The tendency of the lower organisms, such as green algae, to form calcitic skeletons in fresh water is related to their relatively primitive abilities to precipitate skeletal $CaCO_3$. Thus they tend to produce that polymorph which is more likely to precipitate in fresh water, calcite. Higher organisms, such as mollusks, disregard the stability conditions of the environment, and can choose either calcite or aragonite as a skeletal material.

B. Red Algae (Rhodophycophyta) (fig. 5-25)

Solenoporaceae were rock-forming red algae in the Silurian of Gotland and in the Upper Jurassic of the Tethys (KOECHLIN 1949); they survived until the Tertiary. In the Mesozoic era, the Corallinaceae developed from the Solenoporaceae (presumably), and have become increasingly important since middle Cretaceous times. They are subdivided into the articulate Corallineae which build only branched forms (*Corallina, Amphiroa*) and the inarticulate Melobesieae which are represented by *Lithothamnion* and *Porolithon*. The latter form branches and also surf-resistant crusts in many modern reefs. Coralline algae also can form as algal nodules, by which they can survive intense rolling motion. They show the reddish color over the whole living surface (TERMIER & TERMIER 1963).

The red algae are characterized in thin section by a tiny fabric of cells, 10—30 microns in diameter, which are arranged in lines; Corallinaceae contain more pillars than the Solenoporaceae (JOHNSON 1951). The Corallinaceae contain series of larger cavities, sporangia, which are combined into conceptacles in *Lithothamnion*. The skeletons consist mostly of calcite, which contain 5-25 percent (by weight) $MgCO_3$ in the lattice (CHAVE 1954a). The extremely high magnesium contents of *Goniolithon*, however, are partly caused by inclusions of 4—5 percent brucite (SCHMALZ 1965, WEBER & KAUFMANN 1965). MOBERLY (1970) found as much as 53 mole-percent $MgCO_3$ in calcite of *Porolithon*. n_e is oriented perpendicular to the cell walls in recent *Lithothamnion* and *Goniolithon*. In Tertiary *Lithothamnion*, the cell walls are composed of cryptocrystalline calcite, probably the result of early diagenetic micritization (WOLF 1965b).

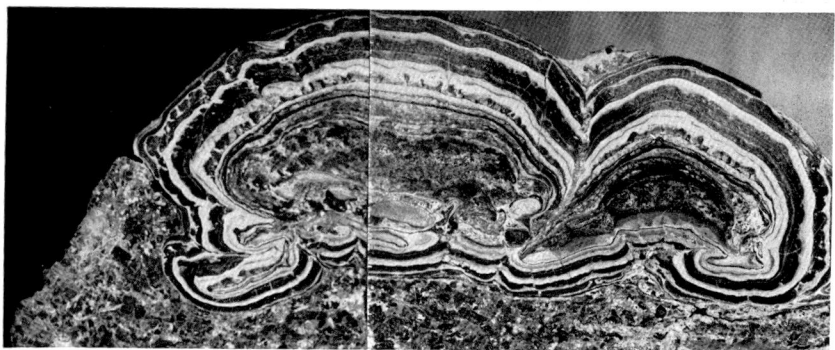

Fig. 5-21b. Transition oncolite-stromatolite. Upper Jurassic. Deister near Hannover (polished section; width of photograph 170 mm).

C. Blue-Green Algae
(Cyanophyceae = Myxophyceae = Schizophyceae)

Together with bacteria, these algae are the oldest preserved inhabitants of the earth, occurring in Precambrian limestones more than 3 billion years old (FENTON & FENTON 1957, KAUTSKY et al. 1960, RUTTEN 1962, A. H. MÜLLER 1964). They have been highly important rock-forming organisms throughout the entire sedimentary record. Moreover, they may have helped create the oxygen atmosphere (BERKNER & MARSHALL 1965). BUTTON (1973) suggested that prior to about 2 billion years, blue-green algal colonies were confined to sparse volcanic lakes, whereas later they started being produced over larger areas of platform seas.

These unicellular or filamentous, multiform masses of asexual algae owe their long history and wide distribution to their ability to tolerate extreme climates and chemical conditions. They grow in salt marshes (for example in the Great Salt Lake, EARDLEY 1938), in marine, brackish, lacustrine, and fluviatile environments and on the surface of moist land surfaces. On the other hand, their life zone extends from cool antarctic lakes to hot springs, such as in Iceland, where temperature can reach nearly 80° centigrade (v. PIA 1926, FRITSCH 1959). The mechanisms making blue green algae resistant to drought, high temperatures and lethal solar radiations have been discussed by MONTY (1971).

Based on growth form, one can distinguish the knobby "oncolites" (HEIM 1916, from Greek onkos = nodule) and the laminated or arched "stromatolites" (KALKOWSKY 1908, from Greek stroma = mat). As a transitional form the plucky "thrombolites" (AITKEN 1967, from Greek thrombos = lump) may be considered.

v. PIA (1928) termed oncolites and stromatolites without cellular structures as "Spongiostromata" whereas they were named "Porostromata" if the tubular filaments of about 0.02 mm diameter were preserved. The latter are represented by *Girvanella, Ottonosia, Somphospongia* and *Osagia*. The encrusting S p h a e r o c o d i u m, which is abundant in the Triassic of the Alps (E. FLÜGEL, pers. commun., SCHMIDT 1928, WAGNER 1931a, GASCHE 1956), is composed of structureless laminations as well as of colonies of *Girvanella* and may contain encrusting Foraminifera (v. PIA 1926, OTT 1966). Species of *Osagia* have been used for the subdivision of the Precambrian (SCHURAVLEVA 1964); structureless stromatolites have been used for the same purpose

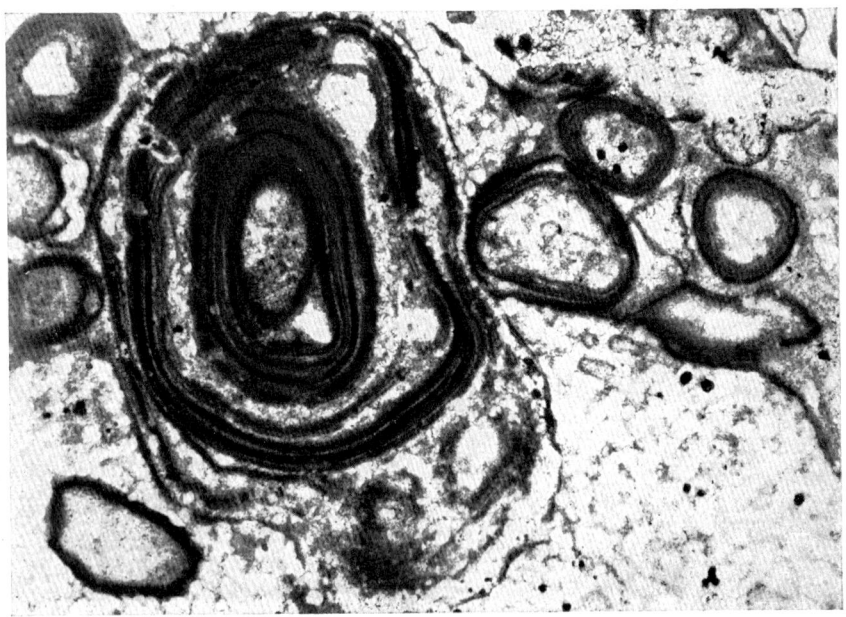

Fig. 5-22. Dolomitic oncolite. The irregular crusts are typical. Upper Permian Zechstein 2, well near Nienburg, Northwestern Germany (width of photograph 2.5 mm).

in the USSR, based on their growth forms and the types of lamination (CLOUD & SEMIKHATOV 1969).

Green algae may participate in the formation of both oncolites and stromatolites.

1. Oncolites (ss = spheroidal structures, LOGAN et al. 1964)

Definition

Irregular shelly, globular to lobate nodules, 0.05 to more than 100 mm in size are termed "oncoids" (figs. 5-5, 8, 21 b, 22, 23, 93, 95, 98). They differ from oöids by the irregular form, the absence of crystal orientation, and the general poor sorting; oncoids of 0.05 and 5 mm in diameter have been observed in one thin section. According to E. FLÜGEL & KIRCHMAYER (1962), defining fine-grained oncolites as "pisolites" is not convenient. Transitions of oncoids and ooids also occur, e. g. in the Upper Permian dolomites of W. Germany. "Oncolites" are rocks composed of oncoids.

Origin

Oncoids may form elastic soft bodies while they are alive as shown by modern (GINSBURG 1960) and fossil material (QUESTER 1964), or they may be slightly hard, such as the oncolites in the Lake Constance (fig. 5-23). With frequent agitation, smaller oncoids show more regular circular laminations, similar to, but not as compact, as oöids. The larger oncoids however are irregularly stratified; in the largest ones the layers are interrupted on the lower side (GINSBURG 1960). The interior structure are often destroyed by boring organisms. Oncoids may incrust pellets, oöids (SCHRAMM 1963), large pebbles (RUTTE 1955), skeletal fragments or whole skeletons ("mummy limestone", GASCHE 1956, PUMPIN 1965). Where hard particles are absent, the oncoids

Fig. 5-23. Oncoids at the sediment surface in shallow water of Lake of Constance (Gnadensee). In the lower right a green algal mat (width 88 mm, from G. MÜLLER 1966 a).

grow around soft intraclasts or even without any perceptible nucleus as in the Upper Permian of Northern Germany (fig. 5-98).

According to LOGAN et al. (1964) marine oncolites are typical to the deeper parts of the intertidal environments with moderate water agitation and to the area immediately below the low tide. Since they commonly occur in areas of lime mud sedimentation, they are rarely associated with oöids. How far they extend into deeper water is not well known. According to their frequent occurrence in stratified limestones at the lower flank of Jurassic bioherms in Southern Germany (GWINNER 1958, BAUSCH 1963a, HILLER 1964), they possibly lived in water as deep as 70 m. FREYTET & PLAZIAT (1965) found smaller oncoids in the sediments of lower energy environments.

In nonmarine environments, oncoids are well-known. Such blue-green algal nodules are found at the bottom of small rivers rich in dissolved $CaCO_3$ (STIRN 1964). In places in Lake Constance, for example north of the island of Reichenau, the sediment consists largely of calcitic oncoids 0.2—100 mm in diameter (SCHÖTTLE & MÜLLER 1968) (fig. 5-23). In certain situations subaerially-formed crusts can look very similar to oncolites (DUNHAM 1969b, THOMAS 1965).

2. Stromatolites

The numerous attributes of stromatolites and the different classifications are discussed by HOFMANN (1969). MONTY (1967) investigated recent algal mats in the Bahamas and found them to be composed of different species and different morphological types depending on the environment:

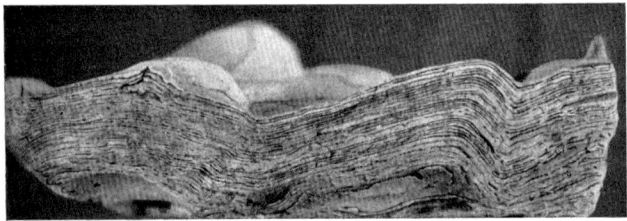

Fig. 5-24. Dolomitic stromatolite. Middle Magnesian Limestone (Permian) bioherm north of Durham, England. (Width of photograph 25 mm.)

a) Supratidal laminated mats form by an interbedding or mixture of the rapid-growing alga *Schizothrix calcicola* (partly calcified thin filaments, preferring moist environments), and the slowly growing alga *Scytonema myochrous* (thick laminated tubes, up to 30 micron in diameter, sometimes calcified, and able to survive arid conditions). *Schizothrix* is able to concentrate in its sheaths magnesium contents 3 to 4 times that of sea water (GEBELEIN & HOFFMAN 1969). This may explain why stromatolites are often preserved in dolomite (see figs. 5-24, 75).

These algae grow alternately upward (in moist periods or when they are buried by mud) and into horizontal mats (in dry times). However, only the Oscillatoriacea *Schizothrix* has the capacity to slip out of its sheath and migrate upwards through a sediment cover and so avoid fatal burial. *Scytonema*, therefore, can only thrive in areas where sedimentation is absent or very infrequent (HARDIE 1973). For this reason, *Scytonema* is the main blue-green alga in the marshes of Andros (Bahamas). HARDIE & GINSBURG (in prepar.) found *Scytonema* molds consisting of fine, granular, crystalline sheaths of Mg-calcite with 12—13 mole % $MgCO_3$ that lightly weld the algal tubes together where they touch. They "seem to be an *in situ* carbonate precipitate generated locally around the filaments, presumably aided by the metabolism of the *Scytonema* colony and/or the bacterial decay of the filaments (see HARDIE 1973). It accounts for at least 30—40% of the total mass of the entire marsh sediment." These earthy layers are interrupted by more or less muddy pellet laminations which presumably were generated during hurricanes, on northwest Andros Island.

Cryptocrystalline-calcified, unicellular blue-green algae or colonies of such coccoid cyanophytes as *Entophysalis deusta* are frequent in the mats, causing a clotted structure. These supratidal algal mats are composed of magnesian calcite with 5—10 mol percent of $MgCO_3$. Many fossil laminated algal mats may be explained in this way (MONTY 1967). An important characteristic of the algal sheaths is their custom of coating sedimentary particles, thus making them sticky, and, at the same time, isolating them to a large extent chemically from the sea water (BATHURST 1971 b).

b) Intertidal. In brackish intertidal environments laminated mats are characterized by organic layers (*Scytonema crustaceum* = 20 micron-filaments, alternating with lime mud layers [BLACK 1933] in sub-millimeter intervals; MONTY 1967). According to HARDIE & GINSBURG (in prepar.), these layers form by flooding during heavy storms. In the tidal flats of Andros Island, 3 depositional events per year were observed as an average, causing a build-up rate of about 1.5 mm/year.

In marine intertidal environments, laminations are less well developed. During neap tides, the mats may become indurated and brecciated. Skeletal particles can be bored and the borings filled with Mg_{5-10}-calcite (fig. 5-66). Small domes occur especially in the lower intertidal areas. They are linked by horizontal mats ("*Schizothrix* laminated domes"; "Collenia"-type; "LLH" = laterally linked hemispheroids, LOGAN et al. 1964) (fig. 5-24). In this environment more lime is trapped than precipitated (see also KENDALL & SKIPWITH 1968).

Similar mats are found in lagoons and in quiet freshwater lakes. In the latter, calcifying forms prevail (GEBELIN, pers. commun.).

c) Subtidal. Globular and domal colonies prevail in the subtidal environment (e. g. *Lyngbya*) ("Cryptozoon"-type; "SH" = discrete, vertically stacked hemispheroids, LOGAN et al. 1964). Scouring of the lower parts during tidal flow and ebb may even form club-shaped masses.

NEUMANN et al (1970), however, have described subtidal algal mats from the Bahamas. These mats are composed of a spongy framework of *Lyngbya* filaments 14—18 µ in diameter and *Schizothrix* filaments 1—1.5 µ in diameter, which are covered by horizontally oriented *Lyngbya* filaments.

The irregular, often ellipsoidal "thrombolites", which form most of the Cambrian and Ordovician algal bioherms, grew in subtidal environments (AITKEN 1967). The same applies to the stromatolites in the lower Triassic north of the Harz Mountains, for which KALKOWSKY (1908) coined the name "stromatolite". Oölites ("Rogensteine") occurring between these stromatolites point to a subtidal environment.

d) In the unsheltered coastal area, isolated pillarlike stromatolite domes of the Cryptozoon type are found (LOGAN et al., l. c. and LOGAN et al., AAPG mem. 13, 1970).

These findings lead to a generalization supported by observations by HOFFMAN et al. (1971) in recent stromatolites: "Where wave action is strong, discrete columns of up to 1 m relief occur. Where wave action is moderate, small branching columns and laterally-linked domes with less than 15 cm relief occur. In quiet water, loferites and stromatolites are stratiform. Stromatolites, with laminated internal structure, accrete by trapping of oöids and foram tests by mats of filamentous blue-green algae. Loferites, with fenestral but poorly laminated internal structure, result from trapping of similar particles by mats of coccoid blue-green algae." Most filaments are found in the lower intertidal zone.

Observations on stromatolites have been published by CLOUD (1942), HAMILTON (1961), MASLOV (1962), SCHURAVLEVA (1964), DAVIS (1966) and GREGOIRE & MONTY (1963, with electron micrographs). DE MEIJER (1969) and HUDSON (1970) found algal filaments in the insoluble residue of limestone stromatolites as old as Cambrian by digesting in 1% HCl.

Modern blue-green algal bioherms are described from freshwater Green Lake, N. Y., by BRADLEY (1929) and from the Great Salt Lake, Utah, up to 3 m below water level (EARDLEY 1938). The importance of blue-green algae as reef building organisms is quoted in ch. 5.18. It is noteworthy that *Schizothrix calcicola* lives in depths down to 390 m, in the Dead Sea (MONTY 1967), see also remarks on p. 214.

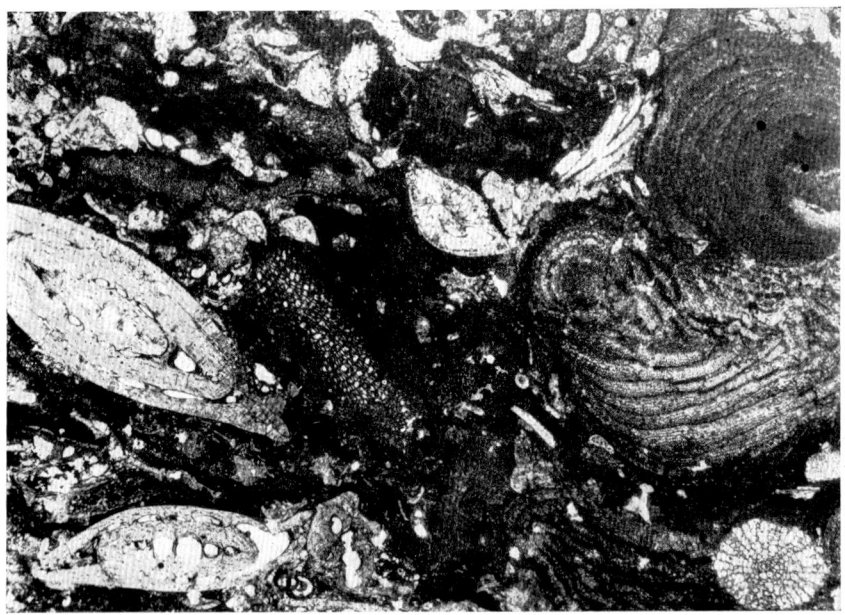

Fig. 5-25. Lithothamnion limestone with two nummulites (left) and several rotaloid foraminifera (center and left). Eocene, Bavaria (width of photograph 6.3 mm).

5.153 Nannoplankton

1. Coccolithophorids

These protista (half plant, half animal) appeared in the Liassic and were particularly frequent in Cretaceous and Tertiary times. They live in the open sea as plankton and consist of unicellular globules composed of tiny rings and discs 2 to 20 micron in diameter, the so-called coccoliths (figs. 5-3 and 5-26 a).

The observation of a mobile stage in the life cycle of the (generally immobile) *Coccolithus pelagicus* may be important for our understanding of micritic limestones: These individuals consist of discs 1 micron in diameter which disintegrate into 0.13 micron-sized isometric rhombohedra after death (GAARDER & MARKALI 1956, PARKE & ADAMS 1960). Such coccolith fragments would not be identified in the sediment.

68 percent coccoliths were found in modern deep-sea sediments, 58 percent in Oligocene deep-sea sediments (SEIBOLD 1964 a). In certain white, porous limestones named chalk ("Schreibkreide"), they can constitute up to 72 percent (MÜNZBERGER 1958). By electron microscope techniques 10—50 percent of coccoliths were found even in such pelagic limestones, which had revealed previously only a few coccoliths in thin sections (fig. 5-3; HONJO & FISCHER 1964). However, they also can be found occasionally in shallow marine limestones (FLÜGEL & FRANZ 1967).

These tiny "nannofossils" consist of calcite and show a distorted black cross under crossed nicols (see BRAMLETTE & MARTINI 1964). Other recent articles which have dealt with this group include those by DEFLANDRE (1952), KAMPTNER (1952), BRAMLETTE (1958), MARTINI (1961), STRADNER (1964), NOËL (1965, 1970), and FISCHER et al. (1967).

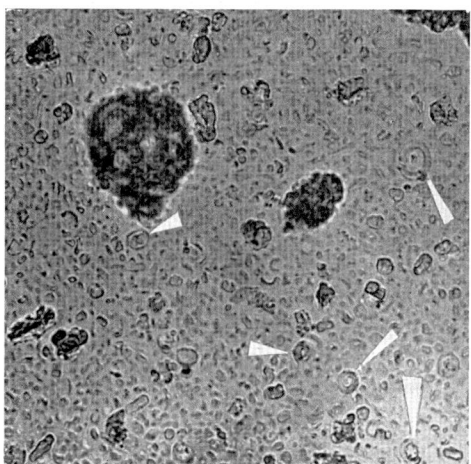 

Fig. 5-26 a. Suspension of chalk with coccoliths (arrows!). Upper Cretaceous, Lägerdorf (width of photograph 0.16 mm).

Fig. 5-26 b. Coccoliths *(Reticulofenestra ornata)* in the bituminous "Bändermergel", Lower Tertiary, Bavaria (width of scanning electron micrograph 0.023 mm; from MÜLLER & BLASCHKE 1971).

226   5.1 *The primary sediments and their origin*

2. Discoasters

Tiny, calcareous, stellate or rosette-shaped plates, known as discoasters, derived from some planktonic organisms presumably related to the coccolithophores (fig. 5-32). They are extinct but were common in some late Mesozoic and Tertiary rocks, particularly in those of open-sea deposition (BRAMLETTE & RIEDEL 1954).

Fig. 5-27. The "Calcisphaera" *Thoracosphaera saxea*, presumably a complete Coccosphaera. Paleocene flysch, Zumaya, Spain (electron micrograph by A. G. FISCHER; width of photograph 0.026 mm).

3. "Calcisphaera" (fig. 5-27)

This name was given by WILLIAMSON (1880) to a group of calcite spheres of unknown origin, 0.03—0.2 mm in diameter. They consist generally of a cryptocrystalline wall engulfing a void which is now filled with larger calcite crystals. As rock formers, they are mostly restricted to Paleozoic limestones (BAXTER 1960). In places they have been interpreted as algal spores (E. FLÜGEL 1966, RUPP 1967), but they may have had very different origin (RICH 1965).

In the Upper Cretaceous, 0.05 mm-sized spheres, named *"Oligostegina"*, are wide-spread, which probably are Coccosphaera of ortholithid calciflagellates (SIEHL 1968, see also fig. 5-27). From their distribution in an Iranian basin, ADAMS et al. (1967) concluded a benthonic habitat exclusively along the shelf edges. As to the taxonomic assignment, these authors stated a relationship with either lagenid foraminifera or siphonate algae.

5.154 Protozoans

1. Foraminifera

These unicellular organisms appeared in the Upper Cambrian, and have been important since the Lower Carboniferous (JOHNSON 1951). Most foraminifera are marine (HILTERMANN 1963); a few species live in brackish environments (ROTTGARDT 1952). According to their mode of life, they may be subdivided into benthonic (i. e. living on the sea floor) and planktonic foraminifera (i. e floating near the sea surface).

A. Planktonic foraminifera did not appear before the Triassic (SCHWARZACHER 1948, OBERHAUSER 1960, LEISCHNER 1968) (fig. 5-28). They mainly consist of

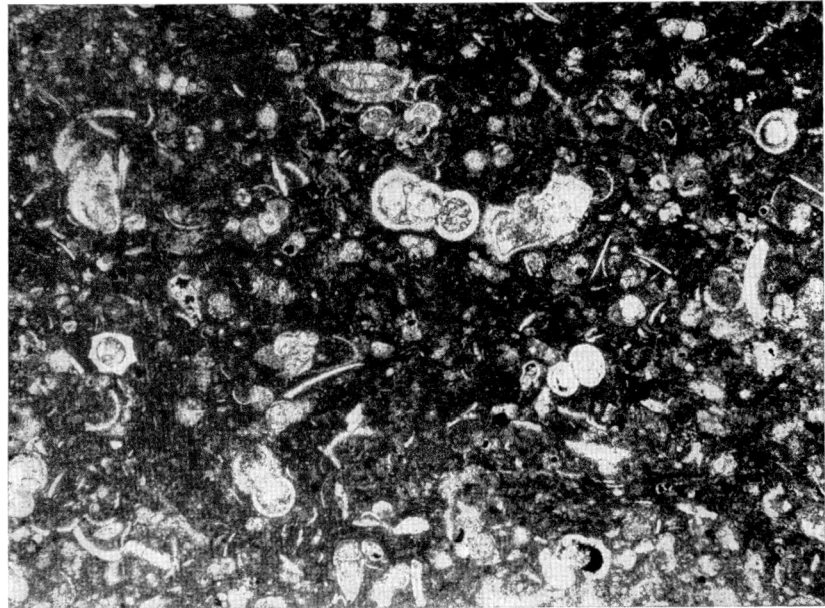

Fig. 5-28. Cryptocrystalline limestone with planktonic foraminifera. Upper Cretaceous, Libya (width of photograph 1.8 mm).

globular, tiny-walled chambers. The diameter of these foraminifera (*Globigerina, Globorotalia, Orbulina, Gümbelina, Hantkenina* and others) rarely exceeds 0.5 mm.

Since they are not attached to the bottom, they do not depend on the facies. This enhances their importance as stratigraphic fossils. They are abundant in many deep sea sediments and sedimentary rocks where benthonic foraminifera are absent (GRIMSDALE & VAN MORKHOVEN 1955), but the abundance of planktonic foraminifera is not necessarily a reliable indication of a deep sea origin (COLE 1957).

B. The **benthonic foraminifera** are quite different in morphology and in size (0.1—100 mm, JONES 1956). They depend strongly on the temperature and sediments at the sea floor where they are living, and are especially common in shallow depths (PHLEGER 1960). Accordingly, most of the foraminiferal limestones supposedly have been deposited in warm shallow seas (CAYEUX 1935, JOHNSON 1951). Encrusting foraminifera are frequent in bioherms (EMERY, TRACEY & LADD 1954, FRITZ 1958, E. FLÜGEL et al. 1963).

NATLAND (1957) observed the following depth zonation characterized by different foraminifera communities:
  I. Lagoonal (0—6 m, temperature fluctuations from 5 to 24° C),
  II. Upper neritic (0—40 m, 13—21° C),
  III. Upper neritic to upper bathyal (40—270 m, 8.5—13° C),
  IV. Upper bathyal (270—600 m, 5—8.5° C),
  V. Middle bathyal (600—1200 m, 3—5° C),
  VI. Lower bathyal (1200—2300 m, 2—3° C) (see also PHLEGER 1964).

A similar depth zonation off Southern California has been applied to Tertiary sediments in the hinterland by NATLAND (1957) and BANDY & ARNAL (1960). PHLEGER (1951) showed that shallow marine foraminifera can be displaced into deep marine areas by turbidity currents. BANDY (1964) investigated the relationships between shell morphology and water depth for different groups of foraminifera (see also BANDY 1967).

On the basis of their skeletal structure, as seen in thin section, the foraminifera can be divided into the following groups (WOOD 1949, JONES 1956):

a) **Agglutinating foraminifera** are composed of granular walls consisting of sand and lime grains. They have prevailed since Paleozoic times but were never major rock-forming organisms.

b) **Porcellaneous foraminifera** are composed of cryptocrystalline carbonate. In thin section they appear dark (figs. 5-29, 5-31), whereas they are white (like porcelain) in reflecting light. They consist of calcite which contains more than 10 mole percent of $MgCO_3$ in the recent foraminifera of warm seas, according to BLACKMON & TODD (1959). The skeletons are not perforated ("**imperforate foraminifera**"). They are represented by the group Miliolidea (e. g. *Quinqueloculina*, fig. 5-31), of which the Miliolidae and Alveolinidae were important as rock-forming organisms in the Lower Tertiary.

c) **Hyaline foraminifera** are composed of tiny fibers or prisms of calcite with n_e perpendicular to the surface. They are light in thin section (figs. 5-25, 28, 30), but

Fig. 5-29. Porcellaneous foraminifera *Rhapydionina liburnica* STACHE, a characteristic inhabitant of marine-lagoonal environments, according to FARINACCI (pers. comm.). Maestrichtian, Monte Turchio, Middle Apennines, Italy (thin section of A. FARINACCI, Rome; width of photograph 3 mm).

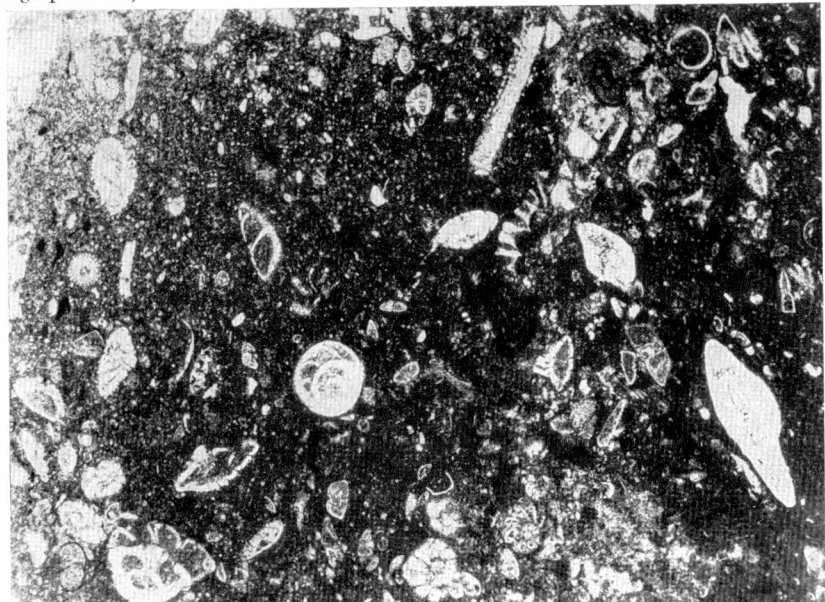

Fig. 5-30. Cryptocrystalline limestone with vitreous rotaloid foraminifera and two sea urchin spines (upper center: longitudinal section, at left: transverse section). Upper Cretaceous, Libya (width of photograph 6 mm).

dark and translucent in reflecting light. Most hyaline foraminifera contain perforations 0.5—15 micron in diameter ("perforate foraminifera"). For more detail see BATHURST (1971b).

Two examples of this group are the Fusulinidae, ranging from 0.5 mm to more than 35 mm in diameter, which were rock-formers in the Upper Carboniferous and Permian, and the Rotaliidea, which include most of the planktonic and many benthonic shallow-water foraminifera. Examples of this latter group are the Rotaliidae (figs. 5-25, 5-30), which became minor rock-formers in the Tertiary, the Orbitoidae, which comprise most of the Tertiary large foraminifera, and the Nummulites (fig. 5-25).

The skeletons of modern planktonic foraminifera are composed of calcite (less than 5 mole percent of $MgCO_3$), whereas hyaline shallow water benthonic foraminifera consist of calcite or Mg-calcite (BLACKMON & TODD 1959). Aragonitic foraminifera are rare (BANDY 1954, WITTHUHN 1968).

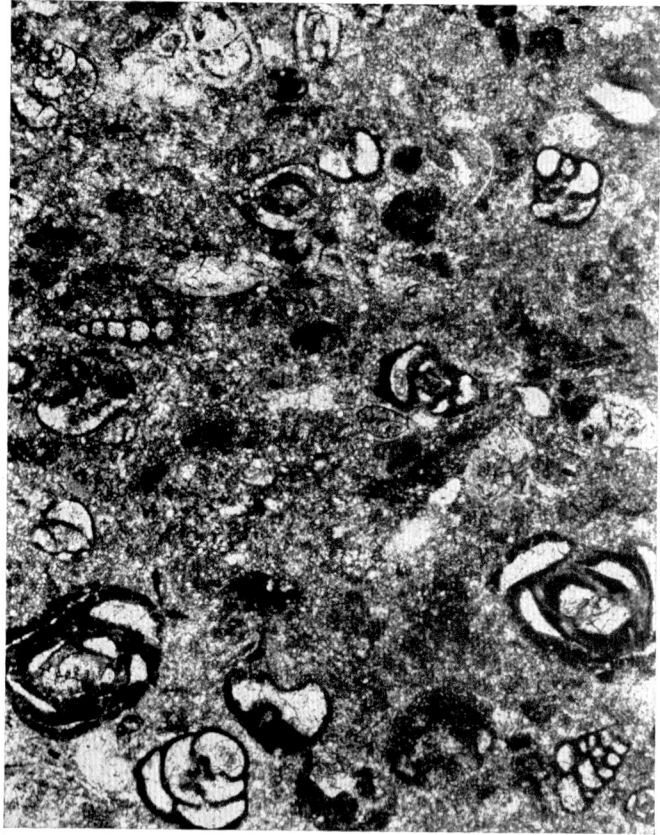

Fig. 5-31. Cryptocrystalline dolomitic limestone, rich in porcellaneous foraminifera (near the right and the left side in the lower part of the photograph are two *Quinqueloculina*). Paleocene, Libya (width of photograph 4.2 mm).

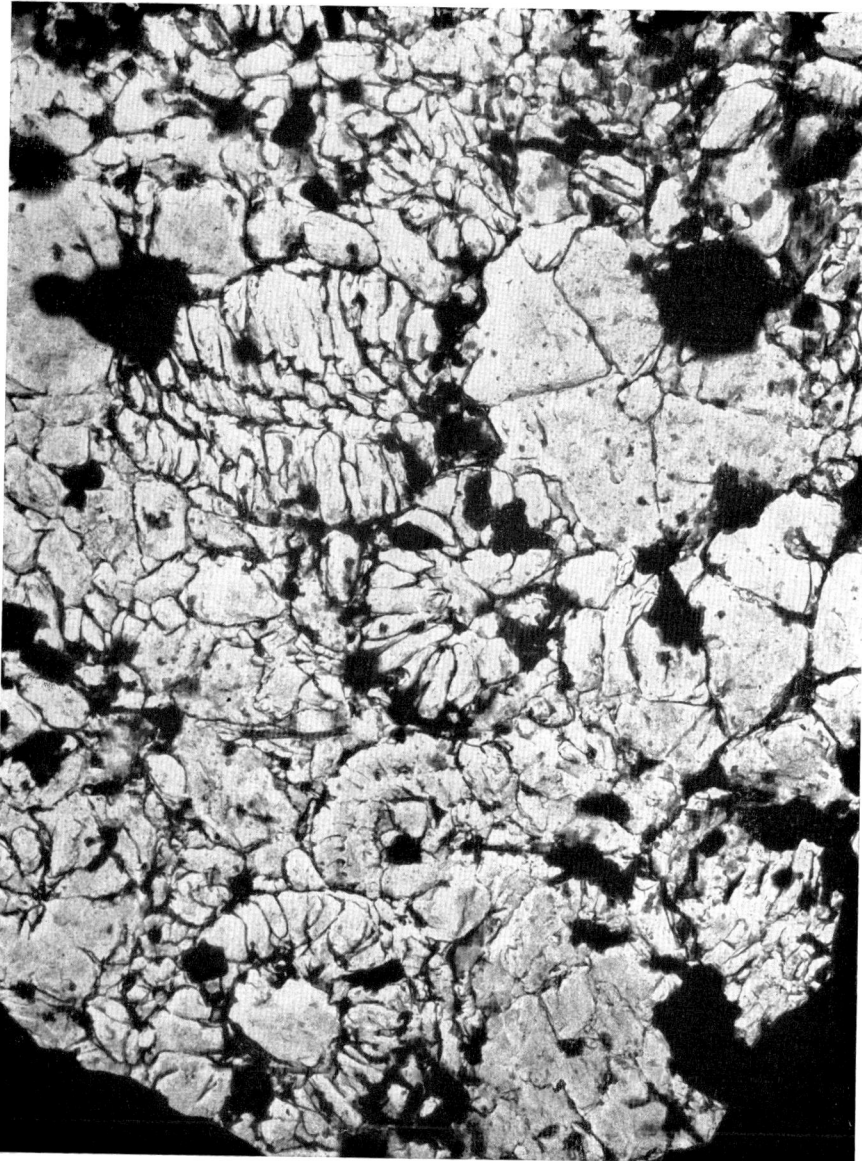

Fig. 5-32. *Nannoconus colomi* (DE LAPPARENT) as a main constituent of the pelagic Calcare Maiolica (Biancone). In the upper right center a discoaster. Neocomian, Monti Martani, Umbria, Italy (electron micrograph by A. FARINACCI, Rome; width of photograph 0.026 mm).

2. Other protozoans

As a supplement, two other groups of protozoa should be mentioned: The Tintinnines (= Calpionellae), planktonic animals with bell-formed skeletons, 0.08—0.15 mm in diameter, occur mainly in the uppermost Jurassic and Lower Cretaceous of the Tethys. These protozoans are thought to have been primarily organic when living, but now are calcified fossils. Similar distributions are seen in Nannoconus, tiny calcite cones 0.005—0.015 mm in diameter (fig. 5-32) (BRÖNNIMANN 1955, A. H. MÜLLER 1958, FARINACCI 1964).

The Upper Cretaceous limestones along the English Channel and the Baltic Sea consist mainly of protozoa. Two rock types are developed; the "Pläner", a dense, more or less marly limestone, containing *Inoceramus* prisms and numerous benthonic foraminifera, which was deposited in a nearshore environment, and the proper "chalk" ("Schreibkreide") facies, a friable white limestone rich in coccoliths and planktonic foraminifera, but poor in *Inoceramus* prisms, which presumably was deposited farther offshore. The water depth in this region has been estimated to about 100—250 m (NESTLER 1965, HUDSON 1967). Primary aragonite skeletons are rare in the chalk (WEGNER 1926, VOIGT 1929, CAYEUX 1935, GEORGE 1957, BLACK & BARNES 1959, HILTERMANN, pers. commun., KENNEDY 1969), suggesting that dissolution of aragonite skeletons (e. g. the mother-of-pearl layer of *Inoceramus*) may have begun at these depths (HUDSON 1967). For hardgrounds see ch. 6.32.

5.155 Sponges (Porifera)

(Cambrian to Recent; maximum development in the Upper Jurassic, Upper Cretaceous and Lower Tertiary); mainly marine.

A. The Calcispongea attained their maximum number of genera in Cretaceous time (A. H. MÜLLER 1958). The Archaeocyathidae, cup-formed calcareous fossils up to 5 m in size, are found in the Lower Cambrian (DEBRENNE 1964) and may have been functional precursors of the Calcispongea (SEILACHER, pers. commun.).

The Calcispongea are inhabitants of the shallow sea, and are most abundant in water depths of less than 4 m (POKORNY 1958). They prefer clear, warm waters (TWENHOFEL 1950a). Their skeletons, which are constructed of calcite, generally do not occur in rock-forming quantities but they are frame-building reef organisms in the Permian of Western Texas (NEWELL et al. 1953) and in the Middle Triassic of the Alps (OTT 1967b).

Two types of fossil remains are found:

1) Complete skeletons with irregular tubes 0.2—1 mm in diameter (fig. 5-33)
2) Spicules, mostly triaxial, about 0.02—0.05 mm in diameter and 0.2—0.5 mm length.

B. The Silicispongea were important frame-building organisms in the Upper Jurassic bioherms of Southern Germany (FRITZ 1958, HILLER 1964). The rectangular skeletons of these hexactinellid sponges are mostly calcified in this occurrence (fig. 5-34). The supporting skeleton, a mesh-like rectangular framework of cells, 0.2 to 0.3 mm in diameter, is often well preserved. The diameters of the sclerites vary from 0.05 to 0.1 mm. The fine spicules from the epidermal layers usually are not preserved

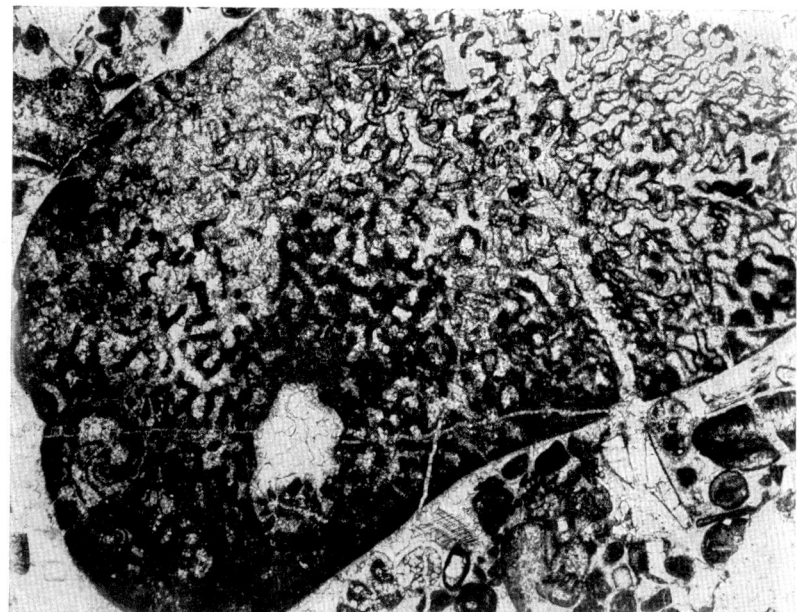

Fig. 5-33. Calcisponge fragment in biocalcarenite. Lower Jurassic, Northern Alps (thin section of F. FABRICIUS; width 10 mm).

(A. H. MÜLLER 1958). In addition to Hexactinellidea, Lithistidea whose skeletons are of similar size but more irregularly formed, are found in the Upper Jurassic. Silicisponges serve frequently as sources of silica (fig. 5-86).

Beginning in the Upper Cretaceous, the Hexactinellidea shifted their habitat into increasingly deeper seas. Although siliceous sponges are found from intertidal to deep sea environments, most are situated in depths between 200 and 500 m. In contrast to other sponges they prefer to live on soft substrates.

The Lithistidea are found mainly between 100 and 350 m at present and supposedly exhibited similar zonations in the past (A. H. MÜLLER 1958).

5.156 Coelenterates

Most Coelenterates are dominantly colonial and can build large bioherms, the best example being coral reefs (ch. 5.18). Coral colonies can be degraded by surf as well as by boring and lime-dissolving organisms [e. g. spongia *(Cliona)*, mollusks, fungi, annelids, and algae (WELLS 1957a, b, GOREAU & HARTMAN 1963)] The debris produced by such degradation, however, usually is confined to the reef environment; limestones composed of coelenterate fragments are rare outside reef complexes.

Important rock-forming organisms occur in the classes Hydrozoa and Anthozoa.

A. Hydrozoa

At present, prominent rock-forming hydrozoans are represented by the orders of Milleporina ("stinging corals") and Stylasterina with aragonitic skeletons, both of which first appeared in the Upper Cretaceous.

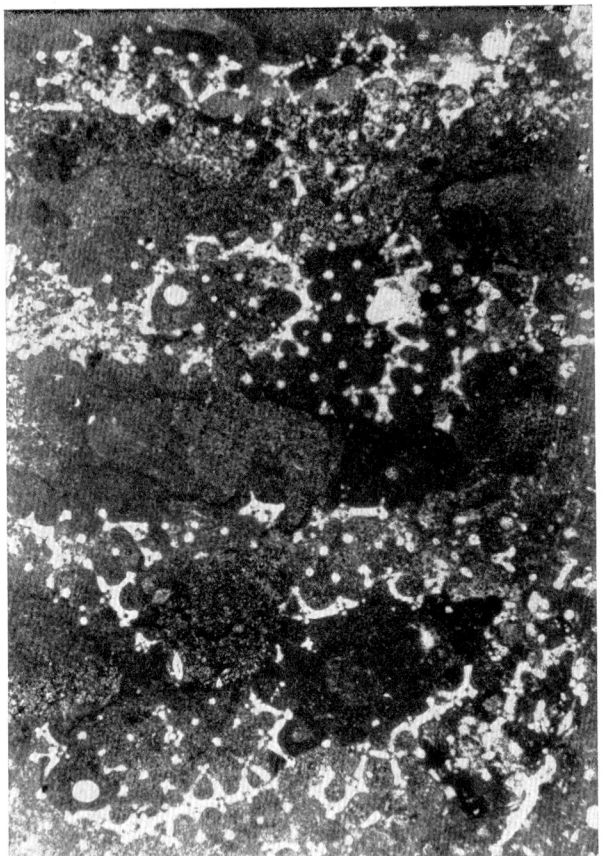

Fig. 5-34. Calcified silicisponge (light). Upper Jurassic, Württemberg. (Width of photograph 4.5 mm.)

In the past, stromatopores were the most important order of Hydrozoa, occuring in rocks from Cambrian to Mississippian age. They attained their culmination as reef builders in Silurian and Devonian time (table 5-4, p. 273). Morphologically similar forms are found in Jurassic and Cretaceous rocks (E. FLÜGEL & SY 1959).

Globular hydrozoans have been observed in arenitic Upper Jurassic sediments, whereas the calcilutites of the same sequence contain very thin and encrusting hydrozoans (DEUTSCH 1972). The growth form of stromatoporoids reflects the energy of the environment. Sub-spherical and bulbous shapes are typical for high energy, tabular and dendroid shapes for lower energy (ABBOTT 1973).

According to MANTEN (1962a), hydrozoa were more sensitive to mud than corals, bryozoa or crinoids.

The stromatopores are nodular, branching or incrusting colonies. They are constructed of delicate laminae which are arranged parallel to the surface of the

The optical orientation is distinctive: whereas n_e is often radially in the border zone, it is oriented parallel to the walls in the interior of the organism, even in early diagenetically dolomitized specimens (fig. 5-37). The latter orientation is found only in bryozoa, serpulae, and recent oöids.

The bryozoa are major rock formers especially in the Paleozoic (bioherms in Carboniferous and Permian), but also in the Cenozoic (such as the peninsula Kertsch and in the Mediterranian, SEIBOLD 1964a). Encrusting bryozoa may be frame-building, in places.

5.158 Brachiopoda

(Cambrium — Recent)

These rock-forming organisms are prominent in Ordovician to Carboniferous rocks. They are exclusively marine and prefer the shallow water where thick-walled types predominate. Today, some species — with thin walls — also occur at greater depths (TWENHOFEL 1950a).

Most of the inarticulate (= hingeless) brachiopoda construct their shells of horn (chitin) layers alternating with layers of hydroxyl-apatite (LOWENSTAM 1963).

The articulate brachiopoda (= with hinge) use only calcite and magnesian calcite (LOWENSTAM 1961) as shown by the good preservation of shell structures with occasional preservation of original colors (A. H. MÜLLER 1950). One occurrence of asphalt-preserved calcitic and aragonitic fossils in the Pennsylvanian of Oklahoma

Fig. 5-38. Brachiopod shells, cross-section (bottom) and tangential section (above the center) with perforations. In the center and at right echinoderm fragments whose tiny pores are filled with chlorite. White = terrigenous grains. Middle Jurassic, Hannover (width 4.6 mm).

contained only calcitic brachiopods (STEHLI 1956). The shells consist of a thin exterior layer with uniform thickness of calcite fibers ($\parallel$ c,) which are oriented perpendicular to the shell surface (WILLIAMS 1956), and an interior layer of irregular thickness consisting of calcite fibers with oblique orientation with respect to the surface. In the "punctate" brachiopoda, the interior layer is perforated at a distance of 0.05—0.1 mm, perpendicular to the surface (fig. 5-38, e.g. Terebratulacea, Terebratellacea). These perforations are not developed in the "impunctate" brachiopoda, e.g. Rhynchonellacea (A. H. MÜLLER 1958). Hollow spines, ca. 0.2 mm in diameter, with tangential lamination, are frequent.

5.159 Worms (fig. 5-39)

(Ordovician — Recent)

Only the permanent dwelling tubes secreted by the serpulids are important as rock-forming materials (subfamilies: the straight Serpulinae and the planispiral Spirorbinae; A. H. MÜLLER 1958). These worms are found in both cold and warm marine environments, mostly in shallow water, often together with other lime-secreting organisms. Occasionally they build entire bioherms, e. g. in the Laguna Madre, Texas (FRIEDMAN 1964).

Prior to the Triassic, serpulids were mostly smooth-shelled; from the Triassic until the Upper Cretaceous however, they became increasingly sculptured. Tube diameter differs from species to species (0.2—10 mm, A. H. MÜLLER 1958). In cross sections of recent serpulas from the Mediterranean as well as of Upper Jurassic serpulas the tubes are constructed of concentric layers of calcite with n_e tangential

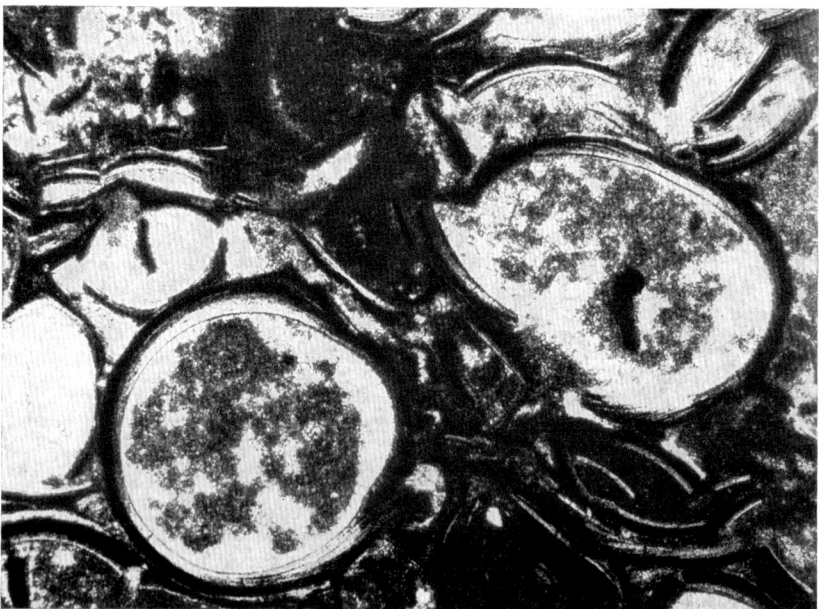

Fig. 5-39. Serpulid limestone, uppermost Jurassic, N.W. Germany (width of photograph 3 mm).

to the surface, perpendicular to the axis of the tube (BØGGILD 1930, p. 16). According to LOWENSTAM (1954), aragonite also may be secreted; the relation of aragonite: calcite may change from layer to layer depending on the temperature of formation (fig. 5-14).

More important from the geological point of view are those worms which modify the fabric of sediments by burrowing (and eroding) into calcareous substrate and by secretion of preservable pellets.

5.1510 Mollusca

Most marine mollusks live at depths of less than 50 m, where their food is predominantly plants and herbivore animals. In the deep sea, mollusks feed mainly upon other carnivores (NATLAND 1957). GIGNOUX (1926) summarized the facies relationships as follows:

Rudistids and oysters build reefs, the latter often in brackish water; *Pecten*, *Avicula*, *Mytilus* and ornamental bivalves live on sand- or calcarenite bottoms; *Astarte* on fine-grained lime mud; smooth bivalves, *Exogyra* and *Gryphaea*, on marly bottoms in a little bit deeper waters; *Pholadomya* on clay bottoms, and *Posidonomya* on bituminous clay (partially after BERGQUIST & COBBAN 1957). According to ZIEGLER (1967) the soft bottoms were presumably colonized by Desmodonts (e.g. *Pholadomya*) the hard bottoms by Dysodonts (= the other bivalves mentioned above, except *Astarte*).

The influence of salinity on Mid-Jurassic forms has been studied by HUDSON (1963), by PAPP (1963) in the Upper Tertiary, and by R. H. PARKER (1959) in modern marine environments. In fresh-water environments, and also brackish waters, mollusks often are characterized by finer shells and smaller sizes than in the marine environment (REMANE 1963). Also in the deep sea, shells are thinner, presumably because of the cold waters. Thin shells are frequent in the limestones of the Tethys and often have been misinterpreted as algal filaments (FABRICIUS 1966).

Shell structure and mineralogy vary strongly with various mollusks. Since their shells are perhaps the most common components within limestones in general, a summarization of the various shell structures is given in the following paragraph, mainly from BØGGILD (1930, see also HUDSON, 1968):

1) H o m o g e n e o u s : The shell seems to be a bent single crystal; this structure is similar to (3).

a) Aragonite. Occurs only in mollusca. $n_x \perp$ shell surface (fig. 5-40).

b) Calcite. Also occurs in foraminifera and trilobites ($n_\varepsilon \perp$ shell surface) and in serpulids ($n_\varepsilon \mathbin{/\mkern-6mu/}$ shell surface).

2) P r i s m a t i c (mostly calcite): delicate polygonal prisms, mostly $\perp$, occasionally $\mathbin{/\mkern-6mu/}$ to the shell surface (in this case often aragonite, and composed of small prisms in feather-like arrangements).

3) B l a d e d

a) Aragonite: "nacreous structure". Occuring only with mollusca, consisting of micron-thick blades parallel to the shell surface, interbedded with thin organic layers; n_x perpendicular to shell surface.

Fig. 5-40. *Cyrena* coquina preserved as aragonite due to oil impregnation. Homogenous shell structure. Lower Cretaceous; oil pool Dalum (Emsland area) (width 8 mm, crossed nicols).

Fig. 5-41. Mollusk shells; primary calcite, blady structure. The lower shell is partly silicified. Matrix: chlorite. Upper Cretaceous, Libya (width 7 mm, crossed nicols).

b) Calcite (more rarely): "foliated structure". Thicker blades, often more or less inclined relative to the shell surface, similar to cross bedding (fig. 5-41); n_e perpendicular to the blades.

4) Crossed-lamellar. In thin section similar to muscle fibers with alternating extinction; occurs in lamellibranchs and gastropods; very common (fig. 5-42).

a) Aragonite. Lamellae $\perp$ or obliquely oriented with respect to the shell surface; in lamellibranchs concentrically arranged around the hinge, in gastropods three layers with different orientations.

b) Calcite. Rare.

5) Complex-crossed lamellar (always aragonite). Fish-bone pattern in all sections perpendicular to the shell surface.

A. Lamellibranchiata (= pelecypods, mussels)

(Ordovician — Recent; rock-forming only since Upper Carboniferous) (TWENHOFEL 1950 a). Maximum: Cretaceous — Recent.

According to their adaptation, two groups are distinguished.

1. Anisomyaria (only one muscle scar or two scars of different size in each valve). They mostly live attached to the substrate (H. SCHMIDT 1935), withstand generally stronger water agitation, and dwell mainly in the shallow sea near the coast. The following shell structure is typical, though others (e. g. crossed lamellae) may occur:

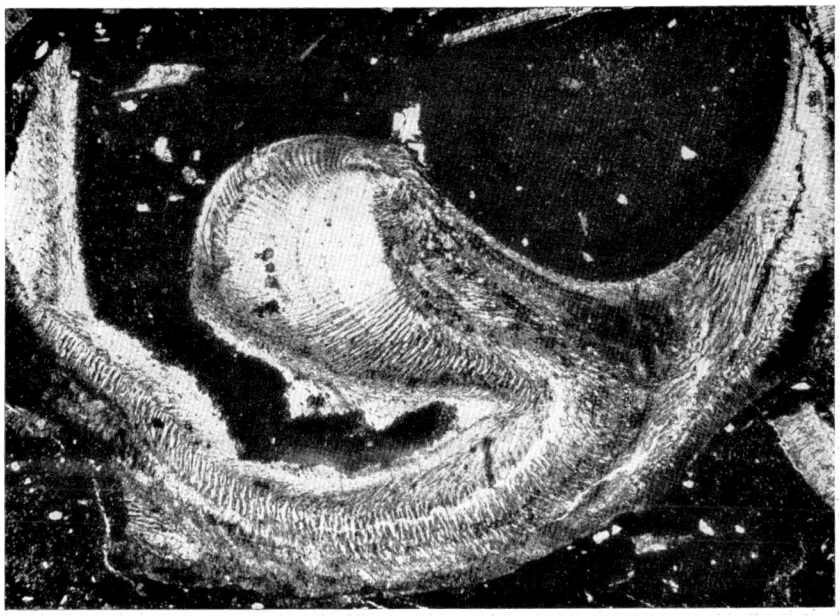

Fig. 5-42. Gastropod, presumably *Glauconia*, preserved as aragonite. Shell structure: crossed lamellae. Matrix: Clayey siderite. Lower Cretaceous; oil pool Dalum (Emsland area) (width 4.5 mm, crossed nicols).

5.1 The primary sediments and their origin

Outer layer: calcite (prismatic or foliated)
Inner layer: aragonite (nacreous layer)

The Ostreae contain only one layer, which is composed of oriented calcite foliae (fig. 5-41).

2. Homomyaria (two muscle scars nearly equal in size)

Most are crawlers or burrowers. Their shells nearly always consist of aragonite (exception: rudistids = calcite, prismatic-chambered). The structure of the inner layer is complex or homogeneous, whereas the outer layer usually consists of crossed lamellae (as fig. 5-42) or, more rarely, of prisms parallel to the layers. In the latter case, a middle layer of crossed lamellae is developed.

It is not always possible to distinguish between the valves of lamellibranchs, brachiopods, and ostracods.

B. Gastropoda (snails)

(Cambrian — Recent; rapid development in Upper Cretaceous and Tertiary).

Pulmonata have been land-dwellers since Upper Cretaceous time but also occur in fresh-water and brackish environments. Most other gastropods live in marine environments; they are found at every depth and on every type of bottom. Gastropods and lamellibranchs are prominent constituents of coquinas (fig. 5-45), and also can form bioherms and biostromes (e. g.*Vermetus* [gastropod] in the tidal area of Florida;

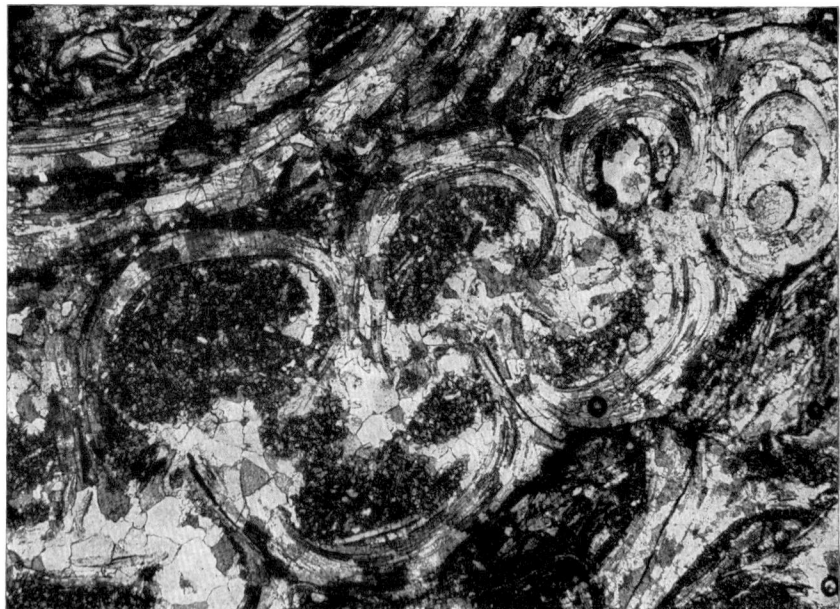

Fig. 5-43. Gastropod coquina, calcitized (previously aragonite). Isometric crystals. Wealden formation (Lower Cretaceous). Water-saturated portion of "Dalum" oil pool, Emsland area (s. fig. 5-40) (width 7 mm, crossed nicols).

Placunopsis [lamellibranch] in the Middle Triassic of Germany; HÖLDER 1961, KRUMBEIN 1963, SEIBOLD 1964 a). The smallest snails are comparable in size to the foraminifera (0.6 mm).

The shells of gastropods are constructed predominantly of aragonite, even in fresh water. Occasionally, an outer calcite shell is developed, mostly consisting of very irregular prisms (BØGGILD 1930). Many snails contain two layers, an outer fine-grained layer, which is homogeneous-prismatic or constructed of crossed lamellae, and an inner nacreous layer. Other gastropods consist of 3—4 layers, mainly composed of alternating oriented crossed lamellae (fig. 5-42). Several gastropods, e. g. *Littorina*, are composed of mixtures of calcite and aragonite, the latter increasing with increasing water temperature. Shells often are covered with color patterns. Such pigments are occasionally preserved in fossil mollusks, presumably only if the original mineralogy of the shells has been preserved, that is if they are not transformed (SCHINDEWOLF 1928, BOWSHER 1957, FÜCHTBAUER & GOLDSCHMIDT 1964).

Pteropoda are planktonic snails (wing snails, Upper Cretaceous — Recent). These snails, 2—20 mm in size, which are naked or have shells, are common components in the surface plankton in typical seas. Their homogeneous aragonitic skeletons can be concentrated in pteropod muds of warm seas.

Tentaculites, 2—20 mm long, conical skeletons, which often show transverse fluting, occur between Lower Ordovician and Upper Devonian, sometimes in considerable quantities. They consist of fine-lamellated, primary calcite shells.

Tentaculites and the similar *Styliolina* (Devonian) are fossils that have been assigned to the Pteropoda, but now are considered as Cephalopoda.

C. Cephalopoda

Rock-forming cephalopods exhibited different growth forms at different times: the orthocone Nautiloidea ("Orthoceras") in the Ordovician and Silurian, the convolute Goniatites and Clymenia in the Devonian and Carboniferous, the Ceratites in the Permian and Triassic, and the Ammonites and Belemnites in the Jurassic and Cretaceous.

Presumably, all of them were marine-stenohaline, as are the modern Cephalopoda. They moved actively both by swimming and creeping, and avoided agitated shallow water. According to GRABAU (1913) the shells were capable of floating over long distances after death. The belemnites preferred shallow, quiet and cool waters, and settled on mud bottoms (BERGQUIST & COBBAN 1957), whereas the ammonites preferred deeper water: According to SCOTT (1940) the strongly sculptured forms lived in depths of 40—200 m (e. g. Jurassic of Swabia), the unsculptured thin *(Desmoceras)* and thick ammonites *(Phylloceras, Lytoceras)* lived at greater depths (e. g. Jurassic of the Tethys). The latter lived presumably nektonic-planktonic; the same is valid for the ammonites covered with spines (BERGQUIST & COBBAN 1957).

On the basis of their irregular-granular structure, the shells of the orthocone Nautiloidea probably were aragonite (BØGGILD 1930). This has been confirmed by preserved aragonite skeletons: according to STEHLI, the outer parts of these animals consisted of aragonite, whereas the walls of the chambers consisted of calcite. The skeletons of the goniatites consisted of aragonite (STEHLI 1956, HALLAM & O'HARA 1962, FISCHER & FINLEY 1949). The same holds true for the ammonites, whose skeletons are often preserved as aragonite and show a colored mother-of-pearl lustre. This outer mother-of-pearl layer is occasionally attached to an inner prism layer. The septae consist only of mother-of-pearl layers, the aptychi of calcite. For more detail of the ammonoid shell structure see ERBEN et al. (1968). The rostra of belemnites are composed of large radial calcite prisms (BØGGILD 1930).

5.1511 Arthropoda

Within this large phylum only the calcitized rock-forming classes are considered.

A. Trilobita

(Cambrian — Permian; occasionally rock-forming in Cambrian and Ordovician).

Trilobites lived on the bottom of shallow seas, below the agitated zone (LOCHMAN 1949). Presumably they were poor swimmers, preferring to feed by burrowing and crawling in the mud bottom (RICHTER 1919). Most of the trilobite remains are products of sloughing.

The exoskeleton consists of chitin with inclusions of calcium phosphate and lime. If the remains are well preserved, they consist (in cross-sections 0.1—0.2 mm in diameter) of an outer pigmented layer and an inner perforated layer (fig. 5-46).

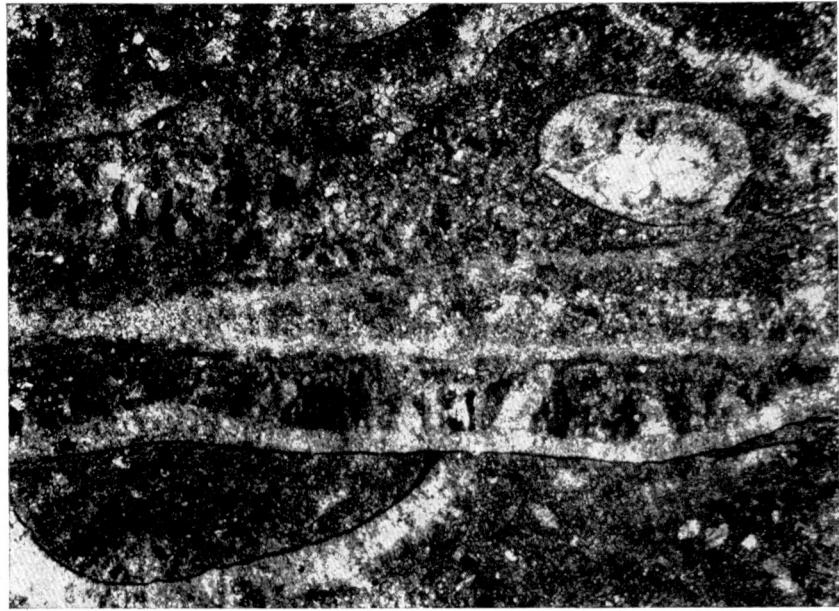

Fig. 5-44. Trilobite; both layers preserved. Cambrian limestone, Comley near Church Stretton E Shrewsbury (England). Width 3.8 mm, crossed nicols.

BØGGILD (1930) postulated that the trilobites built their skeletons with calcite. Remains of preserved fossils described by STEHLI (1956) confirmed this. The observation that the Mg content of the trilobite skeletons is too high for aragonite (5 mole %; LOWENSTAM 1963) also supports such a conclusion. In addition, the color patterns which have been described occasionally on calcitic skeletons (WELLS 1942) point to the preservation of the original mineralogy.

B. Ostracoda (mussel crabs)
(Ordovician — Recent)

The ostracodes are predominantly vagile-benthonic crustaceans. They are found in freshwater, in brackish water and in marine environments. They can live at depths greater than 100 m, but generally are restricted to the shallow sea. Several ostracode species tolerate considerable variations of temperature and salinity, and even can be found in evaporitic environments (e. g. Upper Permian). Ostracodes are very important Tertiary fossils, second in importance only to the foraminifera (POKORNY 1958, II).

The shells are mostly about 0.5—4 mm long and 0.01—0.05 mm thick. Each shell consists of two valves which connect only at the hinge. The outer valve is mostly calcitized; the inner valve is only partially calcitized, on its outer side, mainly near the hinge line (fig. 5-45). Shells possess tiny lamellar structure and are composed of

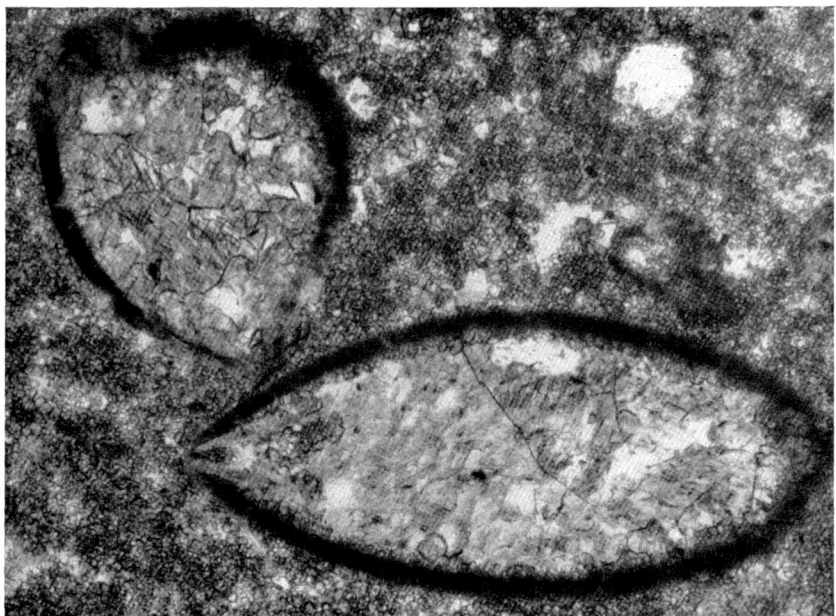

Fig. 5-45. Ostracoda; outer blade micritized, in the lower left with preserved inner blade. Paleocene limestone, Libya (width 0.9 mm).

calcite prisms (n_e perpendicular to the shell surface). They are penetrated by numerous perforations which rarely are preserved. Other crustaceans, e. g. cirripeds (barnacles) and Malacostraca (crabs) do not occur in rock-forming quantities.

5.1512 Echinodermata

Echinoderms are exclusively marine animals. They are found in both shallow and deep marine environments.

Echinoderm skeletons are always formed of single crystals of calcite and therefore are easily recognizable with their uniform extinction under crossed nicols (NISSEN 1963). Viewed in thin section, often the fine pores which characterize the skeletons are found to contain secondary calcite in the same optical orientation or are filled with glauconite and silica (FABRICIUS 1966), or with chlorite (fig. 5-38). Due to the porous nature of the echinoderm skeleton, the specific gravity of recent crinoid columnals is only 1.5 (CAIN 1968); columnals can be transported with quartz grains five times smaller than the stems.

A. Crinoidea

(Ordovician — Recent)

Stem fragments 1—10 mm in size occur mainly from the Upper Ordovician to the Lower Devonian in rock-forming quantities. Although they maintained their importance in later periods, they were rarely rock-fomers (fig. 5-46): Crinoid limestones are found in the Permian and in the Triassic. Blastoidea which are similar in structure to the crinoids, are frequent in the Lower Carboniferous. Generally, the crinoids lived in shallow seas. In modern seas, however, more than half of the stalked crinoids live below the 1000 m line. On the other hand, more comatulid crinoids are found at lower depths (CLARK 1957). During Jurassic and Cretaceous times nectonic crinoids lived *(Saccocoma, Uintacrinus)*.

After death, the columnals tend to disaggregate; entire columns usually occurs only in quiet water, especially in regions with fast sedimentation. According to LOWENSTAM (1949) quiet-water crinoids are more fragile than those living on a reef.

B. Echinoidea (sea urchins)

(Regular echinoidea: Ordovician — Recent, irregular echinoidea: Jurassic — Recent).

Echinoids are common in Pennsylvanian rocks, but are not prominent sedimentary components until the Lower Jurassic (E. FLÜGEL, pers. commun.). They are common components in Cretaceous and Tertiary rocks, but have never attained the importance of the Paleozoic crinoids. Both body plates and spines are present in sediments and can be recognized in thin section (figs. 5-29, 87). Because the smallest echinoids are less than 2 mm in diameter (e. g. Eocene in Northern Africa), the plates may be very small and fragile. Spines are mostly 0.25—0.75 mm in diameter and are covered with ornaments. With the exception of several irregular sea urchins (SEIBOLD 1964a) the sea urchins avoid muddy bottoms and therefore are found in Recent (as

5.15 Skeletal limestones

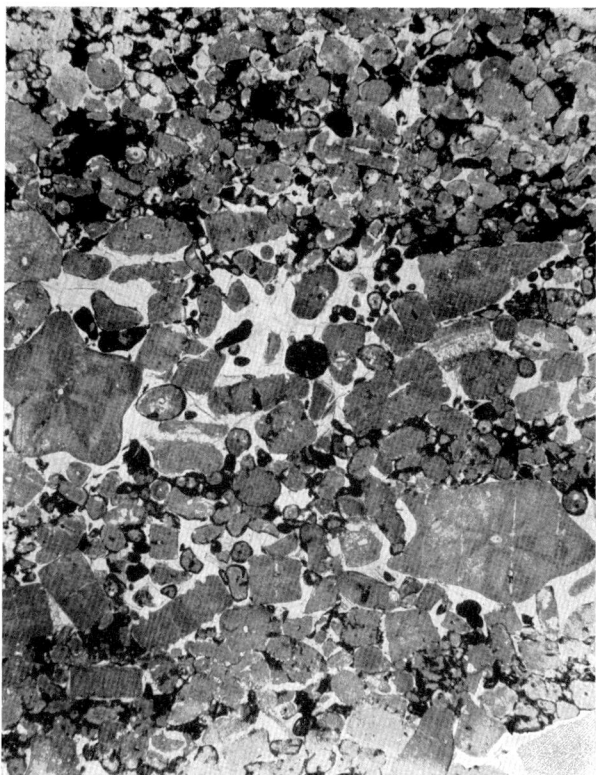

Fig. 5-46. Crinoid limestone, cement mainly calcitic (Liassic-Dogger, Northern Alps). (Width 20 mm; thin section from F. FABRICIUS.)

well as in fossil) deposits mostly on sandy or calcarenitic substrates, deposited in clear water (flat forms) or on rocks (round forms; COOKE 1957a, b). Mainly, they live in the shallow sea near the coast (GRABAU 1913).

C. Asteroidea (star fishes), Ophiuroidea (brittle stars) and Holothuroidea (sea cucumbers)

They are found sporadically in sediments since Paleozoic time, but nearly never are rock-forming (JONES 1956). They form short, blunt spines (1 mm), vertebralike forms and finely ornamented calcite plates and wheels 0.1—0.2 mm in size. Brittle star fragments consist of two differently oriented calcite crystals (FABRICIUS 1966).

D. "Lombardia"

Accumulations of irregular bilateral-symmetrical echinoderm fragments are found in Upper Jurassic limestones. These fragments, however, are distinct from other echinoderm fragments in that they possess linear, dumb-bell-, vertebra- or butterfly-like shapes in thin section. Their maximum length varies between 0.3 and 1.5 mm; their thickness frequently is about 0.03 mm. According to BRÖNNIMANN (1955) they may be interpreted partially as holoturia remains, and partially as planktonic crinoids; according to GROISS (1964) they are fragments of the swimming crinoid *Saccocoma* or of ophiuroids.

5.16 Oölites

Definitions

Oöids are round to oval calcareous grains which possess regular concentric laminations around a nucleus. Rocks composed of oöids are called oölites (KALKOWSKY 1908). Size is not critical in the definition of oöids. If the component grains are larger than 2 mm, the term pisolite has been used. However, because no genetical limit is found at 2 mm (V. SCHMIDT 1961, USDOWSKI 1962), the need to coin a new term seems questionable. The usage of pisolite is all the more dubious considering the fact that oncolites occasionally are also termed pisolites. Spheroids are grains consisting of primary radial fibers.

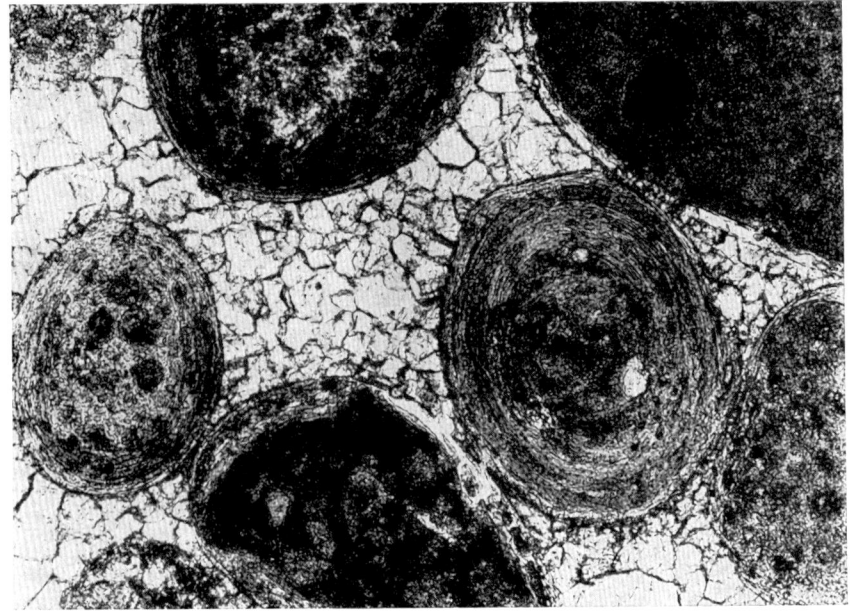

Fig. 5-47. Oölite; top: Superficial oöids. The oöids consist of aragonite, the meteoric cement is calcite. Pleistocene, New Providence, Bahamas (width 1 mm).

Oöid laminae tend to accentuate the roundness of the grains (CAROZZI 1960). Frequently, however, the form of the core is reflected by the form of the oöid. This is especially true in "superficial oöids", in which the nucleus is only coated by a few layers (BERG 1944, ILLING 1954) (fig. 5-47). "Composite oöids" contain several smaller oöids surrounded by a joint coating.

Formation

Recent oöids have been investigated in the occurrences listed below.

5.16 Oölites

Table 5-2. Occurrence of Recent oöids (additional information below the table).

Occurrence	Composition	Orientation of n_e and n_x	Size (mm)	Water depth (m)	Salinity ‰
1. cave pearls	calcite	radial	0.15—15	—	fresh water
2. Lake Aral	aragonite	?	0.1—1	shallow	9.5—10.5
3. Bahamas	aragonite	tangential/disordered	0.25—0.42	<2	37
4. Batabano	aragonite	tangential/disordered	0.125—0.5	shallow	$\sim$ marine
5. Laguna Madre	aragonite	tang./rad./disordered	0.25—0.5	<1	$\geqq$ marine
6. Persian Gulf	aragonite	tangential/disordered	0.07—0.7	1—2	$>$ marine
7. Great Salt Lake	aragonite	(tangential and) radial	0.2—0.5	1—4	203.5

1. References: KIRCHMAYER 1962, 1963, DONAHUE 1965. Cave pearls are not considered as oöids, but as spheroids.

2. In the north-western part of the Lake Aral, the carbonate fraction makes up 10—40% of the sediment and consists of oöids possessing 1—9 laminae; nuclei include quartz, feldspar, carbonate rocks and shell material (BRODSKAJA, cit. in RAUPACH 1952).

3. (Figs. 5-47 and 48). The outer edge of the shallow Great Bahama Banks is rimmed by submarine oölitic dunes. These dunes extend in a 5 to 50 km wide strip; water depths seldom are greater than 2 m. The dunes consist of cross-bedded sets which dip alternating towards the bank and towards the open sea, and are the result of fluctuating tidal currents. As the aerial photograph (fig. 5-48) shows, the predominant current flows onto the bank. The current, with maximum velocities of 0.3 to 1.5 m/sec brings cool CO_2-rich water from the deep straits of Florida (CLOUD 1962, PURDY 1963, PURDY & IMBRIE 1964, SEIBOLD 1964 b). When these cool waters are heated (often to temperatures as high as 30° C), fine layers of aragonite needles are deposited on available nuclei (McCALLUM & GINSBURG 1965). These rod-shaped crystals are 0.5—2 μ long and 0.05—0.1 μ wide, and are oriented tangentially to the grain's surface so that n_x is parallel to the surface (SHOJI & FOLK 1964). Occasionally inclusions of unoriented layers also are found in these oöids.

These oölitic sediments are well to very well sorted ($Q3/Q1 = 1.2—1.8$). The relative thickness of the lamellae increases with decreasing water depth, that is, with increasing water movement (ILLING 1954). Similarly the percentage of oöids within sediment increases with decreasing water depth (NEWELL et al. 1960). Oöids accrete at very slow rates. The outer 10% of the layers from Cat Cay oöids averages 225 ± 100 years old, meaning that the oöids are recent in age; the inner 20% of the same oöids, however, are 2530 ± 100 years (NEWELL et al. 1960). In the area of oöid formation the water is perfectly clear. The salinity is 37‰, a little bit higher than in the open ocean (35‰) (CLOUD 1962 b). Adjacent beds of sea grass may act as a catalyst in the carbonate precipitation by removing CO_2 from the waters during photosynthesis.

The tan color of Recent (and also fossil) oöids is determined by organic substances within the oöid.

4. Oöids in the Gulf of Batabano situated at the South coast of Cuba have been mentioned by DAETWYLER & KIDWELL (1959).

5. The oöids of the Laguna Madre (Texas) are very well sorted ($Q3/Q1 = 1.25$). These oöids possess alternating layers of tangential, radial, and disordered orientation of aragonite needles (RUSNAK 1960).

6. Oöids form on the shallow shore face of the Qatar Peninsula and in the outer part of the tidal deltas in front of the Trucial Coast, Oman, but also within lagoons

Fig. 5-48 a. Submarine spillover lobes composed of oöid megaripples migrating towards the right; near Cat Cay (Bahamas). Bottom: Great Bahama Bank. Top: Straits of Florida (Aerial photograph).

(LOREAU & PURSER 1973). The lagoonal oöids are often unusually big and irregular in shape and are composed of radially oriented aragonite needles. LOREAU & PURSER concluded that all oöids start growing in a relatively protected micro-environment forming layers with "a haphazard or radial orientation, creating a loose fabric. This primary orientation ... is subsequently modified to a secondary tangential orientation on the crest of adjacent bars or beaches where crystals are physically compacted to create a dense fabric" (fig. 5-48 b, 49).

7. In Great Salt Lake, the aragonitic composition of the oöids is the result of the high Mg/Ca ratios (greater than 20/1; EARDLEY 1938). Nuclei consist of quartz, orthoclase, lime pellets, and heavy minerals. In the oöid layers, n_x is oriented tangentially, but more frequently layers of radial fibers occur. EARDLEY assumed that this crystallization is connected with the transformation of aragonite to calcite. But he added that staining tests revealed that the radially structured parts were also aragonitic. This has been confirmed by X-ray analysis: oöids that are frequently radially structured contain only 2% calcite and 4% dolomite; the remainder is aragonite. Because of the prevailing radial structure, they are termed spheroids by FABRICIUS (1972).

An oöid occurrence east of the island of Djerba (Tunisia) described by LUCAS (1955) is caused by reworking of subrecent oölites occurring at the coast, according to FABRICIUS et al. (1970).

As table 5-2 shows, the mineralogical composition of oöids is in chemical equilibrium with the environment. In fresh water calcite spheroids are formed, in sea water aragonite oöids.

5.16 Oölites

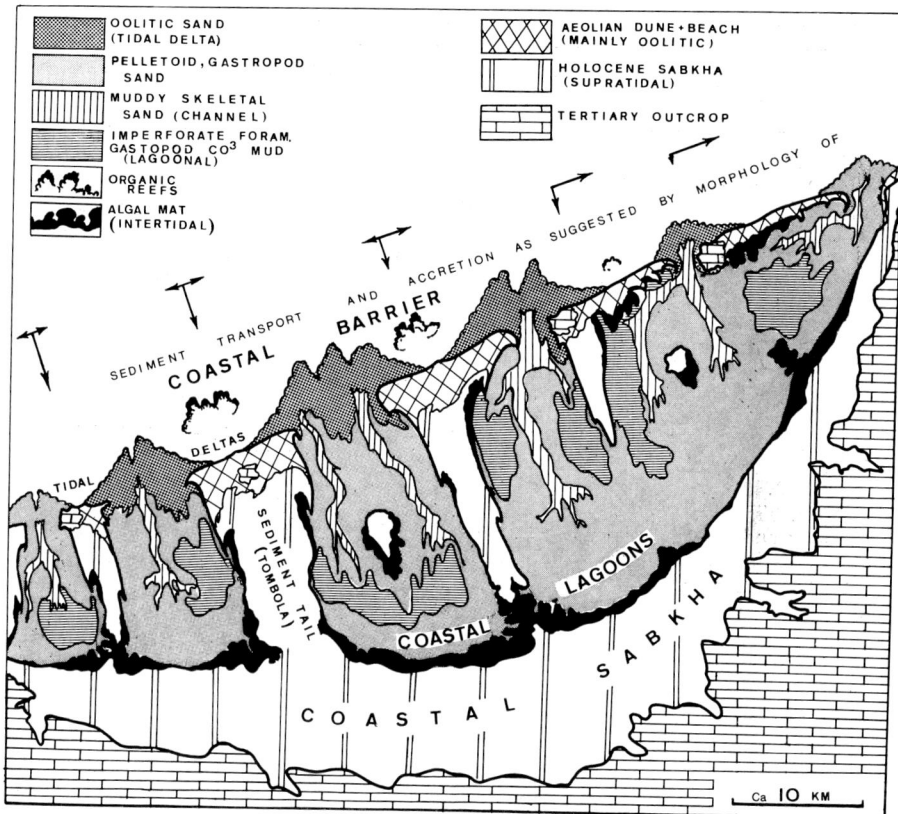

Fig. 5-48 b. Schematic sketch-map of a section of the Trucial Coast (Persian Gulf) showing islands piled around Pleistocene and Tertiary limestones. These islands determine a secondary pattern of protruding oöid "deltas", organic reefs, and protected areas including pelletoid and gastropod sand, lagoonal mud, and intertidal algal mats (from PURSER & EVANS 1973, with permission by Springer-Verlag).

Three general conclusions of geological importance may be drawn from the observations made in modern occurrences:

a) The size of the oöids is between 0.1 and 1 mm, generally.
b) Their sorting is good.
c) They are formed in water mostly less than 2 m deep.

Marine oöids are characterized by alternating layers of aragonite rodlets lacking preferential arrangement or radially oriented, and layers of tangentially oriented aragonite rodlets embedded in a matrix of organic matter which may have influenced the crystallization (LOREAU 1970, FABRICIUS 1972, see below) (fig. 5-49). Coccoid blue-green algae contributing possibly to the oöid growth have been found by dissolving oöids in dilute acid (ROTHPLETZ 1892, NEWELL et al. 1960). Similar fabrics are not found in cave pearls, which form in dark environments in which algae

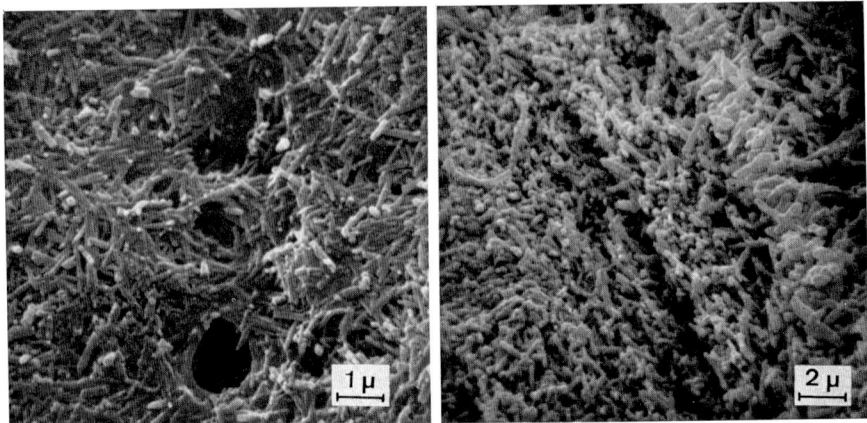

Fig. 5-49. Left: Surface of modern ooïd showing organic mucilage and algal burrows encircled by tangentially oriented argonite rodlets. Abu Dhabi, Persian Gulf (courtesy F. FABRICIUS).
Right: Transected modern ooïd, slightly etched by HCl, showing alternating layers of tangentially and radially oriented aragonite rodlets. Gulf of Suez, Red Sea; coll. JOHANNES WALTHER 1887. Surface is to the right (from FABRICIUS 1972).

generally do not grow; radially structured spherulites are precipitated instead. Spherulites were also formed during diagenesis, e. g. in joints as well as in sedimentary cavities (SCHMIDEGG 1928, CAYEUX 1935). Experimental precipitation of aragonite grains also results in radially structured spherulites (MONAGHAN & LYTLE 1956, LALOU 1957, USDOWSKI 1963a).

According to these observations, the organic coating may be the key to the formation of ooïds (NEWELL et al. 1960). A possible mechanism is the nucleation of $CaCO_3$ which is supposedly caused by aspartic acid found in Recent ooïds as well as in skeletons (MITTERER 1972, see ch. 5.151, end). In the Trucial coast ooïds such coatings are found only in low energy environments. On the Bahama Bank they occur in megaripples which remain motionless for periods of several years, meaning that the ooïds also remain motionless (BATHURST 1967a). In both areas, however, ooïds move periodically from environments of low energy to high energy and vice versa. The alternating laminae in ooïds therefore may correspond to alternating periods of rest and movement.

FABRICIUS (1972), in studying modern ooïds by scanning electron microscopy, found burrows (i. e. excavations for the purpose of living in the excavated space, in contrast to "borings" for the purpose of obtaining food, CARRIKER & SMITH 1969) of blue-green algae perpendicular and tangential to the surface in all ooïds (fig. 5-49). Mucilage of these algae covered the grain surface, and was filled with tangentially oriented rod-like aragonite crystals 1×0.2 μ in size. On the curved parts of an ooïds surface, the ooïd lamellae are thinner, presumably because of a thinner alga mucilage. FABRICIUS found similar elongated 1×0.2 μ rodlets (with hampered growth of the prism faces), but consisting of high-Mg calcite, in the skeletons of miliolids and red algae. His interpretation was that the blue-green algae dissolved the

5.16 Oölites

aragonite in the burrows and reprecipitated it, together with $CaCO_3$ from the seawater, in the outer mucilage. Depending on the agitation of the oöids, the rodlets remained unoriented or became oriented parallel to the surface (fig. 5-49, right).

After decay of the algae, the burrows were cemented with tiny n e e d l e s (with prism faces developed) of aragonite, leading to a cryptocrystalline appearance (fig. 5-51, "micritization", BATHURST 1967a, see also ILLING 1954, NEWELL et al. 1960). The intensity of micritization is controlled by the microenvironment as shown by the alternation of layers of micritized and unmicritized oölites in the Jurassic of Ethiopia (fig. 5-65).

An influence of blue-green algae in the formation of oöids has been suggested already by ROTHPLETZ (1892) and KALKOWSKY (1908). It is corroborated by the unusually high $\delta\ C^{13}$ values (LLOYD 1971).

Though the above observations support a p h y t o g e n e f o r m a t i o n o f o ö i d s, many problems remain (FABRICIUS 1972): Why are oöid-coated skeletal fragments never burrowed by algae, or vice-versa, why are skeletons with micritic envelopes never coated by oöids (fig. 5-51)? Why does the orientation of the aragonite rodlets change from layer to layer in the oöids? What makes the spheroids in the Great Salt Lake different, though they supposedly also are formed by blue-green algae? And what is the difference between oöids and oncoids?

Oöids with oncoid cores occur as well as oncoids with oöid cores; the former occur in upper Permian rocks of Northern Germany (KLIOUMIS, pers. commun.), the

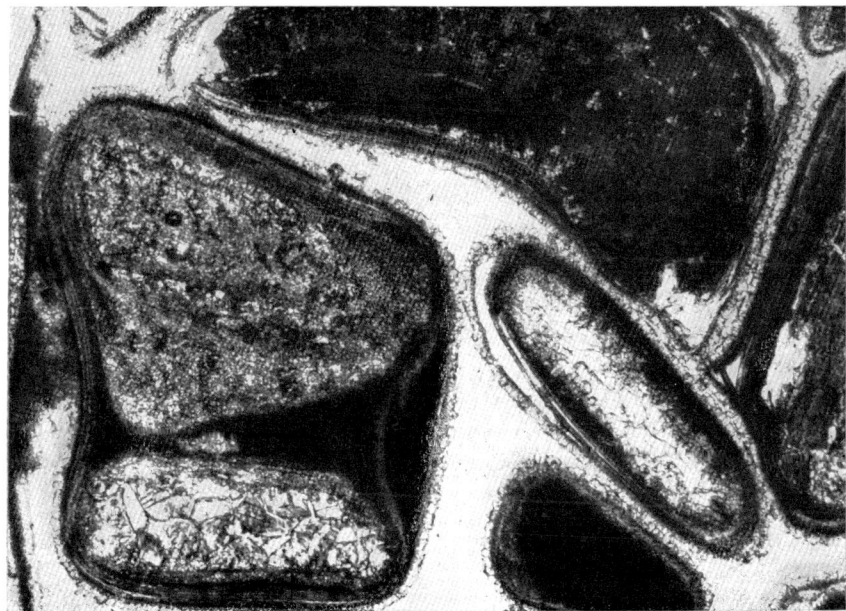

Fig. 5-50. Dolomite-chlorite oöids shrunk after overgrowth of calcite cement A. The shrinkage pores and the normal interparticle pores are filled with anhydrite (white). Upper Jurassic, N.W. Germany (width 1.3 mm).

latter in the Upper Jurassic of northwest Spain (DEUTSCH 1972), see 5.152, C 1.

Blue-green algal mucilage is also responsible for the agglutination of "grapestones" which are considered as the first stage in the formation of "composite oölites" by FABRICIUS (1972).

Presumably oöids grow until an equilibrium is reached between growth and abrasion which increases with increasing weight (BATHURST 1967a). Small nuclei therefore may accrete thicker oölitic lamellae than larger nuclei (fig. 5-51). With higher sedimentation rates, such as when skeletal material is added, the oöids can be buried before this equilibrium is reached. In this case nuclei of different sizes may show equally thick envelopes (CAROZZI 1960).

From the foregoing discussions, the following conditions are considered important in oölite formation:

1. water temperature $> 20°$ C
2. supply of supersaturated (e. g. cold) water
3. marine environment (with somewhat elevated salinity)
4. relatively strong water agitation, at least from time to time
5. very shallow water depth (less than 2 m)
6. supply of potential nuclei
7. low frequency of living organisms which could withdraw carbonate from the sea water: Oöid dunes mostly are "submarine deserts", TWENHOFEL (1950a).

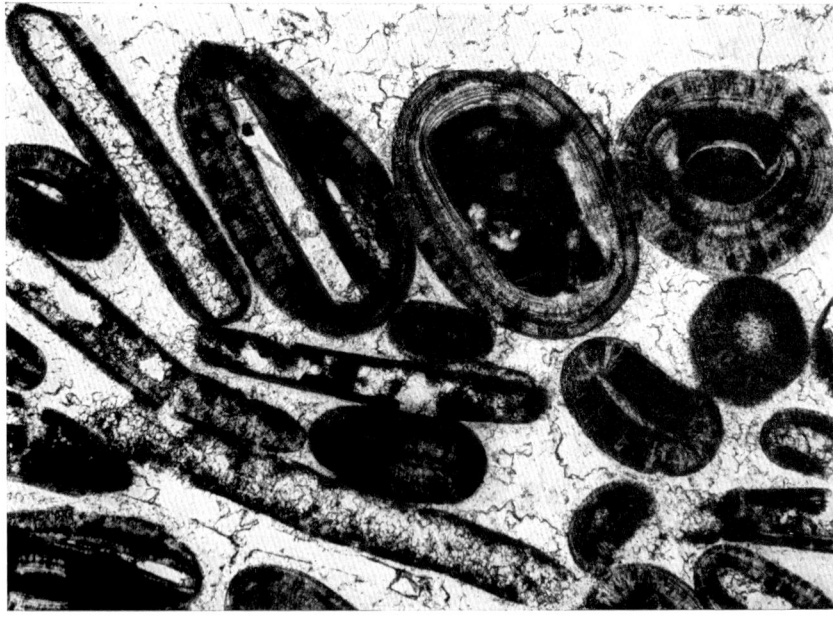

Fig. 5-51. Skeleton oölite. With increasing size of the nuclei the oölitic overgrowths become smaller. These overgrowths are microcrystalline in part (micritization, see below). Upper Jurassic, N.W. Germany (width 2.5 mm).

Due to these limiting conditions pure oölites are limited in thickness and distribution (WELLER 1960), but are more frequent in ancient than in modern deposits.

Although oölitic sediments generally are well-sorted sands, occasionally they contain a calcilutite matrix. This may be caused by burrowing of organisms mixing the sediments. On the other hand, oöids may be transported lagoonward by storms as described by PURDY & IMBRIE (1964) and BALL (1967) from the Bahama Bank. WILSON (1968) has suggested that the oölitic calcilutites of the Upper Jurassic in England may have been deposited in supratidal areas by storms. Such rocks are remarkably frequent as intraclasts in biocalcarenites or oölites (fig. 5-11); presumably a higher energy environment migrated into adjacent calcilutite areas into which already previously oöids had been washed as mentioned above. But oöids may form even at places with only occasional water agitation (FABRICIUS 1972). Oöids also may be washed into deeper troughs; KHVOROVA (1957) found oölitic calcilutites in the Upper Carboniferous flysch deposits west of the Ural Mountains (see ch. 5.53, 2.).

Thus, some oöids found in quiet environments may have been formed in quite different conditions. However, oöid formation can also occur under quiet-water conditions. FREEMAN (1962) found such low-energy oöids in grass-covered sediments of Laguna Madre. These oöids frequently have irregular laminae and dull, pitted surfaces.

Oöids with clay laminae have been found in the Great Salt Lake (EARDLEY 1938) as well as in older rocks (V. SCHMIDT 1961, USDOWSKI 1962). Chloritic oöids behave plastically if packed. Later-on shrinkage may occur as shown in fig. 5-50.

In the course of diagenesis aragonite oöids will be calcitized and receive an unordered or a fibrous structure (n_e radial). The frequency of radially fractured oöids in sedimentary rocks points to an origin of the radial fibers in the uppermost sedimentary layer in which reworking is still possible (CAROZZI 1961).

Normally, the oölites are more or less cemented. Frequently this occurs early in diagenesis, as shown by the abundance of intraclasts with marginal intersected oöids (fig. 5-11; CAROZZI 1964).

5.17 Crustose limestones

Definitions

Crustose limestones are defined as terrestrial limestone crusts without any obvious biogenic origin. They are subdivided as follows:

A. Caliche, a crust occurring at the surface of semiarid, warm areas.
B. Tufa (sinter), a fresh-water limestone occurring near springs or in lakes.
C. Dripstones, a type of "incomplete cementation" occurring in caves.

Origin

A. Caliche

Caliche crusts form in semiarid warm environments, which have long dry seasons and an annual precipitation generally less than 300 mm (KREJCI-GRAF 1960). These crusts generally form by *in situ* evaporation on the ground and in porous surface rocks, which are influenced by capillary migration of ground water or which are moistened at least sporadically by spray, fog, or dew (KNETSCH 1950). Caliche crusts are not found in deserts which have deep ground water levels. Thus caliche formations can serve as important paleoclimatic indicators.

Different types of caliche profiles have been described. BROWN (1956) found double or even multiple layers, each consisting of relatively unindurated chalky material grading upward into an indurated caprock, within the B-horizon of pedocal (= calcareous) soils. BRETZ & HORBERG (1949) described the following caliche profile from New Mexico:

caprock, banded, dense	(1 ft.)
caprock, brecciated, dense	(2 ft.)
caliche, chalky, nodular	(4 ft.)
cupped-pebble conglomerate, calichified	(2 ft.)
substrate: limestone-pebble gravel.	

Brecciation is frequent in caliche and is caused presumably by repeated cycles of desiccation and rainfall. The fragments are generally encrusted irregularly by laminated calcite. These crusts tend to be thicker on top or below a fragment or pebble ("cupped pebble") than laterally (BRETZ & HORBERG l. c., RUTTE 1958, 1961, NÄGELE 1962). The banded crusts which frequently form the surface may be mistaken for algal mats, but MULTER & HOFFMEISTER (1968) presented evidence against this interpretation: Whereas algal mats tend to thicken over domal relief features, subaerial crusts do not, but instead tend to level by dissolution of higher portions and precipitation in depressions (see also RUTTE 1961). MULTER & HOFFMEISTER (l. c.) found such crusts on the top of Pleistocene marine limestones in Florida (see fig. 5-80, and additional information in the caption). They inferred an origin by leaching and reprecipitation processes. pH-values ranging from 6.5 (at night) to 10 (at daytime) have been observed there. Such high pH-values may explain replacements of quartz by calcite which were found even in modern caliche crusts (BROWN 1956, NAGTEGAAL 1969).

RUTTE (1958, 1961) observed multiple crusts forming simultaneously, in the subsurface, especially near escarpments, valley slopes and artificial outcrops; they develop mostly parallel to the bedding.

Four **fabric types** are found in many caliche profiles:

(1) more or less laminated, dense crusts on the top, (2) brecciated crusts with fragments encrusted by irregular calcite laminae, (3) irregular oncolite-like nodules varying in size from 0.1 to more than 50 mm (NÄGELE 1962, DUNHAM 1969b, THOMAS 1965, NAGTEGAAL 1969), (4) cementation of sediments.

In **thin section**, the nodules are composed of cryptocrystalline or equant microcrystalline calcite banded by iron oxides. Encrustations are formed by wedge-shaped crystals oriented perpendicular to the surface on which the encrustation took place. The size of the equant crystals is frequently similar to the size of the clastic grains (e. g. quartz). This is due in part to replacements of silicate or quartz grains by calcite commonly observed in modern and fossil caliche. Microcrystalline quartz occurs intergrown with fine-grained calcite (NAGTEGAAL 1969).

JAMES (1972) found the following microstructures in Holocene and Pleistocene caliche crusts: (1) Alternating laminae of micrite and tangential needle fibers forming horizontal crusts or coated particles identical to marine oöids but composed of calcite.

(2) Spherical to subspherical particles of micrite. (3) 'Flower spar' (similar to the wedge-shaped crystals mentioned above) and random needle fibers forming a void-filling cement.

High porosity is a prerequisite of caliche formation. In the Tertiary Molasse south of Augsburg, caliche (called "Albstein") formed only on the top of fine-grained rocks; presumably coarser-grained rocks did not allow sufficient capillary action for crusts to form (NÄGELE 1962).

Other terms similar to caliche are "calcrete", "duricrust", and "nari" (SANDERS & FRIEDMAN 1967).

B. Tufa and travertine (from tiburtion = rock from Tibur = Tivoli near Rome, Italy).

In more humid areas, such as the European Mediterranean countries, brownish, layered fresh-water limestones are found near springs or in streams and lakes (STIRN 1964).

Dense, banded sinter limestones ("travertines") form inorganically, through CO_2-loss due to warming, whereas spongy, porous "tufas" form with the aid of plant (moss, blue-green algae, water plants) activities (v. PIA 1933; fig. 5-52).

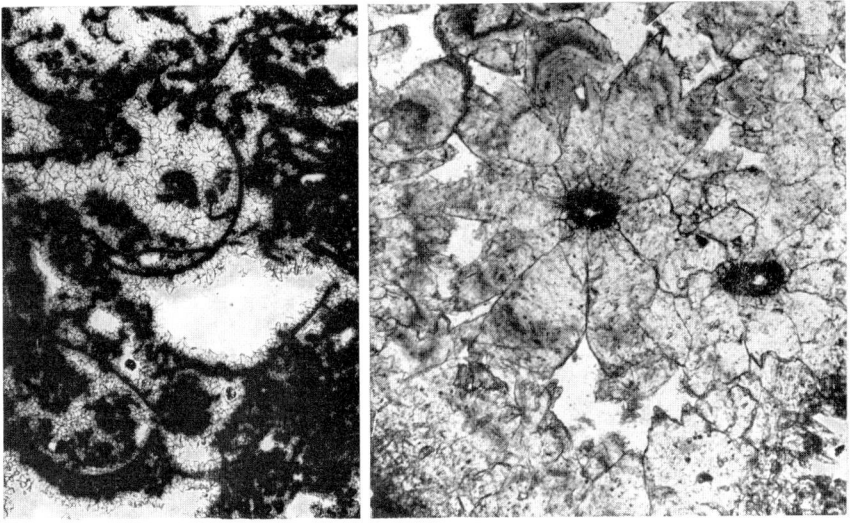

Fig. 5-52. Left: Travertine (tufa), Rome (width 8 mm). Right: Moss tufa near Günningen (Schwäbische Alb). Individual moss stalk, surrounded by radial calcite crystals (width 1.6 mm; from IRION & MÜLLER 1968 b).

As an example, a 20 m-thick tufa near Bad Cannstatt/Stuttgart will be described (REIFF 1955). It consists of thin (mm) layers. Each layer displays a change in color trending from brown at the base to yellow-white at the top: the thicker and lighter parts are interpreted as summer layers. This sequence was deposited from lime- and CO_2-rich mineral waters during an interglacial period which was about 2—3° warmer than present.

According to geochemical studies by WEDEPOHL (pers. commun.), the calcite-tufa of the Leine Valley near Göttingen formed from sulfate solutions.

Aragonite sinter formed in hot springs is partly banded and partly nodular. These sinters can form inorganically (e. g. Karlovy Vary, v. PIA 1933), or with the aid of blue-green algae (e. g. Yellowstone Park, v. PIA, l. c.).

An aragonitic supratidal 'marine tufa' ('coniatolite') is developed on the Trucial Coast (Persian Gulf; PURSER & LOREAU 1973). The distinction between these inorganic vadose encrustations, algal stromatolites, and non-marine tufas may not always be easy. In contrast to stromatolites, marine tufas develop on hard substrates, frequently have a dripstone micro-morphology with downward growth around edges and even from lower surfaces, and do not exhibit desiccation features. In contrast to non-marine tufas, marine tufas are composed of cryptocrystalline or fibrous aragonite, have numerous micro-perforations probably made by endolithic algae or fungi, and encrust generally marine substrates.

C. Dripstones (speleothems)

The walls of limestone caves often are covered with sinter limestones. Stalactites grow downward from the ceiling, while stubby stalagmites grow upwards from the floor. Stalactite growth occurs through the evaporation of $CaCO_3$-rich ground water which slowly drips from the ceiling. Evaporation and CO_2-loss cause deposition of lime at the edge of the drop. The initial deposition results in thin, sometimes monocrystalline tubes, only a few mm thick; later, with more evaporation, the exterior of these tubes accretes and the tube becomes thicker (cf. SANDERS & FRIEDMAN 1967). Occasionally indentations on top of the stalagmites contain cave pearls, up to several mm in size (ch. 5.16).

Dripstones grow inorganically by evaporation of water (v. PIA 1933) or by exsolution of CO_2 (HOLLAND et al. 1964). The fabric is coarse-crystalline and porous. Dripstones are mostly calcitic (e. g. Postojna, Yugoslavia), but occasionally consist of aragonite (e. g. Crystal Cave, South Dakota). Differences in the speed of water flow (MURRAY 1954), or variations in water composition or temperature (SIEGEL 1965) may control the ultimate mineralogy.

5.18 Biostromes, bioherms and their environment

5.181 Nomenclature

This chapter is concerned with *in situ*-accumulations of carbonate skeletons (= biocoenoses) and the interrelations between their depositional environment and their component organisms. According to CUMINGS (1932), biocoenoses are subdivided into

biostromes (from Greek stroma = blanket, sheet): stratified *in situ* skeletal limestones with planparallel upper and lower surfaces, and

bioherms (from Greek herma = reef): massive mound- or lens-shaped bodies which formed through the biologic activities of component organisms.

If these rocks consist primarily of rigid, connected fossil remains, they are termed biolithites (FOLK 1959).

A. Biostromes

They may be classified according to their component organisms. Several characteristics distinguish them from coquinas (NELSON et al. 1962):

1. Fossils are in life-positions.
2. Lamellibranchs often are preserved with both valves.
3. If the valves are separated, biostromes will tend to have equal amounts of left and right valves; coquinas may not.
4. Biostromes may be distinguished from surrounding coquinas by better preservation and poorer sorting of the shells.

Biostromes constructed of algae, bivalves, serpulae or echinoderms presumably originated in very shallow water. Although a sharp limit between biostromes and bioherms does not exist, the former tend to contain only fragile coral colonies (RUTTEN 1958) as well as "fields" ("Rasen") of single corals (table 5-4, no. 11), because environmental conditions (such as water turbidity and food supply) prevent significant colony formation.

B. Bioherms

They are classified here as follows:

(1) "reef bioherms" (or "organic reefs"[2]), forming wave-resistant structures, as indirectly shown by debris accumulations.

(2) "quiet-water bioherms" (or "mud bioherms"), which occur in quiet environments, including deeper water as well as areas with rapid subsidence. They also may form larger morphological structures with slopes up to 50°.

Contrary to the above definition, LOWENSTAM (1950) termed as "reefs" all biocoenoses composed of organisms possessing the ecologic potential to erect wave-resistant topographical highs, irrespective of the present geometry; banks, according to LOWENSTAM, are biocoenoses, which do not contain such reef-builders. NELSON et al. (1962) combined these terms with the terms of CUMINGS, using as criteria the organisms and the morphology of the bildups as follows: bank biostrome, reef biostrome, bank bioherm, reef bioherm.

The definitions adopted in this textbook make clear that "quiet-water bioherms" originate in less agitated water than reef bioherms, for instance in lagoons (e.g. Tavernier Bank, Florida, cf. table 5-4, No. 16). As the table 5-3 shows, typical reef and quiet-water bioherms are easily differentiated, but the transitions often are ill-defined.

Reef bioherms generally rise above the surrounding bottom during their growth (figs. 5-53, 55; "morphological bioherms"). This may not be the case with several smaller reefs and (especially) quiet-water bioherms, which interfinger with the surrounding sediments and are overlain by subsequent sediments without any notice-

---

[2] In navigational terms, any body which protrudes above the bottom and is within 20 meters of the surface can be called a "reef"; thus the prefix "organic" is necessary to distinguish reef bioherms from sand reefs or any rocky shoal.

able deformation ("facies-bioherms"). Morphological bioherms in the geological record probably never achieved the height above the bottom that is suggested by their present thickness; the surrounding normal sedimentation was able to level out differences in relief. This is shown clearly by coral edifices 50—60 m high but only a

Table 5-3. Differences between reef and quiet-water bioherms.

	reef bioherm	quiet-water bioherm
1. organisms		
a) variety	great, especially near sea surface	small
b) zonations (vertical + horizontal)	present	weak to non-existent
c) wave resistance	present in the upper part	non-existent
2. matrix	reef debris or calcilutite	calcilutite
3. debris surrounding the bioherm *	usually present	weak to non-existent
4. surrounding lithologies	biocalcarenites or calcilutite	calcilutite
5. influence on the surrounding environment (e. g. lagoon)	usually great	non-existent or weak
6. resistance to water agitation	strong	weak

* Including only debris from contemporary erosion; debris mantles formed during later uplift may be formed even around quiet-water bioherms. Moreover, debris is formed not only by surf action, but (especially in the interior of the bioherms) also by bioerosion. The surrounding debris dips mostly away from the reef core ("Übergußschichtung").

Fig. 5-53. Reef bioherm of the Steinplatte near Waidring/Tyrol. Rhaetian (Uppermost Triassic). Reef builders include corals, spongiomorpha and green algae. To the left is the stratified basin facies, in the center the fore-reef.

few m in diameter which are enveloped by coral debris and biocalcarenites (Kimmeridge fm. near Lyon, France; ENAY 1966). In fossil bioherms, it is difficult to quantify reef height because of the differential compaction of the bioherm and the surrounding rocks. In soft still-water bioherms, these compaction differences may be minimized.

Net reef growth, according to observations in Pacific coral reefs of Panama, approaches 1 m / 250 years (GLYNN 1971). The ultimate form of a bioherm is governed partially by the relation between reef growth and subsidence of the sea bottom. If subsidence predominates, the reef will die or new populations, more able to withstand deeper water, will evolve, or the reef belt will migrate landwards ("transgressive reef", HENSON 1950, LINK 1950, EINSELE et al. 1967). If the reef growth keeps up with the subsidence, thick reef sequences can develop (Calcareous Alps, MOJSISOVICS 1879, SARNTHEIN 1967). If, however, the reefs grow more quickly than the subsidence of the sea-bottom, the upward growth will cease once the reefs reach the surface. Instead, they will spread laterally, overgrowing their own debris, and may at last develop into atolls. Such a conical enlargement in the upward direction is also typical for the first stages of bioherm growth (ROLL 1934, LINK 1950, HILLER 1964); this is demonstrated excellently by the cup reefs of Bermuda. Typical large-scale examples are the dish-shaped Triassic reef atolls described by OTT (1967b). Such reefs may be confused with the following type, which also overgrows its own debris: If the sea retreats, coastal reefs will move offshore ("regressive reef"), often leaving landward evaporite lagoons (LINK 1950, NEWELL et al. 1953, KEMPER 1966). In this case particularly thick debris breccias originate.

According to GINSBURG & LOWENSTAM (1958) recent reef bioherms are composed principally of frame-builders (e. g. corals), frame-binders (mostly red algae) and the filling (in the reefs around Florida especially *Halimeda* and red algae fragments). A more detailed classification is used below. The frame-builders need not be the prevalent component. For instance in the Funafuti-reef the following order has been found (TWENHOFEL 1950b):

1. *Lithothamnium* (red alga)
2. *Halimeda* (green alga)
3. benthonic foraminifera
4. corals

Compared with the whole reef complex (including the atoll), the frame builders are subordinate (LADD 1950); in fossil reefs it is sometimes difficult even to recognize them, especially in Paleozoic quiet-water bioherms.

5.182 Biologic aspects

In modern reef bioherms (LADD 1950, GOREAU 1963) as well as in their well-preserved fossil counterparts (FRENTZEN 1932, MÄGDEFRAU 1937, LOWENSTAM 1950, LECOMPTE 1958, YOUNG 1959) a vertical zonation of the frame-builders and their growth forms is observed: the upper parts contain the more wave-resistant forms which also tend to be more massive than in lower portions of the framework (cf. ch. 5.156 B). At the same time, a vertical change of the reef dwellers (= organisms using the reef as a substrate only) occurs. Generally, the variety of organisms increases with shallowing depths. Even in quiet-water bioherms slight vertical faunal differentiations may develop, but the wave resistance does not increase upward (cf. LECOMPTE 1958).

EMBRY & KLOVAN (1972) published the following example of vertical zonation from Upper Devonian bioherms in the NW Territories (Canada) (for terms cf. tab. 5-5, ch. 5.2):

top: massive stromatoporoid framestone with debris built of stromatoporoid rudstones and floatstones grading basinward into grainstones to mudstones. Water depth on top of the reef was less than 10 m, presumably.

mid: coral-tabular stromatoporoid bindstone. Depth 10—20 m.

near bottom: coral *(Thamnoporoid-Disphyllid-Alveolites)* bafflestone. Depth 20—25 m.

Thickness, faunal constituents and geometric shapes of the zones were used in combination to place absolute values of water depth for the various reef growth stages. The top of this bioherm corresponds to a reef bioherm, according to the above criteria, whereas the deeper parts are quiet-water bioherms (without a debris mantle).

Bioherms generally show more fossils in living position than the surrounding sediments (fig. 5-54): CROSSFIELD & JOHNSTON (1914, according to AGER 1963) found in Silurian reefs 93% of the corals and stromatopores in living position, whereas in the inter-reef deposits only 16% showed this position. A characteristic of the still-water bioherms is their biological scarcity in species: Modern deep-water bioherms frequently contain only one coral species (TEICHERT 1958). In contrast, more than 100 coral species are found in the uppermost meter, 30 species in 15 m depth of the Bikini atoll reef bioherm (WELLS 1957 a).

On the easterly facing margins of Florida and Bahama platforms optimal conditions for reef growth are found on the seaward side of small islands, which shelter the reefs from unfavorable Florida Bay and Bahama Bank waters (GINSBURG & SHINN 1964). Where these islands are absent, frequently oölites or biocalcarenites are found.

Generally reef growth ceases suddenly as shown by the more or less horizontal upper surface of most ancient reefs. The following events may kill a reef (in brackets the criteria in ancient rocks):

(1) Regressions causing even short-termed emersions, see ch. 5.156 B. (The reef is topped by an unconformity or dissolution surface or buried by nonmarine sediments or soils).

(2) High subsidence rate. (The reef is buried by basin sediments).

(3) Supply of muddy water. (The reef is buried e. g. by shales).

(4) Organic events: epidemics or enemies, e. g. the present starfish *Acanthaster* invasion of Pacific reefs. (Such events mostly will not be recognized in the sedimentary record).

Fig. 5-54. *Montastrea annularis* (M) and other corals (at right) make up about 30 percent by volume of the Pleistocene reef of the Florida Keys (according to L. PRAY, pers. comm.). Quarry Windley Key near Islamorada. The frame-builders have been marked on the quarry wall (height of photograph 180 cm).

5.183 Modern reefs

At present reef bioherms are confined to the tropics and subtropics, generally between 30° N and 25° S latitude, with about 20° C minimum water temperature, but the validity of extrapolating these observations to ancient reefs is unknown (LINK 1950). In modern oceans the following reef types are distinguished (many examples in KUENEN 1950):

1. Fringing reefs grow in front of rocky coasts. They are separated from the shore only by a very shallow lagoon, a few meters in diameter. These reefs may reach several 100 m in diameter. In front of river mouths and other unfavorable places, they are interrupted. One of the longest modern fringing reefs is developed

along the coast of the Red Sea, which is on both sides more than 2000 km (LADD 1950).

2. Barrier reefs are separated from the shore by a deep lagoon. Normally barrier reefs reach about 500 m in diameter, occasionally several km. The Great Barrier Reef stretches along the coast of Australia, for a length of 1600 km and 30 to 250 km from land. Barrier reefs often are composed of smaller reef types: in areas with a prevailing wind- and wave-direction small reef knolls are formed first. These develop into horseshoe-reefs and atolls, which occasionally are filled and become table reefs (FAIRBRIDGE 1950). Barrier reefs are crossed by numerous channels, especially in areas with a high tidal range (e. g. Australia, MAXWELL 1968).

Even sand coasts may be bordered by barrier reefs, for example those east of Madagascar, at a distance of a few hundred meters from land.

3. Atolls are formed by a ring of reefs enclosing a lagoon. DARWIN (1837) suggested that atolls were the final evolutionary stage in coral reef formation: Fringing reefs grew around oceanic volcanic islands; as the volcano sunk, the reef became a barrier reef. If reef growth kept pace with subsidence, an atoll ultimately could form. In detail, the origin of atolls is still open to discussion as they are found also in shelf seas and in barrier reefs (FAIRBRIDGE 1950, DUNBAR & RODGERS 1958) (cf. the above discussion on transgressive and regressive reefs). Many atolls seem to have been influenced by glacially-controlled eustatic fluctuations in sea level (DALY 1910).

Atoll diameters seldom exceed 30 km. Some lagoons contain many patch reefs (reef knolls), generally less than 30 m (but occasionally as great as 1000 m) in diameter. Islands of fine debris may form on the windward side of the lagoon; coarse debris walls are found behind the living reef.

During the last 70 years, the origin of several atolls has been defined by deep drillings (ARMSTRONG et al. 1904, KUENEN 1950, LADD et al. 1950, 1953, EMERY et al. 1954).

4. Table reefs (TAYAMA 1935) are extensive reefs, 1×3 km in size (max.: 25 km; WELLS 1957), which often develop from reef knolls and are found especially in shallow seas, for example in the lagoon behind the barrier reef N. E. of Australia (FAIRBRIDGE 1950). Since shallow seas were more frequent in the geological past than at present, table reefs probably were more important in the past. In the "Dolomiten" mountains (Southern Alps) table-reefs developed during slow sedimentation of the surrounding area (LEONARDI 1967).

EMERY et al. (1954) distinguished the following morphological units within Marshall Island (Pacific Ocean) atolls:

(1) lagoon with reef knolls

The sediment consists mainly of Halimeda fragments and their fine-grained disintegration products. The percentage of foraminifera and coarse reef debris increases towards the main reef. The reef knolls are surrounded by their own debris.

(2) islands (mostly sand-size debris)

(3) reef platform (100—1000 m in diameter)

Dead, back parts of the reef. At low tide the platform is partly above sea level and thus exposed to the chemical dissolution by rain water (KUENEN 1950). Perhaps some of the reefs represent the remnants of reefs that thrived when sea level stood 1 to 2 m higher during the Holocene climatic optimum (FAIRBRIDGE 1970), four to five thousand years ago.

(4) coral-algal zone (up to 100 m in diameter)

Living reef corals and red algae (mostly encrusting) just below the low water line.

(5) lithothamnion (algal) ridge (10—100 m in diameter), only in Indo-Pacific reefs.

It consists of encrusting coralline algae; living algae often display a vivid purple color. These algae are able to survive above water as long as they are sprayed by incoming surf. On the leeside reefs algal ridges are often absent, especially in areas with fluctuating wind direction.

Deep channels perpendicular to the algal ridge (buttress-groove system) intercept part of the wave energy and drain outflowing surf. Apparently these channels represent growth forms rather than erosion (see fig. 5-55).

(6) fore-reef

In the channels mentioned under (5) as well as on the adjacent terrace, reef debris, *Halimeda* plates and benthonic foraminifera are prominent sedimentary components. At a depth of about 15 to 20 meters, the bottom steepens substantially and dips seaward at 40°, locally.

The most important corals in the Great Barrier reef show the following zonation: On the top, in front of the reef, branching *Acropora* is found, in backward parts of the reef, massive brain corals abound *(Favia, Porites)*. Platy *Montipora*, mushroom-like solitary *Fungia*, and various encrusting corals thrive on the fore-reef. Other reef dwellers include green algae *(Halimeda, Acetabularia, Penicillus)*, non-calcareous algae, foraminifera, mollusks, echinids, holothurians, crustacea, and worms (MAXWELL 1968).

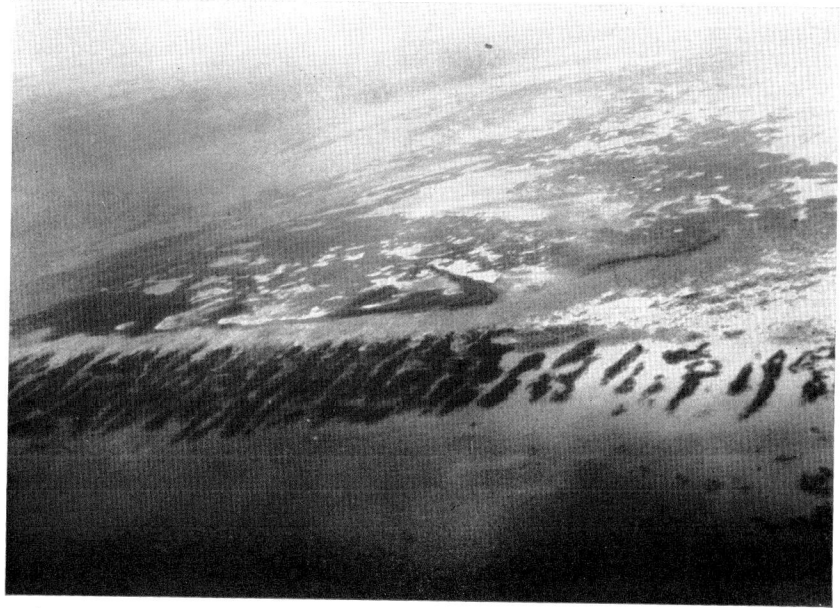

Fig. 5-55. Aerial photograph of a reef seaward of the Florida Keys. In front of the reef a buttress-groove system consisting of corals.

5.184 Ancient biostromes and bioherms

Corresponding units also may be found in fossil reefs:
1. Reef core = the area stabilized by frame builders.

Frame builders usually account for only 5—10% of the reef core (LADD & TRACEY 1949); at most about 40% (fig. 5-54). For this reason it is difficult in thin section to distinguish reef core from other environments, especially fore-reef sediments. The porosity of the reef core generally is lower than the fore-reef sediments. Moreover, the reef core is frequently characterized by layers of algal crusts and encrusting foraminifera of the *Homotrema* and *Nubecularia* type (E. FLÜGEL et al. 1963).

Additional indications may come from geochemical investigations. For instance in the Alpine Triassic the strontium content increases from the back-reef towards the fore-reef (STERNBERG et al. 1959, FLÜGEL & FLÜGEL-KAHLER 1963). According to FLÜGEL & FLÜGEL-KAHLER however, strontium distribution may be only a function of clay content (WEGEHAUPT 1962, BAUSCH 1966), as suggested by TILL (1971) for other elements. Diagenetic phenomena (e. g. filtration) also may play an important role (H. FLÜGEL & WEDEPOHL 1967).

Atoll-like reefs were described from the Upper Devonian of Canada (ANDRICHUK 1958a, EDIE 1961, FISCHBUCH 1962, KLOVAN 1964), and western Germany (KREBS 1966), as well as from the Middle Triassic of the Alps (OTT 1967b).

Fig. 5-56. Coarse debris of the fore-reef: Stromatolite fragments, partly welded by stromatolite crusts. Middle Magnesian Limestone (Upper Permian) N. of Durham (England).

2. Reef flank

In longitudinal reefs exposed to a prevailing wave direction the sediments of the windward side (fore-reef) can be distinguished from those of the lee-flank (back-reef):

Fore-reef sediments begin at the reef proper with coarse, unrounded and unsorted reef debris (HENSON 1950) (fig. 5-56). The bottom dips seaward at 35°, sometimes up to 60—65° in fossil reefs (TWENHOFEL 1950b). These steep slopes are partially explained by differential compaction of the reef core and the flank sediments. The reef debris is cemented or filled with a primary or secondary lime matrix. The latter is often finely graded in the channels of the buttress-grooves as well as within the framework (geopetal fabric, see figs. 5-57 and 5-75; GOREAU 1964, KREBS 1966). Basinward coquinas follow which increasingly contain typical flank dwellers (e. g. thick foraminifera, echinoderms) and may contain calcilutite matrix. They interfinger with normal basin sediments.

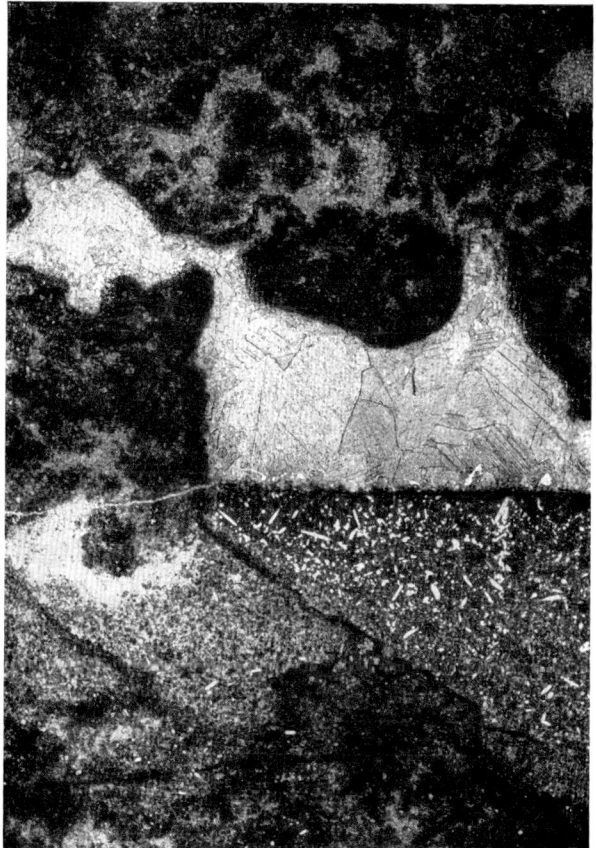

Fig. 5-57. Cavity with internal sediment (calcilutite with authigenic quartz), overgrown by calcite cement. Geopetal fabric (horizontal sediment surface). Bioherm of the Frasne 2h (Upper Devonian, quarry Lion, Frasnes-lez-Couvin, Belgium). (Width 13 mm.)

The back-reef sediments consist mostly of well-rounded, frequently calcitic-cemented biocalcarenites: lagoonward are fine-grained rocks. Typical for this whole area are green algae which may be clearly zonal (OTT 1967a) (nowadays *Halimeda* [fig. 5-20], in earlier times Dasycladaceae etc.), and certain foraminifera, e. g. Miliolidae (HENSON 1950, FLÜGEL & FLÜGEL-KAHLER 1963, KEMPER 1966, KLEMENT 1966). The fauna is typically rich in individuals but poor in species (FABRICIUS 1966). Oncolites are frequent here. Precipitation of calcium sulfate and early diagenetic dolomitization occur in the lagoonal environment. On the landward side, stromatolites are frequent (FISCHER 1965a, BOSELLINI 1967b, GEBELEIN & HOFFMAN 1969), as are corresponding "birds-eye"-structures (p. 342 and 356).

"The reef-complex" includes reef core and reef flank sediments.

3. "Inter-reef" sediments are distinguished in areas rich in reefs by several authors; they may be included in the "reef complex".

In table 5-4 the available data are compiled for 70 modern and fossil biostromes, quiet-water bioherms, and reef-bioherms, each group arranged according to increasing age. The dimensions (length, width, height) refer to the cores of the bioherms. For round bioherms the flank sediments are recorded in the corresponding column, for longitudinal bioherms they were subdivided into back-reef and fore-reef sediments; for the fore-reef sediments the dip (minus the tectonical dip) is added. The marginal dip is also indicated in the case of the reef core.

The following fabric elements are indicated for the reef core, if available:
1. frame-builders and -binders
2. reef dwellers
3. matrix (including reworked frame-builders, -binders and reef-dwellers as well as internal sediment)
4. cement (fig. 5-57)
5. porosity

A gross evaluation of the data compiled in table 5-4 enables the following tentative conclusions:

Quiet-water bioherms

Distinct frame-builders have not been found in the 13 Paleozoic quiet-water bioherms listed. No. 24 and 25 are mounds composed of crinoid ossicles up to 140 m in height with steep flanks. How the ossicles accumulated and remained together is unknown. According to PRAY (1958, p. 270) these mounds are presumably flank sediments of bioherms up to 240 m in height with steep flanks (up to 50°) which are composed predominantly of lime mud (up to 90% according to PRAY & HOROVICH 1959). Differential compaction of such muddy sediments and the surrounding sediments does not seem probable and therefore does not explain the steep flanks. Although real frame-builders may have been absent, sedimentary baffles such as plants or bryozoan tufts may have aided in sediment accumulation (PRAY 1958).

One of the few fabric characteristics of quiet water bioherms is the presence of cavities which are cemented by fibrous calcite crystals which become isometric towards the interior of the cavity. The tops of these cavities frequently are irregular, whereas the bottom is smoothed by internal sediment. Such features are called "stromatactis" (DUPONT 1881), "birds-eye" structures or "fenestrae" (fig. 5-57), but their origin is still unknown. Both inorganic (SCHWARZACHER 1961) and organic processes have been used as an explanation (PRAY 1958, BATHURST 1959b, CAROZZI & TEXTORIS 1963, OTTE & PARKS 1963, LEES 1964, PHILCOX 1971). According to WOLF (1965c)

5.18 Biostromes, bioherms and their environment

Table 5-4. Shape and composition of selected biostromes and bioherms.

No.	period	locality	length (km)	width (km)	max height & (relief) (m)	water depth above top (m)	petrography (F = frame builders, J = inhabitants, M = matrix, C = cement)		flank towards the basin	lagoon towards the shore	open sea	author
A. Biostromes												
1	recent	Niger delta	0.1–2.1		2–17	50–100	cor - thicket	L			S	Allen & Wells 1962
2	Middle Cretaceous	"Golden Lane" (Mexico)	150	1	ca 1200		many intercalations of biostromes, 50m in thickness,1000 m in width, rud, mil, in mil - L, Intragra. por. Dip 30° towards basin					Humphrey 1953, Benavides 1956, Barnetche & Illing 1956
3	Upper Jurassic	Isteiner Klotz, Upper Rhine	>12		45	z.T. <20	flat cor-masses, bra, echinids, Sol oncolite (D). M = onc granulated D. Cross-bedded in part, onc D&L + for, ost crypt; in part crypt - macrocryst. calcitiz.	L D L L	no debris no debris, crypt D&L + for, ost	D and anh	L + silt layers	Kabelac 1955 Fuchbauer 1964 b
4	Upper Permian	NW Germany	180	30	90							
5	Pennsylvanian	3r Aneth Field (Utah) Ismay Field (Utah, Colorado)	22	19	25	few m	F = green alga Jvanova, M = few other fossils, intercry por F = green alga Jvanova, J = bra, ech, ost, mol, bry, Bottom M (calcilutite), top C (calcite). Intergra por	L L	overlain by "oolite" with intragra por biocalcilutite biocalcarenite			Peterson & Ohlen 1963 Choquette & Traut 1963
6	Pennsylvanian		3	1	12	few m						
7	Mississippian	Monroe County (Indiana)	3		18		stratified cri – L with Sh-layers (dip 0°)	L	no debris			Twenhofel 1950
8	Lower Carboniferous	Belgium	50		27		Stromatolite, 1–2 m thick (Collenia-type), alternating with biocalcarenites (Dss)	L	no debris			Monty 1963
9	Upper Devonian	S. Belgium	extensive		25–50		mass & tab str, mass & ramif cor, coastal facies	D,L				Lecompte 1958
10	Middle Devonian	Paffrath (near Cologne)	>10	±2	28		F = tabul & ramif cor, tabul & mass stroms. J = bra, gas, lam. M nearly absent	L	lateral transition into platy skeletal L (many cri in part). No coarse debris			Jux 1960
11	Middle Devonian	Eifel (Germany)	>130	elongate girdle ~3			F = cor, str. M = cor - & bra-coquina. In front of it a biostrome of single cor (,,Rübenriff") M = shL	L	rarely debris	biocalcarenite	Mrl with bra	Struve 1961
B. Stillwater bioherms												
12	recent	Blake Plateau (W-Atl)	0.8		150 (M)	570	F = cor (Lophelia et al)	L			L-mud	Stetson et al 1962
13	recent	Norwegian Shelf	ca 1.1		60 (M)	-200	F = cor (Lophelia et al)	L			Sh	Teichert 1958
14	recent	European Shelf	different		25 (M)	-2500	F = cor (Lophelia et al)	L			Sh	Le Danois 1948
15	recent	Gulf of Mexico	>0.3		55 (M)	450	F = cor (Lophelia et al)	L				Moore & Bullis 1960
16	recent	Tavernier Bank (Florida)	3	1.4	45 (M) (3)	0–1	F = margin: Porites (cor), behind: Goniolithon (red alga), interior: Thalassia (sea weed), green algae. Center: mangroves M = calcilutite, in part disintegrated green algae	D				Baars 1963
17	Miocene	Wairarapa (New Zealand)	0.075		3.4 (M)	1500–2500	cor =,, thicket" (only 1 species: Lophelia parvisepta, slightly convex M: ? Sh with plankton, (& benthon.) for, mol, ost, ech	L	sparse debris (Porites, Goniolithon)		Sh with for	Squires 1964 Vella 1964
18	Upper Jurassic ε – ζ	Swabia (Germany) individual bioherms:	~200 1–3	60	-150 (M) (-60)	>10	F = spo, s, calcified, blue-green algae (crusts, stromatolites) < M = calcilutite, J = for, ser, bry, bra, spoj. (lateral dip 10-25°)	L	no syngenetic debris. In ,,Restlücken & Schüsseln" stratified L with oncoids, mol, ech, bra et al.			Roll 1934, Fritz 1958 Hummel 1960, Gwinner 1962, Hiller 1964 Roll 1934
19	Upper Jurassic β	Wasserdilfingen (Swabia Germany)	0.0015		12 (F) (0)		mostly flat spo, s, calcified. ,,Kleinschwammstotzen"	L				
20	Upper Jurassic	Grand - Salève (Haute - Savoie)	0.02		7 (M)	ca 5–15	F = cor, str, bry M = 20 – 30% granulated L. J = for, ech, ost, das	L				Carozzi 1955
21	Lower Permian	Virgil [Wolfkamp] (New Mexico)	4.8 [4.8]	1.6 [0.5]	30 [18] (M)		F = ? ,Stromatactis" 10–90 Vol% of the rock), top: arched, bottom: flat. Much calcilutite. J = alg, fus, bra, cri, bry, mol	L			Mrl Sh	Otte & Parks 1963
22	Pennsylvanian	San Juan Canyon (Utah)	0.3		18 (M)		calcilutite with calcilutite. Solution pores					Wengert 1951
23	Pennsylvanian	Scurry - Snyder Field (West - Texas)	3.7	6–13	240 (M)		skeletal calcilutite. Solution pores	L			Sh	Levorsen 1956
24	Pennsylvanian	Todd Deep Field (Texas)	4	2	140 (M) 12 (M) (15)		cri – L, mass, with bry, bra. F = ?					Imbt & Mc Collum 1950
25	Mississippian	NE - Oklahoma					cri – L, mass. F = ?		cri - L, platy (dip -45°)			Harbaugh 1957
26	Mississippian	Sacramento Mts. (New Mexico)	0.3–1.6	0.3	100 (M)		calcilutite with < 20% bry & (less) cri (dip -40°). J = bry	L	cri - L			Laudon & Bowsher 1941 Pray 1958
27	Mississippian	Mc Donald Co. (Missouri)	0.1		10 (M)			L	surrounded by shL with cri			Troell 1962
28	Lower Carboniferous	Clitheroe (N - England)	0.005–1		1–150 (M)		basin bioherms, ,,knoll reefs". F = ? Stromatactis, bry. J = bra, cri. M calcilutite (dip 40 – 50°)	L	no debris		Sh, Mrl	Bond 1950
29	Lower Carboniferous	Formoyle N.W. Sievemore Ireland	0.3	0.4	~60 (M) 150 (M)		F = ? bry <<M granulated L with intraclasts ,,Stromatactis." (steep dip)	L	no debris, shL with cri (with current adjustment)			Bathurst 1959 a Schwarzacher 1961

Table 5-4 continued

C. Reef bioherms

No.	period	locality	length (km)	width (km)	max height & (relief) (m)	petrography (F = frame builders, J = inhabitants, M = matrix, C = cement)		flank fore-reef	lagoon and back-reef	open sea	author
30	Lower Carboniferous	Middle Ireland total complex	few 100m >100km, several	0,15	15 (M) 100m thick	F = ?, Stromatactis", J=bry, cri. M=calcilutite (dip -50°)	L	no debris shL with cri, bry, isolated bioherms		Sh,Mrl	Lees 1964
31	Upper Devonian "F2"	S. Belgium	water depth above top:	0,3	80 (M) >10	F= top: tabul cor, mid: mass cor+tabul strom, bottom: tabul cor &,Stromatactis". Additionally bry,spo, J=bra,cri, M=calcilutite	L	no debris, sh layers		Mrl	Lecompte 1958
32	Ordovician	Kullsberg (Dalarna, Sweden)		0,3	40 (M)	F = alg,?,Stromatactis", J=bry,cri,bra,tri,cep. M=calcilutite	L	surrounded by shL with cri, bra, bry		Mrl	s.Thorslund 1960
33	Ordovician	Boda (Dalarna)			100 (M)	F=,Paläoporella,?,Stromatactis",algal crusts, J=bry,cri, bra,tri. M=calcilutite	L	surrounded by shL with cor, cri et al.		Mrl	s.Thorslund 1960
34	Precambrian	Glacier Nat'l.Park(USA)	0,01		≥8 (M)	stromatolite (,Collenia") (dip -50°)					Twenhofel 1950 b
35	recent	Gr.Barrier reef Barr.& fringing reefs (Australia) tabul reef		0,5-1,5	120-150	F = cor, red algae J = mol,cor,for. M = skeletal grains. J,M>>F	L	reef debris behind & in front of the reef	Lime mud & skeletons Near the coast Sh	deep sea sediment	Fairbridge 1950 Maxwell et al.1961
36	recent	Maratoea (Borneo)	atoll	11 5	≥550	F = cor, red algae	L	reef debris (dip 40°)	Lime mud	near the coast S	Kuenen 1950
37	recent	Bikini (Marshall Islands)	atoll	ca 0,8	ca 600	F = cor, red algae C only above low water level M = lutite, for, skeletons (cor,mol,ech-spines)	L	Hal, reef debris (dip 30-40°)	Lime mud, Hal	Globigerina mud	Emery et al.1954 Kuenen 1950
38	recent	unnamed	tabul reef	8	4500					Mrl with fossils	Ladd 1950
39	recent	Shah Allam Shoal (Persian Gulf)	1,8	1,3	≤70	F = cor	L	reef debris. Outer flank: skeletal grains			Houbolt 1957
40	Pleistocene	Florida Keys	220	±5?	54	F = (ca 30 vol%)=cor, M=cor, Hal, calcilutite, few C porosity 40-60%	L				Hoffmeister & Multer 1964
41	Tertiary	Louisiana	11 (with fore-reef and back-reef)	90		F=cor, incrust.alg, Hal, articul.red alg, incrust.for, J = large for, M = debris	L	reef debris angular, at outer flank finer & with lutite matrix reef debris, overgrown by the reef		reef on the higher part of a fault	Forman & Schlanger 1957
42	Oligocene	Bavaria	≥170	30	30	F = Lithothamnion, J = for, bry, ech L with D-flasers. M = few debris of F & J	L		sandy	Mrl	Füchtbauer 1964 a
43	Lower Cretaceous	Edwards Limestone (Central Texas)	"regressive"	0,04	4-17	F = rud, near the bottom cor. M=calcilutite	L (D)		clotted L, chert	calcilutite & biocalcarenite cross-bedded in part	Nelson 1959
44	Upper Jurassic	Swiss Jura	>80 barrier reef	3-8	120	F = cor, (bottom:platy, top:mass) M: bottom:calcilutite, top: coarse reef detritus (cor,ech, biv,gas)	L	biocalcarenite, no debris	biocalcarenite,?ool, onc, chalky in part with Characeae	Mrl with cep	Ziegler 1962
45	Upper Jurassic	Kelheim (Franconia)	5-12	5-6	≈100	F = hyd(18,3 vol%), J=cri, cor, spo, I, bra (1,3%) M= mainly cement, pelletal L (77,3%), Intraclast L (3)%	L	reef debris (,Diceras L"), in the E.1-2 km wide			Bausch & Zeiß 1966
46	Upper Triassic	Sauwand (Steiermark, Austria)	~1	0,2-0,5	150	F = ,,reef buds" (Spo), cor (Thecosmil.), Sol)>>M=reef debris, mostly rounded, in part with C, in part with M, except in the W. J=ech,for,mol, ost, bry. ,,Debris mounds". Onc encrusting skeletons are typical.		coarse debris >1mm (Spo), cor (Thamnast.), Sol), das, for, ech, biv) with fine-grained matrix	biocalcarenite <1mm, cemented (for,mol, ech,alg,bry) ,,Dachsteinkalk"	"Hallstätter" L (red) sea depth 50-200m	Flügel & Flügel 1963 & pers. comm. Schwarzacher 1948
47	Upper Permian	N.E.Durham (England)	34 barrier reef	5	100	F = stromatolite, bry. M = onc; overgrew its own debris basinward (= regressive)	D	reef debris breccia	D - onc	gray D, laminated in part	Smith 1958
48	Zechstein 2	W.of Harz Mts. (Germany)	~2		<20	F = stromatolite J=lam, bra, bry	D	onc in part, D	"Stinking sh" with anh	"Stinking sh" (L with silt layers)	Herrmann 1956
49	Zechstein 1	W.of Harz Mts. (Germany)	>3	1	<20	F = stromatolite, bry. J = bra, lam, mil. C = D	D	fore-reef D-arenite cross-bedded in part	D and anh with onc	black L & D	Herrmann 1956 Füchtbauer 1971
50	Permian	Capitan Texas, New Mexico	many100 barrier	4	~200 550 from top reef to bottom of basin	F= reef buds: (Spo), cor (Thecosmil.), hyd?, J=cri, fus, bra. Cavities. M = fine debris	L	300m reef debris (dip 20-35°) overgrown by the reef	from reef towards coast: 1 biodolomitearenite 2 oncolite, 3 light D 4 anh & eolianite (gas, lam)	dark L and S	Adams et al.1950 Newell et al.1953
51	Lower Permian	Tra-Tau(SE-Russia)	>1,6		~1000?	F = ,,alg"		flanked by black Sh			Twenhofel 1950
52	Carboniferous	Scurry County (Texas)	37	6,5-13	?	biocalcarenite (cri,fus), indurated by early cementation fragments up to 50cm, calcilutite, coarse skeletal debris. Vugs		surroundings: sh in part			Bergenback & Terriere 1953
53	Pennsylvanian	Coffeyville (Oklahoma)	0,03	1	4, (low)	F = Sol, Spongiostroma, cor. C. M=calcilutite	D		biocalcarenite (alg,cri)		Cronoble & Mankin 1963
54	Pennsylvanian	NW-Ellesmere (Arct.Cir.)	>16		650	F = bry, J=cri, bra, bry, cor, gas, amm, tri, fus.	L	reef debris (dip -45°)			Bonham & Carter 1964

5.18 Biostromes, bioherms and their environment

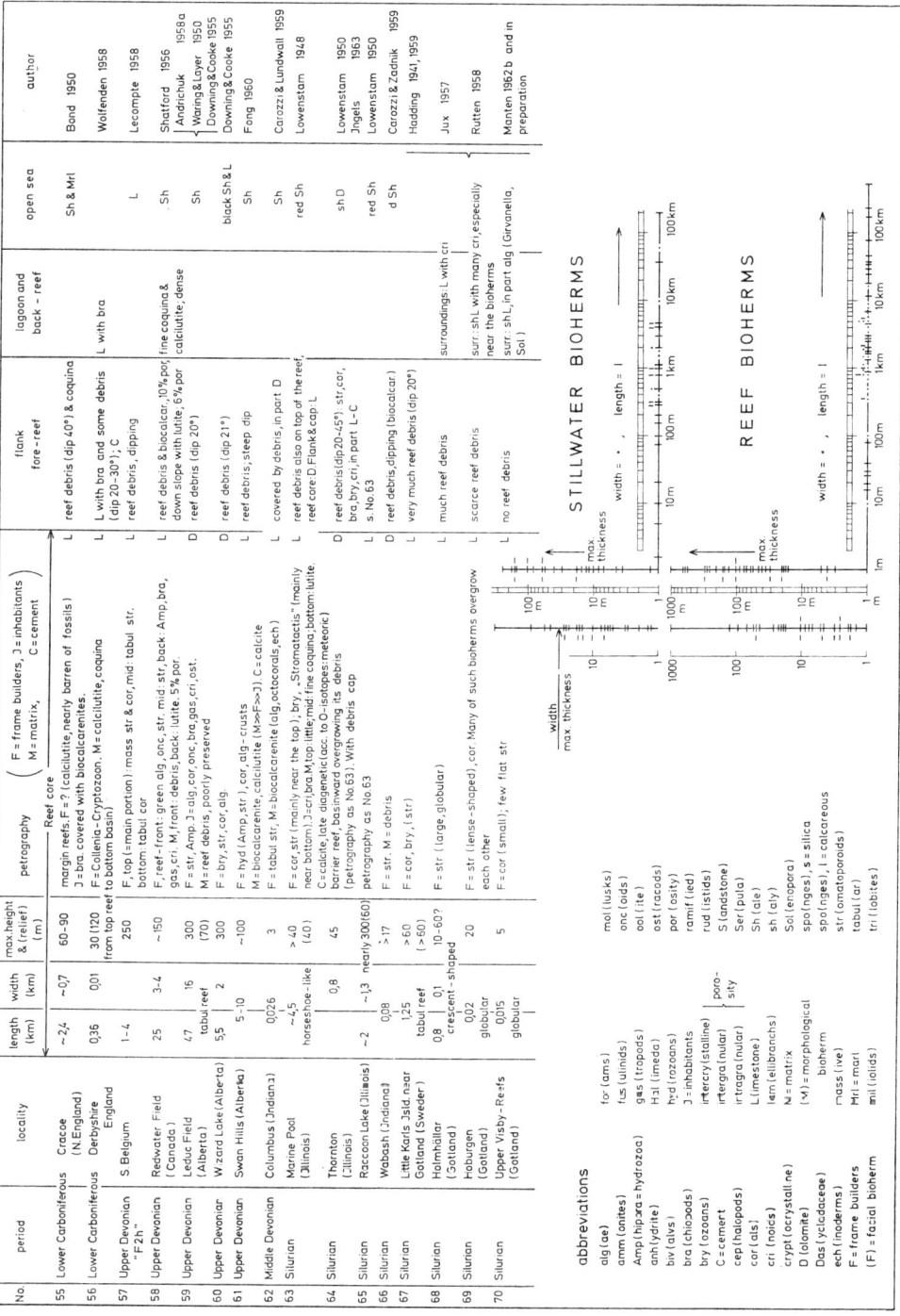

No.	period	locality	length (km)	width (km)	max.height & (relief) (m)	petrography (F = frame builders, J = inhabitants, M = matrix, C = cement) Reef core	flank fore-reef	lagoon and back-reef	open sea	author
55	Lower Carboniferous	Cracoe (N.England)	~2,4	~0,7	60-90	margin reefs. F = ? (calcilutite, nearly barren of fossils) J = bra. covered with biocalcarenites.	reef debris (dip 40°) & coquina		Sh & Mrl	Bond 1950
56	Lower Carboniferous	Derbyshire England	0,36	0,01	30 (120 from top reef to bottom basin)	F = Collenia-Cryptozoon. M = calcilutite, coquina	L with bra and some debris (dip 20-30°); C	L with bra	L	Wolfenden 1958
57	Upper Devonian "F2h"	S.Belgium	1-4		250	F, top (=main portion): mass str & cor, mid: tabul str. bottom: tabul cor	reef debris, dipping	fine coquina & calcilutite, dense		Lecompte 1958
58	Upper Devonian	Redwater Field (Canada)	25	3-4	~150	F, reef-front: green alg, onc, str. mid: str, back: Amp, bra, gas, cri, M, front: debris, back: lutite, 5% por.	reef debris & biocalcar., 10% por, down slope with lutite, 6% por		Sh	Shatford 1956 Andrichuk 1958a Waring & Loyer 1950
59	Upper Devonian	Leduc Field (Alberta)	4,7	16	300 (70)	F = str, Amp, J = alg, cor, onc, bra, gas, cri, ost.	reef debris (dip 20°)		Sh	Downing & Cooke 1955
60	Upper Devonian	W zard Lake (Alberta)	5,5 tabul reef	2	300	M = reef debris, poorly preserved	reef debris (dip 21°)			Downing & Cooke 1955
61	Upper Devonian	Swan Hills (Alberta)	5-10		~100	F = bry, str, cor, alg.	reef debris, steep dip		black Sh & L	Fong 1960
62	Middle Devonian	Columbus (Indiana)	0,026		3	F = hyd (Amp, str), cor, alg-crusts M = biocalcarenite, calcilutite (M>>F>>J) C = calcite	covered by debris, in part D		Sh	Carozzi & Lundwall 1959
63	Silurian	Marine Pool (Illinois)	~4,5 horseshoe-like		>40 (40)	F = cor, str (mainly near the top), bry, "Stromatactis" (mainly near bottom), J = cri, bra, M top: little; mid: fine coquina; bottom: lutite. C = calcite, late diagenetic (acc. to O-isotopes: meteoric) barrier reef, basinward overgrowing its debris (petrography as No.63). With debris cap	reef debris also on top of the reef, reef core: D. Flank & cap: L		red Sh	Lowenstam 1948
64	Silurian	Thornton (Illinois)		0,8	45	petrography as No. 63	reef debris (dip 20-45°) str, cor, bra, bry, cri, in part L-C s. No. 63		sh D	Lowenstam 1950 Ingels 1963
65	Silurian	Raccoon Lake (Illinois)	~2	~1,3	nearly 300(60)	F = str, M = debris	reef debris, dipping (biocalcar.)		red Sh	Lowenstam 1950
66	Silurian	Wabash (Indiana)	0,08		≥17	F = cor, bry, (str)	very much reef debris (dip 20°)		d Sh	Carozzi & Zadnik 1959
67	Silurian	Little Karls Isld. near Gotland (Sweden)	1,25 tabul reef		>60 (>60)	F = str (large, globular)	much reef debris	surroundings: L with cri		Hadding 1941, 1959
68	Silurian	Holmhällar (Gotland)	0,8	0,1 crescent-shaped	10-60?	F = str (lense-shaped), cor. Many of such bioherms overgrow each other	scarce reef debris	surr. shL with many cri, especially near the bioherms		Jux 1957
69	Silurian	Hoburgen (Gotland)	0,02 globular		20	F = str (small), few flat str	no reef debris	surr. shL, in part alg (Girvanella, Sol.)		Rutten 1958
70	Silurian	Upper Visby-Reefs (Gotland)	0,015 globular		5	F = cor (small), few flat str				Manten 1962b and in preparation

abbreviations

alg(ae)	for(ams)	mol(lusks)	
amm(onites)	tus(ulinids)	onc(oids)	
Amp(hipora = hydrozoa)	gas(tropods)	ool(ite)	
anh(ydrite)	Hal(imeda)	ost(racods)	
biv(alvs)	hyd(rozoans)	por(osity)	
bra(chiopods)	J = inhabitants	ramif(ied)	
bry(ozoans)	intercry(stalline) } porosity	S(andstone)	
C = cement	intergra(nular) }	Ser(pula)	
cep(halopods)	intragra(nular) }	Sh(ale)	
cor(als)	L(imestone)	sh(aly)	
cri(noids)	lam(ellibranchs)	Sol(enopora)	
crypt(ocrystalline)	M = matrix	spo(nges), s = silica	
D(olomite)	(M) = morphological bioherm	spo(nges), l = calcareous	
Dasycladaceae	mass(ive)	str(omatoporoids)	
ech(inoderms)	Mrl = marl	tabul(ar)	
F = frame builders	mll(iolids)	tri(lobites)	
(F) = fazial bioherm			

such cavities can originate by blue-green-algal crusts masking the surface of a bioherm, which is irregularly formed or pitted by dissolution. Subsequent dissolution erodes the top of these cavities and smoothes the bottom. This explanation is corroborated by the observations of KAYE (1959) on the coast of Puerto Rico and by PARKINSON (1964). In reef bioherms stromatactis structures are absent or present only in the lower part (LOWENSTAM 1950). SHINN (1968b) compared these structures with boring cavities. According to HECKEL (1972), "stromatactis" cavities can be formed by nonuniform compaction during initial unmixing and segregation of water from an originally partly water-supported accumulation of pure lime mud.

Exceptionally large quiet-water bioherms are the sponge "stotzen" (bioherms) in the Upper Jurassic of Southern Germany (table 5-4, No. 18). Occasionally, they are surrounded by debris, which, however, with rare exceptions (BAUSCH 1963a) are the result of slumpings or mud flows (TEMMLER 1966) or were formed by upheaval of the bioherms after death. The prime criterion for debris formed by the latter process is its absence in the interior of the bioherm (ROLL 1934).

Important from an economic point of view are the leaf-like fragments of red algae forming quiet-water bioherms in the pelletal Pennsylvanian limestones in Utah and Colorado (Ismay-Oil-Field; ELIAS 1963, CHOQUETTE & TRAUT 1963). (They are not included in table 5-4).

Reef bioherms

The oldest bioherms are stromatolites (blue-green algae), which played an important role in the Archean, Algonkian, Cambrian, Pennsylvanian and Permian

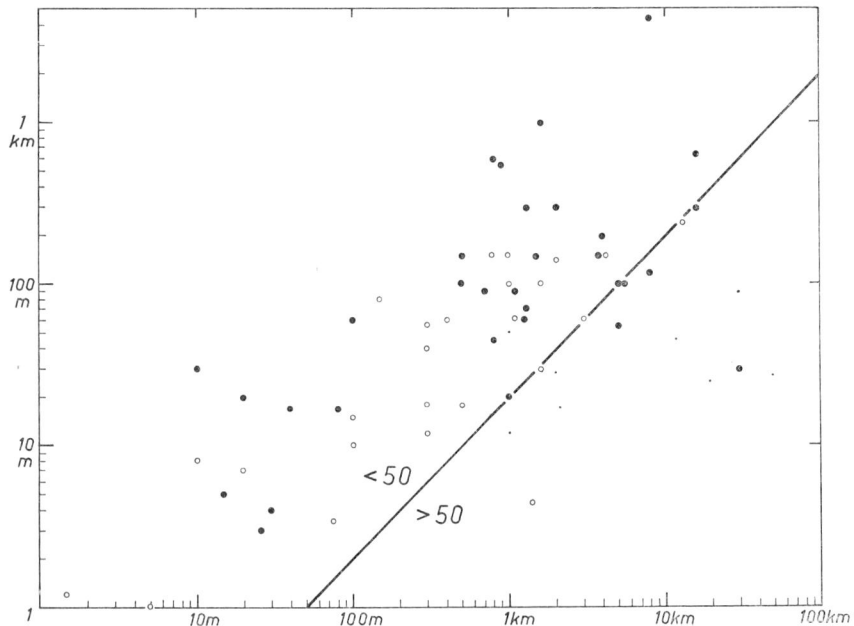

Fig. 5-58. Relationship between maximum width (= abscissa) and maximum thickness (= ordinate) of biostromes (·), still-water bioherms (○) and reef bioherms (•) listed in table 5-4. The line defines rock bodies with a width: thickness ratio > 50.

For several occurrences, the classification given in table 5-4 is applied only with reservation (e.g. No. 16, 34, 51, 52, 55).

periods, whereas since the Tertiary they have built reefs mainly in lakes and some intertidal environments (NEWELL et al. 1953, LOGAN et al. 1964). In the Silurian and the Devonian stromatopores are the main reef-builders, together with tabulate corals and bryozoans. In the Ordovician (and Carboniferous) reef bioherms are rare. Triassic reef bioherms in the Alps are held together by Hydrozoa (Spongiomorpha) and corals as well as by lime sponges and green algae (FABRICIUS 1966). The importance of the corals as reef builders has increased with time, and since the Tertiary, they have been, together with coralline algae, prime reef builders.

Separate developments have not been included in table 5-4: bryozoan reefs in the Cenozoan time, bryozoan-serpula reefs in the Pliocene of Cyprus (W. F. SCHMIDT 1963), serpula reefs in the Upper Jurassic of Northern Germany and in modern seas, e. g. in front of the Brasilian Coast (CORRENS 1939a) and in the Laguna Madre, Texas (FRIEDMAN 1964). Reef bioherms constructed by agglutinating worms on the eastern coast of Florida have been described by KIRTLEY & TANNER (1968). Other bioherms include the rudistid reefs in the Cretaceous, small *Placunopsis* bioherms (bivalves) in the Middle Triassic (HÖLDER 1961, KRUMBEIN 1963). Ostrea reefs in modern somewhat brackish seas (H. & G. TERMIER 1963) and *Vermetus* reefs (snails) at the coast of Florida (SHIER 1965). A recent collection of reef papers in English is included in Geol. Rundschau 61, 2 (1972).

On the third page of table 5-4 and in fig. 5-58 the relations between maximum length and width and maximum thickness (that is, the maximal extensions in all three directions) are used for graphical representations. In the diagram on the table, the abscissa shows the width (dots) as well as the maximum extent (dashes). It can be seen that the reef bioherms are, on the average, larger than the quiet-water bioherms, and that for both the ratio width: thickness is mostly lower than 50, whereas it is higher than 50 for biostromes (NELSON et al. 1962). However, this exact ratio is not necessarily the obligatory limit between bioherms and biostromes.

5.2 Nomenclature of carbonate rocks

In many carbonate rocks three components can be distinguished (in parentheses the terms of FOLK 1959 and 1962, in brackets the terms of LEIGHTON & PENDEXTER 1962):
1. particles (allochems) [grains]
2. matrix (microcrystalline ooze matrix, or "micrite") [micritic limestone]
3. cement (sparry cement, or "sparite") [cement]

Pores can be considered as another component of the rocks, and will be discussed in ch. 5.4.

1. Particles
    a) pellets (in rock terms "pel") [pelletal]
    b) detrital grains [detrital]
    c) intraclasts (in rock terms "intra-") [lump]
    d) skeletal fragments, unrounded } (fossils, in rock terms "bio-")
    e) skeletal grains, rounded   } [algal, bryozoan etc.]
    f) oöids (oölites, in rock terms "oö-") [oölitic]
    g) oncoids (pisolites) [pisolitic]

Rocks composed of these particles are described in ch. 5.13—5.16. Sometimes the rocks consist of two different particle types. Frequent types of such rocks are:

a—c intraclastic-pelletal limestone
a—d pelletal-skeletal limestone
a—e pelletal biocalcarenite
a—f pelletal oölite ⎫ in these instances many of the pellets and the
e—f skeletal oölite ⎭ skeletons are encrusted by oöids (fig. 5-51)

A collective term of all these rocks is: "Particle limestone" or "calcarenite"; if loose, "calcareous sand"; "detrital", "skeletal" etc. may be used as modifiers.

2. Matrix

If the distribution curve of the primary sediment shows two distinct maxima, the maximum in the smaller grain size is called "matrix". Most frequently this fabric is found in skeletal limestones as discussed in the following paragraphs.

3. Cement

Sediment pores formed by grain-grain contact (true of sediments deposited in a high-energy environment) can be filled during diagenesis partly or completely by a chemical precipitation of carbonate cement (ch. 5.321). This crystallization begins from the pore walls and gradually extends into the interior of the pores. Most frequently this fabric is found in calcarenites, biocalcarenites and oölites. A fine-grained matrix in such rocks is sometimes due to welding of pellets.

Whereas the matrix is frequently cryptocrystalline ($< 10\ \mu$), cement is mostly micro- to macrocrystalline ($10—100\ \mu$ and $> 100\ \mu$, respectively), as concluded from the obvious cement in fissures and cavities as well as from the presumable cement of most of the biocalcarenites and oölites.

However, the question often arises (especially in skeletal limestones) as to whether a micro- to macrocrystalline groundmass should be interpreted as a cement, or as a matrix with crystal enlargement (see below). This distinction is important to economic geology, since cemented rocks may grade laterally into porous rocks filled with oil or water, whereas rocks rich in matrix mostly have low porosities (or at least low permeabilities) provided that lateral changes of the depositional environment or of the diagenesis (e. g. late-diagenetical dolomitization or solution porosity) do not occur. The distinction between matrix and cement became critical, since cryptocrystalline Mg-calcite cement (ALEXANDERSSON 1969, AMIEL et al. 1971) and micritization of cement (BATHURST 1971a) have been described (see next chapter).

The question of "cement or matrix" can sometimes be settled by the presence of "grain support" (particles support each other) or "mud support" (particles are separated by a fine-grained groundmass matrix) (DUNHAM 1962). Since most grains are more or less angular, they will support each other at very different percentages by volume. The percentage (by volume) of different particles with grain support is listed below (from DUNHAM 1962). The grain volume includes intraparticle porosity as well as notches on the outside whereas the net volume only comprises the mineral substance.

	grain volume	net volume
porcelain spheres	65%	60%
gastropods	45%	25%
small lamellibranchs	30%	25%
rose corals	70%	20%
branching red algae	20%	15%

5.2 Nomenclature of carbonate rocks

If, however, such loose fabrics are found in compacted rocks, a primary grain support may not be probable, because of the compaction of the matrix. Generally grain support may be assumed in skeletal limestones whose grain volume is higher than 50%. This would correspond to FOLK's (1962) limit between "sparse biomicrite" and "packed biomicrite".

DUNHAM (1962) distinguished

"grainstone" = particles without matrix, with or without cement
"packstone" = with matrix; particles support each other
"wackestone" = particles "floating" in a matrix
"mudstone" = lutite with less than 10% particles

This principle of grouping was extended to a greater variety of rocks:

Table 5-5. Textural classification of carbonate rocks, by EMBRY & KLOVAN (1972).

Components not bound during deposition						Components organically bound during deposition		
less than 10% > 2 mm components				more than 10% > 2 mm components		by organisms		
contains lime mud < 30 μ		no lime mud			> 2 mm component supported	that act as baffles	that encrust and bind	that build a rigid framework
mud supported		grain supported ("grains" = 0.03—2 mm)		matrix supported				
< 10% grains	> 10% grains							
mudstone	wackestone	packstone	grainstone	floatstone	rudstone	bafflestone	bindstone	framestone

Typical criteria for grain support are convex-concave grain contacts as well as cavities which are partly filled with internal sediments (geopetal fabric, SANDER 1936; fig. 5-57).

The term "Offenheitsgrad" coined by SANDER (1950) refers to the number of all particles divided by the number of particles which are in direct contact with one or more other particles in the thin section: This term enables a comparison of the packing density of different fabrics. The "Offenheitsgrad" of the packstones and grainstones illustrated by DUNHAM (1962) is about 1.2. Very bulky artificial aggregates show "Offenheitsgrad" between 1.7 and 3.3. Such fabrics, however, if found in older rocks, genetically may be wackestones because of the compaction which has to be substracted. Until further and exact investigations are done, "Offenheitsgrad" measurements above 2 most likely will infer mud support. For 'contact strength', see ch. 3.314.1.

If mud support is probable or established, the question "cement or matrix" can be answered in favor of "matrix". If grain support is probable or established, this question may be answered by using the criteria described in table 5-7; otherwise, the neutral term "groundmass" may be used.

The rock names can define the character of the three components of the fabric, i. e. particles, matrix and cement. This can be done by terms such as oösparite, biomicrite and biopelmicrite implying a different evaluation of the particles as proposed by FOLK (1959) or by adjective combinations, according to the principles used in the present book (LEIGHTON & PENDEXTER 1962, modified by BISSELL & CHILINGAR 1967):

Table 5-6. Compositional classification of carbonate rocks, shown by one example (BISSELL & CHILINGAR 1967):

percentage of particles (e. g. pellets)	rock name
0 — 10%	calcilutite or micritic limestone (with few pellets)
10 — 25%	pelletal, micritic limestones
25 — 50%	pelletal-micritic limestone
50 — 75%	micritic-pelletal limestone
75 — 90%	micritic, pelletal limestone
90 — 100%	pelletal limestone

Analogous rock names are formed by substituting "micritic" by "sparitic" for cemented rocks and by substituting "pelletal" by other particles e. g. "oölitic", "skeletal", "detrital", "intraclastic", "oncolitic". For "skeletal" the specifications "algal", "bryozoan", "lamellibranch", etc. may be used. Also several particle types can be used together, and "limestone" can be replaced by "dolomite".

If two particle types are present, the following rock names may be used:
20% skeletons, 28% intraclasts, 52% calcite matrix: either skeletal, intraclastic-micritic limestone (or skeletal, intraclastic calcilutite), or, according to fig. 2-1 (p. 11), limestone with skeletons and many intraclasts.
20% skeletons, 28% calcite matrix, 52% intraclasts: either skeletal, micritic-intraclastic limestone, or intraclastic limestone with skeletons and much calcite matrix.
16% pellets, 24% calcite cement, 60% oöids: either cemented pelletal oölite (or cemented pelletal oölitic limestone), or cemented oölite with pellets.

For mixed clastic and calcareous rocks the triangle (fig. 2-1) can be applied as shown by the following two examples:

35% sand ⎫ cemented skeletal sand- | 60% silt ⎫ skeletal
35% skeletons ⎬ stone or cemented very | 14% skeletons ⎬ micritic
 (e. g. bivalves) ⎪ sandy coquina | 26% calcitic ⎪ siltstone
30% cement ⎭ (sandy biosparite) | matrix ⎭ (biomicritic siltstone)

The particle size in mm may be added in parentheses, either the estimated median diameter or in addition the observed deviation, e. g. "(0.05 — 0.2 — 0.4)". The crystal size is indicated without parentheses, e. g. "calcilutite (or limestone) 0.005", corresponding to a median crystal size of 0.005 mm.

The presence of grain support or mud support is indicated mostly by the rock name. "Micritic-skeletal limestones" are mostly grain-supported, whereas "skeletal-micritic limestones" are not. A more genetic nomenclature based on an energy index has been proposed by PLUMLEY et al. (1962).

5.3 Diagenesis

5.31 Introductory note

The transition between sedimentation and diagenesis is not always sharp: micritic or pelletal sediments can be reworked and resedimented. Primary cavities in reefs can be filled by alternating layers of internal sediment and syngenetic cement. Burrowers distort the sediment in which diagenesis already has begun (SANDER 1936, WOLF 1962, BOSELLINI 1964).

Another limit in limestones, however, is relatively sharp: the transition from loose sediment to rigid rock. According to PUSTOWALOW (1940) the changes in loose or soft sediments can be described as early-diagenetic, whereas those in the rigid sediments can be defined as late-diagenetic. "Late diagenesis" can begin quite early (e. g. in subrecent crusts), depending on the time of lithification. Tentatively, the terms "shallow burial" and "deep burial stage" as used in clastic rocks may also be applied to carbonate sediments. At present, however, too little is known about the burial depth of completed lithification.

The diagenesis is "isochemical" if the chemical composition of the sediment as a whole is not changed, and "allochemical" if it is changed. Lithification due to encrusting organisms, e. g. by red algae, blue-green algae, foraminifera, serpulae etc., is not considered here; see ch. 5.152 and 5.17).

"Particles" or "grains" are the smallest sedimented entities (e. g. sand grains, lime grains), notwithstanding whether they consist of one or more crystals. Crystals, however, which probably were not deposited in present form will not be referred to as "grains". As to the crystal fabric the following rock types can be distinguished:

1. idiomorphous rocks which consist of "euhedral" crystals, i. e. terminated by growth faces as observed especially in porous dolomites.

2. hypidiomorphous rocks composed of "subhedral" crystals, i. e. terminated by growth faces only in part.

3. xenomorphous rocks which are composed of "anhedral" (i. e. irregularly terminated) crystals.

FRIEDMAN (1965b) proposed a genetic classification of these fabrics into "idiomorphous" (for magmatic fabrics only), "idioblastic" (for metamorphic fabrics only) and "idiotopic" (for diagenetic fabrics only). In most cases, however, the descriptive terms mentioned above are more convenient.

5.32 Isochemical diagenesis

The most important isochemical diagenetic processes in limestones are

1. Cementation = filling of primary or dissolution cavities within the sediment by chemical precipitation which generally leads to consolidation of the sediment (ch. 5.321). Cone-in-cone (ch. 5.324) and concretions (ch. 5.325) are interpreted as special types of cementation prevailing in marls.

2. Neomorphism (FOLK 1965a) = all processes by which older crystals are gradually consumed and simultaneously replaced by new crystals of the same mineral or a pseudomorph (ch. 5.322). They are subdivided as follows:

a) Crystal enlargement (Sammelkristallisation; included in BATHURST's aggrading neomorphism, 1971b), by which larger (unstrained) crystals grow at the expense of smaller (unstrained) crystals of the same mineral, in wet systems (ch. 5.322.1). The equivalent in dry metallic systems is termed "grain growth" (FOLK 1965).

b) Recrystallization (strain recrystallization, FOLK 1965), by which unstrained crystals form at the expense of strained crystals of the same mineral; this occurs generally in dry systems, in metamorphic rocks only. Crystal size increases during "aggrading recrystallization" and decreases during "degrading recrystallization" (FOLK 1965).

c) Transformation, e. g. aragonite → calcite (ch. 5.322.2), but not magnesian calcite → calcite (see ch. 5.332.2). It occurs in wet systems. The equivalent in dry systems is termed "inversion" (e. g. kyanite → sillimanite).

Table 5-7. Distinguishing features of cementation and crystal enlargement (see also BATHURST 1958, 1964a, STAUFFER 1962).

	cement	crystal enlargement
primary fabric	cavity	particles or matrix
border towards unaltered rock	sharp	partly diffuse, partly idiomorphous, frequently crossing particles
	crinoids are frequently centers of homoaxial	
	cementation ("syntaxial rim cementation")	crystal enlargement ("syntaxial enlargement")
crystal termination	frequently straight; enfacial junctions* are frequent.	coarse crystals irregularly indentated. Frequently inclusions on crystal boundaries
crystal size	uniform in small pores.	fine-crystalline: well-sorted. coarse-crystalline: poor sorting frequent.
	In large pores often fibrous cement A at the walls; towards the center isometrical cement B, sometimes increasing in the center.	Floating relics of micron-sized material surrounded by spar.
special criteria	micrite clusters with an upper boundary parallel to the stratification indicates a geopetal filling of a cavity. In these cases the crystals above this boundary line can be interpreted as cement.	whenever grain support can be excluded, the groundmass cannot be cement, but must be interpreted as a primary matrix which has suffered crystal enlargement

* enfacial junction is defined by BATHURST (1958, 1964b) as the point where the compromise boundary between two crystals runs across the crystal face of a third crystal (fig. 5-57 at right from center) (FENNINGER 1968), showing the discontinuity of cementation processes.

3. **Dissolution processes**, e. g. stylolitization (ch. 5.323) by which carbonates and other minerals are dissolved selectively at clay seams, and formation of secondary porosity (ch. 5.423).

5.321 Cementation

Cementation occurs in two stages which are especially apparent in larger cavities (fig. 5-11).

The **early diagenetic cement A** forms mainly at the expense of the less stable minerals (e. g. aragonite of the skeletons) or, in the supratidal area, by evaporation of the pore fluid, or on the sea bottom. A prerequisite is the decay of the organic mucilage that covers and protects many grains from being coated by cement.

During the **late diagenesis** the main source of **cement B** is supposed to be pressure solution (e. g. FISCHER et al. 1964) (see ch. 5.323). These two stages of cementation are separated also by their fabric:

5.321.1 Cement A and cryptocrystalline cement (micritization)

The cement A (GRAF & LAMAR 1950) forms acicular or serrated seams on the pore walls (fig. 5-59). The longer axis of the fibers corresponds to the crystallographic c-direction which is the direction of the fastest crystal growth. This is explained by a preferred adsorption of colloidal impurities and hydrated Mg-ions on lattice planes perpendicular to the directions with densest packing of the atoms. In the case of calcite (or dolomite or aragonite) the growth perpendicular to c is inhibited by this adsorption. Consequently the fibrous growth $//$ c is promoted (see also MURAVYOV 1970). Due to this faster growth, nuclei with the c-axis perpendicular to the pore walls will be promoted, whereas nuclei oriented in other directions will be hindered ("Keimauslese", SANDER 1950, USDOWSKI 1962). These crystals are frequently brownish colored and show undulose extinction, presumably due to lattice disorder by submicroscopical inclusions (SANDER 1936, BATHURST 1964b, COTTER 1966).

Cryptocrystalline cement (mainly magnesian calcite, but also aragonite) has been observed in many places (fig. 5-59). Only by scanning electron micrographs has the cement A-fabric of such material been established (TAYLOR & ILLING 1969).

As shown by observations in modern sediments, cement A is formed in both the sublittoral and littoral zones ("beach rock"). It represents the first induration of sediments and therefore may be termed "early diagenetic" (ILLING 1954, RUSSELL 1962, GEVIRTZ & FRIEDMAN 1966, TAFT 1967, fig. 9, STODDARD & CANN 1965: beach rock) (fig. 5-60, top). The early formation of this cement can be recognized in limestones by the presence of skeletons covered by cement A and broken by subsequent loading which have not been healed by cement A, but only by cement B (fig. 5-60, bottom). Internal sediment filling cavities which previously were floored by cement A also suggests clearly the early formation of this cement (SANDER 1936, WOLF 1963, TAYLOR 1964). Aragonite cement A transforms with time into a calcite cement poor

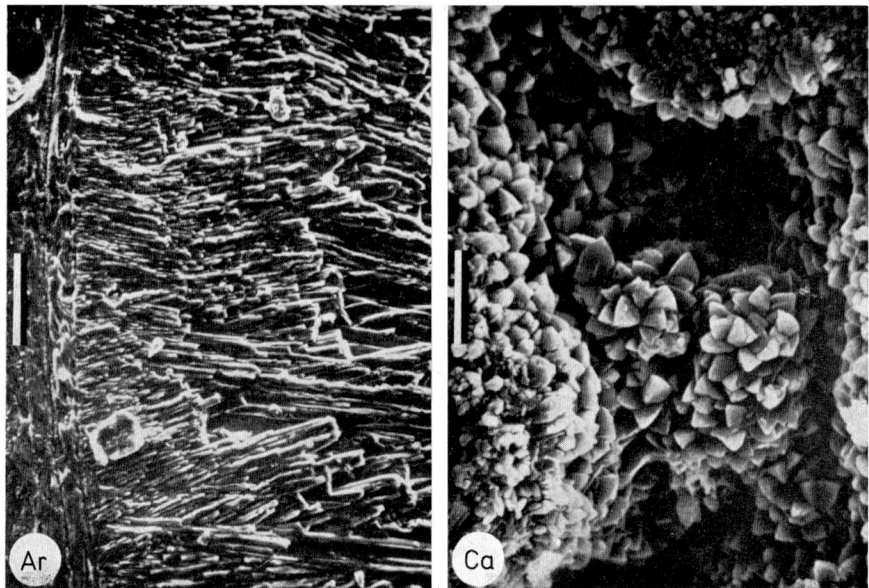

Fig. 5-59. Ar = Fibrous aragonite cement filling gastropod chamber from a Holocene algal reef of Bermuda. Note increase in crystal size away from the substrate (left). Scanning electron micrograph of a polished and etched surface (from GINSBURG et al. 1971, by permission). Bar scale = 50 μ.

Ca = High-Mg calcite cement partly filling interstices between skeletal particles forming a geopetal fill in the gastropod chamber shown above. Increase in crystal size away from particle surface is apparent. Younger crystals show well-developed faces; note also curved faces. Crystals range in size from 0.5—6 μ with a frequency maximum in the 2—4 μ range. SEM-micrograph, as above. Bar scale = 10 μ.

in Fe which retains only the general crystal form and orientation (USDOWSKI 1962, EVAMY & SHEARMAN 1962). Occasionally, a serrated supratidal calcite cement forms directly from solution (FRIEDMAN 1964). Cement A is characteristic of the shallow sea; in deposits of deeper water it is not common (WOLF 1963, 1965 a, MEISCHNER 1964, FLÜGEL & PÖLSER 1965); exceptions, however, are reported (GEVIRTZ & FRIEDMAN 1966, MILLIMAN 1966, WISE & KELTS 1971). Examples of early diagenetic submarine cementation are reported by GINSBURG (1953), EMERY et al. (1954), TAFT & HARBAUGH (1964), HOSKIN (1966), CIFELLI et al. (1966), MILLIMAN (1966), FISCHER & GARRISON (1967), ALEXANDERSSON (1969), SHINN (1969), TAYLOR & ILLING (1969), and many authors in BRICKER (ed.; 1971).

Cementation of grains and internal cementation (hardening) of fecal pellets is inhibited in muddy environments, for two possible reasons: 1) The flux of dissolved carbonate to and within impermeable lime muds is relatively small, and 2) the muddy sediments are generally poorly aerated and relatively rich in organic matter; failure of oxidative decomposition of organic films on carbonate particles may prevent welding of groups of particles or the formation of overgrowths on particles (BOYER 1972). On the other hand, cementation at the sea bottom is most active where adjacent

5.32 *Isochemical diagenesis* 283

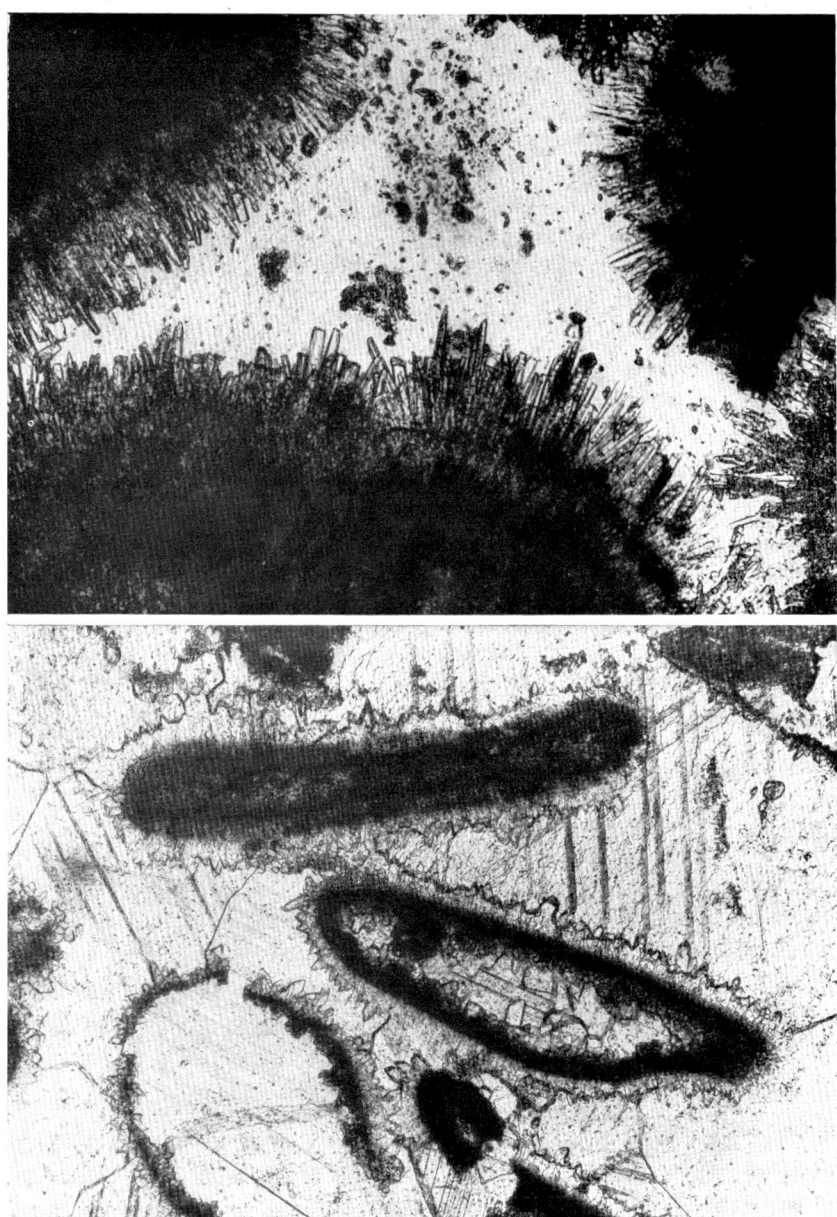

Fig. 5-60. Top: Aragonite cement A in a "beach rock" pebble of oölitic biocalcarenite from the surge of Bimini island, Bahamas (width 1 mm).

Bottom: Biocalcarenite with calcite cement A (engulfing radially the particles) and B (filling the pores as large crystals). Two particles are excavated and filled with cement. The particle in the lower left corner was broken after cementation A. Rhaetian of the northern Alps (width 1.8 mm, thin section by F. FABRICIUS).

grains remain in contact for relatively long periods. These constraints imply that submarine cementation will be most active on sandy floors in moderate energy environments where mud is absent and grains are not agitated (SHINN 1969, PURSER 1973).

In larger cavities, the first generation cement may have a kind of fabric which has been called "radiaxial" by BATHURST (1959 a). The elongate crystals consist of subcrystals radially diverging from the walls of the cavity. These subcrystals, however, show an undulose extinction with the c-axes converging from the walls of the cavity. The origin of this fabric is not known. It is suggested that these crystals, whether they are primary cement or later replacements of cement, grew relatively fast and therefore were able to engulf the numerous inclusions which are so typical of this fabric, and which therefore may cause the undulose extinction. The rapid growth seems to be supported by the large crystal size of radiaxial fibrous mosaics compared with normal cement A. But it does not yet explain the direction of axial divergence. The crystals of the second generation of cement (B) with equant rhombohedral habit and uniform extinction are lower in inclusions, and commonly in lattice continuity with the radiaxial fibres. They presumably grew slowly, i. e., under the normal conditions of late diagenesis. The transition is in places marked by a zig-zag dust line, which presumably marks the final crystal faces of the radiaxial (first generation) crystals which were later buried by the "syntaxial" ($=$ epitaxial) growth of the second generation crystals (BATHURST 1971 b). Generally, a higher rate of impurities and inclusions is typical of early diagenetic compared with late diagenetic cements.

The conditions which favor aragonite or magnesian calcite formation in the marine environment, however, are still not understood.

Aragonite cement has been observed by BATHURST (1971 a) in lagoonal sediments, rich in blue-green algae, of the Bahamas. REZAK et al. (1971) obtained aragonite cement in the presence of organic compounds such as glycine and taurine and magnesium ions. Without the presence of these materials calcium carbonate was precipitated as calcite. According to TRICHET (1971), the decay of cells and mucilaginous sheaths of blue-green algae produces weakly acidic molecules, e. g. amino acids and fatty acids, which take up Ca^{++}. Upon biodegradation of these acids, aragonite precipitates (cf. ch. 5.151).

Serrated magnesian calcite cement A and cryptocrystalline magnesian calcite cement are frequent in the coastal areas of Bermuda (GINSBURG et al. 1967), the Mediterranean (ALEXANDERSSON 1969) and the Persian Gulf (TAYLOR & ILLING 1969, SHINN 1969; figs. 5-59, 5-62) (cf. ch. 5.332.2). The influence of organic matter has been claimed by AMIEL et al. (1971).

Both aragonite and magnesian calcite cement A are observed in the Persian Gulf. The cementation is not confined to the coastal area, but is distributed over an area 70 000 km^2 in size covering the shallow sea of the Southern Persian Gulf to a depth of 30 m (SHINN 1969). In the upper tidal zone, aragonite cement prevails, whereas in lower tidal and subtidal areas, magnesian calcite is more frequent: Aragonite cement seems to be more common in slightly supersaline conditions (FRIEDMAN 1968 b), e. g. in deep sea sediments of the glacial supersaline Red Sea (MILLIMAN et al. 1969; see also ch. 5.331.21, a 7), whereas a low level of supersaturation favors the precipitation of Mg-calcite (DE GROOT 1969).

Intertidal aragonite cementation and polygonal cracks with local buckling into tent-like 'teepee' structures have been described by EVAMY (1973) from the Trucial Coast. According to EVANS et al. (1973), these indurations as well as the submarine hardgrounds described by SHINN (see above) are associated with very slow sedimentation or non-deposition.

In Bermuda GINSBURG et al. (1971) found aragonite cement e.g. lining gastropod skeletons whereas micritic (0.5—8 μ) magnesian calcite cement indurated internal sediments (fig. 5-59). GLOVER & PRAY (1971) observed aragonite cement in the interior of aragonitic fossils and magnesian calcite cement in the chambers of magnesian calcitic fossils. Similar observations were made by AMIEL et al. (1971) and SCHROEDER (1972b).

SCHROEDER (1972a) described green algae filaments with crusts of bladed Mg-calcite crystals penetrating and replacing aragonite cement A in the modern reefs of Bermuda.

A gravitational cement (MÜLLER 1971) which is indicative of alternate wetting and drying has been found by TAYLOR & ILLING ("microstalactites", 1969) in the intertidal area of the Persian Gulf. This cement, which is developed in coarse sediments, consists of elongate magnesian calcite and aragonite crystals forming rims that are thicker on the lower surface of the grains (fig. 5-62).

Cryptocrystalline cement can be mistaken for primary matrix. Possibly similar in origin but different in appearance is the phenomenon of "micritization" which affects mainly the particles, e.g. oöids and the rims of skeletons ("micrite envelopes", BATHURST 1964b, 1966) (Fig. 5-12, 13, 45, 65, 66). EMERY et al. (1954; in cores of Bikini) and PURDY (1963, 1968 in skeletons and oöids) found that the micritization rims of aragonite particles may be aragonitic in modern sediments. Cryptocrystalline envelopes of shells are often preserved in shallow water sediments and rocks. Envelopes consisting primarily of Mg-calcite tend to be preserved, whereas the aragonitic skeletons are dissolved (see fig. 5-67 and ch. 5.322.2; WINLAND 1968; ALEXANDERSSON 1969). On the other hand, mucilagenous envelopes found in the Persian Gulf (SHEARMAN & SKIPWITH 1965) and in the Bimini Lagoon, Bahamas (BATHURST 1967b) may form organic complex compounds with the tiny carbonate inclusions and even with the surface of the skeletal grains. Similar compounds have been identified in micritic envelopes as old as Jurassic in age (SUESS 1968). If rock cementation occurred quickly enough, these envelopes were fixed in their original position (figs. 5-13, 60), otherwise they were compacted and fractured (fig. 5-67).

Through micritization, skeletons, grains and oöids may be transformed into structureless intraclasts (ch. 5.14), the origin of which is no longer visible. On Mediterranean coasts, ALEXANDERSSON (1969) found aragonitic algal crusts *Lithothamnion*-like in appearance, that are more or less replaced by magnesian calcite filling tiny pores in the algae. In other cases, these fillings consist of aragonite (TAYLOR & ILLING 1969, BATHURST 1971a); (see also KENDALL & SKIPWITH 1969).

BATHURST (1964b, 1966) and WILSON (1968) found, as PURDY did, that micritization often is concentrated in slowly agitated lagoonal environments. Frequently these rims are composed of tiny borings possibly caused by blue-green algae such as *Ento-*

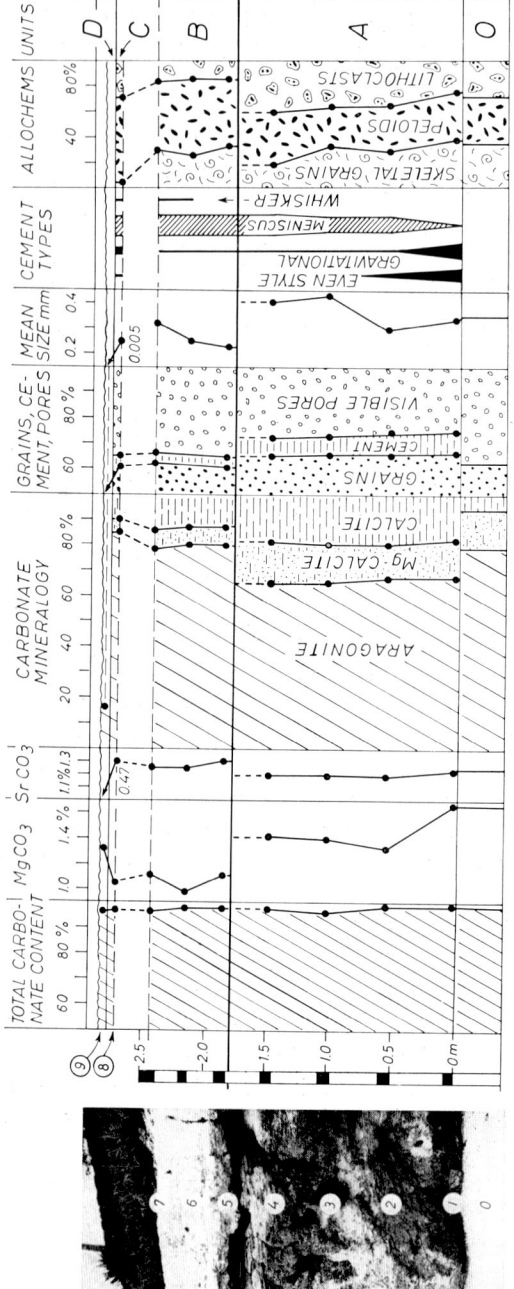

Fig. 5-61. Profile of a very young eolianite cliff at the West-coast of Northern Bimini (Bahamas). Note that the even style "rim cement", as well as the gravitational "microstalactitic cement", increases downward in this profile whereas the "meniscus cement" and the "whisker cement" increase in an upward direction. The even style cement points to a phreatic environment (below water table), at least for part of the time, whereas the meniscus cement is typical of the vadose zone (above water table).

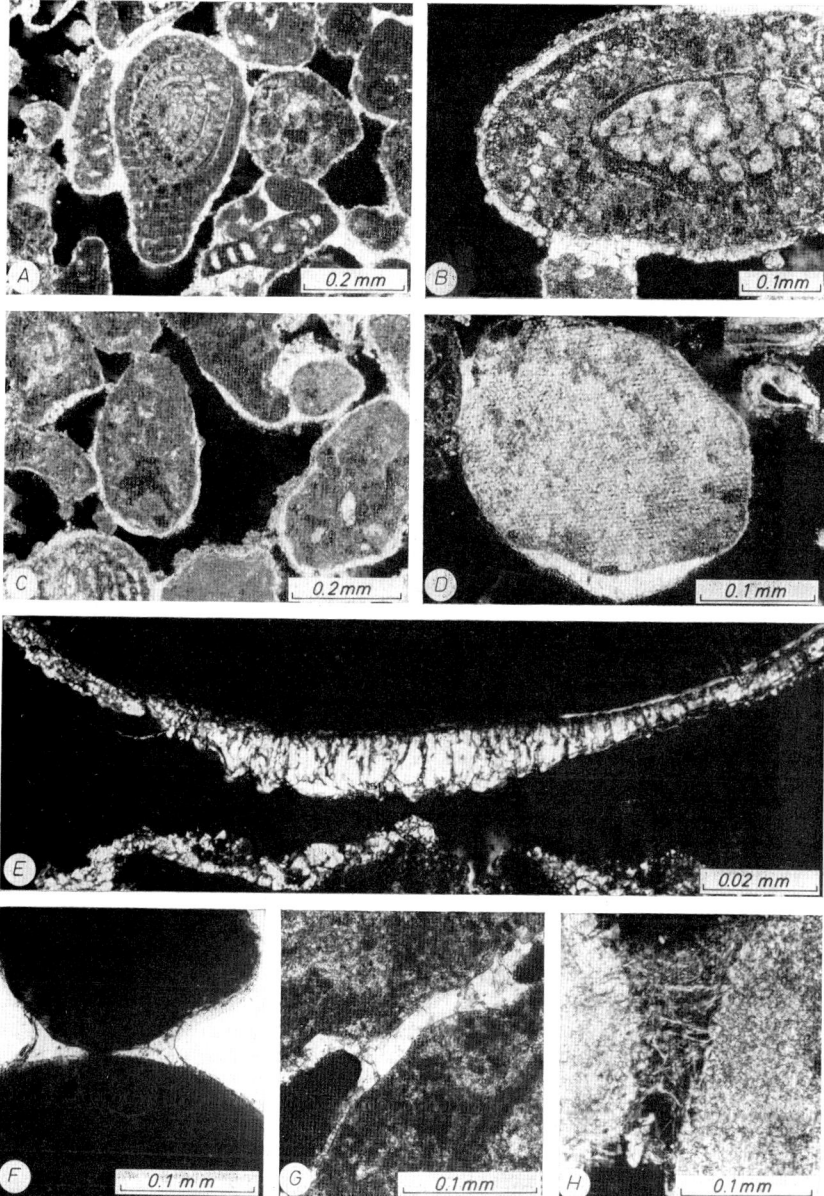

Fig. 5-62. Thin sections from the same profile. A, B = rim cement; C, D, E = gravitational (or "microstalactitic") cement; F, G = meniscus cement. Meniscus and rim cement consist mainly of more or less isometrical (granular) crystals, whereas the microstalactitic "gravitational cement" and part of the rim cement show orientation of c perpendicular to the grain surface. All these cements consist of low-magnesian calcite. The whisker cement, H, consists of aragonite and formed supposedly from sea spray. (All photographs except F with crossed nicols.) (Fig. 5-61 and 62 from G. Müller 1970b.)

physalis (MONTY 1967, p. 78) which subsequently are filled with micrite by cementation (fig. 5-66) (LLOYD 1971). Endolithic fungi and green algae were observed to produce microborings in bottom sediments collected from sites ranging in depth from the intertidal to 750 meters (PERKINS & HALSEY 1971).

Micritization, however, also occurs in deeper water, where algae are absent: FISCHER & GARRISON (1967) found micritized subrecent globigerinae included in crusts, in sites of low sedimentation. Micritization is generally absent in mud-supported limestones. This is accordance with observations by HUGHES CLARKE & KEIJ (1973) in lime muds of the Persian Gulf.

Subaerial weathering also may cause micritization (FRIEDMAN 1964, p. 806).

ALLEN et al. (1969) report on methane-derived cement from the upper continental slope.

In older calcitic rocks, PURSER (1969) found fabrics indicative of different original cementation types (e. g. cryptocrystalline cement, serrated magnesian calcite cement A, fibrous aragonite cement A, and gravitational cement A). Further observations of early cementation from older limestones are published by LINDSTRÖM (1963), PICHLER (1963; early cementation is indicated by differently oriented meniscuses), PRAY (1964), ZANKL (1964), GERMANN (1966) and many other authors.

Submarine rather than subaeric early cementation is responsible for "hardground" formations (see also ch. 6.32 and VOIGT 1959, HUDSON 1967, BATHURST 1971 b). Submarine hardgrounds in Jurassic oölites formed by calcitic or aragonitic cement A, prevented the oölites below from early cementation: they were compacted by stylolitization (PURSER 1971).

Abnormal forms of cement A are the "Groß-Oolithe" of the Northern Limestone Alps, large cavities which are filled by fibrous calcite, dolomite, and iron-dolomite layers in turn (SCHMIDEGG 1928, SANDER 1936, BOSELLINI 1965, GERMANN 1966).

5.321.2 Cement B

In contrast to cement A which generally forms early in diagenesis, cement B forms after the rock has been indurated. Isometrical mosaics of cement B fill the pores which are left by cementation A (figs. 5-11, 60, 65). Frequently the crystal size increases towards the center. This is not apparent in smaller pores in which either an equant cement or a monocrystal cement is observed. The considerable crystal size of cement B indicates that only few nuclei were growing at the same time due to moderate supersaturation (fig. 5-57). The origin of cement B is apparent where this cement develops upon pressure solution welding coated oölites (BATHURST 1971 c).

Cement B frequently consists of iron-bearing calcite as shown by staining (EVAMY & SHEARMAN 1962, 1965 and DICKSON 1966).[3] Reducing conditions during the late diagenesis permit the substitution of Fe^{++} for Ca^{++} in the calcite lattice. OLDERSHAW

---

[3] staining instructions (EVAMY & SHEARMAN 1962, modified by DICKSON 1966):
(A) dissolve 1 g potassium ferricyanid in 100 ccm of n/8 HCl. This solution is not stable and new solutions must be prepared frequently.

& SCOFFIN (1967) found that the iron originated from underlying shales; presumably it was brought with the compaction current into the limestones.

According to observations in modern sediments, cement A is absent in meteoric water environments. Instead, an isometrical low-Mg calcite cement similar to cement B, is formed (fig. 5-47; GINSBURG 1957, G. MÜLLER 1964c, LAND 1967, MATTHEWS 1967) using the $CaCO_3$ of dissolving aragonite particles (FRIEDMAN 1964). Occasionally vadose calcite cement has been detected only by cathodo-luminescence (FREEMAN 1971). The important role subaerial cementation plays in Pleistocene and Holocene limestones is explained by the considerable regressions and transgressions of the sea during ice ages.

By electron microprobe analyses BENSON & MATTHEWS (1971) were able to differentiate between three stages of cementation: (a) phreatic low to medium magnesium calcite, (b) early vadose high magnesium calcite, and (c) late vadose low magnesium calcite.

A special type of meteoric cement is the equant meniscus cement (DUNHAM 1971) filling the corners of pore spaces, e.g. the points of grain contact, thus rounding the pores. This cement develops from evaporation of capillary water and is found in calcarenites in the non-marine vadose zone, i.e. above the ground-water table. In the phreatic zone, i.e. below the ground water table, the cementation generally is more intensive, filling the whole pore space with a granular cement (LAND et al. 1967).

Subaerial stages in early diagenesis may be recognized by "vadose silt" flooring vugs, fractures and interstices as internal sediment postdating cement A and predating cement B. DUNHAM (1969a) in his discussion of a Permian reef, suggested percolating vadose water eroding early cement B crystals 10 to 25 microns in size and leaving them as inclined internal sediment.

Another type of cement is represented by overgrowths on echinoderm fragments (fig. 5-87) or — more rarely — on lamellibranch skeletons and oöids. In these cases, the cement-generations A and B can sometimes be separated by staining (A = iron-free, B = iron-bearing; EVAMY & SHEARMAN 1965). A crinoid limestone of the Mississippian was completely cemented by overgrowths, except those layers in which the crinoids were encrusted by ooids. These layers retained a certain amount of porosity and consequently now contain oil which in turn prevented further cementation (LUCIA & MURRAY 1967). This example demonstrates the high growth-rate of the cement overgrowing the echinoderm single crystals, which is due to the large crystal size and therefore responds to the same principle, as crystal enlargement (ch. 5.322.1), viz. lower solubility of larger crystals.

(B) dissolve 0.1 g alizarin-red S in 100 ccm of n/8 HCl. Mix 40 ccm of solution (A) with 60 ccm of solution (B).

Rinse the thin section in this solution for about 45 sec., afterwards in solution B for another 15 sec. Wash off carefully. By this procedure calcite is stained red, iron-containing calcite (beginning from about 1% Fe) violet-blue, iron-containing dolomite blue-green, whereas dolomite, siderite, magnesite, and rhodochrosite remain unstained. If iron is present in the rock not as carbonate, but in another mineral unstable in HCl, only the simple alizarin-red S staining can be applied. The above staining is weak enough to allow the observation of the rock fabric. For quantitative interpretation see LINDHOLM & FINKELMAN (1972).

5.321.3 Chemical aspects

A hiatus is generally observed between cement A and B; only the latter appears on compaction-fracture surfaces. Thus far, this hiatus is not fully understood. In terms of nucleation rate cement B is lowest and cryptocrystalline cement highest whereas cement A is intermediate. (High nucleation rates may also cause whisker-like cementation in larger pores [fig. 5-62]; see BRICKER, ed. 1971.) Due to the low supersaturation during carbonate sedimentation and diagenesis, epitaxial ("heterogeneous") nucleation is more frequent than neoformation of nuclei ("homogeneous nucleation", BATHURST 1971b).

High nucleation rates point to high supersaturation which occurs mainly in the tropic and subtropic shallow sea. If this supersaturation in the interstitial water is exhausted by different mechanisms, e. g. by precipitation of cement A and cryptocrystalline cement, and if the aragonitic particles are dissolved, the nucleation rate will drop drastically and cementation will stop. This hiatus occurs presumably at most several meters below the sea bottom. Cementation B, which proceeds slowly as indicated by the low nucleation rate, may begin if the saturation in $CaCO_3$ increases again, e. g. by pressure solution.

Occasionally, the formation of cement B may begin quite early in diagenesis. This is shown by the observation of blocky calcite cement (submarine) in recent Bermuda reefs (SCHROEDER 1972b).

In a strict sense, cementation is not always isochemical, as shown by high-magnesium calcite cementation of low-magnesium calcite mud observed by ALEX-ANDERSSON (1969) along the Mediterranean coastal zone (see end of ch. 5.41).

The Mg-content of calcitic skeletons is directly dependent upon water temperature (cf. ch. 5.151 and fig. 5-16). A similar relationship is shown by Mg-calcite cement. Whereas Mg_{12}-calcite — $(Mg_{0.12}Ca_{0.84})CO_3$ — is an approximate mean composition of deep-sea cements, Mg_{18}-calcite is an approximate mean of shallow-sea cements (Bermuda Biological Station, Publ. No. 3, different papers; cf. O. P. BRICKER, ed.: Carbonate Cements. — The Johns Hopkins Univ. Stud. in Geol., *19*, 1971, p. 49).

An interesting observation of (or against) equilibration has been reported by SIBLEY & MURRAY (1972). During micritization of coralline algae in an submarine-cemented layer 35 cm below the sediment surface in the Lac (Bonaire Island), the mineralogy changes from high- to low-Mg calcite, in a marine environment (less than 1 mol-% $MgCO_3$; MURRAY, pers. commun.).

Chemical aspects of many observations described in ch. 5.32 and 5.33 are included in fig. 5-81 (p. 323).

5.322 Neomorphism

5.322.1 Crystal enlargement

Many calcilutites are composed of crystals 5—10(—50) µ in size (FOLK's "microspar") which evidently are not fragments of skeletal particles but are formed at the expense of micritic limestones (1—4 µ) by crystal enlargement.

5.32 Isochemical diagenesis

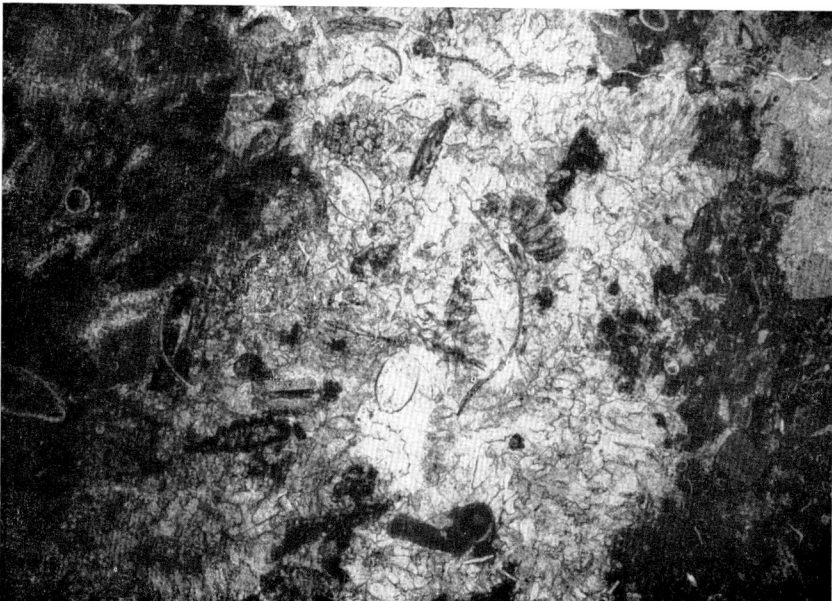

Fig. 5-63. Patches of crystal enlargement (center) in Ordovician Kullsberg reef limestone from Skålberget (Lake Siljan, Sweden). Notice the indented crystal boundaries. The fossils (bryozoa, ostracods, trilobites [left], and crinoids) are well preserved (width 20 mm).

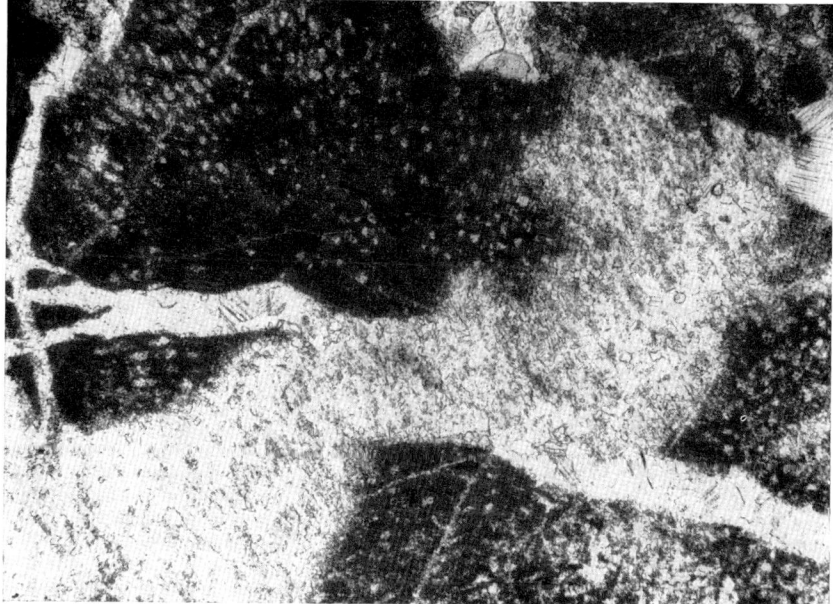

Fig. 5-64. Crystal enlargement transecting a coralline algal fragment. Typical are the blurred transitions between the crystal enlargement zone and the primary particle, whereas the cemented fissures are transecting the particles with sharp boundaries. Rhaetian, Northern Alps (width 2.5 mm, thin section by F. FABRICIUS).

The crystal enlargement of micritic limestones can be explained by the fact that smaller crystals are more soluble than larger ones since the positions of higher "Gibbs free energy" (the areas near edges and corners) are more frequent on smaller crystals. Larger crystals therefore will grow at the expense of the smaller ones, and micro- or macrocrystalline limestones can be formed from cryptocrystalline calcilutites ("micrite enlargement", FRIEDMAN 1964; figs. 5-63, 64).

Because crystal enlargement requires dissolution and reprecipitation, the presence of water is necessary. Tiny communicating capillaries which also are present in dense limestones (as shown by porosity measurements on fragments more than one cm in diameter) play an important role in crystal enlargement. Mechanical stress is not required. This is in contrast to the metamorphic "recrystallization" in which crystals suffer under mechanical stress and are replaced by crystals of the same mineral which are free of stress (VOLL 1960). Water may accelerate metamorphic recrystallization, but it is not required.

The characteristics of normal crystal enlargement, e. g. irregular indentation, are listed in tab. 5-7. Abnormal features are described below.

Crystal enlargement may begin around fossils, especially echinoderm fragments, and advance with a dentated rim towards the adjacent rock ("aggrading crystallization", fig. 5-20) or, vice versa, it starts as a syntaxial continuation of the cement into the grain; more frequently it begins in the matrix and spares the particles. Occasionally crinoids are changed into secondary spherulites by crystal enlargement fibers advancing from outside into the skeleton ("fibrous replacement calcite"; ORME &

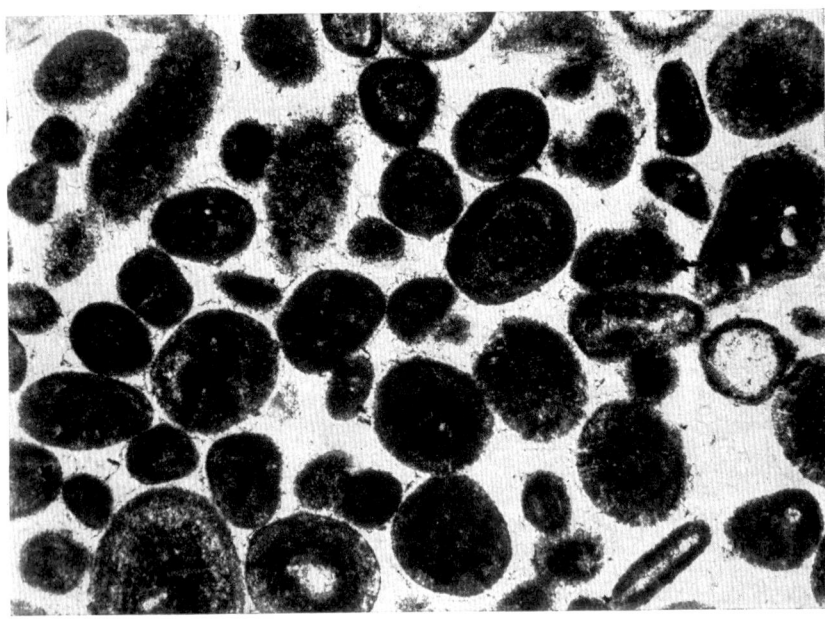

Fig. 5-65. Oölitic limestone, micritized and cemented by calcite (A and B). Jurassic, Ethiopia (width 3 mm).

BROWN 1963). BANNER & WOOD (1964) studied the crystal enlargement of foraminifera shells. Also the origin of the aragonite fibers in skeletal material of the Persian Gulf (SHINN 1969) is due to crystal enlargement. Radial fibrous crystal enlargement occurs also in calcilutites (BATHURST 1971 b). WISE (1973) described selective crystal enlargement of discoasters at the expense of coccoliths in Cretaceous and Tertiary nannoplankton muds of the Pacific. This is difficult to understand because both skeletons consist of calcite.

Finely dissipated clay hinders crystal enlargement. In the Jurassic rocks of southern Germany, limestones with more than 2% insoluble residue are cryptocrystalline (2—10 μ), whereas limestones with less clay are generally micro- to macrocrystalline (50—250 μ; BAUSCH 1968). In the dark dolomitic lutites of the Zechstein 3 in northwestern Germany, the crystal size is about 15 microns, whereas it is higher (50 microns) in the lighter dolomitic oncolites which are lower in insoluble residue (FÜCHTBAUER 1958, 1964b). For these reasons the insoluble residue should be indicated if crystal size is mentioned. Also the observation by NEWELL et al. (1953) and BISSELL (1959) that the crystal enlargement is lowest in the basin sediments, higher in the fore-reef and most intensive in the reef rocks may be explained by an increase in clay content towards the basin (or due to preferential crystal enlargement of aragonitic sediments which are normally more frequent in shallow water than in basin sediments).

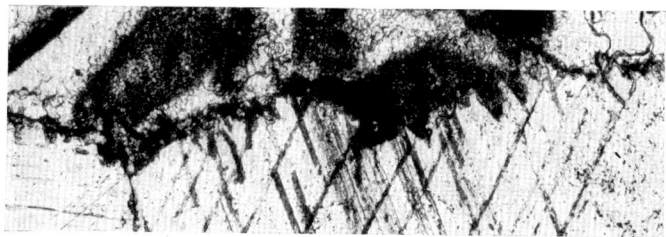

Fig. 5-66. Biocalcarenite. The shell which makes up the lower 2/3 of the photograph has probably been bored by blue-green algae. These tiny tubules (especially left from center) may have been filled by magnesian calcite similar to recent skeletons. Subsequently, the aragonite shell was dissolved and replaced by cement B (with twin lamellae). Rhaetian, Northern Alps (width 1.3 mm, thin section from F. FABRICIUS).

In metamorphic rocks, crystal enlargement occurs also in clayey limestones (BAUSCH 1968), possibly because true recrystallization supersedes crystal enlargement; a direct correlation seems to occur between the degree of metamorphism and the clay content below which the crystal size in the limestone increases suddenly. It may be concluded that the degree of crystal enlargement in limestones is a function of both temperature and time, as well as the amount of impurities. In addition to clay, organic substances and possibly pyrite also may be effective.

The chemistry of the pore fluids and the permeability and porosity of the rock also are important: KARCZ (1964) observed crystal enlargement near fissures; FOLK (1959) found tiny clay layers stimulating crystal enlargement.

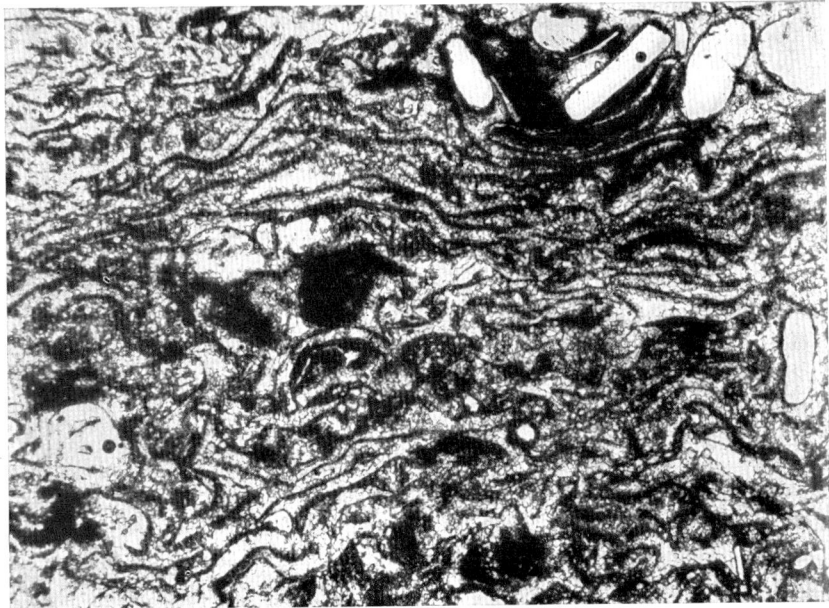

Fig. 5-67. Dolomite coquina. The shells were dissolved during early diagenesis, presumably because they were composed of aragonite. The micritic envelopes are preserved possibly because they consisted of magnesian calcite deposited as a cement in tiny borings of bluegreen algae (WINLAND 1968, ALEXANDERSSON 1969, BUCHBINDER & FRIEDMAN 1970). The envelopes are squeezed together (e.g. above the center); the larger pores are in part filled with anhydrite, in part they are empty (porosity 16%). Upper Jurassic (Emsland area) (width 3 mm).

5.322.2 Transformation aragonite→calcite

Aragonite is unstable at the normal pressures and temperatures found on the earths surface; it becomes stable in metamorphic rocks only, at a depth of about 18 km (SIMMONS & BELL 1963). In shallow water sediments metastable aragonite is formed for kinetic reasons at high Mg/Ca ratios, but it tends to transform into the stable calcite (FYFE & BISCHOFF 1965, ROBERTSON 1964).

Like other diagenetic processes the transformation proceeds by dissolution in and *in-situ* precipitation from an electrolyte. Between the aragonite and the replacing calcite a solution film is present tiny enough to enable *in-situ* preservation of a ghost of the original wall structure. Preservation of aragonite has been limited to strata in which it is embedded in shales of low permeability or in oil shales (ZAPFE 1936, HALLAM & O'HARA 1962, FÜCHTBAUER & GOLDSCHMIDT 1963) or in asphalt (STEHLI 1956) (see figs. 5-40, 42, 43). According to KENNEDY & HALL (1967), amino-acids, adsorbed on the aragonite surface, could prevent the transformation. The loss in Sr during the transformation also points to a wet process (see BATHURST 1971b, p. 347). Meteoric transformation of aragonite into low magnesium calcite was observed in Pleistocene gastropod shells of Bermuda; aragonite cement A needles, however, were not affected, presumably because of their higher strontium content compared with the gastropods (SCHROEDER 1973).

Where embedding in shale prevented an early-diagenetic transformation, later thermal history can influence transformation. For instance, Mesozoic ammonites, which are regionally preserved in aragonite, were transformed into calcite by Upper Cretaceous heating (W. of Osnabrück, Germany) which also altered Lower Cretaceous coals (JORDAN & STAHL 1970).

An impression of the transformation rate is obtained by the following observations: in a deep boring on the Great Barrier Reef (Australia), except for several layers of subaerial calcitization, aragonite was preserved in the biocalcarenites to a depth of 223 m (FAIRBRIDGE 1950). Mineralogic trends of lime muds from other borings in this area were less consistent (MAXWELL 1962). In borings on Pacific atolls, aragonite matrix has been found to depths of 290 m in Upper Miocene sediments (EMERY et al. 1954), and aragonite mud is present at 270 m (LADD et al. 1953). Even in the porous Pleistocene reefs of the Florida Keys which are subjected to frequent rain falls, some corals still contain aragonite (SIEGEL 1960). Possibly the aragonite is stabilized by Mg ions (p. 192) released from magnesian calcites and enriched in the vadose waters. TAFT & HARBAUGH (1964) found no diagenetic changes in the aragonite or magnesian calcite in recent sediments off southern Florida. Accordingly it was still an open question if aragonite is transformed into calcite below the sea bottom. SHINN (1969), however, has observed a transformation of aragonite cement A into calcite cement A in the shallow marine sediments of the southern Persian Gulf. In the same area, TAYLOR & ILLING (1969) found aragonite cement A transformed into cryptocrystalline magnesian calcite cement, while retaining at least some of its needle-like texture.

These two occurrences indicate two different types of polymorphic transformation of aragonite to calcite:

(a) Homoaxial transformation [e.g. of fibrous aragonite to fibrous calcite of (nearly) the same orientation]. A transformation of aragonite needles into calcite needles has been carried out experimentally at 430° C by HATHAWAY & ROBERTSON (1961).

Natural occurrences include spherulitic microtextures in hydrozoans, homoaxially transformed into spherulitic calcite (FENNINGER & FLAJS, in press), and most of the cement A which is now calcite but previously was aragonite.

(b) Heteroaxial transformation (no optical and textural relationship between aragonite and calcite crystals). Fig. 5-68 and BATHURST's (l.c.) fig. 342, p. 488, are examples. Most of the mollusk shells transform heteroaxially.

Transformation of skeletal material occurs mainly in rocks which presumably had initial low permeability, e.g. in shales, marls, or micritic limestones. In such cases, the composition of the local pore fluid probably was between the solubility product of aragonite and calcite. While the interior structures of skeletons are more or less preserved during transformation (V. SCHMIDT 1961, 1965; fig. 5-68), this is not true during the following process: In skeletal limestones, which had high permeability during early diagenesis, the aragonite particles are frequently dissolved selectively without immediate replacement by calcite. This occurs particularly in areas of periodic emersion, under the influence of fresh water (FRIEDMAN 1964, STANLEY 1966): Calcitic cement can fill the cavities eventually.

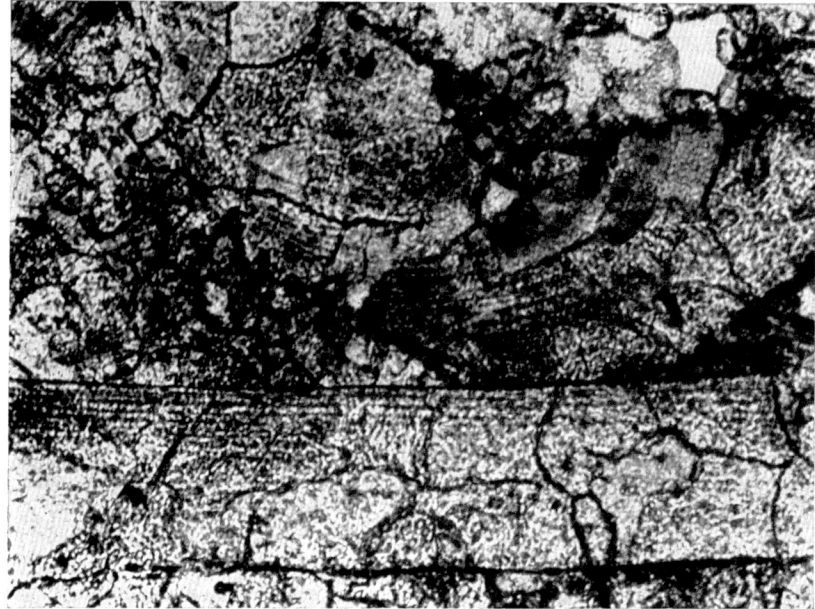

Fig. 5-68. Calcareous coquina. The aragonitic shells retained their interior structure when transformed into macrocrystalline calcite. The crystal boundaries are visible. Upper Jurassic 4 a, Emsland area (width 1 mm).

Whereas such dissolution-cementation fabrics prevailed in the normal-marine middle Triassic limestones, transformation was more frequent in the brackish uppermost layers of this sequence (BACHMANN 1973, p. 62). The occurrence of transformation in brackish or fresh water deposits was also observed by HUDSON (1962). It is suggested that the transformation began soon after deposition in both cases, whereas Mg ions stabilized the aragonite in the marine skeletal limestones, until they were percolated by meteoric water dissolving the whole skeletons.

Many calcilutites are composed of isometric crystals 1—3 microns in diameter (E. FLÜGEL et al. 1968). They presumably developed from acicular aragonite or bladed calcite crystals respectively of about equal length (FOLK 1965 b) which may be partly skeletal in origin.

For the relative stabilities of aragonite, high-magnesium calcite and low-magnesium calcite see ch. 5.332.2.

5.323 Stylolites

Many limestones contain streaks of serrated clay mainly parallel to the bedding, called stylolitic seams. The stylolites (from Greek stylos = column) form teeth-like projections which fit into sockets of similar dimensions perpendicular or, in folded rocks, oblique to the seams (fig. 5-69).

Fig. 5-69. Stylolites in limestone. Thinning of the vertical sections is typical. Upper Permian 2, W. of Nienburg (Germany) (width 2.5 mm).

Stratification of adjacent rocks is interrupted by stylolites; this points to considerable dissolution of the rock at the site of these seams. STOCKDALE (1926) calculated a volume reduction of 13—34 percent for limestones of the lower Mississippian. However, the relatively high concentration of insoluble residues (e. g. quartz grains) at the stylolite surface suggests that a multiple of the "amplitude" of the stylolites may have been dissolved. For a more detailed discussion see MANTEN (1966), TRURNIT (1968) and PARK & SCHOT (1968).

Horizontal stylolite seams have been followed for 500 m (CORRENS 1939a). RIGBY (1953) and WAGNER (1964) found a regular strike of steep to vertical seams pointing to a regional horizontal stress (see also PLESSMANN 1972).

The question as to whether stylolites form before or after lithification is still being discussed. Most authors prefer the latter opinion. The main criteria for a post-lithification origin are as follows:

a) they displace calcitic joints (LANG 1964),
b) they transect crystals formed during crystal enlargement which certainly is a late-diagenetic phenomenon.

Fabric criteria for an origin prior to lithification have not yet been published. However, dissolution may begin before complete lithification on certain irregular clay stringers ("Lösungsschlieren", WEILER 1957, V. SCHMIDT 1961).

According to DUNNINGTON (1954), 79.8 % of the 15569 stylolitic seams observed in Carboniferous to Miocene limestones in Northern Iraq were formed after lithification; a similar origin for the remaining can not be excluded. DUNNINGTON suggests that a minimum depth of burial is required, generally about 500 m.

TRURNIT (1968) claims that dissolution seams are sutured only between particles of equal solubility. Between particles of different solubility straight solution faces develop; the same is true for a more advanced stage if the solution residue is sufficient to act as an insoluble particle. For this reason stylolites are not found on thick clay layers but only on thin clay seams (PROKOPOVICH 1952). PETTIJOHN (1957) also states that the clay seam must be a minor constituent of the rock. Accordingly an optimal thickness of the clay seams exists in stylolite formation. Stylolites do not form if the clay seams are thick enough to allow the water to flow in their interior; the water must be forced to flow at the lower or upper side of the seam, between clay and limestone. Wherever the water flows, the permeability increases; these pathways are not abandoned. If the water moves at the upper surface of the seam, the pathway shifts upwards owing to the dissolution of the upper limestone layer. The opposite is true if the water moves along the lower boundary of the seam. Skeleton particles bordering the seam prevent the water from flowing on this border; it passes to the opposite side of the clay seam. For instance, if the skeleton lies below the seam, a pillar projecting upwards is formed (see fig. 70 in CORRENS 1939a, from WAGNER 1931b). On the flanks of such projections vertical caverns may form which are not closed by the overburden pressure (fig. 1 and 2 in SHAUB 1958).

However, it is also possible that dissolution occurs adjacent to thicker clay layers without being observed, due to the absence of stylolites.

Overburden pressure results in close interlocking of the adjacent rock masses. Vertical stylolitic seams occur only if a horizontal stress is present (see above); they usually begin with joints as the primary paths of the solutions. As a result of dissolution, these joints collect insoluble residue and change into stylolitic seams.

A main problem in stylolite formation is the removal of carbonate. Large quantities of carbonate can be removed during long periods of time. If a flushing system exists, i.e. a supply of undersaturated surface water by fractures, fissures or overthrusts, then a drainage level lower than the stylolites is needed. This is a possible mechanism of subsurface erosion in mountain areas (PFANNENSTIEL, pers. commun.). The removal of large amounts of carbonate from deeply buried strata as described from subsurface investigation of large basins (DUNNINGTON 1954), however, is difficult to understand: the compaction water in limestones is saturated in $CaCO_3$ and therefore is unable to dissolve additional carbonate.

Only the following mechanism seems possible: Pressure solution at the clay seams is stronger than at the crystal boundaries free of clay, possibly because of the easier diffusion of the solutions. The dissolved carbonate reprecipitates in the pressure-shadow, i.e. in the pore space. This would explain why limestones with stylolites frequently have low porosities. The theoretical decrease in porosity by this reprecipitation mechanism can be calculated by the formula

$$T + P - TP / 100,$$

when T = percent thickness decrease by stylolitization and P = present porosity (DUNNINGTON 1967). Such porosity decrease has been shown to cause a stronger compaction below the oil-water table around early oil accumulations, which leads to a reinforcement of the structures (RAMSDEN 1952, DUNNINGTON 1954, 1967).

This explanation is not fully satisfactory since frequently considerably more limestone has been dissolved than could have been precipitated in the adjacent pore space (SCHOT & PARK 1968). On a larger scale, however, much carbonate is needed to account for the nearly complete cementation of most limestones. The time required for such a carbonate transfer is very long, because of the slow motion (if at all) of the subsurface waters and the low solubility of $CaCO_3$ (ca. 0.05 g/l at 25° C, with CO_2; KRAUSKOPF 1967).

From this discussion follows the requirement of pore space available for redeposition of the material dissolved along the stylolites and of a permeability high enough to enable a solution transfer in a reasonable time. Both porosity and permeability decrease during compaction which occurs mainly in clayey limestones and marls (ch. 5.41), whereas in limestones poor in insoluble residue compaction is hampered by early cementation of the grain contacts, thus leaving high porosity and permeability for further cementation. This may explain BARRETT's (1964) observation that limestones rich in silt and clay (but not in sand) are less stylolitic than pure limestones (in which most of the insolubles are concentrated in stylolites). On the other hand it is suggested here that the same process of pressure solution observed in limestones occurs also in marlstones, but on a smaller scale, between the isolated clay flakes and the adjacent calcite crystals.

The apparently important role clay minerals play in pressure solution seems to point to a chemical influence of the clay seams, but an explanation of such a mechanism is lacking. Increased permeability and limestone removal along the seams is a more probable explanation (WEYL 1959).

5.324 Cone-in-cone

Conical aggregates (named cone-in-cone) of an impure fibrous carbonate mineral are found in many marls. They are mostly calcite (but sometimes ankerite) and develop adjacent to carbonate-rich layers (e. g. coquinas or concretions) and diverge towards the clay. For this reason, the axes of the cones frequently are oriented perpendicular to the bedding plane. Consistent association with marcasite and fossil remains or carbonaceous debris (SCHÖNE-WARNEFELD 1962, WOODLAND 1964, FRANKS 1969) implies a reducing environment due to organic decay. Similar to concretions, the production of NH_3 may have triggered the precipitation of carbonate. The insolubles are differently arranged:

(a) cone-in-cone is made visible by tiny cones of impurities collected between the growing calcite bundles (fig. 5-70); (b) these cones are crossed obliquely by less steep and thicker clay cones of tooth-like or zigzag appearance resulting from mechanical replacement in front of larger crystals (USDOWSKI 1963b). For reasons which are not yet understood, the angle of these cones frequently increases from their peak to the base, from about 30° to more than 100° (not developed in fig. 5-70). Presumably

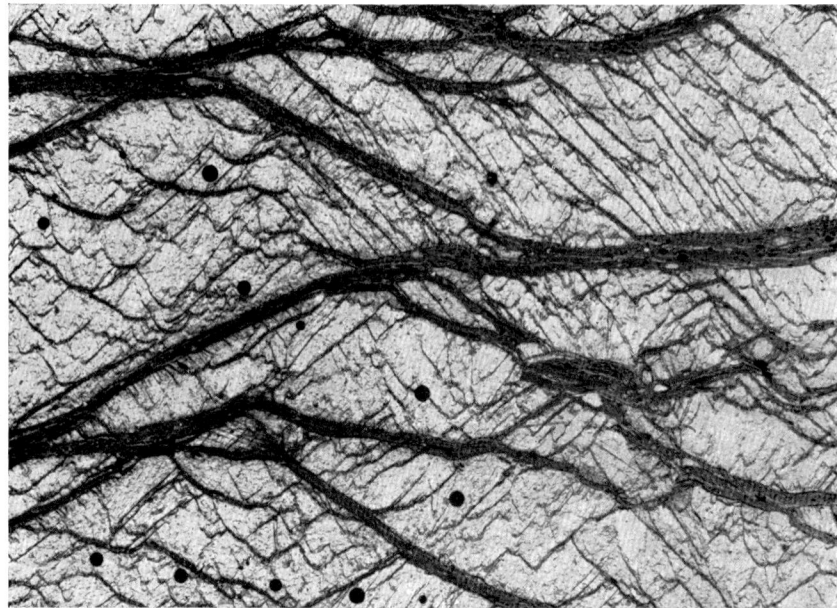

Fig. 5-70. Thin section of cone-in-cone exfoliating a clay layer. Wealden, oil field "Dalum" (Emsland); (black dots = air bubbles) (width 6.3 mm).

different angles depend upon the nucleation rate, as well as on the direction of the section compared with the cone axes; (c) the thickest clay seams are least displaced, but show clearly the expansion due to crystallization (fig. 5-70).

Only these crystallization fabrics are considered as "cone-in-cone"; shatter cones and compaction cones are tectonic fabrics not to be discussed in this book.

Cone-in-cone is considered as a typical form of cementation of marls. Their occurrence around concretions points to an origin subsequent to the formation of the concretion. On the other hand, the features shown in fig. 5-70 and many other occurrences (FRANKS 1969, WOODLAND 1964) suggest an origin before lithification. This means that cone-in-cone features normally form earlier than stylolites and later than concretions. A relative late formation has been shown for cone-in-cone in the Dalum oil pool (Emsland area; FÜCHTBAUER & GOLDSCHMIDT 1964): coquinas of pelecypods, ostracods and gastropods embedded in Lower Cretaceous shales and oil shales are cemented by cone-in-cone below the oil-water table only; therefore cone-in-cone must have grown after formation of the structure and oil migration into the trap.

The large crystals point to a low nucleation rate and slow growth which again agree with their relatively late origin. Clay content is necessary, according to ch. 5.41, to prevent these rocks from early induration and thus to enable the formation of cone-in-cone. The size of the crystal fibers frequently increases towards the base of the cones (i. e. during formation) due to a decrease of nucleation rate. At the same time, the quantity of inclusions decreases (WOODLAND l. c.), pointing to a decrease of growth rate. Both observations suggest decreasing supersaturation during cone-in-cone formation.

It is not clear whether the cone-in-cone fibers were formed as calcite or as aragonite and subsequently transformed into calcite fibers of the same orientation (as reported above for oöids). Primary calcite is suggested for veins parallel to the bedding composed of teeth-like micro-cone-in-cone crystals, the c-axes orientation of which shows 3 maxima:

(1) perpendicular to the bedding
(2) at an angle of about 25—30° from this direction
(3) at an angle of about 65—70° from this direction.

(2) and (3) can be explained by interference of calcite crystal pairs growing at the same time with a steep rhombohedron $02\bar{2}1$ and the main rhombohedron $01\bar{1}2$ of calcite as planes of interference, respectively. This points to primary calcite (FÜCHTBAUER 1971). A contact angle of 25—30° has been reported also by GRIGGS et al. (1960) in recrystallization experiments. — Such veins are believed to have formed later than cone-in-cone when the clay layers were more consistent and less permeable so that larger cone-in-cone fibres could no longer form (WOODLAND l. c.).

5.325 Concretions

Concretions are rounded nodules of diagenetic origin, many of which are flattened parallel to the bedding plane. Whereas "concretions" begin growing in the center, "geodes" grow from outside towards the center (PETTIJOHN 1957), possibly as replacement of earlier concretions (HAYES 1964). "Septaria" are concretions with shrinkage fractures which become broader towards the interior and are filled with different minerals (LIPPMANN 1951, EINSELE & MOSEBACH 1955, SCHMIDT 1965). Frequently concretions are accumulated in layers.

Concretions are most frequent in marls (for instance in loess). They consist of calcite, dolomite or siderite (spherosiderites). The carbonaceous Permian of Workuta contains limestone concretions in the marine layers, whereas spherosiderite concretions are found in the fresh water layers (MAKEDONOV 1957). The same has been observed in the Liassic of Yorkshire (HALLAM 1967a). In limestones concretions of chert and phosphorite are found. Concretions of iron hydroxide can form by weathering of spherosiderites.

Chemically, concretions form by migration of ions due to concentration and saturation differences: bacterial generation of NH_3 during organic disintegration raises the pH leading to a local supersaturation in and precipitation of $CaCO_3$. In many concretions, however, visible organic remains are absent (CORRENS 1939a, SEIBOLD 1962). In such cases, alkaline putrefaction of decaying organisms rich in basic proteins, such as fish, may be an important mechanism (BERNER 1968b): In the presence of excess free fatty acid, several calcium soaps are more stable than $CaCO_3$. If, however, all fatty acid is used up, the Ca-soap becomes thermodynamically instable relative to $CaCO_3$ + hydrocarbon (BERNER 1971a).

Also sulphate reduction by bacteria can result in the formation of excess $HCO_{3aq.}^-$; since sulphide ion is a relatively strong base (H^+-acceptor) it readily reacts with bacteriogenic $CO_{2aq.}$ to form the much weaker base $HCO_{3aq.}^-$:

$$S_{aq.}^{--} + 2\,CO_{2aq.} + 2\,H_2O \rightarrow 2\,HCO_{3aq.}^- + H_2S_{aq.}$$

Similarly the above reaction may be written

$$NH_{3aq.} + CO_{2aq.} + H_2O \rightarrow HCO_{3aq.}^- + NH_{4aq.}^+$$

Increase in the concentration of $HCO_{3aq.}^-$ may cause the equilibrium ion product for calcium carbonate to be exceeded, resulting in precipitation of $CaCO_3$ (BERNER 1971 b).

Geometrically, concretions may form by

1) filling of pore space
2) mechanical replacement due to growth pressure in the soft sediment
3) metasomatic replacement

Only in the first type (pore-filling concretions) is it possible to calculate the porosity during formation and the approximate time of formation (by comparing the volume of carbonate and non-carbonate with that in the rock outside the concretion). In type (1) — concretions the included non-carbonate material generally is prevented from further diagenesis (ILLIES 1949 a, FÜCHTBAUER & GOLDSCHMIDT 1963).

The three-dimensional preservation of fossils in concretions compared to deformation in the surrounding shales indicates an early formation of the concretions. Some siderite concretions contain trace fossils of benthonic oxygen-consumers, which suggests an origin after sediment burrowing since siderite forms only in reducing environment (ILLIES 1949 a). WEEKS (1957) described concretions with fish inclusions in Post-Pleistocene marine muds. Also LIPPMANN (1955), PANTIN (1958) and SEIBOLD (1962) postulated an origin only a few meters below the sediment surface. Theoretical growth times of 12 000 years for concretions of 5 cm radius have been calculated by BERNER (1968 b), based on ionic diffusion only. Active ground water flow would lead to higher growth rates.

However, sometimes concretion growth can continue for longer periods, as shown by the continual bending of the penetrating clay lamellae (TOMKEIEFF 1927) as well as by the continual change of clay minerals, away from the center (FÜCHTBAUER & GOLDSCHMIDT 1963). Based on the orientation of clay mineral flakes in clay-ironstone concretions compared with the enclosing shales, OERTEL & CURTIS (1972) concluded an early period of rapid growth followed by a second, protracted phase which extended effectively until the sediment achieved its present degree of compaction.

In several cases clay lamellae which penetrate without being bent point to a very late origin of the concretion (TAYLOR 1964).

The initial stages of certain nodules, such as those in the thin red nodular Jurassic limestones in the Alps, must have formed at the very surface of the sediment. This is shown by their partial corrosion, including parts of embedded ammonites ("halmyrolytical subsolution", HOLLMANN 1962, 1964, FABRICIUS 1966). Dissolution of calcite in the marly stringers and precipitation in the nodules intensified the nodular structures subsequently. Presumably the deposition rate of such sediments must have been very slow: For instance in Greece, the Middle and Upper Triassic is condensed into a few meters. The cephalopods have been coated with black films of manganese oxide (AUBOUIN 1965, p. 128; see also MENSINK 1960).

5.33 Allochemical diagenesis

5.331 Dolomitization

5.331.1 The processes

Below 1075° C calcite and dolomite are separated by an immiscibility gap which begins directly at stoichiometric dolomite (GOLDSMITH & HEARD 1961). Metastable calcium dolomites (commonly termed "protodolomites"), however, are frequent. Though they are known from Paleozoic rocks, they transform quickly into stoichiometric dolomites at temperatures of about 200° C (GRAF & GOLDSMITH 1956). Like-

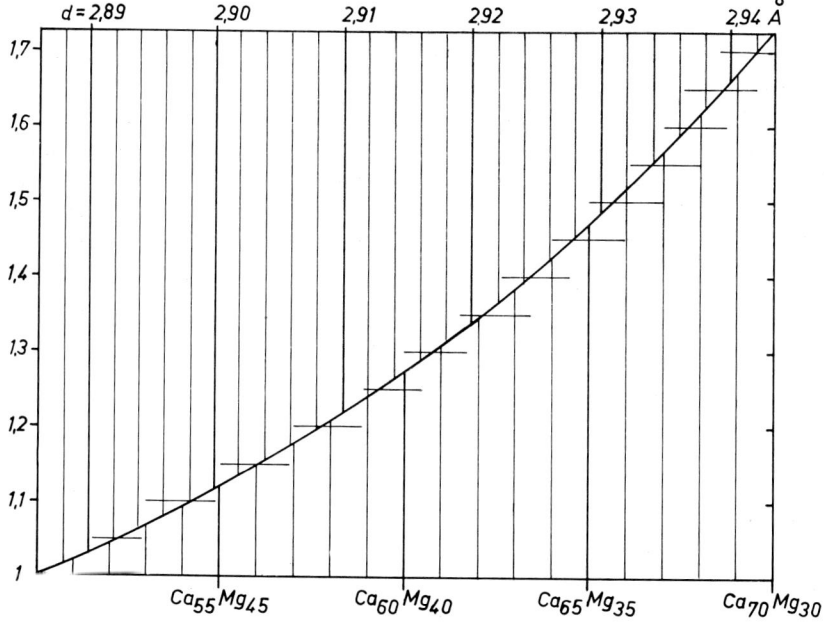

composition of dolomite

Fig. 5-71. Graph for the calculation of the composition of calcium dolomites (lower abscissa) from the d-value of the strongest X-ray reflexion (upper abscissa) according to GOLDSMITH et al. (1955). The ordinate indicates the factor by which the dolomite content calculated on the assumption of stoichiometric composition is to be multiplied in order to obtain the real dolomite amount (from FÜCHTBAUER & GOLDSCHMIDT 1965).

wise metastable Mg-contents are found in many calcites. This magnesian calcite turns out to be much less stable than the calcium dolomite; from pre-Tertiary rocks only two occurrences have been described (see ch. 5.332.2). The gap between stable calcite (= up to 4 mol-% $MgCO_3$) and dolomite, however, has been filled completely by natural occurrences of metastable phases. The transition from Mg-calcite to Ca-dolomite is defined by the appearance of the ordering line (01.5) = (221) which reflects the mixed layering of Ca and Mg ion layers perpendicular to the c-axis. The shift towards the diffraction peak of calcite enables an X-ray determination of the Ca-excess in the dolomite lattice (fig. 5-71) with an accuracy of ± 0.5 mole percent $CaCO_3$.

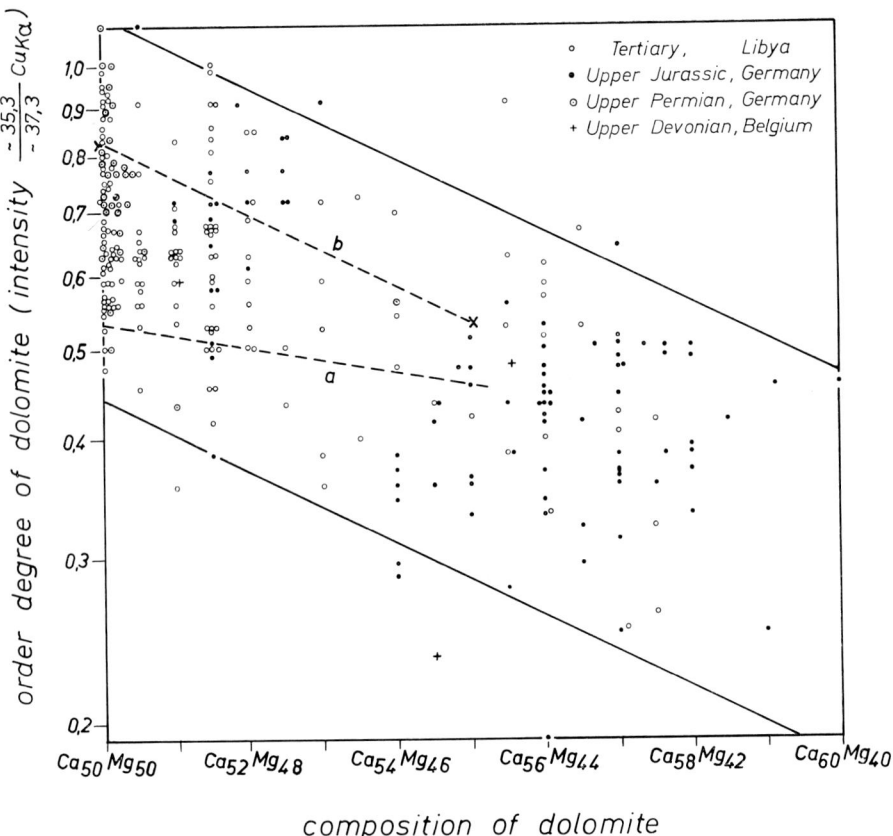

Fig. 5-72. Relationship between composition and ordering degree of natural dolomites. The hatched lines are from Goldsmith & Graf (1958, table 3): a) indicates the relationship calculated from the chemical composition, b) indicates two natural dolomites; line b) is steeper than a) presumably due to an increasing irregularity of the mixed layering of Ca and Mg planes with increasing Ca-content (Graf & Goldsmith, l. c.). This is also true for the other occurrences shown in the diagram (Füchtbauer & Goldschmidt 1965).

On the other hand the ordering degree can be calculated by comparing the intensity of the ordering line (01.5) = (221) (which is influenced by the mixed layering of Ca and Mg in the c-axis) with the diffraction peak of (11.0) = (101) (GOLDSMITH & GRAF 1958). The higher this quotient the better ordered is the dolomite (fig. 5-72).

A comparison of the theoretical equilibrium constant of dolomite with the product of the concentrations of the free ions Ca^{++}, Mg^{++}, and CO_3^{--} present in sea water (both values are about 10^{-17}) indicates that sea water is approximately saturated with respect to dolomite (BLATT et al. 1972, p. 482 f.). This explains why dolomite, though stable in marine environments (KRAMER 1959), will not form at considerable rates by primary precipitation. Upon evaporation, it frequently forms by homoaxial replacement of magnesian calcite present as metastable mineral in marine sediments. This replacement has been confirmed by isotope investigations (EPSTEIN et al. 1963, FRIEDMAN & HALL 1963, see p. 311) as well as by the trace-element take-up (BRÄTTER et al. 1971). True nucleation seems to be rare. Instead, homoaxial dolomitization occurs, in which epitaxis is substituted for nucleation. Or older dolomite crystals are used as nuclei (RICHTER 1972a).

However, it cannot be excluded that **dolomite cement** may form within the pore space of carbonate sediments not only during late diagenesis, but even soon after deposition. This is suggested by investigations reported at the end of ch. 5.331.21.

The kinetics of dolomite nucleation at low temperatures are hampered for the following reasons:

a) the simplicity principle (GOLDSMITH 1953): minerals composed of two or more kinds of cations at non-equivalent, but energetically similar lattice places, form slowly.

b) the strong hydratation of Mg ions.

On the basis of **geological occurrence**, dolomitization may be subdivided as follows:

1. early diagenetic (syngenetic) dolomitization, which takes place in unconsolidated sediments exposed to super-saline sea water (concentrated by evaporation): High nucleation rate.

2. late diagenetic (epigenetic) dolomitization, occurring mostly in lithified sediments under the influence of the interstitial water: Low nucleation rate.

1. **Early diagenetic dolomitization** generally takes place at elevated **temperatures** (above 30° C, YANAT'EVA 1957), for the kinetic reasons discussed above, though in many modern occurrences, only temperatures of 15—20° C are observed. Most syngenetic dolomites also require relatively high **Mg/Ca ratios**, up to 15—30 as opposed to a ratio of 5 for normal sea water. This ionic increase occurs through evaporation and the precipitation of $CaCO_3$ or $CaSO_4$. The third requirement of dolomite formation are **high volumes of water** passing through the sediment and providing the necessary number of Mg ions. Based on these requirements, three mechanisms of dolomitization are possible:

306 5.3 Diagenesis

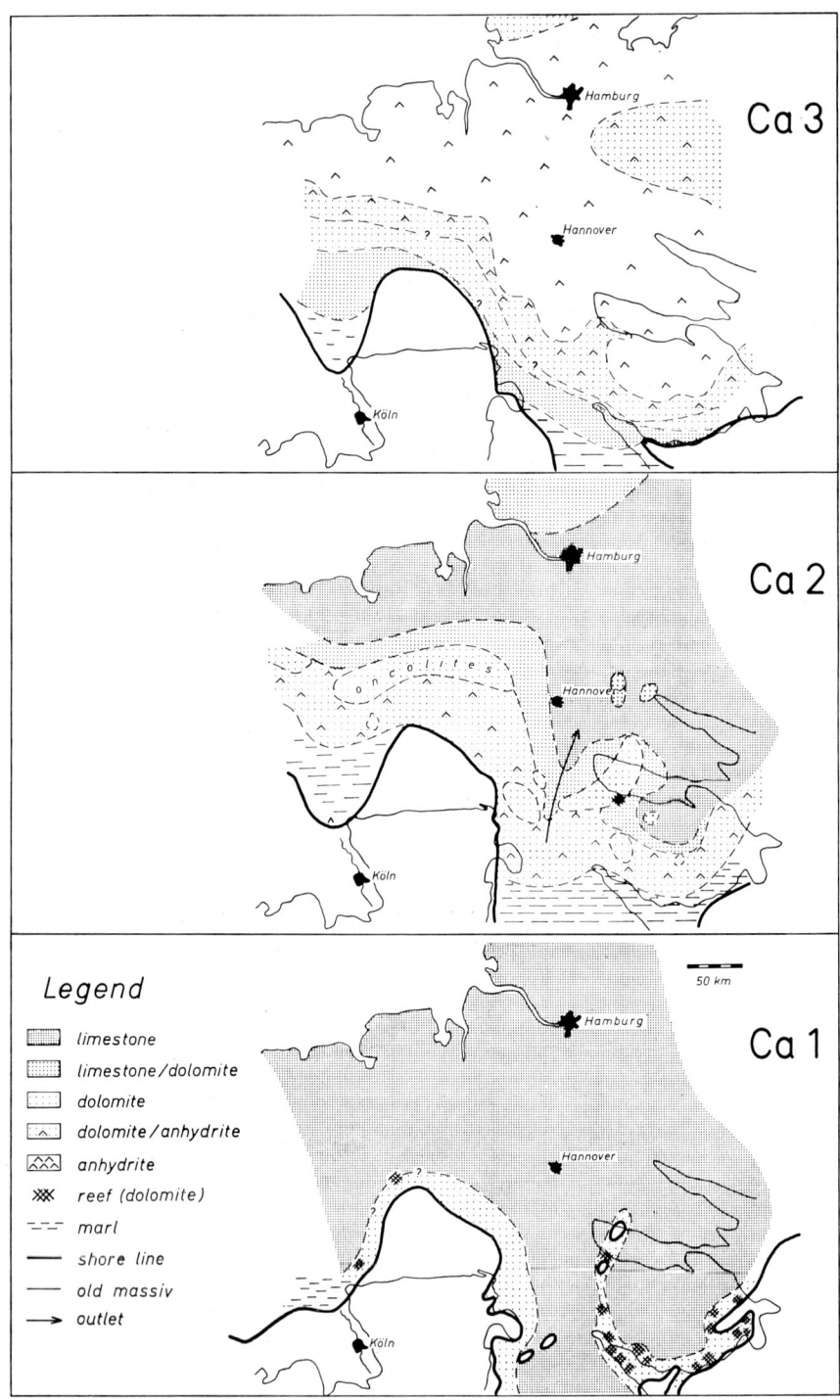

(1) Evaporation of lagoons or separated bays may cause precipitation of gypsum and increase the Mg/Ca ratio: Subaqueous dolomitization can occur, because sufficient Mg ions are provided from the brine. Examples of such "subevaporitic" and "evaporitic environments" are the dolomitic sediments of the Kara Bogas Gol (a bay of the Lake Kaspi, see below) and the Upper Permian dolomites (figs. 5-73, 74), as well as dolomite forming in modern lakes (MÜLLER et al. 1972).

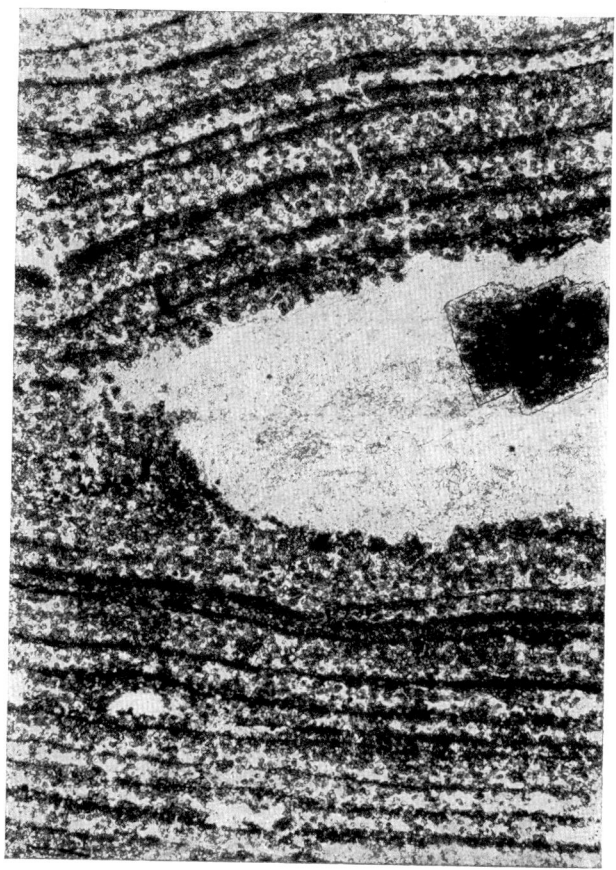

Fig. 5-74. Lamination of anhydrite (light) and dolomite (dark). According to the geologic setting, these strata have been formed at the bottom of the sea. A fluorite crystal is present in the center of the anhydrite lens. Zechstein 1, Northern Germany (width 3.3 mm).

◀

Fig. 5-73. Facies distribution in the carbonate rocks of the Zechstein formation, Germany. During Ca 1 and Ca 2, an early submarine dolomitization took place near the border of the basin ("subevaporitic basin"). In the Ca 3, carbonate sediments were deposited only near the border (limestones grading into dolomites towards the basin), whereas in the interior of the basin anhydrite was deposited ("evaporitic basin") (FÜCHTBAUER 1972).

(2) Evaporative reflux may occur in supratidal areas which are flooded occasionally by the sea. The evaporated brine percolates the sediment on its way back to the sea. This model was termed "seepage refluction" by ADAM & RHODES (1960) and was suggested also by KING (1947), CHILINGAR & BISSELL (1961) and FRIEDMAN & SANDERS (1967).

For the most-cited example, however, recent observations by LUCIA (1967) are at variance with the seepage refluction theory: Borings made in the Pekelmeer (Island of Bonaire, see below), show that no dolomite has formed in the subsurface. Moreover, in view of the low permeability of the fine grained muds found in these environments, it is difficult to understand how this mechanism can provide enough magnesium for the formation of thick sequences of early diagenetic dolomites.

(3) For these reasons HSU & SIEGENTHALER (1969) proposed another model of formation for such dolomites which they named "evaporative pumping": In an arid coastal area the interstitial water from the uppermost layer of sediment evaporates and is supplemented by water seeping e.g. from the lagoon through the

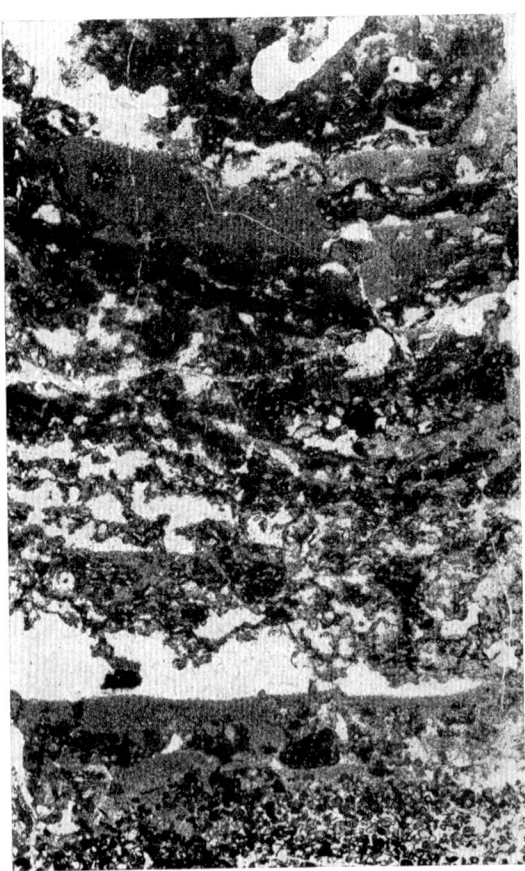

Fig. 5-75. Dolomitic blue-green algal mats (black; cryptocrystalline) with caverns containing a geopetal fill of internal calcilutitic sediment (gray). The rest has been closed by more or less equant calcite cement (light), when the rock already had been dolomitized. The strong magnesium uptake of some algal sheaths observed by GEBELEIN & HOFFMAN (1969) may have triggered this early dolomitization. Dachsteinkalk (Lofer facies; Nor, Upper Triassic), Lattengebirge near Berchtesgaden (width 8 mm).

sediment. In this way, an uninterrupted flow of Mg-containing water from the sea into the adjacent areas of dolomitization is possible.

The Ca-excess of the dolomites is influenced by the chemistry of the solutions, especially by the salinity: with increasing aridity and salinity the Ca-excess in the dolomite lattice decreases as shown in the following modern and fossil examples:

a) Florida: $Ca_{0.58-0.67}$-dolomite (humid climate)
   Bonaire: $Ca_{0.54-0.56}$-dolomite (slightly arid)
   Persian Gulf: $Ca_{0.50-0.55}$-dolomite (strongly arid)

b) in the lakes of Southern Australia, with increasing salinity, magnesian calcite, calcian dolomite, stoichiometric dolomite and magnesite are formed (see below).

Even in rocks, a decrease of Ca-excess in the dolomite lattice has been found with increasing evaporitic character, which is shown by increasing dolomite content, increasing thickness of overlying anhydrites and by the character of fossils (FÜCHT-BAUER 1964b, FÜCHTBAUER & GOLDSCHMIDT 1965, MARSCHNER 1966) (fig. 5-77). During diagenesis,. however, the calcium excess decreases. This decrease is greater in porous rocks than in rocks with low porosity and permeability (fig. 5-78). A zonal distribution of the calcium excess in the rhombohedra has been observed by KATZ (1968).

Possible bacterial assistance in dolomite formation has been suggested by the experiments of LALOU (1957), NEHER & ROHRER (1958) and OPPENHEIMER & MASTER (1963).

In early diagenetic dolomites the nucleation rate is high due to the easy access of magnesium ions through the soft sediment. For this reason the crystal size generally is small, but it may increase by subsequent crystal enlargement (LAPORTE 1967, QUESTER 1964).

2. In late diagenetic dolomitization which proceeds relatively slowly (in consolidated limestones), concentrations of 0.1 mole Mg + Ca in 1000 moles H_2O are sufficient, according to calculations by USDOWSKI (1967). (In comparison, the

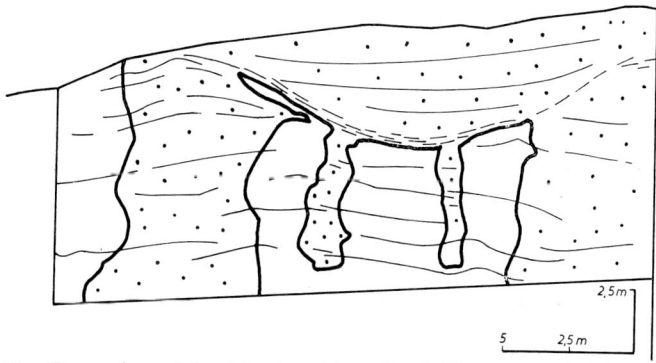

Fig. 5-76. Descendent dolomitization (dotted) of blue-green algae-silicisponge quietwater bioherms in the Upper Jurassic, S.W. Germany. Magnesium-rich solutions were prevented from penetrating clayey layers (dotted lines) (modified from LANG 1964).

concentrations in sea water are 1 mole Mg and 0.2 moles Ca in 1000 moles H_2O). Moreover, the Mg/Ca molar ratio must be higher than 0.37 at 50° C and higher than 0.16 at 120° C (for comparison: sea water 5.26).

Though a higher Mg/Ca ratio is required in order to compete with the negative free energy for dolomitization of calcite, sea water or similar interstitial water is more than adequate to allow for late diagenetic dolomitization (KRAMER 1959). Only the quick early diagenetic dolomitization, for reasons of reaction kinetics, needs evaporation and subsequent Ca^{++}-precipitation in order to provide an increase of the absolute and relative Mg^{++}-concentration.

Due to the low concentrations, the nucleation rate in late diagenetic dolomites is low, resulting in large crystals.

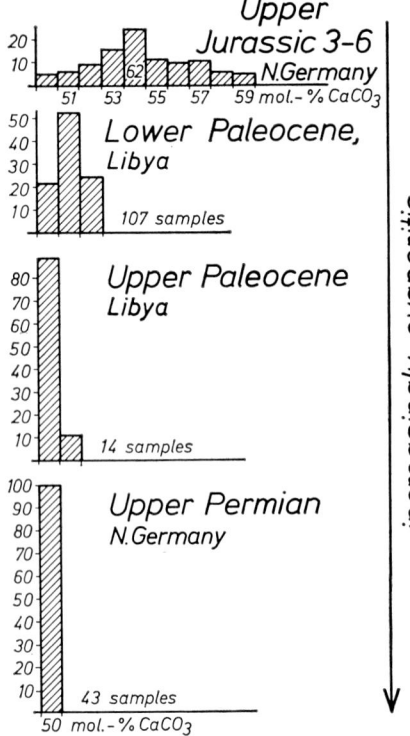

Fig. 5-77. Influence of environment on the calcium uptake in the lattice of early diagenetic dolomites. The upper two examples contain normal marine fossils and are capped only by a small sequence of anhydrite. The two lower examples, especially the lowest, contain mostly fossils from a restricted environment and are capped by thick anhydrite sequences. The evaporitic tendency defined as the amount of sulphate and chloride sediments in the sequence increases from the upper to the lower examples (FÜCHTBAUER & GOLDSCHMIDT 1965).

Whereas sea water is the main source of magnesium for early diagenetic dolomites, interstitial water is the most important source of magnesium for late diagenetic dolomites. This can be concluded from the considerable decrease in magnesium in the interstitial water compared to the sea water: the Mg/Ca mole ratio in sea water is 5.26, in interstitial waters of Mesozoic sandstones (Northern Germany) 0.3 (v. ENGELHARDT 1960). This marked drop in Mg may be due to the diagenetic formation of dolomite and chlorite.

Occasionally the interstitial water of underlying early diagenetic dolomites supplies the necessary magnesium. A similar mechanism may be operative at the surface as shown by the following example (D. K. RICHTER 1974): Water percolating through Devonian dolomites in the Eifel Mts. (Germany) was collected in playas forming after the Variscian orogenesis, during Permotriassic times, resulting in dolomitization of the underlying Devonian limestones as well as of Permotriassic conglomerates.

The magnesium ions adsorbed by clay minerals may be another source of magnesium; these ions become available as they are replaced by Na and especially Ca, during diagenesis (KAHLER 1965). This may explain stringers in silt layers (e. g. Buntsandstein). In other cases the clay content has also triggered dolomitization (FAIRBRIDGE 1957, RUDOLF 1959, V. SCHMIDT 1965). Skeletal material consisting of magnesian calcite loses its magnesium during the early diagenesis and therefore probably is not a major source of magnesium during late diagenetic dolomitization; however it may be one of the most important sources of magnesium during early diagenesis and in the stage between early and late diagenesis (see no. 2 below table 5-8, ch. 5.152 C 2 a, and fig. 5-72; LAND & EPSTEIN 1970).

The relatively small magnesium concentrations required for late diagenetic dolomitization necessitate a high volume of solutions flowing through the rock. This is supported by isotope investigations of late diagenetic dolomites: the O^{18}/O^{16} ratio in dolomites differs from the limestones from which they derived (JAVOY & FAYARD 1964, 1966). Contrary to this, the C- and O-isotopes in early diagenetic dolomites generally are similar to those in the adjacent calcilutites (EPSTEIN et al. 1963).

Principally two chemical modes of dolomitization are possible:

a) exchange of Ca and Mg without supply of CO_3
b) supply of $MgCO_3$

a) From the lack of an oxygen isotope fractionation between dolomite and calcite in young sedimentary deposits, EPSTEIN et al. (1963) concluded that the dolomite resulted from addition of magnesium to, and subtraction of Ca from, existing crystalline calcite without an isotopic change in the oxygen of the precursor $CaCO_3$.

According to WEYL (1960), interstitial waters are poor in bicarbonate ions compared with their contents in Ca and Mg. For this reason, late diagenetic dolomitization also may take place by a mole-by-mole exchange of Mg for Ca:

$$2\,CaCO_{3\,solid} + Mg^{++} \rightarrow CaMg(CO_3)_{2\,solid} + Ca^{++}$$

Because dolomite is more dense than calcite, porosity should increase, provided that the rock does not compact during dolomitization. The theoretical increase in porosity occurs through the transition of

    aragonite into dolomite: 5,5 % by volume
    calcite into dolomite: 13 % by volume

(see also ch. 5.423.2).

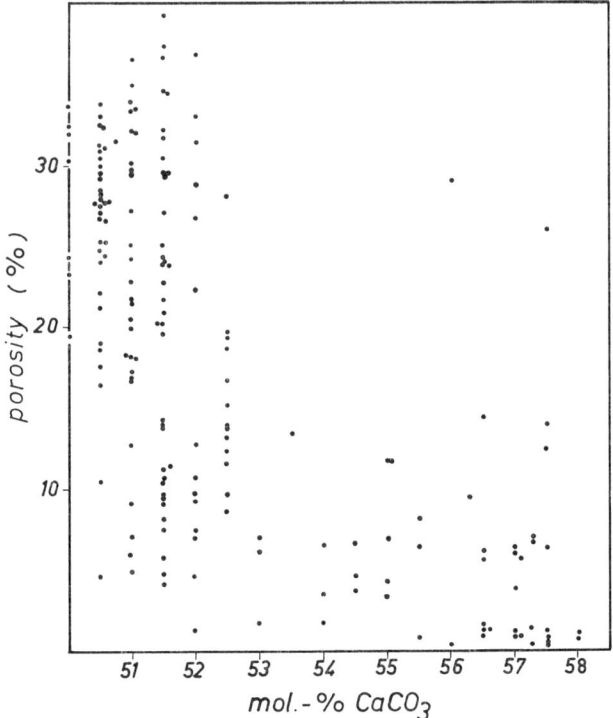

Fig. 5-78. Relationship between $CaCO_3$-content in the dolomite lattice and porosity (Paleocene, Libya): In porous samples, the dolomite has lost most of its calcium excess, whereas it has been preserved in the samples with low porosity (FÜCHTBAUER & GOLDSCHMIDT 1965).

b) many late diagenetic dolomites, however, have less than 5% porosity, which suggests compaction or addition of carbonate (e. g. from stylolites, ch. 5.323). LIPPMANN (1967), on the basis of his synthesis of norsethite $BaMg(CO_3)_2$ at low temperatures, suggested an uptake of CO_3 as the normal way of dolomitization:

$$CaCO_{3\,solid} + Mg^{++} + CO_3^{--} \rightarrow CaMg(CO_3)_{2\,solid}$$

This, however, would require a porosity of more than 40% during dolomitization, which will be realized only occasionally during early dolomitization, but neither in the walls of most skeletons nor during late dolomitization.

The petrographic differences between early and late diagenetic dolomites are listed in table 5-8.

Comments on table 5-8:

1. An increase in crystal size with depth has been found in several early diagenetic dolomitization cycles, e. g. in the Paleocene of Libya (V. SCHMIDT, pers. commun.; FÜCHTBAUER & GOLDSCHMIDT 1965). This trend suggests a transition into the conditions of late diagenetic dolomitization in deeper portions of the sediment.

5.33 Allochemical diagenesis

Table 5-8. Criteria for the distinction of early and late diagenetic dolomites.

	early diagenetic	late diagenetic	
time of dolomitization	before consolidation (except crusts)	mostly after consolidation	
		early	late
1. crystal size	< 0.01—0.02 mm (if no crystal enlargement occurred)	> 0.02 mm	
2. skeletons	minute structures preserved because the dolomitization generally is homoaxial. Selective dolomitization of high magnesian calcite skeletons	only coarse crystalline structures (echinoderms) preserved; other skeletons destroyed.	
3. characteristic fossils	yes (e.g. restricted environment)	no	
4. relationship to paleogeography	yes (e.g. dolomites near basin border)	yes	no
5. relationship to bedding plane	yes (e.g. interlayering of dolomite and limestone)	no	
6. connection with $CaSO_4$	yes	no	
7. indications of desiccation	yes (brecciation)	no	
8. iron content	no	frequently yes	

2. Examples of homoaxial dolomitization are reported by V. SCHMIDT (1961), LUCIA (1962), MURRAY (1964b), FÜCHTBAUER (1967b). SCHLANGER (1957) found coralline ("red") algae preferentially dolomitized in subrecent sediments, due to the high Ca-excess and specific surface of coralline algae (G. MÜLLER, pers. commun.). The same was true of micritic envelopes observed by BUCHBINDER & FRIEDMAN (1970) in Miocene reef talus rocks. Red algae and micritic envelopes in modern environments are mostly high magnesian calcite. Workers have inferred that magnesium released during early diagenesis can serve as a source of magnesium in early dolomitization. The susceptibility of some organisms to dolomitization and the resistance of others certainly depends on their porosity, permeability, crystal size, and crystal surface area. A selective (early) dolomitization of aragonitic compared with calcitic skeletons, as observed by RUDOLF (1959) and V. SCHMIDT (1961, 1965), may be caused by higher surface area of aragonitic compared with calcitic mollusks reported by ANDERSON et al. (1973). In addition, the mineralogy of the skeletal particle at the time of dolomitization is very important as calcite, aragonite, and high magnesium calcite have different solubilities (BLATT et al. 1972).

3. Typical of early diagenetic dolomites are oncolites, stromatolites, green algae, gastropods and ostracods (EDIE 1958, DEGENS et al. 1961, LAPORTE 1967). In the Lower Tertiary of Libya only Miliolidae and Dasycladaceae are found in dolomite layers, whereas in the surrounding limestones a rich fauna of foraminifera, mollusks, echinoderms and bryozoans is found. — Since there is no genetic relationship between late diagenetic dolomitization and depositional environment, no characteristic fossils are found in late diagenetic dolomites.

4. Early diagenetic dolomites generally display characteristic paleogeographic zonations in which the mostly highly evaporitic rocks occur near the edge (RICHTER-

BERNBURG 1955, RONOV 1956, FIEGE 1938) or the center of the basin (RICHTER-BERNBURG 1955, RONOV 1956, IMBRIE 1957, BRIGGS 1958). This point has been discussed earlier in this chapter (fig. 5-73) shows examples for both distributions; FÜCHTBAUER 1972).

5. Alternating thin layers of dolomite and limestone (fig. 5-72; HERMAN & BARKELL 1957, CHOQUETTE & TRAUT 1963, BAUSCH 1963b) or dolomite and anhydrite (fig. 5-74) only can be interpreted as early diagenetic (SANDER 1936, SCHWARZACHER 1947, HOBBS 1957, KEPPER 1966, LAPORTE 1967). This assumes, however, that the dolomite layers are not reworked detrital grains (SANDER 1936, SARIN 1962). Late diagenetic dolomites frequently form dome-shaped, mushroom-like or cedar-formed complexes (fig. 5-79; INST. FRANÇ. DU PÉTROLE, 1959) and parallel the bedding plane only if this corresponds to a major petrographic change. Frequently, a connection with faults and joints is observed.

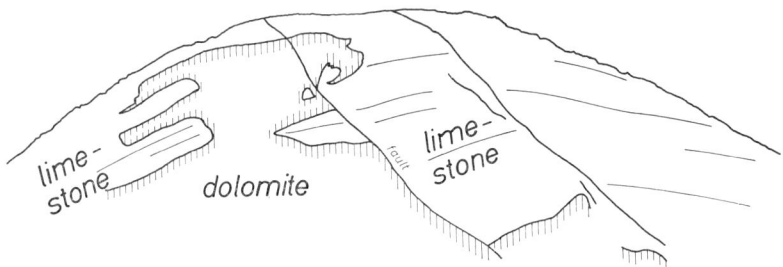

Fig. 5-79. Ascendent dolomitization, late diagenetic. Schematic view of the Massif of Robin (Provence; France). The exposure is 150 m high (from Inst. Franç. du Pétrole 1959).

6. The calcium sulphate that was connected with the dolomite must not be preserved. D. K. RICHTER (1971) was able to find traces of gypsum and anhydrite, which had disappeared from the rock, as inclusions of authigenic quartz crystals.

7. Dolomitization cycles frequently are capped by a sedimentary breccia which is overlain by anhydrite (EDIE 1956): In several, but not all, modern examples, the dolomitization is related to the formation of a crust. During either flooding or desiccation this crust may be fractured. Such structures are extremely frequent in early diagenetic dolomites (SANDER 1936, GREENMAN 1951, FARINACCI 1965).

8. The connection between iron content and late diagenetic dolomites (DEININGER 1964, FRIEDMAN & SANDERS 1967) parallels the iron-calcitic late diagenetic cement B. This may be due to higher temperatures during late diagenesis or higher Fe-concentrations in these solutions. The Devonian dolomites of the Eifel Mts. (Germany) may serve as an example (D. K. RICHTER 1974): Late diagenetic dolomite rims above and below early diagenetic dolomite bodies are enriched in Fe compared with the latter.

5.331.2 Occurrence of dolomite

5.331.21 Early diagenetic dolomites

a) Recent

1. Caribbean Sea (DEFFEYES et al. 1964)

On the island of Bonaire (off the coast of Venezuela), areas of wet soft sediments and small supratidal ponds nearly 1 m above the mean high water level are separated from the sea by a rampart of coral boulders. Sediments in these areas consist of gypsum, aragonite, calcite and dolomite. In a crust consisting of about 95% dolomite, a C^{14}-age of 1480 ± 140 years has been found, indicating recent dolomitization. Dolomite rhombs of about 0.002 mm size, a relatively well-ordered $Ca_{0.54-0.56}$-dolomite, replace carbonate mud and even shell fragments.

The lakes and ponds and the interstitial waters of the wet sediments are hypersaline, but not excessively so. These areas are periodically flooded by high tides and storms; presumably the outlet of brines is through subsurface strata, because the lake is below sea level. According to chemical analyses the water looses only $CaCO_3$ and $CaSO_4$ during evaporation; gypsum precipitation is observed. This causes an increase of the Mg/Ca molar ratio from 5 (in sea water) to as high as 30. DEFFEYES et al. (1964) suggested that as these heavy brines seep through the strata to the open sea, they cause dolomitization. This theory, "seepage reflection", however, is at variance with new observations by LUCIA (1967) (see above). "Evaporative pumping" has been proposed instead (HSU & SIEGENTHALER 1969).

2. Florida Keys and the Bahamas (SHINN & GINSBURG 1964, FRIEDMAN 1964, SHINN 1964, SHINN et al. 1965)

Recent dolomitization has been observed on Sugarloaf Key in the Florida Keys and also on the western part of the island of Andros; sediments consist mainly of aragonite mud together with pellets, oöids and skeletal grains. Both areas are situated several cm above the mean high water level and are covered with blue-green algal mats as well as by mud cracks. High dolomite concentrations (more than 25% on Sugarloaf Key and up to 80% on Williams' Island W. Andros) are found only in the hard crusts, especially beneath the cracks. ATWOOD & BUBB (1970) observed on the Florida Keys that dolomite concentration is highest near the shore line and within topographic lows where sediments are under water at least once a day. The crystals are 1—3 microns in size and consist of $Ca_{0.58-0.67}$-dolomite (FÜCHTBAUER & GOLDSCHMIDT 1965). Partially dolomitized gastropod shells and pellets prove that this is not a primary dolomite precipitation but an early diagenetic dolomitization. Alternating drying and flooding increase the salinity of the interstitial water. In this humid climate, however, gypsum is not precipitated. Mg/Ca ratios of 40 have been found in the interstitial water.

3. Persian Gulf (WELLS 1962, CURTIS et al. 1963, SHEARMAN 1963, EVANS & SHEARMAN 1964, ILLING & WELLS 1964, KINSMAN 1964b, ILLING et al. 1965, EVANS et al. 1969) (see fig. 5-48 b).

The lagoons bordering the arid southwestern coast of the Persian Gulf, such as the Peninsula Qatar and the Trucial Coast, pass into a tidal belt which on the landward side is bordered by blue-green algal mats (see ch. 6). Landward of the algal mat zone is an extensive (several km-wide) very flat area, the so-called s a b k h a, which lies only a few inches above the mean high water level and is covered partly by a crust of fine anhydrite and halite crystals. In the underlying soft aragonite mud dolomitization begins about 15 cm below the surface and increases downward until as much as 100 percent dolomite is present. Crystals of $Ca_{0.50-0.55}$-dolomite, 1—5 microns in size, are developed. The dolomitization comes to an end 60 to 120 cm below the surface at an irregular face. This distribution of dolomite points against a primary precipitation. Age determinations of two dolomite samples 25 and 45 cm below the surface resulted in 2670 and 3310 years, respectively. The sediment remains unconsolidated during dolomitization. Locally, however, rigid lumps are formed with increasing dolomite content in which the aragonitic shells are dissolved. The Sabkha mud contains large gypsum crystals with small aragonite inclusions and traces of magnesite (KINSMAN 1965, EVANS et al. 1969).

The Sabkha is flooded only periodically, during favorable winds or by storm tides. Evaporative pumping may also supply some solution (see above). Consequently evaporation prevails. The water table is as much as 75 cm below the surface. The temperatures within this interstitial water can be as high as 35—40°, the pH between 6 and 6.7. The Mg/Ca ratio in the seaward parts of the Sabkha is as much as 3—4 times that of sea water, due to $CaSO_4$ precipitation; this ratio decreases landward because of the strong dolomitization. "Evaporative reflux" and "evaporative pumping" have been proposed as the mechanisms of dolomitization.

4. Southern Australia (MAWSON 1929, ALDERMAN & SKINNER 1957, SKINNER et al. 1963, ALDERMAN & VON DER BORCH et al. 1964)

The Coorong is a series of small coastal bays S. E. of Adelaide. During the evolution of the Coorong, the inner bays have been progressively isolated from the open ocean. Thus these inner bays receive much of their water supply from seasonal rains rather than from the ocean. During photosynthesis, water plants extract CO_2, with the subsequent increase of pH (up to 10) and precipitation of lime. With increasing isolation (age) of the lakes, the following sediments are found:

Coorong:	aragonite + magnesian calcite	Mg/Ca = 5
	magnesian calcite	6
Lakes	magnesian calcite + calcian dolomite ($\sim Ca_{0.56}$)	8
landward (east)	ordered dolomite	10
of Coorong	dolomite + magnesite	16
	aragonite + hydromagnesite	20

(Figures indicate Mg/Ca ratio in the water at the time of the highest pH).

It is supposed that the Mg/Ca ratio, which strongly increases during precipitation of $CaCO_3$ (see above), causes the transformation of aragonite or calcite sediments into magnesium-rich minerals (ALDERMAN 1965). The dolomite is of recent age according to C^{14} analyses; its "sedimentation rate" is 0.2—0.5 mm/year (FRIEDMAN & SANDERS 1967).

5. Kara Bogas-lake (STRAKHOV & ZWETKOV 1946, v. RAUPACH 1952, KAZAKOV et al. 1959, TEODOROVICH 1961)

This is an eastern bay of the Caspian Sea which has attracted much interest because of the salt equilibria in the sediments and the brines. In the areas of lowest salinity, calcite with traces of dolomite is formed. At higher salinities (7—18%) near the inlet, the mud contains dolomite and occasionally calcite. In the interior of the bay, with increasing salinities, dolomite, gypsum, and calcite + hydromagnesite + gypsum are formed. This latter sedimentary facies (together with mirabilite [$Na_2SO_4 \cdot 10H_2O$]) prevails in the main part of this bay. The water contains 2.905% SO_4, 1.427% Mg and only 0.0459% Ca. This is an example of subaqueous dolomitization by evaporation.

6. A dolomite layer 2300—4300 years old which formed approximately 5—7 m below sea level was found in the Baffin Bay (Texas) by BEHRENS & LAND (1972).

7. Etoscha pan, S. W. Africa (GEVERS 1930)

This depression is about 100 km in diameter, has no outlet and is filled with water during the rainy season. The dolomite that is formed is not reworked, but a newly formed dolomite. Fossil dolomites are covered by limestone crusts. The dolomite content of the surface limestone increases towards the Etoscha pan with decreasing annual precipitation. The intensity of evaporation determines the dolomite content.

Another occurrence of this type is the Deep Spring Lake (PETERSON et al. 1963). Glacial lakes with dolomite deposits include Lake Bonneville (GRAF et al. 1961), the Salt Flat Graben (Texas; FRIEDMAN 1966) and the Tuz Gölü (Turkey; MÜLLER & IRION 1969).

8. Subrecent dolomites in lakes with low salinity

Dolomite has been found in several lakes with low salinities and low Mg/Ca ratios. This dolomite has formed in subrecent if not in recent times under conditions dissimilar to normal early diagenetic dolomites discussed above. A genetic relationship with the occurrence described below in ch. 5.331.22 a is suggested.

a) In the mud of the lake Neusiedl (Austria) calcian dolomite and (below the sediment surface) dolomite has been found. The salinity of the water is 1.5‰, with 104 mg/l Mg, 21.9 mg/l Ca, and a pH of 9 (SCHROLL & WIEDEN 1960, WIEDEN 1964).

b) Subrecent calcian dolomite also has been found in the sediments of Lake Balaton (Hungary); the Mg/Ca ratio in the interstitial water is lower than 2 (MÜLLER 1970a) (fig. 5-93).

c) Lake Balkash (Kazakhskaya SSR), which is less than 25 m deep, has muds which, according to chemical analyses, contain up to 57.4% dolomite. Dolomite-rich layers alternate with Mg-free layers of diatom and ostracod skeletons. The water contains 0.169% SO_4, 0.067% HCO_3, 0.011% CO_3, 0.111% Cl, 0.137% K + Na, 0.028% Mg, 0.0015% Ca; the pH is 9.2 (according to GRAF 1960), (TEODOROVICH 1946, SAPOZHNIKOV 1951, ZALMANSON 1951, RUCHIN 1958).

Three environmental mechanisms of dolomitization are active in the above examples:

A. subaquatic

1. Evaporation (I a 1 in part, I a 4, I a 5, I a 6, fig. 5-73)

B. subaeric

2. Evaporation (I a 2, I a 3 in part)
3. Evaporative pumping (I a 1 in part, I a 3 in part)

The fourth mechanism, evaporative reflux, occurs only where permeability and pressure are high enough and the distance small enough to allow reasonable rates of percolation. An example has been investigated by MÜLLER & TIETZ (1966): Supratidal calcarenites in the Canary Islands are, from time to time, washed by high waves and spray. Some of the sea water remains in pools on the surface of the calcarenite and evaporates, causing the precipitation of gypsum and halite. The rest of the water percolates down through the calcarenite altering the high-magnesium calcite grains of coralline algae to cryptocrystalline dolomite. Seepage reflux may have been operative also in example II a 1 (below).

b) Fossil

Only a few occurrences of early diagenetic dolomite rocks shall be discussed; extensive descriptions are found in FRIEDMAN & SANDERS (1967).

Some terrestrial sediments contain evaporite cycles which end in dolomite. An example are the cycles gravel-fine sand-marl-dolomite (top) in the Upper Freshwater Molasse (Tertiary) which are about 1 m thick, and may be formed in small lakes (SCHMEER 1962). In the same strata NÄGELE (1962) found a layer of clayey-sandy cryptocrystalline $Ca_{0.54}$-dolomite as the last phase of regression of the Miocene sea, below an exsudation limestone named "Albstein".

On the basis of the excellent preservation of included plant cells, ECKHARDT & v. GAERTNER (1955) concluded that the so-called "peat dolomites" of the Upper Carboniferous are early diagenetic in origin. The magnesium in these iron-containing $Ca_{0.5-0.55}$-dolomite possibly was supplied from the overlaying marine horizons.

The uppermost parts of the quiet-water bioherms in the Swabian Upper Jurassic are dolomitized. This has been observed in many subrecent reefs (CULLIS 1904, REU-

LING 1934, FRIEDMAN & SANDERS 1967). This dolomitization, however, is not strictly parallel to the bedding (fig. 5-76). Nevertheless, it is presumably early diagenetic. This is indicated also by isotope investigations (P. FRITZ 1966) which are in harmony with those by EPSTEIN et al. (1963). Such a descendent dolomitization as shown in fig. 5-76 is characteristic of early diagenetic dolomitization. However, it is found also in the early phases of late diagenetic dolomitizations (see below: Bonaire) and even in later phases, after erosion of the overlaying strata and subsequent marine transgression. This has been claimed for the Dolomite Mountains (N. Italy) by LEONARDI (1967) (see also below, No. b, 5). Ascendent dolomitization with larger vertical migrations is found only in the realm of late diagenesis (see below, No. b, 4), as a consequence of the compaction water squeezed out of the strata.

In limnic limestones of Miocene age in the Steinheim impact crater (southern Germany), dolomite was formed presumably below the lake surface, according to WOLFF (in press); four stages of decreasing dolomitization are developed on the slope of a subaquatic ridge, the upper part of which is covered with algal bioherms: (1) near the original lake surface, the bioherms are completely dolomitized; (2) downslope, only the surfaces of the bioherms are dolomitized; (3) farther downslope, spherulitic dolomite cement is lining the cavities of the bioherms; (4) in the sediments at the base of the slope, single dolomite crystals 0.02 mm in size are attached to the walls of the interstices of aragonitic and calcitic biocalcarenites as a cement.

5.331.22 Late diagenetic dolomites

a) First stage

The following examples belong to the late diagenetic dolomites on the basis of apparent low nucleation rates, although they are formed either in the soft sediment or soon after lithification.

1. In the northern part of Bonaire (see above, a 1), dolomitized Plio- to Pleistocene normal-marine biocalcarenites dip slightly towards the coast. Dolomitization is especially prominent landward in large zones crossing the stratification. The crystal size is as large as 0.075 mm, and composition averages $Ca_{0.54-0.56}$-dolomite. An origin by subsurface evaporative reflux towards the sea is possible. Presumably this descendent dolomitization occurred (and still occurs) when the sediments were already lithified. This as well as the large crystal size point to a late diagenetic type of dolomitization.

2. Deep-sea occurrences of dolomite

In the Mediterranian (BØGGILD 1912), in the Gulf of Guinee (CORRENS 1937, 1939 b), in the Peru-Chili-Trench (ZEN 1959), and at several guyots (flat-topped submarine mounds; BISCAYE 1964), isolated 0.01—0.02 mm dolomite crystals have been found on the deep sea floor. The exact mode of formation is difficult to evaluate; it is possible that they have been transported by turbidity currents or by wind, and do not represent in-situ precipitation (FAIRBRIDGE 1957). Occasionally, however, they enclose glauconite grains pointing to formation in deeper water.

More recent observations come from about 4500 m depth, 100 km S. E. of the Bermuda Islands (FRIEDMAN 1964): 0.2—2.3 m below the sea floor, Pleistocene unlithified biocalcarenites and calcilutites containing isolated flat dolomite rhombohedra, 0.05—0.07 mm (rarely 0.2 mm) in size, have been found. A detrital supply from the Bermudas can be excluded since no dolomite occurs on these islands; therefore they must have formed within the sediment or on the surface.

In the preparatory borings of the Mohole project off Baja California, layers of nearly unlithified dolomite have been found in a late Tertiary sequence 160—180 m below the sea floor (RIEDEL et al. 1961). These pure dolomite sediments certainly have been formed below the sea. Possibly all these cases correspond to a very slow dolomitization under the conditions of late diagenesis.

b) Second stage

Typical examples of late diagenetic dolomite rocks follow below:

1. GEORGE (1954, 1956) described a sequence of normal marine Carboniferous skeletal limestones about 15 m in thickness which borders on dolomite with a sharp boundary crossing the bedding planes; the boundary is definitely not a fault, and is a common occurrence in late diagenetic dolomites.

2. The lower Carboniferous limestones of the Pennines (England) are dolomitized along faults, joints and bedding planes (GEIKIE 1903, HATCH et al. 1936).

3. In the Arbuckle limestone (Oklahoma) dolomitization is found along a fault crossing a 360 m-thick sequence and penetrating into the adjacent rocks for nearly 5 km. Ankerite is formed here (with max. 11.5 mole % $FeCO_3$; HAM 1951). The direction of migration of the Mg solutions along faults or joints has been investigated using the gradient of the Mg isotopes (DAUGHTRY et al. 1962).

4. Many anticlines in Upper Jurassic limestones of southern France were dolomitized in an upward direction (ascendent) during the Alpine or Provencial folding phase, according to investigations by the INSTITUT FRANÇAIS DU PÉTROLE (1959): The dolomitization penetrates with sharp borders upwards into stratified calcarenites and calcilutites; in places it follows the bedding planes (fig. 5-79). The dolomite crystals are 0.05—0.2 mm in size; according to chemical analyses these are pure dolomites with a composition of $Ca_{0.51-0.58}$. The mean porosity in the dolomite is 3.2%, in the adjacent limestone 2.8%.

Near Caucanas (Cevennes) a dolomite body 800 m in length, 500 m in width and 170 m in height has replaced clayey pelagic calcilutites of Lusitanian age. This dolomite rises from an early diagenetic dolomite of Bathonian age. The borders of the late diagenetic dolomite body are sharp and frequently impregnated with iron oxide. The dolomite crystals are 0.03—0.06 mm in size, the mean porosity is 3.7%, in the adjacent limestone 1.7%.

5. On top of Devonian biostromes consisting of stoichiometric dolomites, D. K. RICHTER (1974) found a thin layer of normal marine limestones which were altered into Ca-dolomite presumably by ascendent dolomitization by the compaction water percolating the dolomite rocks below, previously.

6. A descendent dolomitization which is independent of the age of the dolomitized rocks has been described from the Eifel Mountains (Germany) by REULING (1931) as "Geländedolomite". D. K. RICHTER (1974, 94) concluded that ground water released from dolomite areas evaporated in depressions of exposed limestones and various conglomerates and dolomitized these rocks.

7. A predominant dolomitization of fractured anticlines has been described by CAYEUX (1932) from the Cretaceous of the Paris Basin and by JODRY (1955) from the Middle Devonian of the Michigan Basin which is underlain by an evaporitic sequence. However, not all dolomitizations connected with anticlines belong to this group. Examples are known in which such structures are syn-sedimentary; local emersion caused early diagenetic dolomitization.

8. NECHAEV (1959) described ascendent dolomitization in a limited, strongly fractured area from the Devonian and Carboniferous of the Donetz Basin. He suggested a hydrothermal Mg-supply from local intrusions of Devonian to Carboniferous

age. Normally, however, volcanic activity prevents rather than triggers dolomitization (LEONARDI 1967).

9. FREEMAN (1965) reported dolomite rhombs from the Ordovician of Arkansas, which grew on stylolites (late-diagenetically deformed clay seams) and developed in the cement. These crystals therefore did not grow before lithification.

10. The Turner Valley formation (Mississippian, Alberta) was dolomitized only where it was deposited as a carbonate mud. Mud-free crinoidal sands are preserved as limestones. This has been tentatively explained by MURRAY & LUCIA (1967) as an early cementation commonly found in crinoidal sands which inhibited subsequent dolomitization.

11. Redolomitization, a dolomitization of calcitized dolomites, also belongs to this group.

5.332 Mg-loss

5.332.1 Dedolomitization

Dedolomitization includes the selective dissolution of dolomite rhombs in limestones (H. MÜLLER 1958, BAUSCH 1965), as well as the replacement of dolomitic rhombohedra or rocks by calcite (LANG 1964, v. MORLOT 1847, LUCIA 1961).

Two types of dedolomitization can be distinguished (KHVOROVA 1957, SHEARMAN et al. 1961), between which some transitions are observed (MATTAVELLI 1966):

1. Microcrystalline. Dolomite rhombs are replaced by an accumulation of small calcite crystals. This replacement often begins in the interior of the rhomb, so that only a small envelope of dolomite remains or in the extreme instance only the rhombohedral form is left. Occasionally, fabric traces of the original limestone may reappear in such "dedolomites" (EVAMY 1967). In other cases, dedolomitization begins at the outer rim of the rhomb (D. K. RICHTER 1974).

2. Macrocrystalline. Calcite crystals, up to cm in size, form. Frequently they are filled with dolomite relics some of which are euhedral (see p. 326).

Disseminated dolomite rhombohedra in limestones generally are calcitized microcrystalline, whereas large-scale dedolomitization tends to be macrocrystalline (SHEARMAN et al. 1961, P. FRITZ 1966, D. K. RICHTER 1974).

The strange centrifugal growth of microcrystalline calcite (1) may be explained by the zonal structure of many late diagenetic dolomite rhombs (BANNER & WOOD 1964): In the first stage of dolomitization, the rhombs include many small calcite crystals. Due to the volume contraction sometimes connected with the dolomitization, a porous zone can form around the growing dolomite crystal (MURRAY 1964b), which enables the crystal to replace mechanically the limestone and to form a marginal zone without inclusions. Calcitizing solutions reaching the interior of such dolomite rhombs, will find available nuclei. This agrees with the observation by KATZ (1968), that calcian dolomites are preferentially attacked by dedolomitization; moreover, he found that the calcium content in such calcian dolomite crystals is arranged in a zonal manner.

In a more general way, the occurrence of the two different types of dedolomitization may be explained by the different degrees of previous dolomitization:

(1) With low magnesium supply, only sparse dolomite rhombs can form. These contain relic-calcite inclusions providing lots of nuclei for micritic calcitization.

(2) With high magnesium supply, the whole rock is dolomitized. No calcite relics are left to provide nuclei for calcitization: Macrocrystalline limestone is formed.

In most cases dedolomitization is related to the action of meteoric waters (Tatarsky 1949, Khvorova 1957, Lang 1964, de Groot 1967). This has been confirmed by combined thin section and isotope investigations of the Upper Jurassic limestones in Southern Germany: The dedolomitized areas are impoverished in O^{18} and C^{13} compared with the dolomites and with the primary limestones (P. Fritz 1966). This points to an isotope exchange with the meteoric water from the surface and the organic CO_2 dissolved in such water (see ch. 5.333). This exchange increases with increasing calcitization. In several cases, the observations suggest an influence of $CaSO_4$-containing solutions (Chilingar 1956, Friedman & Sanders 1967).

During dedolomitization, Fe is separated from the lattice, forming zonal rims of iron oxide. This is because a complete solid solubility exists between $MgCO_3$ and $FeCO_3$ at temperatures at least as low as 300°C, whereas no solid solubility is found at 300°C between $CaCO_3$ and $FeCO_3$, from 5 to 97 mole-% $FeCO_3$ (Goldsmith et al. 1962). This miscibility gap is due to the fact that the ionic radii of $Mg^{\cdot\cdot}$ (0.66 Å) and $Fe^{\cdot\cdot}$ (0.74 Å) are closer to one another than those of $Ca^{\cdot\cdot}$ (0.99 Å) and $Fe^{\cdot\cdot}$ favoring an uptake of $Fe^{\cdot\cdot}$ into the dolomite rather than into the calcite lattice. Generally, substitution is not possible if ionic radii differ by more than 15 percent.

Due to the high Mg quantities freed during dedolomitization, these calcites may be significantly higher in Mg (by 1—5 mole percent) than the cement calcite in the same rock (D. K. Richter 1974).

Occasionally, dedolomitization occurs very early as shown by dedolomitized intraclasts in dolomites (Katz 1968).

Some rocks have been dolomitized and dedolomitized several times (Mattavelli & Tonna 1967). Dedolomitization is typical to r a u h w a c k e s (= cellular dolomites, ch. 5.423.2).

5.332.2 Magnesian calcite → calcite (incongruent dissolution)

Magnesian calcites with more than 4 mole-percent $MgCO_3$ are metastable (Chave 1964). Magnesian calcite is found in the skeletons mainly of shallow marine organisms (Stehli & Hower 1961), as well as in shallow marine early diagenetic cement. Such magnesian calcites are preserved for geological periods of time only in bituminous or in rocks with very low permeability. Most Mg loss occurs during early diagenesis.

Occasionally whole skeletons are dissolved (Chave 1964); mostly the magnesium excess is released without destroying the structure, even if the rocks were exposed to the influence of meteoric water soon after their formation (fig. 5-80; Friedman 1964, Gross 1964). Incongruent dissolution (exsolution, Land 1967) of Mg as well as replacement of Mg by Ca are the main processes. Additionally an exchange of some of the CO_3^{--}-ions occurs as shown by the enrichment in C^{12} and O^{16} (Land & Epstein 1970, Bathurst 1971b).

While calcite ("low-magnesian calcite") is stable under present atmospheric conditions as well as during diagenesis, controversal observations and experiments

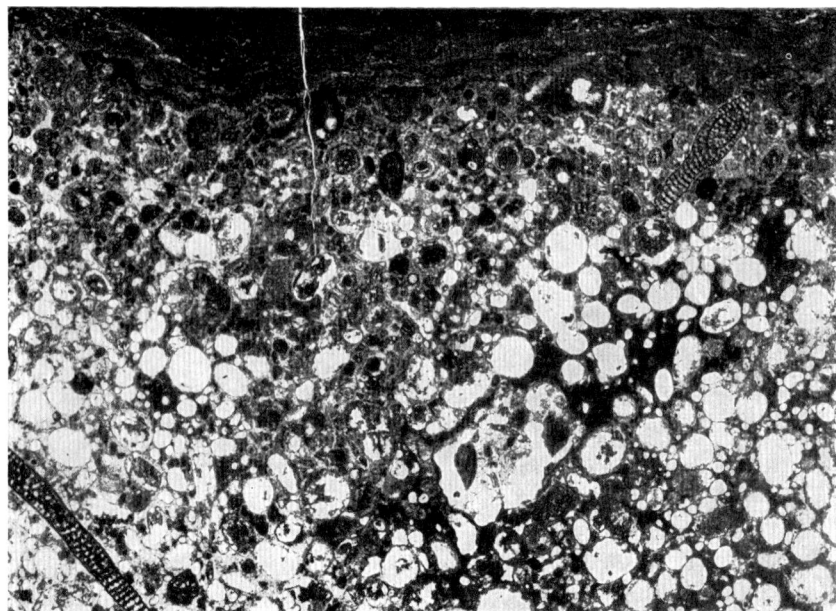

Fig. 5-80. Miami-oölite. The aragonite oöids have been dissolved by meteoric water; the foraminifera originally consisting of magnesian calcite are preserved (lower left, upper right), presumably because the water was saturated only with respect to calcite but not to aragonite. — The oölite is topped by a Mg-calcitic caliche (Pleistocene, Sugarloaf Key, Florida; width 10 mm).

have been published as to the relative stability of magnesian calcite ("high-magnesian calcite") and aragonite: Experiments by CHAVE (1962) and BERNER (1967) point to a lower stabilitiy of magnesian calcite under marine conditions. On the other hand, no differences in stability between aragonite and magnesian calcite are observed in shallow marine (FRIEDMAN 1964, ALEXANDERSSON 1969) as well as in deep marine sediments (MILLIMAN 1966, FISCHER & GARRISON 1967; see ch. 5.321.1). Both carbonates are more stable in submarine than in meteoric interstitial waters. Under meteoric conditions, magnesian calcite quickly transforms into calcite, whereas aragonite can survive longer as shown by the Pleistocene reef limestones of the Florida Keys, as well as by observations of LAND et al. (1967) in the Bermudas.

The lower stability of magnesian calcite during diagenesis is also shown by the fact that fossil aragonitic skeletons are ubiquitous whereas only a few occurrences of pre-Tertiary magnesian calcite are reported (LOWENSTAM 1963, FÜCHTBAUER & GOLDSCHMIDT 1964). In the realm of metamorphism, magnesian calcite is stable above 600° C (GOLDSMITH 1960).

Additional information on the stability of the three minerals discussed above is reported by SCHMALZ & CHAVE (1963) and FRIEDMAN (1965 a). Stability relationships are also discussed in fig. 5-81.

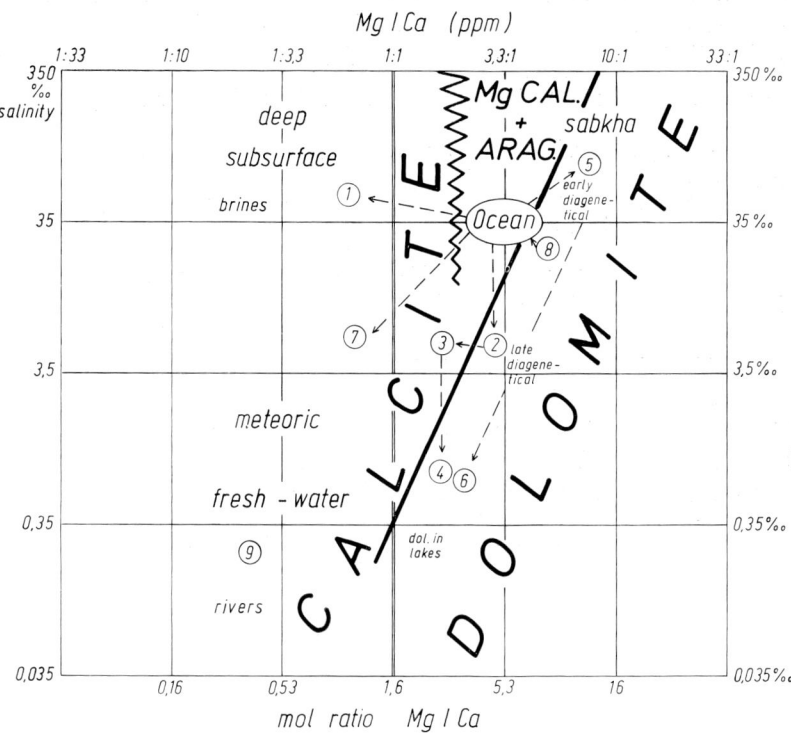

Fig. 5-81. Approximate relationships between water composition and carbonate minerals (after R. L. Folk, pers. comm. 1972) summarizing many of the observations discussed on the preceding pages, and showing areas of formation, not stability fields! According to Lippmann (pers. comm.), the diagram is approximately valid for a solubility product of dolomite $(Ca^{2+}) \cdot (Mg^{2+}) \cdot (CO_3^{2-})^2 = 10^{-17}$. Because of the influence of the salinity on the activities, the boundary of calcite and dolomite is tilted (Lippmann 1973). Fibrous Mg-calcite and fibrous aragonite form in highly supersaturated shallow marine waters because the hydrated Mg-ions and organic compounds hinder the growth perpendicular to c (cement A). During diagenesis, seawater loses Mg, and equant calcite crystals can form (cement B) (1). Mingling of seawater-like interstitial water with meteoric water can produce late-diagenetic dolomite (2) (Land 1972), which can be dedolomitized (3), if clay minerals capture Mg from the interstitial water. By later mingling with meteoric water redolomitization can occur (4). Evaporation can lead to dolomitization only by precipitation of $CaCO_3$ or $CaSO_4$ (early-diagenetic dolomite) (5). Crystal enlargement of dolomite can occur by mingling of brines of sabkha-like environments with water of continental origin high in HCO_3^- and CO_3^{--} during diagenesis (6). Transformation of aragonite to calcite and Mg-loss plus crystal enlargement of Mg-calcite can occur by Mg-loss of the interstitial water (1) or by mingling with meteoric water poor in Mg (7) lowering the supersaturation and favoring growth of larger crystals. Slow growth of large dolomite rhombs can also occur at the sea bottom and late-diagenetically in interstitial waters of seawater composition (8). Large equant crystals of calcite grow during meteoric water cementation (9).

5.333 Isotopic composition of carbonate sediments and rocks

The isotope ratios O^{16}/O^{18} and C^{12}/C^{13} are indicative of sedimentary and diagenetic environments. O^{18} and C^{13} are enriched in marine compared with non-marine environments due to fractionation during evaporation and equilibration with the atmosphere. Isotopic compositions resulting from these phenomena are shown in fig. 5-82.

According to CLAYTON et al. (1966), many interstitial waters, even oil field brines, are in isotopic equilibrium with meteoric water (brines: δO^{18} at $10°C$ about -8, at $70°C$ about $+4$; meteoric water: δO^{18} -4 to -9). Also the pre-Tertiary

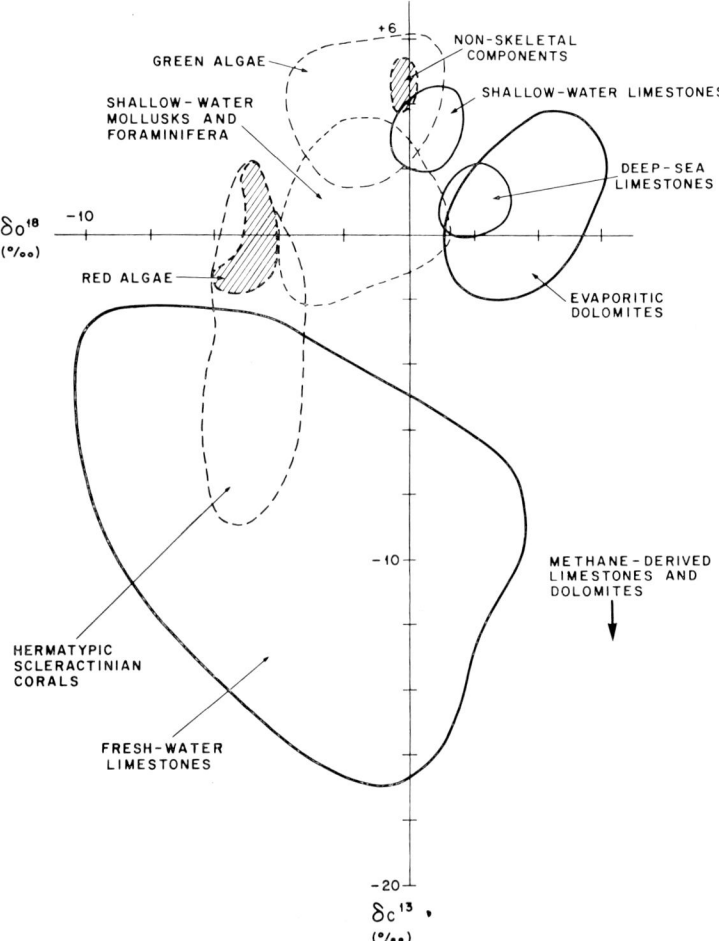

Fig. 5-82. Distribution of δO^{18} and δC^{13} values in various types of carbonates (from MILLIMAN 1974, fig. 19, reproduced with permission by the author and by Springer Verlag). For comparison: δO^{18} of rain water is -4, of sea water $+1$.

abundant nuclei (e. g. sand grains). In sandstones the solutions mostly are supersaturated only with respect to quartz.

B. Silicification may occur in the marine realm early in diagenesis as shown by reworked silicified intraclasts. Again chalcedony prefers carbonate rocks whereas sandstones are silicified by authigenic quartz.

The conditions under which chalcedony or quartzine form are not known. MILLOT (1960) has suggested that the presence of gypsum favors quartzine formation; also FOLK & PITTMAN (1971) found that all occurrences of quartzine cited in the literature are connected with evaporitic environments. Opal is rare; according to MILLOT (l. c.) it occurs especially in shales.

In carbonate rocks silica occurs mostly as flat nodules parallel to the bedding. They display various colors, show conchoidal fractures, and frequently merge laterally in beds. These nodules are called "chert" or "flint" (french: "silex"). They originate as chalcedony or quartz, with traces of opal (CAYEUX 1931, PETTIJOHN 1957). With time, they transform into quartz without changing their fibrous texture (see below). Different criteria suggest a diagenetic formation of these nodules (PETTIJOHN 1957): they can merge from layer to layer, sometimes through joints. Moreover, the fabric of the adjacent rock (e. g. skeletons) may be found in the silica nodules (DUNBAR & RODGERS 1957, p. 251). Carbonate inclusions in the silica, however, are not in themselves proof of replacement, as was shown experimentally by CORRENS (1924).

Obvious sources of silica include highly soluble SiO_2 particles, such as silica skeletons of organisms (sponges, radiolaria, diatoms) and volcanic glass. First of all, however, terrigenous supply must be taken into account. Inorganic precipitation can occur when SiO_2-rich river water reaches the ocean. Solutions need not be excessively concentrated, for example the Mississippi River (BIEN et al. 1958). HARDER (1965) observed X-ray amorphous hydroxides of aluminum and iron to coprecipitate with silica.

One example of chert is the Upper Cretaceous flint (A. H. MÜLLER 1956). These nodules are mostly brownish or dark grey with a white crust and consist of chalcedony (see also FLÖRKE 1962, MICHEELSEN 1966). Inclusions of bryozoans, echinoids, bivalves, belemnites, and foraminifera (in the flint nodules) which are replaced by silica, in part by quartz, suggest that the flint is a replacement (at least in part; ILLIES 1949b, GRIPP 1954). The time of formation of flint nodules in the chalk is still under discussion. Soft microfossils (O. WETZEL 1933) with preserved organic material (W. WETZEL 1957) point to an early diagenetic origin, but the filling joints or replacing crushed fossils by flint, on the other hand, suggest some occasional late silicification (GRIPP 1954). The silica of the flint is supplied mainly by the numerous sponge needles that occur in the chalk and are often calcitized in the rock adjacent to the flint (CAYEUX 1935, see also LOWENSTAM 1949, HEALD 1952, NEWELL et al. 1953, p. 160 ff., SUJKOWSKI 1958, FABRICIUS 1966). Similarly in the Lower Campanian, horizons which are rich in flint nodules often are adjacent to chalk containing a minimum of silica (G. ERNST 1964). A volcanic origin of SiO_2 is also possible in view of the tuffite layers found by VALETON (1960).

Generally the mechanism of silica enrichment in such replacement fabrics is unknown. SIEVER (1962) suggested an influence of organic material adsorbing silica which in turn could form the nuclei for further enrichment. At the same time the carbon dioxide that formed during the disintegration of the organic material could dissolve the $CaCO_3$. Accordingly, skeletons sometimes can serve as "nuclei" of silicification as described by A. H. MÜLLER (1956) from Upper Cretaceous flint and by FABRICIUS (1966) from Liassic siliceous limestones: this is especially true of echinoderm fragments (fig. 5-86). Silica may move through the sediment either by diffusion or by compaction (ILLIES 1949b, EMERY & RITTENBERG 1952). Whether it accumulates as a gel is unknown. Opal certainly forms by ageing of silica gel. The same is suggested for the concentric silica rings on the surface of many silicified fossils (WETZEL 1957, JUX 1959).

DAPPLES (1967b) compared the tendency of silicification of different fossils from seven occurrences of Ordovician to Permian strata. He observed a small preference of bryozoans, brachiopods and corals. These are followed by (in order of decreasing silicification) mollusks (fig. 5-41), echinoderms, foraminifera, calcareous sponges, and dasyclads (NEWELL et al. 1953). The reason for such selective silicification is unknown. The resistance of dasyclads may be caused by the absence of organic material in the calcitic remains which only wrap the tissues without being true skeletons.

C. In the course of later diagenesis, the unstable forms of silica, e. g. glass, opal, chalcedony, and quartzine, gradually give way to the stable quartz. According to MIZUTANI's (1967) diagram, amorphous silica is transformed quantitatively into quartz after $0.8 \cdot 10^6$ years at $100°C$, $10 \cdot 10^6$ years at $50°C$, or after $100 \cdot 10^6$ years at $20°C$. In a manner analogous to the devitrification of volcanic glass, birefringent chalcedony fibers can form within opal. Whereas further transformation of chalcedony to quartz took place quickly at $250-300°C$ in experiments by WHITE & CORWIN (1961), in nature it requires more time but needs less temperature, as indic-

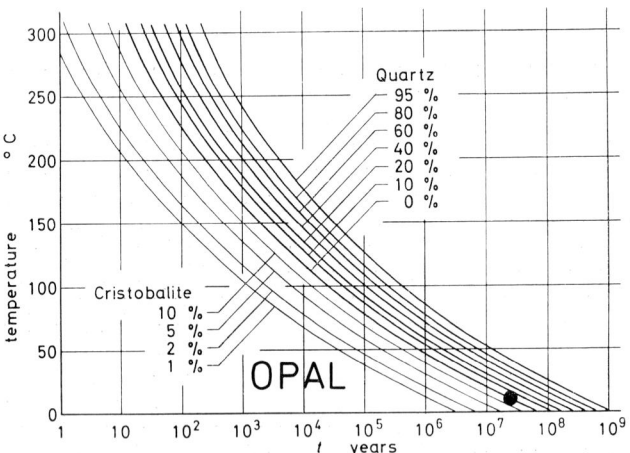

Fig. 5-85. Transition opal-cristobalite-quartz related to temperature and time, based on observations and calculations by MIZUTANI (1970). The maximum values for cristobalite are found in the area of beginning quartz development. At 95% quartz, traces of opal are still present. The dot indicates chert samples from Upper Oligocene deep sea cores (PIMM et al. 1971).

ated above. Simultaneously, the refractive index moves towards that of quartz. In a later phase of diagenesis, possibly near the transition to metamorphism, the fibrous or cryptocrystalline "chalcedony quartz" reorganizes by crystal enlargement into quartz with homogeneous extinction (MUKHOPADHYAY & CHANDA 1972). Under appropriate preservation unstable forms of silica can be maintained for long periods; e. g. in Jurassic limestones of Ethiopia plenty of volcanic glass (n = 1.49—1.52) is included.

X-ray investigations indicated the following transitions: First opal is converted to low-cristobalite, which in turn is transformed into low-quartz. The duration of this transformation depends on the thermal history. In one natural example it was completed after about 15 million years in sediments, whose depth of burial increased at the same time from 0 to 1500 m, a figure which is comparable to the above mentioned transformation (MIZUTANI 1970). Observations in deep sea cores (PIMM et al. 1971) confirm the diagram of MIZUTANI (1970), showing the influences of age and temperature on the transformation opal-cristobalite-quartz (fig. 5-85).

Only quartz formation is observed in late diagenetic silicifications since solutions are only supersaturated with respect to quartz. In anhydrite, halite and carbonate rocks, authigenic bipyramidal quartz crystals are frequent (figs. 5-57 and 81b; GRIMM 1964). Such crystals, however, may form early as well as late diagenetically.

In druses the sequence opal-chalcedony-quartz has been observed. (For opal-chalcedony see fig. 3-47). Interparticle pores of silicified oolites contain radial chalce-

Fig. 5-86. Limestone with sponge spicules consisting of chalcedony or fibrous quartz (upper right); in the lower left echinoderm fragment replaced by silica. Liassic-Dogger, Northern Alps (crossed nicols, width 3.6 mm; thin section from F. FABRICIUS).

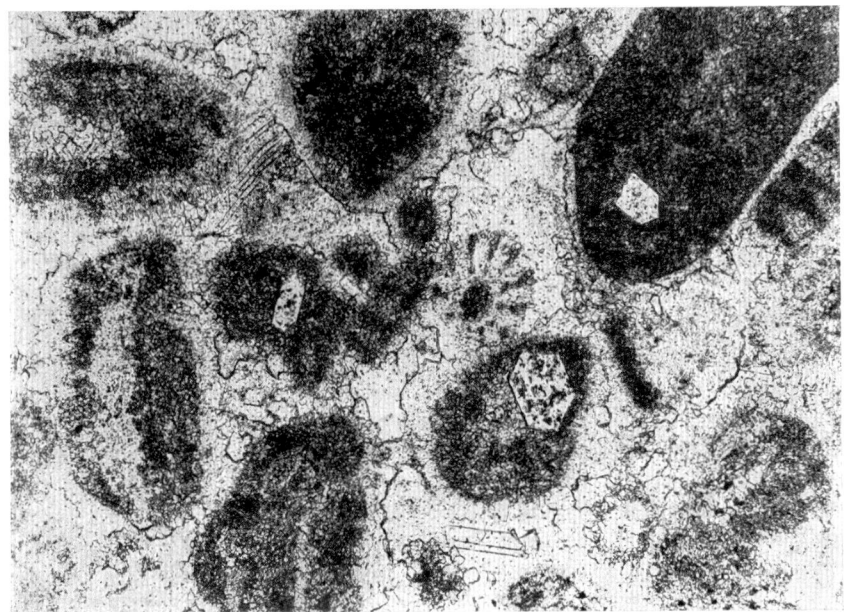

Fig. 5-87. Pelletal limestone with cement A and B, the latter in part nucleated by an echinoderm spine (center). In three cryptocrystalline grains, inclusions of authigenic quartz have formed. Rhetian, Northern Alps (width 1.3 mm, thin section from F. FABRICIUS).

dony fibers perpendicular to the pore walls and isometric quartz crystals in the center of the pores (CHOQUETTE 1955). This fabric is similar to the cement generations A and B calcite (ch. 5.321).

According to SIEVER (1962) the solubility of quartz increases by the factor of 10 during a temperature increase from 25 to 120° C. The salinity increasing with burial partially counteracts the temperature effect but cannot completely neutralize the increased solubility. With respect to the carbonate solubility, the influences of increasing pressure and temperature compensate each other; however, increasing salinity raises the carbonate solubility. The system SiO_2—$CaCO_3$ is therefore rather complex with increasing burial. If the salinity increases at constant pressure and temperature, e. g. during a long burial at a shallow depth, SiO_2 should precipitate and $CaCO_3$ should dissolve. The silicification of limestones generally takes place in shallower depths than the calcitization of quartz and silicates; this possibly is caused by the increasing pH with depth. According to WALKER (1962) both replacements are reversible.

For late diagenetic silicification large-scale changes of pressure, temperature and salinity may be less important than short-term chemical transport by diffusive or advective processes. An example is shown in the photographs of etched radiolarian chert by SCHWARZ (1928): the radiolaria obviously are dissolved at the top and bottom sides in contact with clay stringers; the dissolved silica precipitated in the clay-free part of the same rock as cement between the radiolaria.

The filtering of the compaction waters by semipermeable membranes (e. g. clay layers) may play a role in the late diagenetic enrichment of silica (SIEVER 1962).

5.336 Authigenic silicates

Minerals which have been formed after the deposition of the sediment, are termed "authigenic" (from Greek authigenes = formed at this place). Although this term is somewhat in disuse, it is applicable in designating diagenetic minerals that differ from their surroundings considerably.

One of the most frequent authigenic silicates is glauconite, which is found mainly in sandstones and shales, and has been treated there (see Part II/2). It is found also as internal molds of planktonic foraminifera, thus indicating authigenic formation.

Chlorite and chamosite also occur in limestones, especially in oölites (fig. 5-50). They can form as individual layers (V. SCHMIDT 1965) or compose entire oöids, which often are plastically deformed (BERG 1944). Chlorite and chamosite also fill skeleton pores, e. g. in echinoderms (fig. 5-40). For further discussion, see Part II/2.

The main authigenic silicates in limestones are in order of decreasing frequency (TOPKAYA 1950):

1. sodium feldspar
2. potassium feldspar
3. tourmaline
4. muscovite
5. zircon

The following observations indicate an authigenic formation:

a) euhedral outline
b) pseudomorphs after sedimentary textures such as oöids and fossils (VAN STRAATEN 1948)
c) abundant occurrence of one mineral species which cannot be explained by transport-sorting phenomena
d) correlation of crystal size and abundance (CAROZZI 1953a)
e) inclusions which correspond mineralogically to the surrounding rock (e. g. carbonates, opaque minerals)
f) differences in habit and color as compared to detrital grains of the same species
g) those minerals with isomorphous miscibility are usually confined to the small chemical variability near the end members, due to the low temperature of formation
h) for feldspars: optical properties contrasting to those of detrital grains.

1. authigenic sodium feldspars rarely are larger than 0.2 mm. Smaller crystals generally are tabular parallel to (010), the larger ones parallel to (001). "Roc-Tourné" twins (twice twinned combinations of albite and X-Carlsbad twins) are typical (fig. 5-88; FÜCHTBAUER 1948, KASTNER & WALDBAUM 1968).

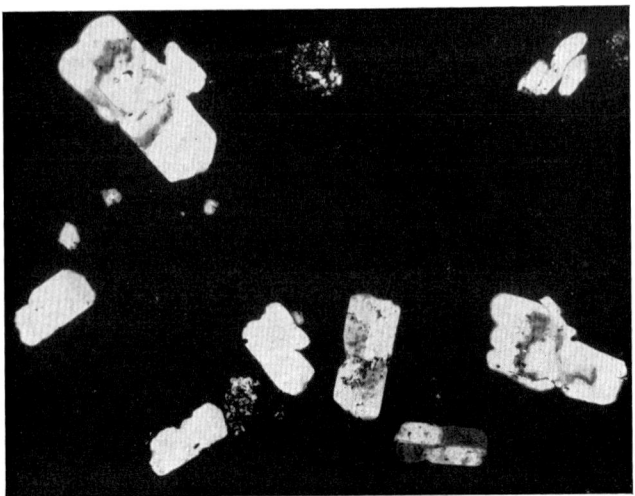

Fig. 5-88. Authigenic albites; "Roc-Tourné" twins (twice twinned) with typical lateral notches on the (010) faces. These twins are best shown in the lower right. Insoluble residue of Upper Permian 2-dolomite, well "Norddeutschland 6" (Emsland area) (crossed nicols; width 0.9 mm).

According to refractive index and chemical analyses these albites correspond to the end member ($<$ 3 mole-% potassium feldspar and $<$ 1 mole-% anorthite, BASKIN 1956). The axial angle $2 V_z$ generally is enlarged compared with detrital albites (FÜCHTBAUER 1948, 1950, H. MÜLLER 1958, MILTON et al. 1960, SCHÖNER 1960). Based on X-ray analyses, BASKIN explained this as an incomplete Al-Si-order corresponding to a slight shift towards the high-temperature albite ("analbite", see table 5-9). Authigenic feldspars are highly ordered compared with high-temperature feldspars, but are not as ordered as low-albites and microclines from low-metamorphic rocks and pegmatites (KASTNER 1971).

Table 5-9. The optical axial angle (FÜCHTBAUER 1950) and the angle γ^* in the reciprocal lattice (BASKIN 1956) for different sodium feldspars.

	$2 V_z$	γ^*
albite (low-temperature feldspar)	77°	90° 30'
authigenic albites (Switzerland)	ca. 85°	90° 30' — 90° 05'
authigenic albites (Middle Triassic)	90°	90° 00' — 89° 12'
analbite (high-temperature feldspar, SCHWARZMANN 1956)	140°	88° 00'

In this table the albites from Switzerland possibly were exposed to higher temperatures. In albites from (nonmetamorphic) sediments, the $2 V_z$ is mostly between 85 and 105°, instead of 76—80° for low-albites.

Authigenic albites may attain several percent of a limestone. They occur mainly in marine rocks (MILTON et al. 1960, FÜCHTBAUER 1972).

2. most authigenic potassium feldspars are smaller than sodium feldspars and may develop different combinations of faces. The adularia type with completely or asymmetrically developed 110 planes is frequent. Rectangular (?) microclines are more frequent (fig. 5-89). Twins are reported from authigenic microclines (albite-pericline-double twin; PERRENOUD 1952, BASKIN 1956). Larger authigenic potassium feldspars frequently are not euhedral and contain impressions of surrounding mineral particles on their surfaces.

According to refractive indexes and chemical analyses, these potassium feldspars correspond to the end-member ($<$ 2 mole-% albite, BASKIN 1956). The optical axial angle $2V_x$ is about $43°$ (FÜCHTBAUER 1950, H. MÜLLER 1958, MELLIS 1960, SCHÖNER 1960); in the "adularias" found in alpine joints, which underwent elevated temperatures, this angle varies between 22 and $64°$.

According to optical investigations, both monoclinic and triclinic K-feldspars occur (FÜCHTBAUER 1956). The latter has been confirmed by the splitting of the 130 and 1$\bar{3}$0 as well as of the 131 and 1$\bar{3}$1 lines in X-ray diffraction patterns (D. K. RICHTER 1974).

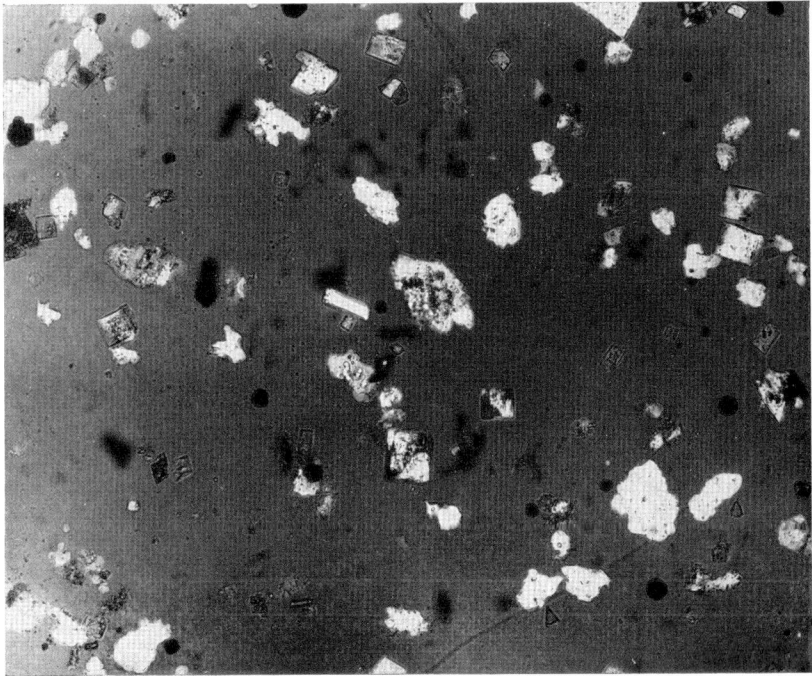

Fig. 5-89. Authigenic potassium feldspars (rectangular, in part complex twins), authigenic quartz (lobate outlines) and one authigenic tourmaline (small column near center). Insoluble residue of Upper Permian dolomite, Northern Germany (nearly crossed nicols; width 1.1 mm).

The parameters controlling the distribution of authigenic Na- and K-feldspars are not clear. The following discussion may provide a basis for understanding the different distributions of these two feldspar-types.

The silica demand is high, but this seems to be no problem since most rocks with authigenic feldspars also contain authigenic quartz, suggesting a supply of silica in a more soluble form, e. g. opaline skeletons (CALVERT 1968). The silica contents in the pore fluids of modern ocean sediments are considerably higher than in sea water (SIEVER et al. 1965).

The solubility of aluminum is low in the carbonate environment; a long transport therefore is improbable (LIPPMANN & SCHLENKER 1970). Clay minerals present in the rock are a possible Al-source for the feldspar formation. This applies especially to illite which is poorly crystallized in most sediments and therfore more soluble than muscovite.

Sodium in the interstitial water of marine sediments is inherited from sea water and may, with sediment compaction, migrate even into non-marine deposits. Potassium, on the other hand, is 1/30 as abundant as sodium in sea water; however, its concentration can change considerably within the sediment. According to SIEVER et al. (1965) and MARCHIG (1970), potassium can be enriched by a factor of 2—5 within modern oceanic clay sediments (due to filtration, base exchange and decomposition of K-feldspars) whereas during diagenesis its content can decrease (due to adsorption by illites). Part of the potassium adsorbed by the clays may be subsequently available for diagenetic formation of K-feldspar; at the same time the aluminum from the illites is taken up by the feldspars. According to ion-exchange data and thermodynamic calculations by KASTNER (1971), the $K / (K + Na)$ ratio in sea water is in equilibrium with both albite and potassium feldspar. Therefore both feldspars have been formed in many marine carbonate sediments, in which the above-mentioned enrichment of K did not take place. If, however, the Na-content of the interstitial water was not sufficient or if the K-supply was especially high, only potassium feldspar was formed. The first alternative applies to the fluvial sandstones of the Lower Triassic Buntsandstein (FÜCHTBAUER 1967b), the latter, as discussed above, to clays and illitic limestones (SWETT 1968). A positive correlation between the quantity of authigenic K-feldspars and the (illitic) clay content has been reported from Triassic limestones (FÜCHTBAUER 1956).

A well-known origin of authigenic feldspars is the replacement of zeolitic precursors in tuffaceous sediments of alkaline lakes. Whereas glass shards are replaced by montmorillonite in fresh-water environments, diagenetic sequences such as volcanic glass — phillipsite or clinoptilolite (from basaltic or rhyolitic glass, respectively) — analcime — potassium feldspar are developed in increasingly saline environments; minerals of decreasing water content are formed, because the H_2O-activity decreases during evaporation and subsequent diagenesis (SHEPPARD & GUDE 1969). Feldspars replacing analcime have also been observed in tuffaceous members of the Green River Formation (Eocene). According to IIJIMA & HAY (1968), "the distribution of albite and K-feldspar may reflect the K^+/Na^+ ratio of the intrastratal solutions. K-feldspar predominates in most beds containing sodium-bearing saline minerals, whereas albite is most abundant in the outer part of the transition zone, where saline minerals are absent or rare. Probably the K^+/Na^+ ratio was elevated in the more saline facies by crystallization of sodium-bearing salts, thus favoring K-feldspar. Albite would be favored over K-feldspar by low K^+/Na^+ ratios."

In smaller quantities, authigenic feldspars are present in various carbonate rocks, and in anhydrites of non-tuffaceous Upper Permian evaporite sequences (FÜCHTBAUER 1972). Albite is found in the less evaporitic lowermost strata (anhydrite "1" and "2"), whereas K-feldspar is present in anhydrite and dolomite "2" and "3" associated with the main rock salt deposits, analogous to the occurrence of both feldspars in the

Green River Formation. Since the volume of sodium salts (halite) exceeds that of the potassium salts in the Upper Permian evaporites as a whole, the corresponding brines and interstitial waters were enriched in potassium (RICHTER-BERNBURG 1955). During compaction and crystal enlargement of the halite, K^+-rich solutions were pressed into the younger sediments and triggered the formation of K-feldspar. BRAUN & FRIEDMAN (1969) also mentioned that evaporation may be a prerequisite for K-feldspar formation.

3. Authigenic tourmalines are mostly colorless, green, blue, or blue-green. In sandstones, they form small columnar overgrowths on detrital tourmalines. In limestones, long, prismatic tourmalines with small detrital cores are found (fig. 5-82b).

The refractive index of the secondary overgrowth is mostly lower than that of the detrital core, for the same reason as mentioned for feldspars. TOPKAYA (1950) found $n_\varepsilon \sim 1.615$ and $n_\omega = 1.633$ in colorless species, whereas for magmatic tourmalines $n_\varepsilon = 1.620-1.657$ and $n_\omega = 1.639-1.692$ are reported (CORRENS 1949). However, authigenic tourmaline with $n_\omega = 1.672$ has also been described (SCHÖNER 1960).

Authigenic tourmalines are particularly frequent in evaporitic rocks (TOPKAYA l. c.). In anhydrites small tourmaline needles occasionally form the main residue mineral (LOHSE 1957). In an Upper Permian dolomite abundant tourmaline has stained the stylolites green.

4. Photographs of presumably authigenic muscovite are published by TOPKAYA (1950); they form hexagonal leaflets up to 0.2 mm in size. Authigenic muscovites replacing a brachiopod shell by a fibrous aggregate were described by SOROTCHINSKY (1954). In high-diagenetic sand- and siltstones muscovite formation is common.

5. Authigenic zircons together with other authigenic minerals are reported by SOROTCHINSKY (l. c.), as fillings of bryozoan chambers in the Devonian of Belgium.

5.337 Authigenic pyrite

Pyrite is a frequent authigenic mineral in sediments.

"Framboidal pyrite" aggregates are found in Devonian to Cretaceous marls and limestones. FABRICIUS (1961) observed this species associated with benthonic foraminifera and ostracods and interpreted it as an indicator of "gyttja" (semi-sapropelite). He found framboidal pyrite in two varieties:

1. "fine pyrite": spheres < 0.03 mm in diameter, composed of pyrite crystals $< 1 \mu$, which aggregate into globules of second order, 0.15—0.2 mm in size (fig. 5-90).
2. "coarse pyrite": spheres < 0.12 mm in size composed of pyrite crystals, 5 to 10μ in size. They are nearly never arranged into larger globules.

These regular forms suggest an organic origin; they have been interpreted as caused by sulfur bacteria (SCHNEIDERHÖHN 1923, NEUHAUS 1940). LOVE & AMSTUTZ (1966) upon dissolving pyrites found an organic spongelike matrix body of unknown origin, sometimes surrounded by a sack; they interpreted this as not being bacterial in origin, although bacteria may be at least partially involved in pyrite formation.

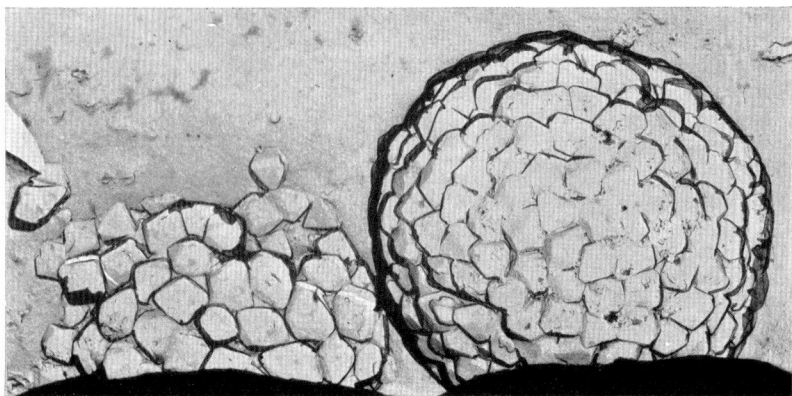

Fig. 5-90. Framboidal pyrite in radiolarian chambers. Upper Jurassic (Tithon) near Unken, Salzburg (electron micrograph from HONJO et al. 1965; width of photograph 12 μ).

5.4 Pore space

5.41 Porosity of calcilutites

The argonitic muds in the Florida Bay have a porosity of 87% at the surface, and 75% at a sediment depth of 3.4 m (GINSBURG 1957). In the Gulf of Oman (Persian Gulf), 70% porosity has been measured at the sediment surface, 60% in 1—3 m depth (KÖGLER 1967): The porosity and its decrease in the uppermost meters is similar to that of modern clay muds. Further compaction, however, is very different in lime and clay muds. Whereas the porosity in clays decreases from 80% at the surface to about 20% at 1000 m depth, and 4% at 3000 m depth (v. ENGELHARDT 1960), the porosity of calcilutites decreases much more rapidly: the porosity of many Tertiary limestones with shallow depths of burial, for instance, is less than 1%. The lack of compaction in such limestones, however, is demonstrated by the following observations:

1. even fragile fossils retain their form in limestones, whereas they are mostly fractured in shales and marls.

2. internal sediments in cavities have as low porosities as the surrounding calcilutites even though the internal sediments were not subjected to loading pressure.

3. oölites and biocalcarenites frequently contain resedimented fragments of the same sequence ("intraclasts") with oöids in a cryptocrystalline matrix. The frequency of transected oöids at the margin of such intraclasts is surprisingly high (fig. 5-11). This indicates that the matrix has been more lithified than (or as lithified as) the primarily rigid oöids at the time of formation of these intraclasts.

Is the early lithification of lime muds caused by neomorphism (crystal enlargement, transformation aragonite-calcite) or by cementation or both?

Crystal enlargement generally is observed at higher overburden pressures. Such pressures, however, are excluded by the observations listed above (1.—3.).

The transformation aragonite-calcite may play an important role in lithification, as suggested by the following experiments by ROBERTSON (1964): Aragonite mud deposited in sea water transformed to calcite and became lithified after a few days exposure at 400° C and a pressure of less than 100 bar. A calcite mud from the deep sea did not change its consistency at the same conditions, even up to pressures of 1000 bars. This suggests that aragonite muds are easier to lithify than calcite muds. Two examples of presumably primary calcite sediments seem to confirm this model:

(1) Limnic limestones of Upper Tertiary age in the Swiss Jura, buried by a few hundred meters of sediment, still have porosities of about 40 percent (KÜBLER 1962 a, b).

(2) Upper Cretaceous chalk retained porosities of 30—40 percent at a burial depth as great as 1500 m (v. ENGELHARDT 1960). As an explanation, NEUGEBAUER (1972 and pers. commun.) suggested an interstitial water highly supersaturated in calcite, so that pressure solution did not occur. A supersaturation in calcite but not in magnesian calcite was concluded from crinoid layers in the chalk of Kansas. These, originally consisting of magnesian calcite, were broken down to micritic calcite by overburden pressure and subsequently underwent crystal enlargement.

Danian chalk of the Ekofisk oil pool (Norwegian North Sea) retained a porosity of 30%, when oil-saturated, at a depth of 3300 m. The primary permeability can be lower than 1 md in some sections. However, due to secondary fracturing, the average permeability is 12 md, allowing for individual well potentials of 10,000 bopd from 200 m of limestone (WORLD OIL, May 1971, 51—52).

On the other hand, shallow marine calcilutites of Tertiary and Cretaceous age in the Persian Gulf area which presumably were rich in aragonite when deposited, also have retained porosities of 10—30% even in the subsurface. Thus it can be concluded that the transformation aragonite-calcite may influence the lithification of lime mud, but this cannot be the only mechanism.

Calcilutitic rocks consist frequently of isometric calcite crystals 1—3 μ in size (E. FLÜGEL et al. 1968). In the same order is the crystal size of modern lime muds, e.g. the aragonite needles forming the sediments on the Great Bahama Bank (p. 196). Transformation and subsequent cementation are required for the transition of this mud into the dense calcilutite. It is suggested that transformation alone was not able to form a splintery limestone; instead, presumably a spongy mass of tiny calcite crystals formed which then enlarged by overgrowth (= cementation) to form a low-porosity rock (FOLK 1965 b).

Both transformation and cementation of lime mud depend on the composition of the electrolytes. For this reason, they can begin at different depths below the sediment surface. Early cementation and transformation were observed mainly in the shallow sea (SHINN 1969, TAYLOR & ILLING 1969). Breccia components lithified presumably in less than 15 to 30 m below the bottom (NEWELL et al. 1953, SNYDER &

Odell 1958). Low-magnesium calcite mud forming internal sediment along the Mediterranean Sea coastal zone lithifies by high-magnesium calcite cementation (Alexandersson 1969). Late cementation and transformation (> 270 m below the bottom) is inferred from observations in Pacific atolls (Ladd et al. 1953). Late cementation of calcitic nannoplankton oozes also was reported from the Deep Sea Drilling Project (1969): The uppermost rigid though porous limestone layer was observed at different depths, between 100 and 450 m below the deep sea bottom, in Miocene sediments.

Two observations are important in this context: According to Zankl (1969), a clay content as low as 2% is sufficient to cause compaction of a calcilutite. At lower clay contents, no compaction has been observed. On the other hand, a clay content of 2% can hinder a crystal enlargement in the Upper Jurassic (Bausch 1968). Similarly, the presence of clay seams hinders cementation in sandstones. This suggests that both crystal enlargement and cementation may be influenced by the clay content in limestones (Ballance & Nelson 1969); but the role of clay is not fully understood. Limestones investigated by Daley (1971) reacted as marls and remained plastic during early diagenesis only if their clay content was in excess of 20%. Such mobile bands tended to fill cracks similar to sand dikes in the overlying more cohesive limestones.

Perhaps the most critical stage is the initial cementation during which the crystals composing the calcilutite are welded. Once the pores are "frozen" by initial cementation, the complete filling may take a longer time. In interbedded marls and limestones, cementation of grain contacts is hampered in the marls by the deficiency of calcite-to-calcite contacts. Instead, calcite solubility may increase at the calcite-to-clay contacts due to pressure solution. The dissolved $CaCO_3$ is used to fill the pores in the limestones by homoaxial overgrowth of calcite crystals. This may lead to an increase in the difference of clay contents of the limestone and the marl layers, thus leading to a more distinct layering (Hallam 1964).

Though in many limestones textural evidence seems to exclude compaction, limestone sequences generally decrease in thickness during diagenesis. This is mainly due to porosity decrease by cementation, because most of the carbonate cement is derived from the same limestone sequence, e.g. by stylolitization. For this reason, compaction water streaming upward through the sequence is found not only in shales and sandstones, but also in limestones.

5.42 Porosity of particle limestones and dolomites

5.421 Pore types

The following classification of pore types has been modified from Choquette & Pray (1970) by grouping the 15 types proposed by these authors into two categories with 4 groups each. The intention is to classify and combine pore types with similar geometric properties (see fig. 5-91).

5.42 Porosity of particle limestones and dolomites

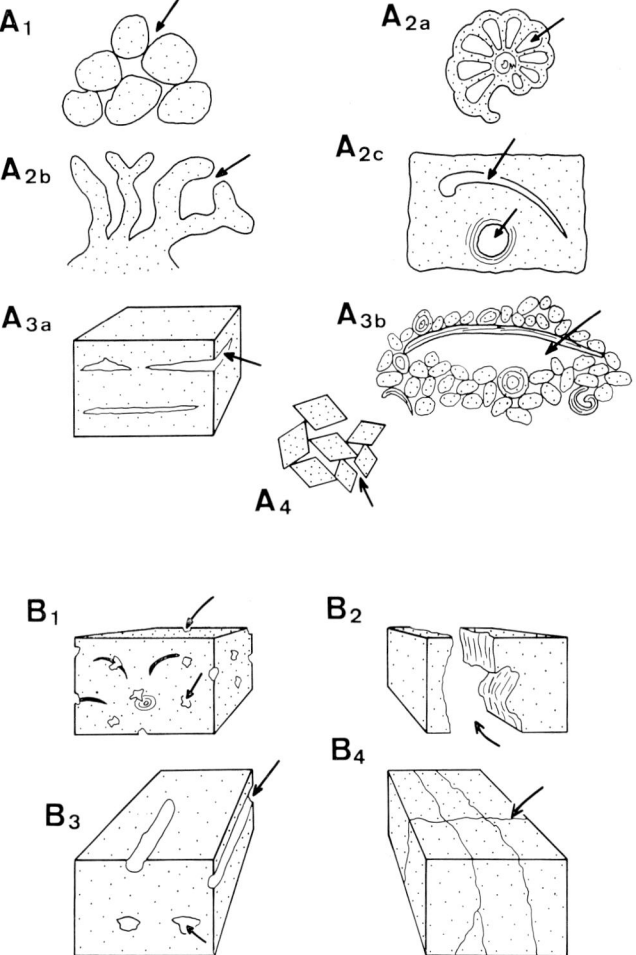

Fig. 5-91. Classification of pores, modified from CHOQUETTE & PRAY (1970). The letters correspond to the subdivisions of ch. 5.421.

A. fabric selective

   1. **Interparticle pores** (fig. 5 52, 60, 92)

   They include all pores between packed particles (e. g. in depositional breccias, sandstones, and in chalk), and are synsedimentary in origin.

   2. **Intraparticle pores**

      a) "chambers" or other openings within skeleton organisms ("presedimentary"; figs. 5-25, 35).

      b) "growth framework" created where elements of the growth structure, such as the platy arms of the colonial coral *Acropora palmata* or irregular sheets of the alga

Fig. 5-92. Interparticle pores.
Left: Very sandy dolomite-arenite with calcite cement. Tertiary Molasse north of the Alps (Bausteinschichten, well Schwabmünchen 1). 35% sand (white; median 0.19 mm), 12.0% calcite, 52.4% dolomite grains. Porosity 26.1%, air permeability 2500 millidarcies (width of photograph 0.9 mm, crossed nicols, pores are black).

Right: Capillary pressure curve of the same sample. This curve indicates the portions of the pore space filled with mercury as pressure is gradually increased from 0 to 100 atmospheres (left ordinate). The right ordinate shows the capillary diameters calculated from the formula $r = 2\sigma \cos\delta/P$ (r = capillary diameter; σ = surface tension, in the present example 480 dynes · cm^{-1}; δ = marginal angle, here 140°; P = pressure). The capillary diameters are theoretical figures for comparison purposes, for the capillaries of rocks are never exactly circular in cross-section. The diagram shows that 12% of the "capillary diameters" are below 1 µ, 11% are between 1—5 µ, 27% are between 5—20 µ and 50% are larger than 20 µ (pers. comm. by Dr. MIESSNER, Gewerkschaft Elwerath, Erdölwerke Hannover).

Lithoporella or *Archaeolithothamnium* intergrow in such a manner as to isolate voids from sedimentation (synsedimentary) (fig. 5-57).

c) "molds" formed by diagenetic solution of a former sedimentary constituent, such as a shell, oöid or anhydrite crystal (postsedimentary) (figs. 5-17, 80, 93).

3. **Fenestral** and **shelter pores** (synsedimentary)

a) "fenestrae" are primary or penecontemporaneous gaps in rock framework, larger than grain-supported interstices. They form by decay of sediment-covered algal mats, shrinkage during drying, or accumulation of pockets of gas. Fenestrae commonly are somewhat flattened and parallel with the bedding planes. This term is preferred to "birds eye" (fig. 5-75).

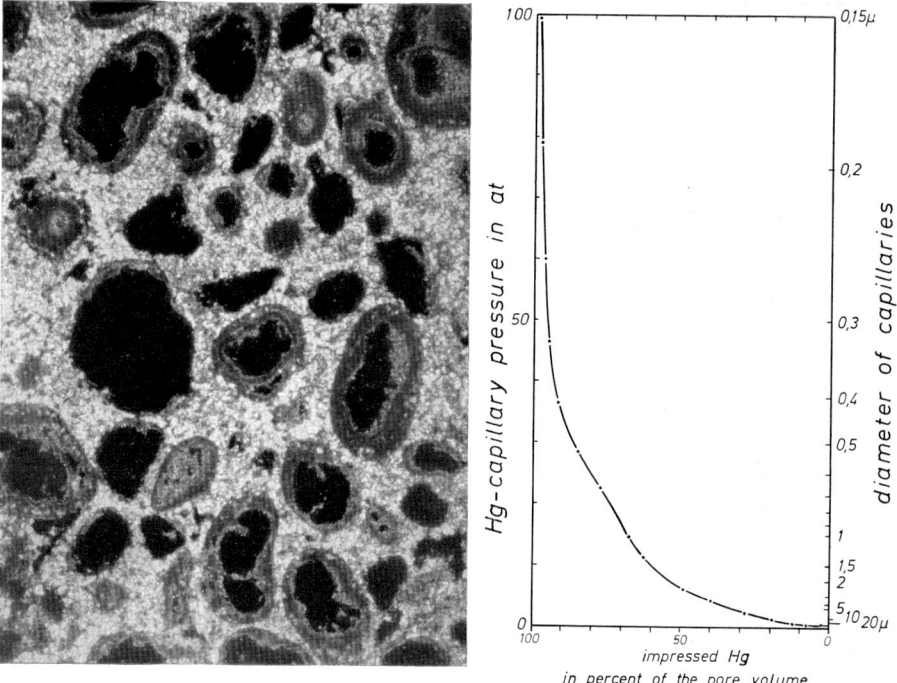

Fig. 5-93. Molds.
Left: Dolomitic oncolite; oncoids are cryptocrystalline but largely dissolved. Upper Permian 2 - dolomite W. of Nienburg (Germany); crystal size of the matrix ~ 35 μ. Porosity 30.3%, air permeability 11 millidarcies (width 2.3 mm, crossed nicols, pores black).
Right: Capillary pressure curve of the same sample (for explanation s. fig. 5-92). Capillary radii: 32% < 1 μ, 34% 1—5 μ, 17% 5—20 μ, 17% > 20 μ.

b) "shelter pores" are created by the sheltering effect of relatively large sedimentary particles, e. g. shell fragments which prevent the filling by finer particles of adjacent pore spaces.

4. **Intercrystalline pores** (syn- or postsedimentary depending on mineral species, see below) (fig. 5-94).

B. not fabric selective (postsedimentary)

1. **Vugs** (fig. 5-95) (occasionally fabric-selective)
Pores which are not markedly elongate. The are formed by the solution of indistinct areas of the rock rather than distinct particles.

2. **Caverns**
Man-sized openings.

3. **Channels**
Markedly elongate pores in one or two dimensions (ratio of length to average cross-sectional diameter exceeds 10). They may form inorganically during diagenesis,

Fig. 5-94.   Intercrystalline pores.
Left: Euhedral, pelletal dolomite. Paleocene, Libya. Mean crystal size 20—25 μ. Porosity 36.0%, air permeability 170 millidarcies (width 0.4 mm, crossed nicols, pores black).
Right: Capillary pressure curve of the same sample (for explanation see fig. 5-92). Capillary radii: 4% < 1 μ, 32% 1—5 μ, 62% 5—7.5 μ, 2% > 7.5 μ.

by boring or burrowing organisms or by shrinkage (FISCHER 1964) thus constituting transitions to fenestral pores.

4. Fractures

They may originate in diverse ways, such as by collapse related to solution, slumping, or various kinds of tectonic deformation including joints. Fracture-breccia pores form if rotation of the fragments has occurred.

The most important types are discussed in detail below.

Ad A1) Interparticle pores are well-known from sandstones. They form if particles are packed without matrix and are prevented from later cementation. They may also form if the cement or matrix is dissolved.

The difference between the narrowest and the widest pore openings is slight in this pore type as shown by the following example: in the tightest packing of spheres 1 mm in diameter, i. e. between 3 spheres, a sphere with a maximum diameter of 0.1554 mm would find a place whereas the loosest packing (i. e. in octahedral openings, between 6 spheres) would allow a sphere with a maximum diameter of 0.4142 mm to find a place. Generally, the difference in pore openings is larger because

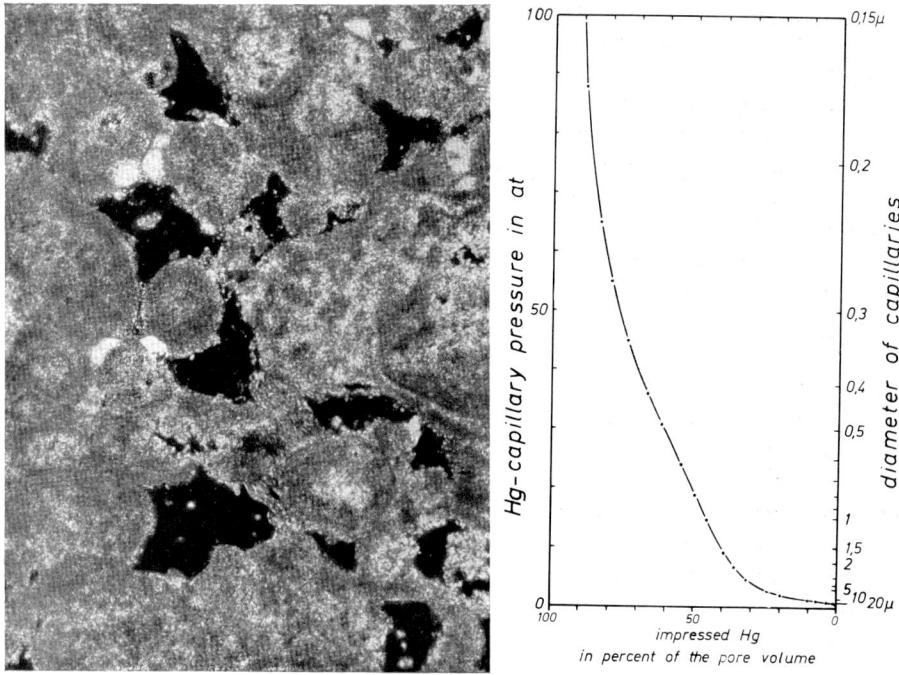

Fig. 5-95. Vugs.
Left: Dolomitic oncolite. The pores represent a transition between vugs and interparticle pores re-opened and enlarged by solution. Matrix: Dolomite and calcite. Upper Permian 2 - dolomite W. of Nienburg (Germany). Oncoids 0.05—0.5 mm in size, cryptocrystalline. Porosity 23.9%, air permeability 160 millidarcies (width 0.9 mm, crossed nicols, pores black).
Right: Capillary pressure curve of the same sample (for explanation see fig. 5-92). Capillary radii: 54% < 1 µ, 19% 1—5 µ, 27% > 5 µ.

natural packings are more bulky so that considerably larger pores can occur. The relatively good "sorting" of the pores also results from the capillary pressure curve (fig. 5-92) which can be translated into pore diameters under certain assumptions (v. ENGELHARDT 1960, STOUT 1964).

The relatively high proportion of large capillaries causes a high permeability-porosity ratio which is typical for particle limestones with primary porosity (see also fig. 5-96, a). For oil reservoir rocks, such pore types are especially favorable.

Ad A2) Intraparticle pores alone do not provide an effective porosity. In most cases they are combined with a system of very small intercrystalline pores connecting the larger intraparticle pores; such combinations of large and small pores (macro- and microporosity) are frequent. In fig. 5-93, the macropores (molds up to 0.7 mm in size) communicate via a capillary system of micropores about 0.5 µ in size (QUESTER 1964). The quotient of largest to smallest pores is therefore at least 1000. This corresponds to a very poor sorting of the pores, which is indicated also by the capillary pressure curve (fig. 5-93). The result is a low ratio of permeability/porosity (figs. 5-96, b and 97, b).

Ad A 4) **Intercrystalline pores** occur mainly in dolomites. In cryptocrystalline dolomites they can be recognized in thin section by euhedral crystals. Coarse-crystalline dolomites with intercrystalline pores reflect light in a similar manner to sugar crystals. Therefore they are named "sucrose dolomites".

The porosity of such rocks may exceed 30%. The ratio between largest and smallest pore diameter is similar to that found in interparticle pores. Correspondingly, the capillary pressure curve indicates a relatively low dispersion of the pore diameters (fig. 5-94).

Ad B 1) **Vugs** are irregular pores formed mostly by solution processes. With decreasing size they become intercrystalline pores. Vugs as well as intraparticle pores can form only if another pore system is available to provide communication. Therefore the capillary pressure curves of both pore types are similar (figs. 5-93 and 95). Both can be described as capillary systems in which the diameter is extremely variable.

Ad B 4) **Fracture pores** include unhealed or enlarged (by solution) tectonic fractures or joints. They generally make up less than 1—3% of the rock and in most cases considerably less (Murray 1960, Drummond 1964). With some exceptions (Adams 1953) this pore type does not contribute to the reservoir, but it is very important for the drainage towards oil wells. The relation between porosity and permeability is particularly favorable in this pore type: relatively high permeability corresponds to low porosity. For example, in gas-bearing Mesozoic dolomites of the Vienna basin Hawle et al. (1967) found a mean permeability between 1—5 millidarcies and only 1% porosity.

It is well-known that dolomites are more intensely broken by joints than limestones (v. Engelhardt 1960, Hawle et al. 1967). This was confirmed by Handin & Hager (1957) who demonstrated that dolomites are more brittle than limestones under increased unidirectional pressure.

Whereas the above classification, though genetical, is arranged according to descriptive aspects, the following shorter classification is based on genetic subdivisions. It is slightly modified from an exhibition prepared by Gulf Oil of Canada for the International Association of Sedimentologists' meeting in Heidelberg, 1971, and corresponds, except for A 3, to the system used in the German edition of this textbook.

A. Primary porosity
  1. interparticle pores
  2. intrafossil pores (including "growth framework")
  3. sheltered pores (including "fenestrae")

B. Secondary porosity
  1. intercrystalline pores (e. g. sucrose dolomite)
  2. leached (vuggy, moldic porosity)
  3. fractures

For other classifications of the pore types in reservoir rocks, see Robinson (1966) and Amer. Assoc. Petrol. Geol., Memoir 1.

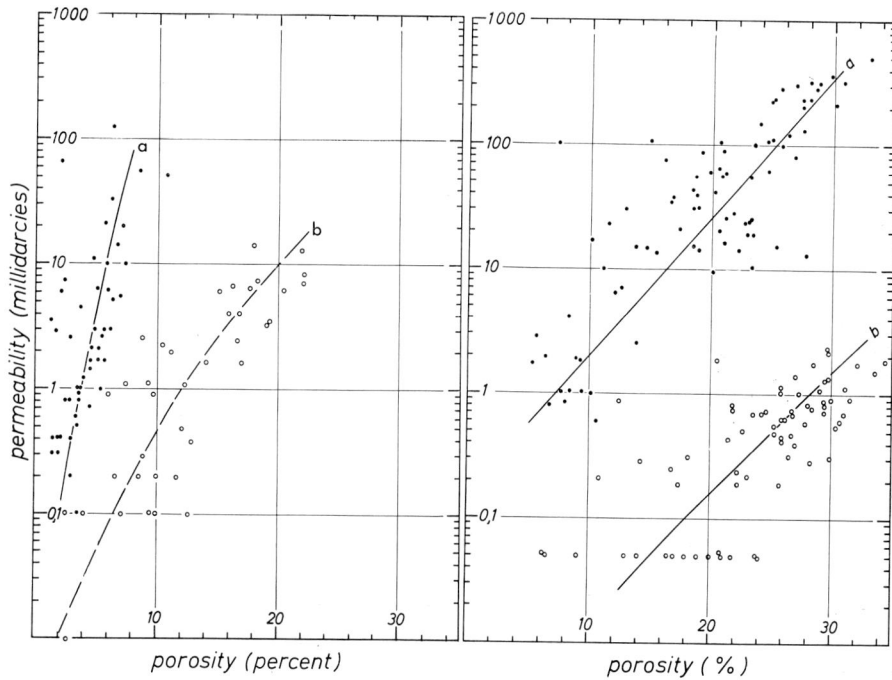

Fig. 5-96. Correlation between porosity and permeability in partly dolomitized limestones of the Paradox formation (Pennsylvanian), modified from CHOQUETTE & TRAUT (1963).
a) Interparticle pores (calcarenites and calcilutites of the green alga *Ivanovia*);
b) molds (dissolved skeletons of *Ivanovia*);

Fig. 5-97. Correlation between porosity and permeability in dolomites of the Upper Permian (Zechstein) 2.
a) Intercrystalline pores between rhombs, 0.02—0.05 mm in size (gas pool Rehden, BAUSCH & WIONTZEK 1961);
b) molds (dissolved oncoids) and intercrystalline pores between rhombs, 0.002—0.006 mm in size (see QUESTER 1964, and fig. 5-93).

5.422 Examples of depositional porosity

A considerable amount of data on distribution and properties of pores has been published (see LEVORSEN 1956), especially from oil and gas reservoirs. Generally each rock contains more than one pore type. The following examples describe rocks containing pores which were present immediately after deposition (interparticle and intraparticle pores) and which are categorized as primary or depositional porosity (MURRAY 1960).

Even if their shape is modified by pressure solution, by cementation or by other diagenetic events, their origin remains clearly visible.

Oölites with interparticle pores form oil reservoirs in the Jurassic (Smackover) Magnolia pool, Arkansas. In favorable zones, the porosity amounts to 20%, the permeability to 1000 millidarcies.

Biocalcarenites composed of dasyclads (Chlorophyta), intraclasts, oncoids, and stromatopores, form the Upper Jurassic Arab-D-formation in the giant oil pool Ghawar (Saudi-Arabia; POWERS 1962). The porosity at a depth of 2000 m is between

20 and 30%, the permeability varies from 100 to more than 1000 millidarcies. Slight degrees of cementation are frequent.

Coquinas consisting of lamellibranchs and gastropods are productive in the Lower Cretaceous of the small oil pool Dalum (Emsland area). These coquinas are preserved as aragonite in the oil-impregnated area whereas they are calcitized below the water table. This suggests an early oil migration (FÜCHTBAUER & GOLDSCHMIDT 1964). The porosity in the reservoir rock at a depth of about 1000 m varies between 15 and 35%; the permeability varies between 20 and more than 2000 millidarcies (MARSAL 1959). Only interparticle pores are developed. The cyrena shells are in part very closely packed; cementation is absent.

Numerous oil pools have been developed in rudistid and miliolid biostromes of Cretaceous age along a narrow belt in Mexico (No. 2 in table 5-4).

Much of the porosity is the result of fossil chambers interconnected by fissures. In the Mississippian of Utah and New Mexico, oil occurs in biostromes consisting mainly of skeletal fragments of the green alga *Ivanovia* (No. 5 and 6 in table 5-4, and fig. 5-96, a).

In bioherms, depositional porosity is developed if the matrix consists of biocalcarenite (skeletal particles) rather than calcilutite. This applies especially to reef bioherms, e. g. to the Upper Devonian of Swan Hills, Alberta (No. 61 in table 5-4, FONG 1960, THOMAS 1962). The deposits of the fore reef (and occasionally the back reef) are frequently more porous than the reef core: Within the Upper Devonian reef of the Redwater oil field (Alberta) the porosity is only 5%, compared with 10% in the adjacent fore reef (No. 58 in table 5-4). Similarly, pore space in the Kirkuk oil pool (Tertiary, Iraq), is largely restricted to the fore reef (DUNNINGTON 1958).

5.423 Origin of secondary porosity

5.423.1 Secondary porosity in limestones

Secondary pores are formed diagenetically by solution, chemical reactions, or mechanical fractures ("post-depositional porosity"; MURRAY 1960). They are at least as important as primary pores in the accumulation of oil and gas and therefore their origin is an important problem.

In many oil pools solution pores are connected with unconformities (MURRAY 1930, HOHLT 1948, LEVORSEN 1956, p. 242 ff.). Disconformities generally infer subaerial erosion, at which time carbon dioxide-rich meteoric waters percolated the permeable layers of the outcropping strata, and dissolved carbonate, thereby increasing the porosity. If the strata crop out down dip, meteoric water will migrate through the rocks and occasionally will flush oil and gas accumulations.

Because of the absence of overlying unconformities and for geometrical reasons, meteoric migration in many instances is not possible. Dissolution processes in such strata may be due to the upward movement of compaction water. This water, however, generally is saturated in $CaCO_3$, and therefore cannot remove considerable quantities of carbonate.

An additional mechanism is required for releasing the dissolved carbonate (HOHLT 1948).

Differential solubilities of various sedimentary components can but must not provide secondary porosity: for example carbonate is dissolved at stylolites without creating porosity; on the contrary, the dissolved carbonate precipitates in adjacent pores diminishing the porosity. In order to provide porosity, spots of higher solubility must be distributed properly in order for the rock to support itself around the secondary pores (HOWARD & DAVID 1936). Two cases may be distinguished genetically:

1. The sediment consists primarily of minerals with different solubilities. This would apply to a sediment consisting of calcite and aragonite, of which only the latter would dissolve. This has been claimed as an explanation for the high porosity of chalk (BØGGILD 1930, p. 244). In a similar manner, vugs are formed by dissolution of anhydrite or gypsum inclusions in carbonate rocks. Frequently, the dissolved carbonate of skeletal or non-skeletal grains is used in the formation of calcitic cement indurating the matrix between the grains, thus leaving molds, e. g. oömoldic porosity (fig. 5-80, see also FRIEDMAN 1964 and ROBINSON 1967).

2. At the time of dissolution the sediment may have been homogeneous mineralogically but not texturally. For example, the Upper Permian oncolites are frequently porous as a result of the dissolution of the cryptocrystalline oncoids; the supporting framework was provided by a microcrystalline matrix (figs. 5-93, 98). The dolomitic lutites of the same sequence are not porous. Similarly, in the Devonian crinoid limestones of the Andrews South Oil pool (W. Texas) the matrix-free regions of the rocks are tightly cemented whereas the regions with primary matrix display

Fig. 5-98. Solution molds (white) concentrated in layers. Dolomitic oncolite, Upper Permian 2, W. of Nienburg (Weser). Cryptocrystalline oncoids, which are not dissolved, or only partly so, are black (width 14 mm).

vuggy or interparticle porosity (up to 7%). This is the result of dissolution of the crypto-crystalline matrix (LUCIA 1962, LUCIA & MURRAY 1967).

Particle limestones (and dolomites) generally contain a more suitable distribution of different solubilities or crystal sizes than calcilutites, and therefore contain more extensive secondary porosity (CHOQUETTE & TRAUT 1963). Once such pores have formed, they can grow because their higher permeability is conducive for the migration of compaction water (fig. 5-98) (see also ch. 5.423.2).

Quiet-water bioherms in most cases contain only secondary porosity, e. g. the Pennsylvanian bioherm of the Scurry-Snyder Pool (W. Texas; no. 23 in table 5-4).

The observation that secondary porosity frequently is connected with structural domes (e. g. bioherms) has been explained by HOHLT (1948) through the accumulation of tectonic fractures in such domes. Perhaps more important is the fact that compaction water is directed strongly into such structures (v. ENGELHARDT 1967). Fractures may even drain the water and thus prevent the adjacent rock from secondary solution. This might be responsible for the observation that the secondary porosity in the Upper Permian dolomites in the Emsland area is considerably lower than in the Weser area in which the same dolomites are less fractured (FÜCHTBAUER 1964b).

Joints represent a type of secondary pores which are not connected primarily with solution processes. As mentioned above, they are rarely important as reservoir rocks, but are important as drainage systems. The large Persian Oil pools in the Tertiary Asmari limestone are connected with fracture zones due to folding. Outside such zones, the permeability is lower than 0.5 millidarcies, the porosity is between 2 and 15%. Similar conditions are found in the Mara pool (Venezuela): The Cretaceous limestone contains oil over the whole thickness of 550 m; because of the low permeability (< 0.1 millidarcy) it can be pumped only in areas of fracture zones (LEVORSEN 1956, p. 120). Fractures also play important roles in the Iraquian oil pools, Kirkuk (Tertiary) and Ain Zalah (Cretaceous) (DANIEL 1954). Joints or fissures or even

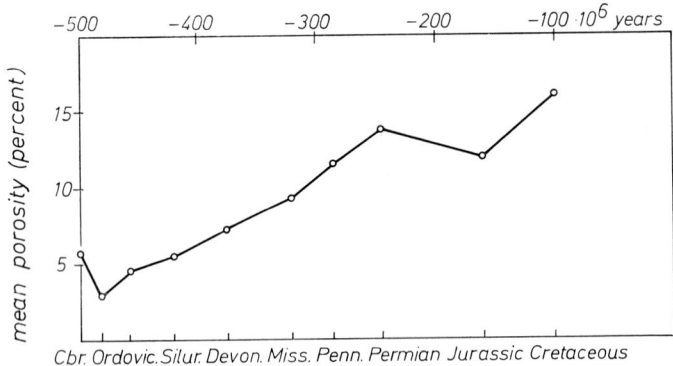

Fig. 5-99. Mean porosity of American carbonate rocks, mostly from oil pools (based on data from MANGER 1963).

bedding planes may be modified by solution processes into larger channels or even caverns (LEVORSEN l. c., p. 115—116).

The porosity-increasing processes contrast with the pore-destroying processes: cementation, crystal enlargement and stylolites. Due to these latter processes, the mean porosity generally decreases with increasing geological age (fig. 5-99).

5.423.2 Secondary porosity in dolomites

The formation of intercrystalline porosity during late diagenetic dolomitization is one of the most important examples of secondary porosity.

Porosity in carbonate rocks frequently increases with increasing dolomitization. For instance in S.W. Persia a porosity of 0—4% is found in rocks with 0—20% dolomite, 4—8% porosity in rocks with 10—35% dolomite, 8—12% porosity in rocks with 20—60% dolomite and more than 12% porosity in rocks with 30—75% dolomite (GASKELL, in CHILINGAR et al. 1967, table XI, see also MURRAY 1960).

As discussed in section 5.331.1, a late diagenetic mole-by-mole dolomitization of calcite theoretically can produce a porosity of 13%. In many cases, the porosity of such dolomites is more than 13% higher than the porosity of the adjacent limestones or dolomitic limestones (MURRAY 1960: 30% in the dolomites, 10% in the limestones; SCHMIDT 1961, 1965: 10—30% in the dolomites, less than 10% in the limestones; POWERS 1962: 20% in the dolomites, 5% in the limestones; LUCIA & MURRAY 1967: 27% in the dolomites, 7% in the limestones).

Several possibilities may explain these large differences:
a) more calcite was dissolved than dolomite was formed
b) preferential dolomitization of the more porous and permeable zones (e. g. Elkton, see below)
c) the dolomites became less cemented than the adjacent limestones, perhaps because an early accumulation of hydrocarbons prevented further diagenesis.

Mole-by-mole dolomitization may be responsible for the preference of dolomite to act as a replacing material rather than as a cement. Thus, an oölite may retain its interparticle porosity during dolomitization (MURRAY 1960). Dolomitization in particle limestones generally begins in the matrix. A potential volume contraction frequently is balanced by dissolution of the later dolomitized constituents of the fabric. This may explain why the skeletal material in many late diagenetic dolomites is dissolved (V. SCHMIDT 1965). An additional dissolution of calcite may increase this moldic porosity.

Early diagenetic dolomites occasionally retain a high porosity (SPENCER 1964) if no crystal enlargement has occurred. Such a preservation of the original crystal size is observed in rhythmical alternations of the crystal size in dolomitization cycles, such as the Paleocene dolomites of Libya which retain porosities of 17—38% (FÜCHTBAUER & GOLDSCHMIDT 1965). Sometimes an early oil accumulation may have hindered the crystal enlargement. Generally, early diagenetic dolomite lutites are poor in porosity as are many late diagenetic dolomites. This may be due to crystal enlargement combined with compaction, or to cementation at the expense of stylolitic pressure solution or of unstable grains.

Three examples of porous dolomites will be discussed:

In the Lima-Indiana pool (Ohio and Indiana), oil is produced from Ordovician porous dolomites which are surrounded by dense limestones. Presumably the dolomites formed late diagenetically from the dense limestones (LEVORSEN 1956).

In the "Elkton" carbonate cycle (Mississippian, Alberta) porosity and oil accumulation also are restricted to the dolomites. The late diagenetic coarse-crystalline dolomitization did not affect the dense, calcite-cemented biocalcarenites, whereas the uncemented biocalcarenites were replaced by sucrose dolomite with 28% porosity and 300 millidarcies permeability. The calcilutites were dolomitized early-diagenetically (in a lagoonal environment) into crypto- to microcrystalline, dense dolomites. Where these calcilutites were rich in skeletal material, the latter was dissolved during dolomitization, leaving a microcrystalline dolomite with up to 40% porosity but only 4 millidarcies permeability (G. E. THOMAS 1962).

Many of the large Canadian oil pools produce partly from Upper Devonian dolomitized reef bioherms. In the Leduc pool, the mean porosity, which is only 6.8%, consists of moldic and intercrystalline pores drained by fissures and joints (WARING & LAYER 1950). More information on oil-bearing carbonate rocks is published by HARBAUGH (1967).

"Rauhwackes" (cellular dolomites) have to be mentioned in connection with the secondary porosity in dolomites. These vuggy, mostly brownish rocks consist mainly of calcite, but evolved by calcitic healing of tectonic fractures in dolomites (MAYRHOFER 1955, LEINE 1968). The dolomite domains between the fractures were in part calcitized and in part dissolved by meteoric water. Thus, rauhwackes are mostly a phenomenon of surface weathering. The fracturing of the rigid dolomites can be caused by motions or dissolution of adjacent evaporites (STANTON 1966), which may provide Ca. But also a sedimentary breccia of early diagenetic dolomite embedded in limestone may be changed into a rauhwacke.

5.5 Environmental indicators

5.51 General aspects

Many indications of depositional environments have been mentioned in ch. 5.1—5.3; these are summarized here. References already cited in the above chapters are not quoted again. A quantitative investigation of modern carbonate sediments in the Persian Gulf including illustrations, synsedimentary diagenetic features, and relations to environment has recently been published by WAGNER & V. D. TOGT (1973). It illustrates a number of criteria useful in identification of depositional environments in ancient carbonate sediments.

The application of uniformitarian principles, that is relating modern and fossil environments, has several severe qualifications. One difficulty is presumably that shallow seas were considerably more extensive in the past than at present. During non-glacial times, when there were no ice-caps at the poles, seas were considerably higher than at present. The possible current and tidal systems in such epicontinental seas are not known.

Further difficulties arise from diagenetic alteration of primary sedimentary structures, e. g. the obliteration (by welding) of pelletal muds. On the other hand, fabrics caused by early diagenesis, such as cementation and replacements, may elucidate the depositional environment. For instance, meniscus and gravitational cements are indicative of environments of alternate wetting and drying, e. g. of the vadose and the supertidal zones. In view of the chemistry of interstitial water during early diagenesis, paramorphous calcitization of skeletal fragments originally consisting of aragonite is to be distinguished from replacements by dissolution and later cementation of the mold or from dissolution and subsequent compression.

Especially informative are regions in which the same environmental conditions persisted from the geological past to the present. This is true of the deep-sea, and the Deep-Sea-Drilling Project will bring to light important information. The same is true of the well Andros 1 (Bahamas) which penetrated environments similar to the present one down to the lower Cretaceous at a depth of more than 4000 m (SPENCER 1967). In the Persian Gulf WOOD & WOLFE (1969) found carbonate cycles of the sabkha type in the oil pool of Umm Shaif (Upper Jurassic).

Most limestones (except detrital limestones) generally form in environments of relatively high temperature and low terrigenous supply. From these conditions four possible depositional environments for carbonate rock result:

  a) shallow marine environments, large distances from the coast
     (e. g. Bahama Banks)
  b) adjacent to land with low morphology (e. g. Florida)
  c) adjacent to land with low precipitation (e. g. Persian Gulf)
  d) in seas and lakes whose drainage area is formed exclusively by carbonate rocks (see ch. 5.52).

The environmental criteria compiled in the next paragraphs should be used with reservation only (V. SCHMIDT 1965, WILSON 1968).

5.52 Continental environments

1. Rivers

With the possible exception of dead water basins, only detrital carbonate sediments accumulate in rivers (calcarenites, dolomite arenites and conglomerates as well as marls). Pure carbonate sediments except tufas are rarely deposited in alluvial environments.

2. Lakes

In addition to loss or extraction of CO_2 and evaporation, mixing of different water bodies may cause precipitation of carbonates in lakes: G. MÜLLER et al. (1972)

investigated highly alkaline (pH 9—10) Anatolian lakes (Turkey) into which enters river water rich in Ca- and HCO_3-ions and a pH of about 7.5. In the mixing zone aragonite precipitates, forming "whitings" due to the low solubility of calcium carbonate at elevated pH.

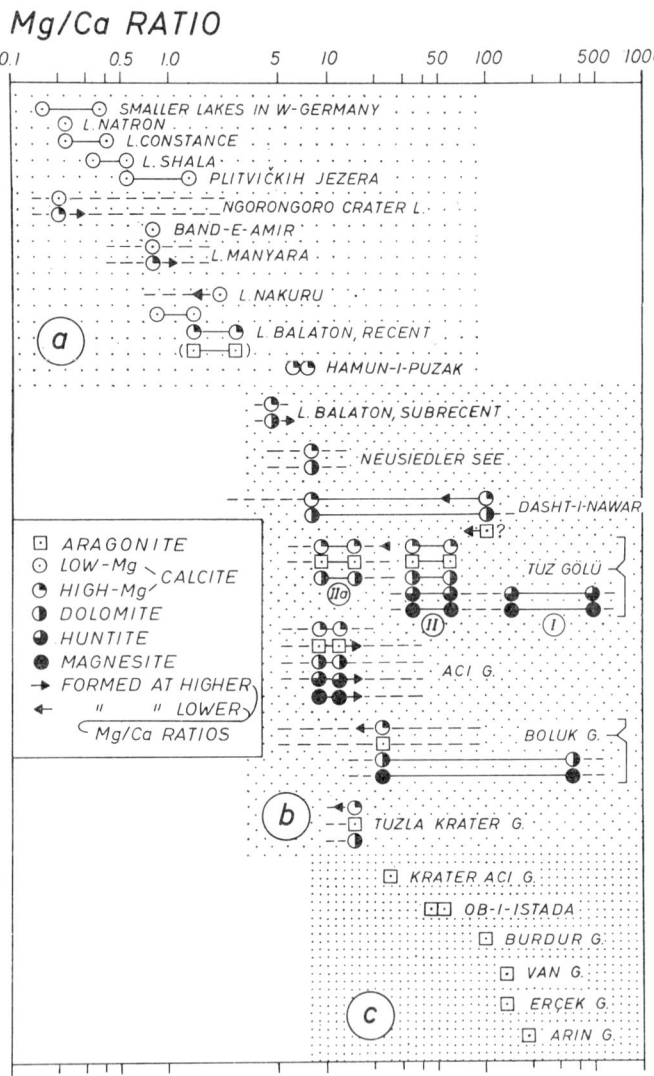

Fig. 5-100. Carbonate mineralogy in different lakes as related to the Mg/Ca-ratio of the lake or interstitial water (from G. MÜLLER et al. 1972).

The primary composition of the precipitates in the lakes investigated by MÜLLER et al. (1972) depends on the Mg/Ca-ratio at the time of formation (fig. 5-100):

Mg/Ca < 2 low-magnesian calcite
Mg/Ca 2—12   high-magnesian calcite
Mg/Ca > 12 aragonite.

Diagenetic carbonates [dolomite, huntite $CaMg_3(CO_3)_4$ and magnesite $MgCO_3$] form only in high-magnesian calcite sediments and at Mg/Ca-ratios between 7 and (15) for dolomite, and more than 40 for huntite and magnesite. Generally, dolomite acts as a precursor of huntite and magnesite formation. Fluctuations of the volume and chemistry of lake water produce different primary and diagenetic minerals in the same lake.

The fauna generally is poor in species but occasionally rich in number of individuals. Oncolites and stromatolites are found, but they are not exclusive to the limnic environment. Clotted limestones also are formed in lakes (e. g. in Lake Constance, G. MÜLLER 1966a).

3. Eolianites

This name was coined for cemented eolian dunes composed of calcareous grains, mostly biocalcarenites or oölites supplied from submarine environments. Generally, they show large-scale, relatively steep-dipping cross-bedding (LAND et al. 1967). They are cemented by granular calcite (cement B).

5.53 Transitional environments

1. Brackish water

Few species. Thin-shelled pelecypods and gastropods; ostracods.

2. Supratidal environment and emersions

"Primary" anhydrite forms only in dry, hot (40° C) supratidal environments. Early diagenetic dolomites, in part connected with algal mats, form in supratidal environments (BOSELLINI 1965, LAPORTE 1967, MATTER 1967). Vertical and horizontal shrinkage cracks (sheet cracks, e. g. FABRICIUS 1966, FARINACCI 1967) are frequent; during sporadic inundations such material may be brecciated. The minor displacement of fragments implies gentle transgression. Skeletal particles consisting of magnesian calcite release the magnesium without destruction of the fabric, whereas aragonite particles (e. g. oöids) are dissolved. A typical cement of this environment is a granular calcite cement which may be modified towards meniscus and gravitational fabric. Intraclasts of oölitic micrites may point to transgressions (p. 257).

Fenestrate ("birds-eye") structures are characteristic of this environment and of the intertidal areas: vugs which are elongated parallel to the bedding and mostly cemented by calcite or anhydrite in limestones and especially in dolomites (LAPORTE 1967, SHINN 1968a). These structures have been interpreted as shrinkage pores by

FISCHER (1965a). CLOUD (1962b) observed that such structures originated as gas bubbles in lagoons. WOOD & WOLFE (1969) also suggested a lagoonal formation for fenestrae and D'ARGENIO (1966) discussed the formation of such pores.

3. Intertidal flats

Blue-green algal mats and LLH-stromatolites indicate lakes or sheltered intertidal flats. Algal filaments and birds-eyes (decaying organic matter) are typical of this environment (P. HOFFMAN, pers. commun.), as are mud cracks and intraclast chips. Additional information is given by BLACK (1933), MONTY (1967) and recently by HARDIE & GINSBURG (in prepar.), see also ch. 5.152, C2.

5.54 Marine environments

WILSON (1969) recommended the following terminology of major marine carbonate environments:

a) Evaporite platform (including sabkhas and salinas)
b) Restricted marine platform (including lagoons and ponds)
c) Open marine platform (unwinnowed)
d) Tidal shelf of moderate depth (wide spread open marine seas)
e) Winnowed platform edge sand (tidal bars, offshore bars, eolianite dunes)
f) Organic reef of platform margin (includes boundstone framework reefs and entrapped lime mud accumulations)
g) Foreslope of carbonate platform (marine talus)
h) Basin margin (base of slope)
i) Basin (including starved basins)

A simpler subdivision is used on the following pages.

1. Shallow marine environment

Rich benthonic communities are typical of this environment, provided the water near the bottom is aerated. Characteristic are certain pelecypods, e. g. oysters, megalodonts, as well as gastropods, thick-shelled brachiopods, calcareous sponges, serpulae, regular sea urchins, irregular sea urchins (on soft bottoms), benthonic foraminifera, green algae, blue-green algae (oncoids and SH-stromatolites as isolated domes). Boring algae are restricted to water depths less than 50 m according to extensive investigations by NADSON (1927). Plants generally indicate a water depth of less than 150 m.

These depth limits and the definition of 'shallow marine' depend essentially on the light penetration which is controlled by the suspended sediment in the water column. HUGHES CLARKE & KEIJ (1973) used the following criteria in their study of the Persian Gulf:

"Shallow marine": recognized by the presence of one or other of the following groups: micro blue-green algae, responsible for most of the finely bored peripheries on skeletal grains. Calcareous, and other algae. Hermatypic corals (which include most reef-building corals). Large perforate Foraminifera and certain imperforate groups, e. g. Peneroplidae.

"Deep marine": areas with the sea bed below the well-lit zone and therefore lacking the elements listed above (unless they are transported or relic, which is often the case in the Persian Gulf).

Luxuriant growth of flora, and also that fauna directly associated with plants, reaches down to more than 40 m in the Pacific and Caribbean, but is limited to depths less than 20 m in the Persian Gulf.

Paleontological criteria enable further differentiation at least qualitatively of such environmental factors as light, temperature, salinity, substrate, and oxygen content (e. g. ZIEGLER 1967). Several organisms are restricted to normal marine environments: corals, echinoderms, red algae, brachiopods and cephalopods. Proximity of the coast is indicated by plant remains (wood), spores and pollen.

Needles of aragonite and steep rhombohedra of magnesian calcite are precipitated as a cement A on the walls of large pores. Cryptocrystalline magnesian calcite cement also forms in pores (ALEXANDERSSON 1969, TAYLOR & ILLING 1969), and can be mistaken for primary matrix. During subsequent diagenesis, cement A may be replaced by rhombohedra of calcite cement in the same orientation. The rest of the pores can be filled by coarse-crystalline cement B.

1a. Calm shallow sea (lagoonal in part)

A thin carpet of mucilagenous algae is an important characteristic of calm shallow sea bottoms (BATHURST 1967b). It prevents soft sediments from being eroded and may promote aggregation of calcilutites and calcarenites into clotted muds and into grapestones and crusts, respectively. Also pellets are protected and possibly gain time for induration and preservation. Calcilutites with intraclasts, according to FREEMAN (1962), form on bottoms stabilized by seaweed. Oncoids which also form in this environment need occasional motion in order to grow from all sides. Green algae prefer calm environments and depths of less than 10 m, but are found as deep as 50 m. The partial or complete micritization of particles seems to be characteristic of this environment, and is caused by small tubes formed by boring organisms (e. g. blue-green algae) and filled by cryptocrystalline cement. The micrite envelopes prove to be more resistant to diagenesis than most skeletal material.

Restricted environments can be characterized by certain associations, such as Miliolidae (foram.) and Dasycladaceae (green algae) in the Paleocene of Libya (BRÖNNIMANN, pers. commun.). Geologists from Shell Research (Rijswijk, The Netherlands) have, in past years, investigated Recent sediments of a protected bay south of Bahrain, Persian Gulf (HUGHES CLARKE & KEIJ 1973).

This shallow-water area is characterized as a 'restricted marine environment' by its high salinity (more than 50‰) and the strong temperature fluctuations (more than 33° C in summer and 15° C or less in winter). Sedimentation rates are low, and many skeletons are destroyed by boring algae and sponges. Corals, echinoids and red algae are absent. Pelecypod and oyster shells are smaller and less colored than in the adjacent Persian Gulf. Blue-green algae, *Acetabularia* (green alga), miliolids and peneroplids (imperforate foraminifera), *Lingula* and a limited number of species of gastropods *(Cerithium)*, ostracods, encrusting bryozoans and star fish are present (see also WAGNER & V. D. TOGT 1973, pp 136, 140).

Highly restricted environments (salinity exceeding ca. 70 ‰), which have been investigated in the Persian Gulf, are faunal deserts, containing little more than cyprideid ostracods. These lagoons are sites of gypsum precipitation.

1b. Agitated shallow sea (except reefs)

One of the best environmental indications of water agitation is the occurrence of clean (e. g. matrix-free) oölites. In modern seas they form only in water depths of less than 2 m, but only under special conditions (see ch. 5.16). Well-sorted biocalcarenites (in part with thin micrite envelopes) are characteristic of this environment, too. The occurrence of calcilutite, however, does not necessarily argue against agitation, although it may be hindered locally by a strong current.

2. Biostromes, bioherms and their environment

Frame-building organisms in life-position indicate bioherms or biostromes. The latter differ from the bioherms only by the quotient of lateral to vertical extension (ch. 5.18). Organisms, however, are not always preserved in fossil bioherms.

The abundance of colonial organisms, such as corals, within a sediment, generally indicates the proximity to a reef. Water depths and exposure to surge in reefs can be estimated from growth form e. g. of stromatopores and corals. In geologically young reefs (e. g. Pleistocene) the depth of formation can be compared with modern reefs. In the Caribbean Sea, the following zonation was described by GOREAU (1959) and MESOLELLA (1967):

        0— 5 m depth: *Acropora palmata*
        5—15 m depth: *Acropora cervicornis*
       $>$ 15 m depth: *Montastrea annularis,*
                        Siderastrea, Diploria (coral head zone).

The portion of frame-builders in a reef is generally between 5—40%. In Triassic reefs it is very low (ZANKL 1964), in modern reefs relatively high. Encrusting coralline (red) algae are an important, modern-day frame-builder.

Competition seems to be an important factor in bioherm growth. Examples are the alternations of bryozoans and stromatolitic crusts in the Upper Permian bioherms, and the quiet-water bioherms in Upper Jurassic rocks of Southern Germany, constructed of silica sponges and stromatolitic crusts. Such alternations may be a criterion of bioherms.

Reef debris with a matrix of fine-grained skeletal debris and/or with cement suggest fore-reef, whereas well-sorted cemented biocalcarenites are typical of back-reef environments, especially if they interfinger with calcilutites, dololutites or sulphates in a landward direction. Certain biota are typical of back-reef, especially green algae. The interpretation of any rock as reef (or as quiet-water bioherm) should be based also on the adjacent rocks. In ideal cases it is possible to trace a continuum of environments: from the lagoon through the back reef, reef, and fore-reef area down to the basin.

3. Deeper water

The absence of plants and the frequency of calcilutites are characteristic; both criteria, however, are not necessarily indicative of deep water since modern calcilutite also is deposited at shallow depths. Calcisiltites and fine pelletal calcarenites, usually showing small-scale grading or ripple cross-lamination, are relatively common. The same is true of dark colors although in places pink and red limestones occur. Moreover, very even planar 10—30 cm limestone beds interbedded horizontally with much thinner intercalated shales are typical in this environment (WILSON 1969).

Animal populations tend to decrease with water depth; this generally is due to decreasing water temperature, but also may be related to light intensity. Such a relation has been investigated for corals by WELLS (1957b); brachiopods, trilobites and trace fossils also have been discussed in the same GSA volume (memoir 67a). ZIEGLER (1967) published many observations on mollusks: In the Upper Jurassic shallow seas, ammonites, pelecypods and gastropods are nearly equal in quantity; among the ammonites, *Perisphinctes* and *Aspidoceras* prevailed. In deeper water, ammonites *(Phylloceras* and *Lytoceras)* are the prevailing mollusks. Pelagic pelecypods are *Posidonia, Pseudomonotis,* and *Halobia* from the Triassic and Jurassic, and *Buchiola* from the Devonian (WILSON 1969).

The foraminifera are the most suitable depth indicators of the microfossils. The ratio of planktonic to benthonic foraminifera increases in deeper water, as benthic populations decrease and planktonic foraminifera populations increase or remain constant towards the open sea (GRIMSDALE & VAN MORKHOVEN 1955, SMITH 1955). Precaution is necessary, however, because some shallow marine bottoms also are sparsely populated due to unfavorable conditions. In the Gulf of Mexico, the coastal area also is poor in species of benthonic foraminifera, due to fresh water influences (WALTON 1964); the size variability of several foraminifera is larger in this area than in the open sea (PHLEGER 1964). Other physical and chemical influences on the distribution of foraminifera are discussed by FUNNEL (1967). STEHLI & CREATH (1964) used the ratio planktonic : benthonic foraminifera to delineate barriers of oceanic circulation. For more recent periods (beginning with Upper Cretaceous) one can correlate species distributions of benthonic foraminifera with recent analogues (BURNABY 1961) for reasons of depth zonation.

Coccoliths are suitable for a depth estimation. They are common mainly in deeper water (more than 100 m) sediments, but they occur also in shallow-marine environments. Examples are the lagoonal, coccolith-rich carbonate muds of British Honduras which surround coralgal pinnacle reefs (SCHOLLE & KLING 1972) and the Solnhofen limestone in southern Germany.

With regard to early diagenesis, fibrous aragonite cement A is missing because the environment is not supersaturated with respect to aragonite. Aragonite particles can dissolve, as shown by FISCHER & GARRISON (1967) at a depth of 280—440 m near Barbados. This explains the absence of aragonite shells in chalk (HUDSON 1967). The "Aptycha" limestone of the Alpine Jurassic contains only the calcitic aptycha but not the aragonitic ammonites, suggesting deposition at a depth below aragonite saturation. Most deep-water limestones contain only calcitic skeletons of planktonic foraminifera (e. g. Globigerinae) and Calpionellae (= Tintinnidae). Micritization also can occur at great depths as shown by FISCHER & GARRISON (1967) in recent sediments at 280—440 m depth.

As FISCHER & GARRISON (1967) stated in their discussion, "the petrographic approach to bathymetry becomes feasible only, when the samples are viewed in their stratigraphic context: when sedimentary sequences representing tens of millions of years are uniformly composed of sediments of deep-water aspect, or contain only exceptional beds of shallow-water derived materials showing evidence of resedimentation. When rocks of deep-water aspect in turn show a logical temporal sequence of increasing or decreasing depth appearance (such as our sequence of mixed faunas, ammonite-dominated faunas, coccolith oozes, radiolarian ooze [= deepest] and back to coccolith-calpionellid ooze), the evidence for a "deep" rather than "shallow" origin becomes compelling."

4. Deep Sea

At greater depths carbonate sedimentation is absent. Instead, radiolaria are found. In the present oceans, aragonitic pteropods disappear at about 3500 m depth (FRIEDMAN 1965 a). Down to 4000 m, Globigerina mud is found. Beneath 5000 m depth calcium carbonate is absent. The exact compensation depth may have been different during earlier geologic periods (HUDSON 1967).

Limestone turbidites are found in the Tongue of the Ocean (Bahamas), at a depth of about 1000 m (RUSNAK & NESTEROFF 1964), and at 3600 m depth in the Gulf of Mexico (DAVIES 1968).

Since the ocean is only supersaturated in $CaCO_3$ in the upper few hundred meters (PETERSON 1966), the absence of $CaCO_3$ at depths exceeding 5000 m is not caused by a chemical inversion but by a change in the velocity of dissolution.

The globigerina mud is calcite-cemented in different places of the ocean bottom (MILLIMAN 1966, FISCHER & GARRISON 1967).

Additional information on pelagic sediments and sedimentary rocks, especially on carbonates and chert, is contained in the Special Publication No. 1 (HSÜ & JENKYNS, Ed.) of "Sedimentology" (1974).

6. Cyclic Sedimentation

6.1 Definitions and methods

Rocks containing two or more alternating sedimentary types are more frequent than petrographically uniform rock series. The sequence may be irregular (acyclic), but recurring sequences can be described as "rhythms" or "cycles" depending on whether only two or several rock types participate in the sequence, respectively (WANLESS & WELLER 1932, FIEGE 1952, BERSIER 1959).

The distinction between rhythm and cycle is minor, since many rhythms prove to be cycles when investigated more closely. These terms refer to either the phenomenon or the time of the alternation. When referring to the sediments involved, the terms "rhythmite" (SANDER 1936) and "cyclothem" (WANLESS & WELLER 1932) are used.

Rhythmites may involve alternation of two different components ("alternation rhythmite"), or a constant or slowly varying supply superposed by another supply varying over shorter periods ("superposition rhythmite").

Distinctly bedded, petrographically homogeneous rocks, e. g. limestones or sandstones, may be rhythmites: sedimentation alternating with minor unconformities at the upper surface would be "aOaO". If limestone beds are interrupted by marl layers, the rhythm could be expressed as "abab", but if the marl layer originated by dissolution of the limestone at the contact surface, this would then be a diagenetic rhythmite.

Cyclothems can be symmetrical (abcba) or asymmetrical (abcabc), to mention only two extreme cases. Cyclothems characterized by beds with a continual changing of grain size may be fining upward or fining downward (ALLEN 1965 a). Such sequences have been termed "positive" and "negative", respectively (LOMBARD 1956). BRINKMANN (1929) termed those cyclothems in which the most resistant unit is at the base "sole unit (Sohlbank) cyclothem", whereas those sequences in which the most resistant unit in weathering profiles is on the top were named "top unit (Dachbank) cyclothems".

In order to investigate the origin of such alternations, their properties must first be defined:

1. In studying r h y t h m i t e s one tries to determine (by chemical or microscopical investigation) whether alternation or superposition rhythmites are present (see above). Included fossils may indicate the depositional environment.

For further characterization of the rhythmite the thickness distribution of both rock types is defined: for each rock type the frequency of thickness intervals is plotted in cumulative frequency diagrams, the abscissa of which is the logarithm of thickness and the ordinate is the cumulative frequency on a probability scale. These curves are usually nearly linear (PETTIJOHN 1957, MCBRIDE 1962, SCOTT 1966). Median sorting or other statistical measures are easily determined.

2. In cyclical alternations involving several rock types the following three types are distinguished (DUFF & WALTON 1962):

a) the modal cycle is the most frequent grouping of rock types observed in the sequence.

b) the composite sequence combines all rock types present in the sequence into their most frequent order.

c) the theoretical cycle (model cycle) is an artificial order of the rock types present based on a theoretical model.

The modal cycle is derived from the frequency distribution of all occurring sequences, or it is the cycle from which the observed sequences deviate least (MARSAL 1967, p. 36; see also PEARN 1965).

The composite cycle is found by arranging the rock types into the most common sequence. For this purpose, a transition matrix indicates the pairs of the highest transition frequencies, which are then combined into a composite cycle. MARSAL (1967, p. 39) found the following composite cycle in a profile of the Pennsylvanian rocks of Illinois:

(marine calcareous) shale (= top)
coal
(under-) clay
sandy shale
sandstone (= bottom)

Deviations from the composite cycle or from the modal cycle can be used as an additional characteristic of cyclic sequences.

Fourier analysis has been used to investigate cyclic sequences in electrical well logs by PRESTON & HENDERSON (1965). Similar investigations are published by SEIBOLD & WIEGERT (1960) and by ANDERSON & KOOPMANS (1963). For additional statistical methods see DUFF et al. (1967).

3. The next investigation is a lateral comparison of the cycles. This requires correlation of the sections by paleontological or petrographic methods. If this is possible, the lateral thickness variation of single beds can be investigated. If an exact correlation is not possible, the regional thickness-variance may be investigated for each rock type independent of the cycles. Such investigations, although mostly not utilizing modern statistical techniques, have suggested several generalities in paralic carbonaceous cyclothems: coals and especially marine clays show little variation in horizontal thickness, whereas the sandstones may thicken and thin locally, forming "clastic wedges" (WANLESS 1965). Such findings may be described quantitatively by the variance as well as by mapping the trends (READ & DEAN 1967).

4. The next task involves a comparison of the superimposed cycles with respect to their thickness and petrography (see DUFF & WALTON 1962, and KRUMBEIN

1965). The thickness of cyclothems may change continually (JESSEN 1957) or one of the rock types may increase and decrease rhythmically in the vertical succession, pointing to larger rhythms or cycles.

5. Finally the absolute time span of each cycle should be investigated. This may be done by a comparison of the number of cycles with the duration of the formation (e. g. FISCHER 1965a), or by counting varves if they are assumed to be annual layers (RICHTER-BERNBURG 1957, VAN HOUTEN 1965, ANDERSON 1965).

Annual varves are composed of a thicker and a thinner (mostly darker) layer. The latter frequently contains organic substances either formed in the autumn or in the winter (BRADLEY 1957), or as in the Black Sea in the summer (SEIBOLD 1958). The thickness of such a pair is between 0.1 and 1 mm (ANDERSON 1965), except in evaporites (see below). The typical environment of varves is the interior of basins. Glacially layered clays ("Bänderton") have been used by DE GEER (1912) for chronological investigations. Similar varves have been found by HÜBNER (1965) in glacial clays of the Permo-Carboniferous in Africa. Further discussions of varves are given in BRINKMANN (1930), v. BUBNOFF (1947) and LOMBARD (1956).

In many cases it cannot be proven whether such varves are annual layers. For example in the varves of the Upper Permian evaporites in Northern Germany, RICHTER-BERNBURG (1955) found the following rhythmites:

a) limestones interbedded with silt layers, in thicknesses of about 0.05 mm ("stinking shales"),
b) anhydrites interbedded with dolomite layers, in thicknesses of about 0.5 mm,
c) rock salt interrupted by anhydrite layers in thicknesses of about 50 mm.

If these rhythms actually were annual layers, the entire late Permian would be less than 1 million years (even assuming longer periods of non-sedimentation between the different series) compared with the more probable figure of 25 million years.

In such evaporitic rhythms, the chemical layer, e. g. the limestone in the "stinking shale" (see above), may represent one year or part of one year only, whereas the silt layer could represent several years. Observations in the Dead Sea suggest such possibilities (NEEV & EMERY 1967, p. 107). Generally, fine-grained clastic rhythmites (e. g. silt and clay) are deposited more continuously and therefore are more suitable for development of annual rhythms than homogenous clastic sediments with intercalations of chemical precipitations which may or may not correspond to annual events. Other requisites of the formation of typical annual layers are quiet sedimentation, absence of burrowing, and slow deposition. Larger sequences of varved sediments can be preserved only if the rate of deposition is lower than the basin subsidence.

A rhythmicity corresponding to the 11 years sun-spot period has been used by RICHTER-BERNBURG (1955) as an argument in favor of annual layers (fig. 6-1).

A control of the absolute time span of geological units is possible in well-sections exhibiting a steady subsidence for a long time (PHILIPP 1961, FÜCHTBAUER 1964b).

By counting varves ANDERSON (1965) found cycles varying in length between one year ("1st order") and 1 million years ("7th order") which were partially superimposed on each other. The carbon cyclothems of the interior of the United States were grouped into the 6th order, to which the main glacial-interglacial cycles of the Pleistocene also belong (see 6.21a).

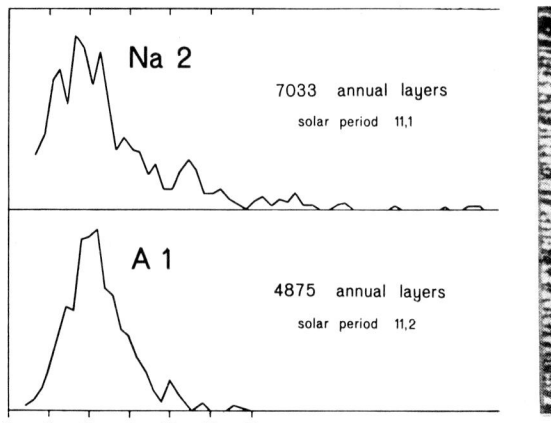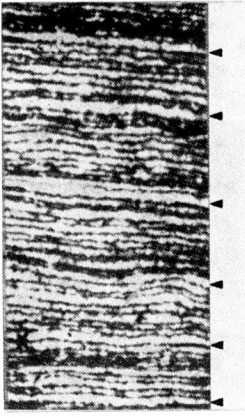

Fig. 6-1. Left: Frequency of varve anomalies in rock salt (Na 2) and anhydrite (A 1), Upper Permian, Germany. Abscissa = number of varves between two anomalies (from RICHTER-BERNBURG 1955). — Right: Dolomite varves (dark) in anhydrite (white), with thicker anhydrite layer (arrows) at a distance of about 11 varves (solar period?). Mean distance between dolomite layers is 0.4 mm (from RICHTER-BERNBURG 1960).

6.2 Possible causes

Alternations form principally in two ways:

a) by periodic influences (6.21—6.26)
b) by accumulating processes (6.27)

Periodic influences involve the quantity or composition of sediment, or cyclic influences on material with different sedimentation characteristics (e. g. dissolved and suspended matter). Cyclical influences of one or more factors may be reinforced or diminished by other factors.

Rhythmites may be accentuated or formed diagenetically. Examples are alternations of limestone and early diagenetic dolomite, nodular limestones, and sandstones with concretionary enrichments of calcite in distinct layers (DUFF et al. 1967). According to HALLAM (1964) many limestone-marl alternations may be at least partly diagenetic (see also EDER 1971).

In the following paragraphs the main causes of cyclic sedimentation are discussed briefly; in most cases more than one influence may have operated, but rarely are such influences proved. One of the few criteria aiding in interpretation is the lateral distribution of the cyclothem.

6.21 Fluctuations in climate

a) Temperature

Large temperature variations influence the ice balance of the earth markedly and cause eustatic sea level fluctuations. This mechanism has been suggested for

Pennsylvanian cycles by WANLESS & SHEPARD (1936), BEERBOWER (1961), and VAN SICLEN (1965). The coincidence of the frequency of such cyclic deposits with periods in which glaciations are known (e. g. Lower Carboniferous to Lower Permian) may be meaningful (DUFF et al. 1967). Temperature fluctuations also influence chemical and biological equilibria: a general warming leading to a transgression in the coastal area (due to eustatic sea level rise) may simultaneously cause calcareous sedimentation in the shallow sea offshore. Such "eustatic" cycles are very extensive in lateral directions.

b) Precipitation

Fluctuations in rainfall influence the intensity of erosion, in part indirectly, by modifications of vegetation (BEERBOWER 1961).

In lakes, fluctuating rainfall may influence the chemical equilibria and even the mechanical conditions: In the Azov Sea (U.S.S.R.), clay layers correspond to stages of high sea level caused by increased rainfall, whereas skeletal layers form by re-working during lowered sea level. Similar observations are reported from the Caspian Sea (KUKAL 1970).

The large-scale sandstone-shale rhythms in the Triassic of Southern Germany grading into a shale-limestone rhythm towards the basin (v. FREYBERG 1957) may have been caused by fluctuating rainfall. Lower precipitation may have resulted in clay sedimentation near the coast and limestone sedimentation in the basin.

c) Wind velocity

Fluctuations in wind velocity are responsible e. g. for the coarse storm sand layers in the shallow sea (REINECK & SINGH 1971) and for the tidal flat lamination of the carbonate muds in the Bahamas (HARDIE & GINSBURG, in prepar.) which are laminated by three storm layers per year as an average (see ch. 5.152, C 2). It is remarkable that big storms do not leave any thicker layers than do small storms. This is caused by the sticky *Schizothrix*-type algal mats acting like a fly-paper. The result is a uniform thickness of the millimeter laminations in modern and ancient carbonate tidal flats (HARDIE & GINSBURG l. c.).

d) Complex climate fluctuations

In most cases it is not possible to attribute cycles unequivocally to fluctuations in temperature or in precipitation, since these are interconnected. An example is the annual layers (varves), which are interpreted as complex climatic-biological rhythms (ch. 6.1).

e) Fluctuations in water chemistry

Such fluctuations may or may not have climatic origins. In epicontinental seas or in gulfs they may be caused by rhythmic formation of reducing conditions and lowered pH in the bottom water, combined with dissolution of carbonate (EXON 1971).

6.22 Large-scale tectonic movements of the ocean bottom

Such movements may have led occasionally to eustatic sea level movements causing "world-wide" transgressions and regression. The regressions at the end of the Jurassic period may have been caused by the downwarping of the deep-sea bottom, as reported by HEEZEN & FISCHER (1971). On the other hand, in times of increased sea floor spreading, which seem to coincide with orogenies, the areas of mid-oceanic ridges possibly were uplifted (I. G. JOHNSON 1971), resulting in "worldwide" eustatic transgressions, e. g. in late Cretaceous time (see also RONA 1973).

6.23 Different subsidence rates of basins
(STOUT 1931, EDWARDS & STUBBLEFIELD 1948, DELMER 1952)

In order for thick sequences of shallow marine sediments to accumulate, the sea bottom must subside over a long period. Since such movements in nature are never strictly constant, statistical fluctuations of the subsidence velocity are probable. During such irregular fluctuations, a threshold velocity should be surpassed at approximately equal intervals of time. Such thresholds may cause rhythmic transgressions of the sea e. g. into coal swamps. On the other hand, a threshold decrease of subsidence may periodically lead to the separation of a lagoon and to the precipitation of chemical sediments. Such events may have produced many marine cycles (see also LOMBARD 1956).

This type of cycle should have a smaller lateral extension than 6.21 (a).

6.24 Fluctuations in the rise of the hinterland
(HUDSON 1924, RUTTEN 1952, WELLER 1956)

Uplift increases the intensity of erosion and may cause a coarsening of the sediments in rivers and deltas. If this rise also involves the marine depositional environment, it can cause a regression. For example, the fluvial channel sands of the Pennsylvanian have eroded as much as 30 m into the underlying marine deposits (WANLESS & WELLER 1932). Such cycles generally have limited horizontal extent, e. g. thinning or wedging out may be observable within one outcrop.

6.25 Episodic shifting of rivers and deltas
(MOORE 1950, BERSIER 1959, SCHIEMENZ 1960)

Cyclothems formed by such events have the smallest horizontal extent. Examples are the "clastic wedges", sandstones and siltstones intercalated more or less locally in the sequence of coals, marine shales, and limestones, in Illinois; they may even subdivide coal measures (WANLESS 1965).

6.26 Biological factors

Rhythmic intercalations of siliceous limestone layers (consisting of radiolaria and *Calpionella*, embedded in a matrix of coccoliths) in Upper Jurassic shales of the Alps have been interpreted as the result of fluctuating organic productivity upon the deep-sea sediments (GARRISON & HONJO 1964).

Another example has been described by WAAGE (1965): Brackish-marine siltstones of late Cretaceous age contain rhythmic layers of fossiliferous calcareous concretions at 3—5 m intervals. Each of these layers contains only a few species of mollusks. A periodic mass-mortality is suggested as a probable cause. Such biological events may be typical of areas of changing salinity (lagoonal environment), thus suggesting chemical fluctuations which, in turn, were caused by climatic variations.

6.27 Cycles caused by steady processes

The steady accumulation of sediments in a delta near the shelf margin (e. g. Mississippi) periodically results in the breaking-off of the outer parts of the delta by slumping or turbidity currents. This separation may be caused either by the steadily increasing load or by periodic earthquakes (VAN DER KNAAP & EIJPE 1968). The suspensions may lead to cyclic deposits (e. g. turbidites) on the deep sea bottom.

Rhythms of marine clay and pebbly mudstones may point to glaciation on the adjacent continent, the pebbles being transported by icebergs (KUKAL 1970), if they are not caused by intermittent slumps.

6.3 Examples of rhythms and cycles

Many examples are described and compiled in the symposium on "Cyclic Sedimentation" (Kansas State Geol. Surv. Bull. 169, Vol. I and II, 1964; D. E. MERRIAM, editor) and in the textbook by DUFF, HALLAM & WALTON on "Cyclic Sedimentation" (Dev. in Sedimentology, 10, Elsevier 1967).

6.31 Deep sea

The turbidites as mentioned above are deposited from suspension currents in deep water (KUENEN & MIGLIORINI 1950); frequently they are developed as ideal fining-upwards cyclothems (BOUMA 1962):

(top)
e. pelagic clay (not a turbidite in origin)
d. upper layer with parallel laminations (silt and clay)
c. layer with current ripple laminations (sand and silt); convolute bedding frequent
b. lower layer with parallel laminations (sand)
a. graded layer (sand or gravel, fining upwards)
(bottom)

Layer "e", which contains planktonic foraminifera in younger formations, corresponds to the continual, very slow pelagic sedimentation (KUENEN 1964 d). This sedimentation is interrupted periodically by suspension currents, depositing in a short time layers "a"—"d", or parts of them.

The intervals between the suspension currents in a recent deep sea trough (Tongue of the Ocean, Bahamas) vary between 460 and 1000 years (Rusnak & Nesteroff 1964, Kuenen 1964c): in older rocks between 500 and 100,000 years (Kuenen 1953, Kimura 1967). Turbidites contain chiefly benthonic fossils, testifying to their shallow marine origin.

The turbidite activity along the present continents was higher in the Pleistocene than in the Holocene, due to the periodic coincidence of sea level with the upper margin of the continental slope during periods of glaciation. According to Carson (1971), Pleistocene deposits in the Cascadia basin, northeast Pacific Ocean, accumulated 10 times more rapidly than their Holocene equivalents.

An important difference between the layer "a" in turbidites and fining-upward cyclothems of different origin is the content in fines generally present also in the coarse basal layer of turbidites (fig. 6-2). Because the term "graded bedding" (Kuenen & Menard 1952) frequently is used for fining-upwards cycles in general, the above-mentioned special fining-upward cycle with fines included at the base may be termed "flysch-type grading", according to Kuenen (pers. commun.).

Bouma's divisions "a" and "b" belong to the upper flow regime, whereas divisions "c" and "d" belong to the lower flow regime, according to Harms & Fahnestock (1965).

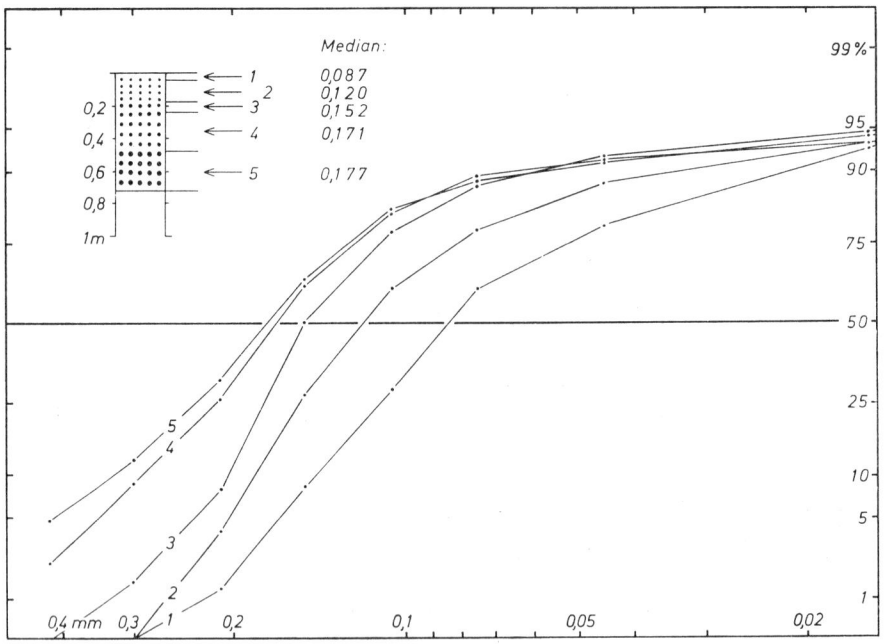

Fig. 6-2. Flysch-type graded bedding. Average grain size analyses from a modern turbidite of the submarine Hudson fan in 4810 m depth (adapted from Kuenen 1964c). Typically, coarse and fine sand layers contain nearly the same percentage of fines.

Examples of graded bedding have been published recently by ANGELUCCI et al. (1967) and MUTTI & RICCI LUCCHI (1972). The fining upward may have a concave (at first rapidly, then slowing down), or a convex form, depending on the source material (SCHEIDEGGER & POTTER 1965). For the same reason, layer "a" is not always graded in all turbidites. Layer "c" is topped frequently by current ripples with less than 20 cm wave length and less than 5 cm amplitude.

The thickness of turbidites is on the order of cm to m. Frequently grain size increases with increasing thickness. This rule, which was reported as early as 1936 and 1937 from the Lower Carboniferous turbidites of the Harz Mountains by FIEGE, has also been observed in other sandstones (PETTIJOHN 1957). In addition to sandstone turbidites, limestone turbidites also exist (sometimes with a "pre-phase": MEISCHNER 1964; RUSNAK & NESTEROFF 1964, FLÜGEL & PÖLSLER 1965, TEMMLER 1966). The extensive literature on turbidites has been compiled by KUENEN & HUMBERT (1964).

The proximal facies of suspension currents is characterized by conglomerates which are sometimes lenticular or show erosional channelling at their base. Occasionally this facies is referred to as "fluxoturbidite" (DŻUŁYŃSKI et al. 1959); it may extend into the slope facies consisting of shales interrupted by slumps (MUTTI & RICCI LUCCHI 1972).

The upper divisions of the BOUMA cycle (see above) are lacking and groove casts are more frequent than flute casts in the proximal facies (PAREA 1965). Erosion of the substratum of turbidites has been described also by SEILACHER (1962) and EINSELE (1963b).

In the current direction, the turbidites thin out and become finer grained, until they are nearly invisible in the pelagic clays (KULICK 1960a, PLESSMANN 1961, RADOMSKI 1961, BOUMA 1962, DŻUŁYŃSKI & WALTON 1965, MUTTI & RICCI LUCCHI 1972). Such waning turbidites have been termed "laminites" (LOMBARD 1963). Burrows and trails are the most common sole marks in this distal facies of turbidites (PAREA l. c.). A statistical criterion on proximity of a turbidite is the A B C index as defined by WALKER (1970): the percentage of turbidites beginning with the "a"-layer (graded) plus half of the percentage of turbidites beginning with the "b"-layer (parallel laminated sand). A simple statistical treatment of the thickness of the different turbidite divisions has been published by R. G. WALKER (1967).

Carbonate turbidites with a complete grain size spectrum and indistinct internal structures are described by ENGEL (1970).

6.32 Shallow sea

Purely detrital fining-upward cyclothems, about 30 m in thickness, have been described by NIEHOFF (1958) from the Devonian shallow sea of the Variscian geosyncline:
        shale (= top)
        "shaly graywacke" (siltstones, in part)
        graywacke and quartzite

THOREZ (1964) found fining-downward cyclothems in sandstones of the Upper Famenne in Belgium.

One of the most frequent cyclothems of shallow marine environments is composed of
limestone (= top)
marlstone
claystone

The limestone is frequently terminated by a hiatus which may or may not correspond to a short-termed regression of the sea.

KLÜPFEL (1917) found such cycles in the Liassic (Lower Jurassic) of Lorraine (France) and similar cycles, about 10 m in thickness, in the Dogger (Middle Jurassic; KLÜPFEL 1916). The surface of the limestones was bored, dissolved or reworked. In the Helvetian chains of the Northern Alps, CAROZZI (1953) defined depth zonations on the basis of planktonic and benthonic organisms and found megacycles of about 300 m thickness. He concluded that they indicated a steady shoaling. A similar result was obtained by SCHMASSMANN (1944) in the Dogger of the Swiss Jurassic, in which the cycles begin with marls overlain by calcilutites grading into fossiliferous limestones topped by an unconformity.

Shoaling due to increasing carbonate deposition may serve as an explanation of such cycles. Another, and perhaps more reasonable, explanation is a decreasing rate of subsidence. During this shoaling the basin was presumably subdivided into lagoonal areas; supply and transport changed drastically: the steady clay sedimentation ceased, and finally sedimentation stopped all together. At the same time, the subsidence of the basin continued or increased (see above). These processes caused a sudden deepening of the sea, and a new cycle began (see also BRÜCKNER 1953).

Whereas these cycles generally are asymmetric, the following 20 m-cycle from the lowermost Jurassic ("Rätolias") of the Calcareous Alps is symmetric (OTTE 1972):
Top

1. dark gray calcareous shales
2. greenish calcareous shales
3. reddish calcareous shales
4. yellowish marls interbedded with nodular calcilutites
5. medium gray, bedded calcilutites with corals
6. light gray calcilutite rich in skeletons including coral colonies
7. micritic and cemented pelletal limestone with intraclasts and skeletal grains
8. cemented particle limestone (intraclasts, pellets, skeletal grains, oncoids which frequently form coatings of the other grains)

7.—1. as above
Bottom

This cycle reflects the depth of deposition which becomes increasingly shallow from 1 to 8 and deepens from 8 to 1, according to the interpretation given by OTTE.

Added insight into certain cycles is possible if one considers the recent investigations on early diagenetical, submarine and meteoric cementation in the Persian Gulf (TAYLOR & ILLING 1969, SHINN 1969). Taking into account these modern analogs, PURSER (1969) offered a new interpretation of the thick (up to 100 m) cycles in the Dogger of France:

6.32 Shallow sea

"hardground" with burrows (= top)
pelletal limestone
biocalcarenite
clayey calcilutite

PURSER's interpretation was as follows: The sea shoaled for unknown reasons, reaching the intertidal environment during the biocalcarenite phase (gravitational cement A). The hardground corresponds to a time of slow (or even interrupted) sedimentation (hiatus?) connected with submarine cementation A. During this time, the sea deepened by continued subsidence and the next cycle began.

In the Cenomanian chalk of the Pas-de-Calais, 0.4 m cycles are found which begin with marl and grade through marly chalk into chalk ending with an indurated and partly bored surface. These "hardgrounds" also have been interpreted as a hiatus or an even longer term subaerial exposure (VOIGT 1959). However, paleontological evidence points to a considerable water depth at least immediately before and after the formation of these hardgrounds. Moreover, encrusting organisms, iron and manganese incrustations, glauconitization and phosphatization of the hardground surfaces point to non-deposition and cementation under the influence of sea-water as the most probable mechanism of hardground formation. This may have happened on top of submarine swells where tidal currents could remove the newly deposited fines and occasionally rework the surface layers of the hardground (BATHURST 1971b).

A rhythmical sequence of cemented and non-cemented calcarenites is found in the bottom of the modern shallow sea in the Southern Persian Gulf (DE GROOT 1969, SHINN 1969).

From the Lower Muschelkalk (Middle Triassic), SCHÜLLER (1967) described the following cycles, about 5 m in thickness (see also FIEGE 1938):

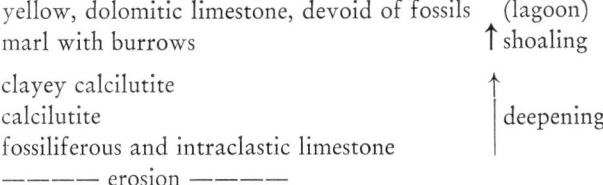

yellow, dolomitic limestone, devoid of fossils   (lagoon)
marl with burrows   ↑ shoaling

clayey calcilutite   ↑
calcilutite | deepening
fossiliferous and intraclastic limestone |
————— erosion —————

A rhythmical alternation of limestones and marlstones is explained best by warming and cooling, possibly connected with fluctuations of the fresh-water supply (KLÜPFEL 1916, BRÜCKNER 1951, 1953).

The Upper Jurassic limestone-marlstone sequence investigated by SEIBOLD (1952) is a superposition rhythmite; both calcite and clay phases are present in each layer. The boundaries of the deposits are sharp, suggesting rapid changes in the depositional environment. Since the supply of fine detritus (clay or marl) probably was rather sluggish, the rhythmicity can only be explained by a sporadic precipitation of carbonate (see also SANDER 1948), whereas a generally constant clay supply must be assumed. This model allows one to compare the rates of limestone and marl formation if diagenetic migration of calcite from the marlstone into the limestone is neglected. In the Jurassic example, this is possible since the calcite content of thin marl layers is very similar to that of thicker marl layers. In the case of diagenetic migrations, a stronger impoverishment in calcite of the thinner marl layers would be expected.

Excluding diagenetic migrations, the relative sedimentation rate of limestones and marl layers in the above model follows the equation:

$$S_l/S_m = P_m/P_l$$

with S_l and S_m = sedimentation rates of limestone and marl layers, respectively, and P_l and P_m = percentage of insoluble residue in limestone and marl layers, respectively. Because the rate of carbonate deposition is higher than that of the clay, the carbonate plays the active role in defining this rhythm.

Rhythmic alternations of siderite bands and shale in the Liassic of England have been explained by decreased and increased rates of sedimentation, respectively (SELLWOOD 1971).

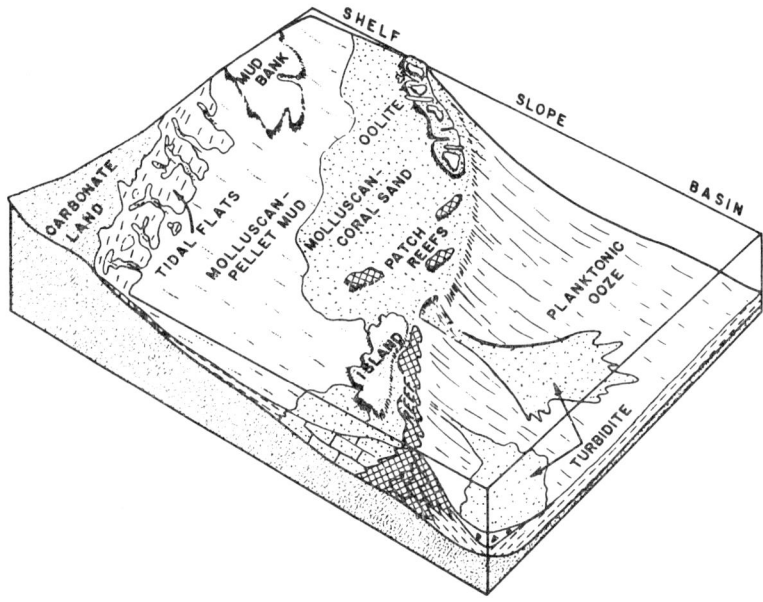

Fig. 6-3. Generalized framework of carbonate environments on the shelf, on the slope, and in the basin. The land and shelf sections are comparable to the Trucial Coast area (from A. H. COOGAN 1969, with permission).

Fig. 6-4. Lithological diagrams of idealized regressive carbonate cycles:
A = modern carbonate cycle;   C = Cretaceous reef cycle;
B = nonreef Cretaceous cycle;   D = Permian reef cycle;
                                   E = Devonian reef cycle.

In the left portion, the time sequence is vertical and the horizontal zonation is from left (= land) to right (= sea) (from A. H. COOGAN 1969, with permission).

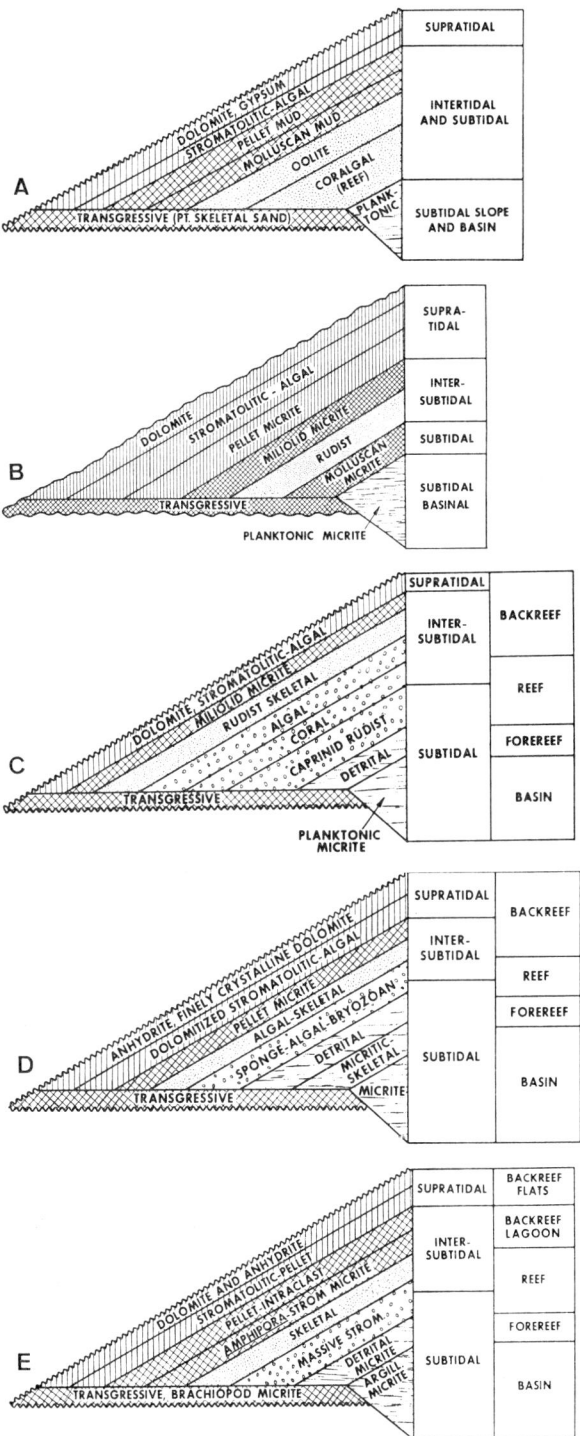

In the Callovian of England, BRINKMANN (1929) described cycles of less than 1 m thickness:

coquina (= top)
light clay
bituminous clay

Similar cycles, 1—2 m in thickness, have been described from the brackish-marine Wealden (Jurassic-Cretaceous transition) in N. Germany (MARTIN & WEILER 1963, FÜCHTBAUER & GOLDSCHMIDT 1964).

In the Middle Devonian of the Eifel Mountains, KREBS (1969) found the following cycles, 2.3—2.8 m in thickness:

5. calcilutite with vertical burrows (= top)
4. calcilutite with mm-laminations   } brackish
3. calcilutite with Leperditia (ostracod)   } ("regressive")
   and coquina layers

2. calcilutite, dark-gray, with globular stromatopores, } marine
   corals and marl seams   } ("transgressive")
1. black marlstone with corals

Pure calcitic cyclothems have been found in the Middle Triassic of the Northern Calcareous Alps (SARNTHEIN 1965):

arenitic calcilutite } 15—25 m, and { laminated dolomite
calcarenite   }   { calcarenite

In the Upper Triassic of the same area, SCHWARZACHER (1954) and FISCHER (1965a) found the following cyclothems:

limestone with megalodonts, several m in thickness
partly dolomitic stromatolite, about ½ m thick

The massive megalodont limestones show influences of early dissolution from the surface of the beds down as far as 10 m into the rock. If these "karst" formations were meteoric, emergences up to 10 m must have interrupted the subsidence of the sea bottom every 50.000 years; perhaps these sea level fluctuations were eustatic (FISCHER 1965a). The thickness of the sequence changes periodically every 5—8 cycles. These "megacycles" may be related to oscillations in the velocity of subsidence, assuming the period of the cycles was constant (FISCHER l. c.).

In similar cycles in the S. Alps the stromatolites were dolomitized early-diagenetically (BOSELLINI 1967). The stromatolites suggest an intertidal environment, the megalodont limestones and dolomites a subtidal one.

Typical regressive cycles are shown in fig. 6-4. The cycle of fig. 6-4 A is illustrated by the facies model of fig. 6-3. These cycles are similar to the "transition land-sea" cycles (next chapter).

6.33 Transition land — sea

In this environment, the most differentiated cyclothems are found. Perhaps the best studied are those in paralic coal areas. Two examples are compared in table 6-1. A more thorough discussion is found in the textbook by DUFF et al. (1967).

The Bavarian cycles contain more limestone and less sandstone than the Westphalian cycles. The Upper Carboniferous cycles in England, France, Belgium, and Westphalia are similar (see fig. 6-5).

Table 6-1. Suggested composite cycles of the Tertiary Lower Marine Molasse (Bavaria; STEPHAN 1965) and of the Upper Carboniferous molasse (Westphalia; JESSEN 1957).

Bavaria (ca. 15 m)	Westphalia (ca. 10 m)
8. coal measure (= top)	8. coal measure (= top)
7. underclay (dark green-gray marl, limnic)	7. underclay, limnic
6. green-gray marl, brackish	6. missing
5. calcareous marlstone, limnic	5. clay-silt-sand mixtures, limnic
4. sandy calcareous marlstone	4b. sandstone, nonmarine
	4a. clay-silt-sand mixtures
3. green-gray marlstone, brackish	3. claystone, limnic
2. green-gray marlstone, brackish-marine	2. claystone with spherosiderite layers and nodules, marine
1. brownish, bituminous marlstone, limnic (= bottom)	1. bituminous claystone with pyrite or spherosiderite (= bottom)

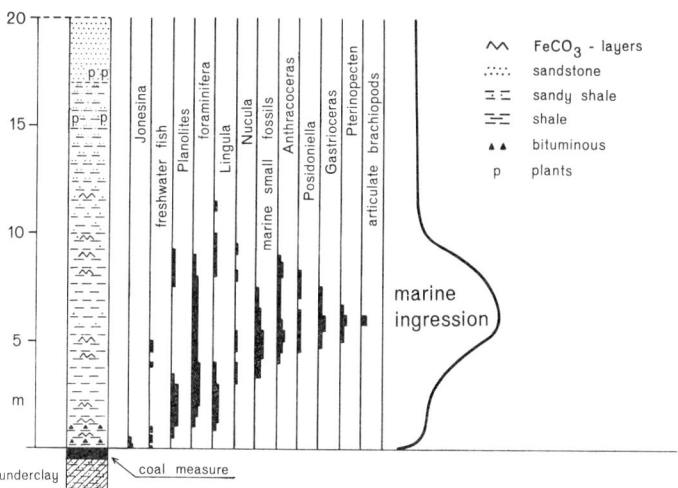

Fig. 6-5. Scheme of an "ideal" cyclothem in the "Sarnsbank" group, Upper Carboniferous (W. Germany). The distribution of fossils indicates the environment. Most important from a practical point of view is *Planolites ophthalmoides* JESSEN, a worm burrow indicative of brackish environment (from BÖGER 1964).

The classic cyclothems of the Pennsylvanian in the Eastern Interior Coal Basin (Illinois) frequently show an erosional unconformity beneath the sandstone (fig. 7-6). These carbon cycles are 1.5—29 m thick (WELLER 1931, WANLESS & WELLER 1932). The demarcation between such complicated cycles is more or less arbitrary. The deviations of the real cycles from the composite sequences mentioned above have not been investigated statistically. The composition of these cycles (fig. 6-6) is complicated, because

a) more than one genetic factor is involved;
b) these factors influence the different rock types in different ways.

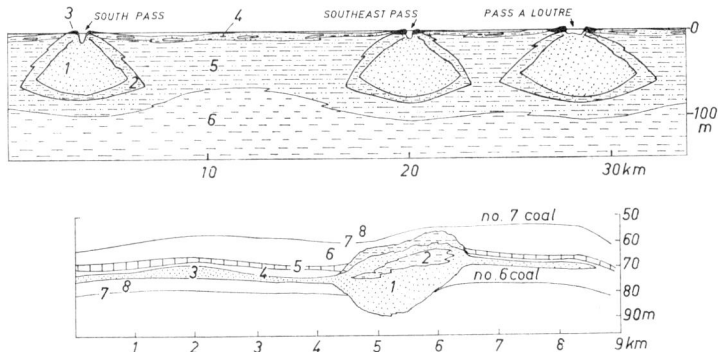

Fig. 6-6. Below: Channel and blanket facies of the Anvil Rock sandstone (Pennsylvanian) on the stable shelf of the Eastern Interior Coal Basin (Illinois); adapted from Hopkins (1958) by projection on a profile line perpendicular to the channel.

Legend: 1 = fluvial channel sand, median diameter ca. 0.3 mm; 2 = transitional facies (sandstone and siltstone); 3 = marine sheet sand, median diameter ca. 0.15 mm; 4 = siltstone and (top) claystone, carbonaceous; 5 = limestone, barren of fossils; 6 = claystone and siltstone alternations, carbonaceous, grading upwards into underclay; 7 = coal; 8 = shale with marine fossils, occasionally with a fossiliferous limestone bank.

Above: Cross section through three bar-fingers of the Mississippi River (in part schematic, scales three times smaller than Anvil Rock) from Fisk (1955, 1961).

Legend: 1 = fluvial channel sands (median diameter ca. 0.1 mm); 2 = fluvial transition zone (sands and silts); 3 = natural levee (silt); 4 = delta plain = topset (silt; relics of sheet sands. The latter are no longer formed at present, because the main part of the Mississippi sediments is lost towards the deep sea. In earlier times, however, the sheet sands were widespread over the shelf); 5 = marine delta-front silt (clayey) = foreset; 6 = pro-delta clays = bottomset.

Shepard & Lankford (1959) modified this picture to some degree. The vertical scale in both sections has been expanded 40 times.

The occurrence of a sand body tends to be more localized (for example a delta bar) than the formation of coal layers and marine clays, which are influenced by regional factors such as irregular subsidence of the sea bottom or eustatic sea level oscillations (Teichmüller 1955 a).

Several marine levels of the Upper Carboniferous which tentatively have been correlated from the USA to N.W. Europe, suggest eustatic sea level fluctuations (Duff & Walton 1962). For the majority of cycles, however, such correlations are not valid; instead, the number of cycles depends upon local subsidence as shown by a positive correlation of the bulk thickness with the number of cycles (Duff & Walton 1962, and Read & Dean 1967).

The distribution of number and thickness of single carbon cycles (as well as bulk thickness) forms a "basin pattern". This is opposite to the thickness of the coal measures which increases towards the "coast", i.e. away from the "basin center" ("wedge pattern", Read & Dean 1967). The latter may also be partly valid in sandstones which result from distributary meandering.

Surprisingly, in spite of these various influences, the composite sequence frequently is recognized. This can be explained only by the interaction of the different factors. For instance, sand wedges occasionally form the substrate of local coal measures (GRIBNITZ 1952).

The understanding of such processes, especially in the Pennsylvanian coal measures, is facilitated by a consideration of the Mississippi delta (LOWMAN 1949, FISK 1955, 1961): The strongest sedimentation occurs in the channels frequently formed by erosion and filled with sand. Similar is the case in the Pennsylvanian in Illinois (fig. 6-6). In both areas the grain size of the sands increases upwards (FISK 1960). Only during floods does sedimentation occur in the adjacent swamps (silt). Periodically every 400—1000 years the Mississippi shifts its whole delta system laterally into a deeper region which previously was marine. Subsidence in the abandoned delta continues but it is not compensated by sedimentation; the result is a marine trans-

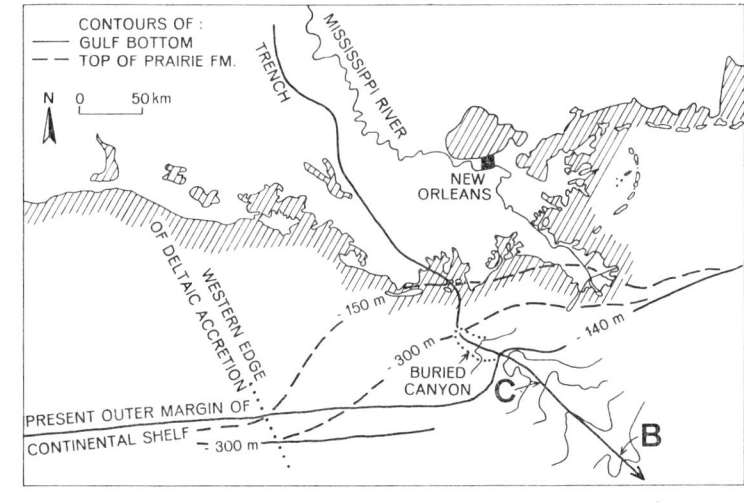

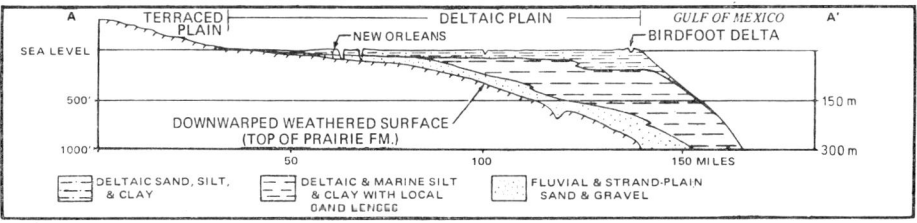

Fig. 6-7. Above: Mississippi delta region. The dotted 300 m-line is the outer margin of the deltaic accretion complex. Since the bottom layers within this complex consist of fluvial and strand plain deposits (s. section 'Below'), which were deposited near or above sea level, the present depth of 300 m (exceeding the lowest sea level by about 150 m) must be caused by downwarping. The arrow connecting the Mississippi trench with the canyon (C) and the bulge (B) is the axis of the delta complex, to which the present birdfoot delta is in a marginal position.

Below: Longitudinal section through the deltaic mass showing two transgressive units topped by the present regressive phase of delta protrusion (modified from GOULD 1970 and FISK & MCFARLAN 1955).

gression ("destructive phase"). Some additional subsidence could be provided by contemporary compaction of peat and fine-grained sediments (TEICHMÜLLER 1955b).

The developmental history of the Mississippi deltaic complex began with the last glacial stage when sea level stood about 140 m lower than at present. At that time, the Mississippi carried its load directly to the head of the present-day submarine canyon through which the sediments were flushed by turbidity currents and deposited in the Mississippi submarine bulge on the continental slope (fig. 6-7). With the subsequent rise of the sea to its present level, a thick onlapping (= transgressive) sequence (see fig. 6-8) was laid down, grading upward and seaward from fluvial and strand-plain sands and gravels (= topsets) to deltaic and marine silts and clays (= foresets) with local sand lenses.

After the decrease in rate of sea-level rise (4 to 6 thousand years ago), the Mississippi built its present birdfoot delta forward to form the thick offlapping wedge of deltaic sands, silts, and clays ("constructive phase", fig. 6-7). Only half of the thickness beneath the deltaic plain, however, can be attributed to the eustatic rise. The remaining thickness, up to 150 m, results from regional downwarp of the underlying Pleistocene Prairie surface beneath the deltaic mass (GOULD 1970) (fig. 6-7).

Fining upward as well as fining downward cycles occur in the transitional zone between land and sea.

Fining upward cycles include bays, tidal channels, and regressive tidal flat cycles.

Fining downward cycles include transgressive tidal flat cycles, (regressive) marine deltas and regressive coastal and barrier island cycles.

A rock sequence composed of several regressive cycles can form a transgressive or a regressive series depending on the relationship between subsidence and sedimentation (YAPAUDJIAN 1972).

Tidal flat carbonate cycles begin with bioturbated unlayered subtidal and intertidal sediments behind protecting barrier islands. These sediments are capped by (algal) laminites deposited by the most severe onshore storms (HARDIE & GINSBURG, in prepar., see also pp. 223, 356, 365). For clastic tidal cycles see p. 117.

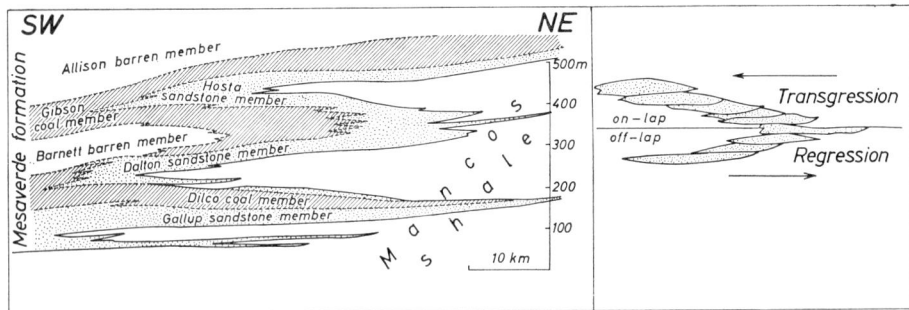

Fig. 6-8. Cross-section of the Upper Cretaceous coast line in New Mexico (adapted from MOORE 1949). The fluvial to littoral Mesa Verde formation, composed mainly of sandstones, interfingers with the dark marine Mancos shales, as a result of minor regressions and transgressions. On the right, the corresponding off-lap and on-lap of the sandstone bodies are shown schematically (see also HOLLENSHEAD & PRITCHARD 1961).

The sabkha-type regressive cycle (Persian Gulf) is caused by an alternation of subtidal, tidal and supratidal environments (fig. 6-4 A):

>dolomite + calcium sulphates (= top)
>stromatolites
>pelletal limestones
>oölites
>fossiliferous limestones

BULLOCK (1965) described the following cycles in the Jurassic of Utah (USA). These have been investigated geochemically by BORDINE (1965):

>calcilutites (= top)
>calcareous oolites
>red beds (sandy shales)
>calcarous clays (paper shales)

These cycles have been interpreted being due to changes in water depth.

6.34 Lacustrine cycles

An example is the Triassic of New Jersey (VAN HOUTEN 1965 a). In this area, "detrital cycles" about 5 m thick, and "chemical cycles" about 3 m thick have been found:

1. detrital cycles:
>silty marlstone (with shrinkage cracks) (= top)
>clayey marlstone
>black claystone

2. chemical cycles:
>analcite-claystone (brecciated, possibly by synaeresis) (= top)
>claystone and marlstone alternating (with synaeresis cracks)
>dolomitic marlstone (with shrinkage cracks which are in part sun cracks, in part synaeresis cracks)

The predominance of one or the other cycle type forms megacycles. Whereas the above-mentioned cycles represent about 21,000 years each, corresponding possibly to precession movement of the earth axis (VAN HOUTEN), the megacycles of 20—30 m correspond to ~ 100,000 years and those of ~ 100 m correspond to ~ 500,000 years. These cycles have been explained climatically. The sediments were probably deposited in a lake which had no outlet during chemical sedimentation.

Regressive sequences which also may be caused by climatic variations are reported from the Great Salt Lake (cf. KUKAL 1970):

>soils (= top)
>calcitic and aragonitic crusts
>oölites
>skeletal calcareous sediments
>calcareous dolomitic clay
>calcareous clay
>clay

6.35 Fluvial cycles

BERSIER (1959) described fining-upward cycles in the fluvial Lower Miocene Molasse (north of the Alps), 2—5 m in thickness:

sandy marlstone (= top)
clayey calcareous sandstone
calcareous sandstone

He interpreted these cycles as being due to lateral shifting of rivers: the calcareous sandstones represent the channel facies, the marlstones correspond to the flood facies deposited adjacent to the channels. The same origin has been postulated by SCHIEMENZ (1960) who investigated the same formation closer to the Alps; for this reason his cycles are thicker (10—40 m, VOLLMAYR 1958) and coarser:

sandy marlstone (= top)
calcareous sandstone
conglomerate, fining upward

The origin of these cycles, according to the above-mentioned authors, can be explained as follows: Initially the rivers eroded a channel. The shifting or splitting of the stream channel decreased the current velocity, and thus gravel and sand were deposited in marginal portions of the river bed. Similarly, during the continuous shifting of the river channel, the grain size decreased upward in the bed. This differs from the graded bedding of turbidites, since there is a deficiency of fine matrix in the basal layer. Finally, the area became dry and accumulated sediments only during floods, when fine-grained material was deposited in the flood plain.

Similarly constructed, but finer grained, cyclothems (5—9 m in thickness) have been described form the Old Red Sandstone (Devonian, England) by ALLEN (1964 b).

Fining-upward, clearly non-marine, coal-bearing cycles are reported from the Gondwana system (India) by CASSHYAP (1970):

coal (= top)
siltstone
fine-grained sandstone
medium-grained sandstone
coarse-grained sandstone

6.4 Cycles of higher order (megacycles)

6.41 Symmetrical megacycles

In the previous paragraphs, facies oscillations which modulate the cycles have been mentioned repeatedly. Such "megacycles" or "cycles of higher order" may have small dimensions. An example has been published by DE RAAF et al. (1965, fig. 12): In the English Upper Carboniferous, which is composed of an alternation of small sand, silt, and clay layers, 50 cm-megacycles are defined by the predominance of one of these rock types.

On the other hand, JESSEN et al. (1952) described megacycles 400—500 m in thickness from the Westphalian Upper Carboniferous. These megacycles are characterized in the lower part by a symmetrical increase and decrease of the sandstone percentage and in the upper part by a varying frequency of marine transgressions. FIEGE (1960) found an alternation of similar magnitude between series rich and poor in coal measures.

Eight megacycles, 600 to nearly 4000 m thick and characterized by nearly symmetrical alternations of marine clays with beach and fluvial sands, are developed in the Tertiary and Pleistocene of the Gulf Coast (LOWMAN 1949). These are transgression-regression cycles connected with the coastal zone (see also fig. 6-8).

The steeper the coastal plain and the sea bottom dip, the smaller is the area covered by such cycles. In the case of steep coastal plains the facies zones are mostly arranged in strips whereas they may have a very irregular distribution in the case of flat morphology. In the latter case hydrological factors may influence the sedimentation (LAPORTE & IMBRIE 1965).

Shallow marine and coastal sediments formed during transgression and regression may be removed by erosion. Accordingly, transgressions and regressions have been classified as "depositional" or "erosional" (CURRAY 1964). Erosional surfaces interrupting the sediments of the Holocene transgression in the Gulf of Paria were produced by temporary regressions, according to VAN ANDEL & SACHS (1964).

6.42 Asymmetrical megacycles

It is remarkable that fining-downward cycles prevail in most asymmetrical detrital cycles of higher order; several examples are listed in the next paragraphs.

DE RAAF and others (1965) have described clastic fining-downward cycles of about 70 m thickness from the Lower Westphalian of England. The depositional environment changed from deep water (with isolated turbidites) to coastal plains.

Table 6-2. Comparison of cycles and megacycles from the Rheinisches Schiefergebirge (FIEGE 1936, 1937, KULICK 1960a).

	Lower Carboniferous ("Kulm")	Upper Carboniferous ("Namurian")
fining-upward cycle	a few meters	ca. 10 cm
fining-downward megacycles	10—30 m	4—5 m

The principle "cycles fining-upward, megacycles fining-downward" is realized in deep water (Kulm) as well as in shallow water sediments (Namurian) (table 6-2). WENTZLAU (1960) described fining-downward megacycles of 300 m thickness from the Upper Devonian.

According to WOLBURG (1961), the Buntsandstein of N. Germany is divided into 6 more or less distinct fining-downward megacycles (silt-sand), each on the order of 100 m thick.

This type of megacycle also is frequent in larger sandstone bodies embedded in shale sequences, e. g. the Valangian sandstone in the Emsland area, Germany (ca. 30 m in thickness).

6.4 Cycles of higher order (megacycles)

In many cases such fining-downward cycles may correspond to deltaic regressive or even transgressive cycles (OOMKENS 1970).

The Tertiary Molasse north of the Alps can be subdivided into 7 fining-downward megacycles, the thickness of which decreases from the margin of the Alps towards the north from 200—1100 m to 0—150 m, respectively. These megacycles frequently are topped by a transgression or a sudden increase in marine character, and can be correlated with tectonic movements and increase of the slope in the Alps as well as in the Molasse basin (LEMCKE 1962, FÜCHTBAUER 1967a).

References and Authors' Index

The numbers in square brackets at right indicate pages on which the publication is cited.

ABBOTT, B. M. (1973): Terminology of stromatoporoid shapes. — J. Paleont., 47, 805—806. [234]
ACKERMANN, E. (1955): Zur Unterscheidung glazialer und postglazialer Fließerden. — Geol. Rdsch., 43, 328—341. [82]
ADAM, K. D. (1950): Über Windtransport in Wüstengebieten I. Beobachtungen in Nordost-Afrika. — N. Jb. Geol. Paläont. Mh., 289—294. [103]
ADAMS, J. E. (1953): Non-reef limestone reservoirs. — Bull. Amer. Assoc. Petrol. Geol., 37, 2566—2569. [346]
ADAMS, J. E. & FRENZEL, H. N. (1950): Capitan barrier reef, Texas and New Mexico. — J. Geol., 58, 289—312. [272]
ADAMS, J. E. & RHODES, M. L. (1960): Dolomitization by seepage refluxion. — Bull. Amer. Assoc. Petrol. Geol., 44, 1912—1920. [308]
ADAMS, T. D., KHALILI, M. & KHOSROVI SAID, A. (1967): Stratigraphic significance of some oligosteginid assemblages from Lurestan Province, northwest Iran. — Micropaleontology, 13, 55—67. [227]
ADAMS, W. L. (1964): Diagenetic aspects of Lower Morrowan, Pennsylvanian, sandstones, Northwestern Oklahoma. — Bull. Amer. Assoc. Petrol. Geol., 48, 1568—1580. [149]
AGER, D. V. (1963): Principles of paleoecology. — McGraw Hill, 370 pp. [264]
AHRENS, W., STADLER, G. & WERNER, H. (1960): Beitrag zur Genese der Westerwälder Tertiärquarzite. — Z. dt. geol. Ges., 112, 253—258. [141]
AITKEN, J. D. (1967): Classification and environmental significance of cryptalgal limestones and dolomites, with illustrations from the Cambrian and Ordovician of Southwestern Alberta. — J. Sediment. Petrol., 37, 1163—1178. [220, 223]
ALDERMAN, A. R. (1965): Dolomitic sediments and their environment in the south-east of South Australia. — Geochim. Cosmochim. Acta, 29, 1355—1365. [316]
ALDERMAN, A. R. & VON DER BORCH, C. C. (1963): A dolomite reaction series. — Nature, 198, No. 4879, 465—466. [316]
ALDERMAN, A. R. & SKINNER, H. C. W. (1957): Dolomite sedimentation in the south east of South Australia.. — Amer. J. Sci., 255, 561—567. [195, 316]
ALEXANDERSSON, T. (1969): Recent marine high-Mg calcite lithification in the Mediterranean. — Sedimentology, 12, 47—61.   [276, 282, 284, 285, 290, 294, 322, 340, 357]
ALLEN, J. R. L. (1962): Petrology, origin and deposition of the highest Lower Old Red Sandstone of Shropshire, England. — J. Sediment. Petrol., 32, 657—697.   [64, 73, 136]
— (1963): The classification of cross-stratified units, with notes on their origin. — Sedimentology, 2, 93—114. [76]
— (1964a): Primary current lineation in the Lower Old Red Sandstone (Devonian), Anglo-Welsh basin. — Sedimentology, 3, 89—108. [90]
— (1964b): Studies in fluviatile sedimentation: Six cyclothems from the Lower Old Red Sandstone, Anglo-Welsh basin. — Sedimentology, 3, 163—198. [90, 380]
— (1965a): Fining upwards cycles in alluvial successions. — Liverpool and Manchester Geol. J., 4, 229—246. [361]
— (1965b): A review of the origin and characteristics of recent alluvial sediments. — Sedimentology, 5, 89—191. [103]
— (1966): On bed forms and palaeocurrents. — Sedimentology, 6, 153—190. [90]
— (1968a): Current ripples, their relation to patterns of water and sediment motion. — North-Holland Publ. Co. Amsterdam, 433 pp. [76]

— (1968 b): The nature and origin of bed-form hierarchies. — Sedimentology, 10, 161—182.
[73, 90, 101]
— (1970 a): Physical processes of sedimentation. An introduction. — G. Allen and Unwin Ltd. London, 248 pp. [77, 78]
— (1970 b): A quantitative model of climbing ripples and their crosslaminated deposits. — Sedimentology, 14, 5—26. [78]
ALLEN, J. R. L. & BANKS, N. L. (1972): An interpretation and analysis of recumbent-folded deformed cross-bedding. — Sedimentology, 19, 257—283. [78]
ALLEN, J. R. L. & WELLS, J. W. (1962): Holocene coral banks and subsidence in the Niger Delta. — J. Geol., 70, 381—397. [271]
ALLEN, R. C., GAVISH, ELIEZER, FRIEDMAN, G. M. & SANDERS, J. E. (1969): Aragonite-cemented sandstone from outer continental shelf off Delaware Bay: Submarine lithification mechanism yields product resembling beachrock. — J. Sediment. Petrol., 39, 136—149. [288]
AMIEL, A. J., FRIEDMAN, G. M. & MILLER, D. S. (1971): Lithification of modern Red Sea reefs (Abstr.). — Progr. VIII Internat. Sediment. Congr., Heidelberg, 3.
[198, 276, 284, 285]
ANDERSON, C. A. (1933): The Tuscan Formation of northern California, with a discussion of the origin of volcanic breccia. — California Univ. Publ., Geol. Sci., 23, 215—276.
[180]
ANDERSON, E. M. (1951): The dynamics of faulting and dyke deformation with applications to Britain. — Oliver and Boyd, Edinburgh, 206 pp. [84]
ANDERSON, R. Y. (1965): Varve calibration of stratification. — In: D. F. MERRIAM (ed.), „Symposium on cyclic sedimentation". — Kansas Geol. Surv. Bull., 169, 1—20. [363]
ANDERSON, R. Y. & KOOPMANS, L. H. (1963): Harmonic analyses of varve time series. — J. Geophys. Res., 68, 877—893. [362]
ANDERSON, T. B. (1965): Turbidites whose sole markings show more than one trend — a further interpretation. — J. Geol., 73, 812—814. [96]
ANDERSON, T. F., BENDER, M. L. & BROECKER, W. S. (1973): Surface areas of biogenic carbonates and their relation to fossil ultrastructure and diagenesis. — J. Sediment. Petrol., 43, 471—477. [313]
ANDRESEN, M. J. (1961): Geology and petrology of the Trivoli sandstone in the Illinois basin. — Ill. State Geol. Surv. Circ., 316, 31 pp. [105]
ANDRICHUK, J. M. (1958): Stratigraphy and facies analysis of Upper Devonian reefs in Leduc, Stettler, and Redwater areas, Alberta. — Bull. Amer. Assoc. Petrol. Geol., 42, 1—93. [273]
ANGELUCCI, A., DE ROSA, E., FIERRO, G., GNACCOLINI, M., LA MONICA, G. B., MARTINIS, B., PAREA, G. C., PESCATORE, T., RIZZINI, A. & WEZEL, F. C. (1967): Sedimentological characteristics of some italian turbidites. — Geol. Romana, VI, 345—420. [128, 136]
ARAMAKI, S. & YAMASAKI, M. (1963): Pyroclastic flows in Japan. — Bull. Volc., 26, 89—100.
[172, 175]
ARISTOTLE [160]
ARMSTRONG, H. E. et al. (1904): The Atoll of Funafuti. — Roy. Soc. London, 428 pp. [266]
ATHY, L. F. (1930): Density, porosity and compaction of sedimentary rocks. — Bull. Amer. Assoc. Petrol. Geol., 14, 1—24. [4]
ATWOOD, D. K. & BUBB, J. N. (1970): Distribution of dolomite in a tidal flat environment, Sugarloaf Key, Florida. — J. Geol., 78, 499—505. [315]
AUBOUIN, J. (1964): Réflexions sur le faciès «ammonitico rosso». — Bull. Soc. géol. France (7), VI, 475—501. [125]
— (1965): Geosynclines. — Elsevier, Amsterdam, 335 pp. [125, 127, 129, 303]
AXELSSON, V. (1967): The Laiture Delta. A study of deltaic morphology and processes. — Geograf. Ann., Stockholm, 49, Ser. A, 1—127. [111]

BAAK, J. A. (1936): Regional petrology of the Southern North Sea. — Diss. Leiden, 114 pp.
[43]
BAARS, D. L. (1961): Permian blanket sandstones of Colorado Plateau. — In: J. A. PETERSON, J. C. OSMOND (ed.), Geometry of sandstone bodies.. — Amer. Assoc. Petrol. Geol., 179—207. [99]
— (1963): Petrology of carbonate rocks. — In: R. O. BASS (ed.), Shelf carbonates of the Paradox Basin. — Four Corners Geol. Soc. 4. Field Conf., 101—129. [271]

BACHMANN, G. H. (1973): Die karbonatischen Bestandteile des Oberen Muschelkalkes (Mittlere Trias) in Südwest-Deutschland und ihre Diagenese. — Arb. Inst. Geol. Paläont. Univ. Stuttgart, N. F. 68, 99 pp. [198, 296]
BACHMAYER, F. & FLÜGEL, E. (1961): Die Hydrozoen aus dem Oberjura von Ernstbrunn (Niederösterreich) und Stramberg (ČSR). — Palaeontographica A, 116, 122—143. [235]
BAGNOLD, R. A. (1941): The physics of blown sand and desert dunes. — Methuen, London, 265 pp (reprinted 1965). [60, 89]
— (1953): The surface movement of blown sand in relation to meteorology. — In: Desert Research (UNESCO Sympos.), Jerusalem, 89—96. [101]
BAKER, G. (1962): Detrital heavy minerals in natural accumulates, with special reference to Australian occurrences. — Austr. Inst. Min. and Metallurgy, Melbourne, Monogr. No. 1, 146 pp. [27]
BALL, M. M. (1967): Carbonate sand bodies of Florida and the Bahamas. — J. Sediment. Petrol., 37, 556—591. [257]
BALLANCE, P. F. (1964): Streaked-out mud ripples below Miocene turbidites, Puriri Formation, New Zealand. — J. Sediment. Petrol., 34, 91—101. [72]
BALLANCE, P. F. & NELSON, C. S. (1969): Differential cementation in the Waikawau limestone (Waitemata group), West Auckland, New Zealand. — J. Geol. Geophys., 12, 67—86. [340]
BALTZER, F. & LE RIBAULT, L. (1971): Néogenèse de quartz dans les bancs sédimentaires d'un delta tropical. Aspect des grains en microscopie électronique et optique. — C. R. Acad. Sci., Paris, 273, 1083—1086. [141]
BANDY, O. L. (1954): Aragonite tests among the foraminifera. — J. Sediment. Petrol., 24, 60—61. [230]
— (1964): General correlation of foraminiferal structure with environment. — In: IMBRIE & NEWELL (ed.), Approaches to paleoecology. — Wiley & Sons, 75—90. [228]
— (1967): AGI short course lecture notes, New Orleans. — Amer. Geol. Inst. [228]
BANDY, O. L. & ARNAL, R. E. (1960): Concepts of foraminiferal paleoecology. — Bull. Amer. Assoc. Petrol. Geol., 44, 1921—1932. [228]
BANERJEE, I. (1964): Size-roundness relation in the Barakar sandstones of the South Karanpura Coalfield, India. — Sedimentology, 3, 22—28. [68]
BANNER, F. T. & WOOD, G. V. (1964): Lower Cretaceous — Upper Jurassic stratigraphy of Umm Shaif field, Abu Dhabi marine areas, Trucial Coast, Arabia. — Bull. Amer. Assoc. Petrol. Geol., 48, 191—206. [293, 320]
BARNETCHE, A. & ILLING, L. V. (1956): The Tamabra limestone of the Poza Rica oilfield. — XX Congr. Geol. internac., 38 pp. [271]
BARRETT, P. J. (1964): Residual seams and cementation in Oligocene shell calcarenites, Te Kuiti Group. — J. Sediment. Petrol., 34, 524—531. [299]
BASKIN, Y. (1956): A study of authigenic feldspars. — J. Geol., 64, 132—155. [334, 335]
BATES, C. C. (1953): A rational theory of delta formation. — Bull. Amer. Assoc. Petrol. Geol., 37, 2119—2162. [109, 110, 113]
BATHURST, R. G. C. (1958): Diagenetic fabrics in some British Dinantian limestones. — Liverpool and Manchester Geol. J., 2, Pt. 1, 11—36. [280]
— (1959a): Diagenesis in Mississippian calcilutites and pseudobreccias. — J. Sediment. Petrol., 29, 365—376. [190, 271, 284]
— (1959b): The cavernous structure of some Mississippian Stromatactis reefs. — J. Geol., 67, 506—521. [270]
— (1964a): Diagenesis and paleoecology: A survey. — In: Approaches to Paleoecology; ed.: IMBRIE & NEWELL. Wiley and Sons, 319—344. [280]
— (1964b): The replacement of aragonite by calcite in the molluscan shell wall. — In: Approaches to Paleoecology; ed.: IMBRIE & NEWELL. — Wiley and Sons, 357—376. [281, 285]
— (1966): Boring algae, micrite envelopes and lithification of molluscan biosparites. — Geol. J., 5, 1, 15—32. [206, 285]
— (1967a): Depth indicators in sedimentary carbonates. — Marine Geol., 5, 447—471. [254, 255, 256]
— (1967b): Subtidal gelatinous mats, sand stabilizer and food, Great Bahama Bank. — J. Geol., 75, 736—738. [204, 285, 357]
— (1970): Problems of lithification in carbonate muds. — Proc. Geol. Assoc., 81, 3, 429—440. [199, 200]

— (1971a): Grapestone and submarine cementation, Berry Islands, Bahamas (Abstr.). — Progr. VIII Internat. Sediment. Congr., Heidelberg, 6. [276, 284, 285]
— (1971b): Carbonate sediments and their diagenesis. — Developm. in Sedimentol., **12**, Elsevier, Amsterdam, 620 pp.
[191, 192, 195, 197, 209, 223, 230, 280, 284, 288, 290, 293, 294, 321, 371]
— (1971c): Two generations of cement. — In: O. P. BRICKER (ed.), Carbonate cements. — Johns Hopkins Stud. in Geol., **19**, 296, The Johns Hopkins Press, Baltimore, Md. [288]
BAUSCH, W. M. (1963a): Der Obere Malm an der unteren Altmühl, nebst einer Studie über das Riff-Problem. — Erlanger Geol. Abh., H. 49, 38 pp. [223, 274]
— (1963b): Geologisches Erscheinungsbild eines Dolomitisierungsprozesses. — Geol. Bl. NO-Bayern, **13**, H. 2, 89—92. [314]
— (1965): Dedolomitisierung und Recalcitisierung in fränkischen Malmkalken. — N. Jb. Miner., Mh., 3, 75—82. [320]
— (1966): Strontiumgehalte in süddeutschen Malmkalken. — Geol. Rdsch., **55**, 86—96. [268]
— (1968): Clay content and calcite crystal size of limestones. — Sedimentology, **10**, 71—75. [293, 340]
BAUSCH, W. M. & WIONTZEK, H. (1961): Petrographische Untersuchungen am Hauptdolomit von Rehden. — Erdöl u. Kohle etc., **14**, 686—692. [347]
BAUSCH, W. M. & ZEISS, A. (1966): Zur Zusammensetzung des Kelheimer Riffkalks. — Geol. Bl. NO-Bayern, **16**, 240—242. [272]
BAUSCH VAN BERTSBERGH, J. W. (1940): Richtungen der Sedimentation in der Rheinischen Geosynkline. — Geol. Rdsch., **31**, 328—364. [89]
BAXTER, J. W. (1960): Calcisphaera from the Salem (Mississippian) limestone in southwestern Illinois. — J. Paleont., **34**, 1153—1157. [227]
BEAL, M. A. & SHEPARD, F. P. (1956): A use of roundness to determine depositional environments. — J. Sediment. Petrol., **26**, 49—60. [68]
BEALES, F. W. (1958): Ancient sediments of bahaman type. — Bull. Amer. Assoc. Petrol. Geol., **42**, 1845—1880. [198, 202]
BEALL, A. O. & FISCHER, A. G. (1969): Initial report of the Deep Sea Drilling Project, Vol. 1, 521—593. [323]
BEAUDOIN, B. & GIGOT, P. (1971): Figures de courant et traces de pattes d'oiseaux associées dans la molasse miocène de Digne, Basses Alpes (France). — Sedimentology, **17**, 241—256. [92]
BEERBOWER, J. R. (1961): Origin of Cyclothems of the Dunkard Group (Upper Pennsylvanian — Lower Permian) in Pennsylvania, West Virginia, and Ohio. — Geol. Soc. Amer. Bull., **72**, 1029—1050. [365]
BEHRENS, E. W. & LAND, L. S. (1972): Subtidal Holocene dolomite, Baffin Bay, Texas. — J. Sediment. Petrol., **42**, 155—161. [316]
BENAVIDES, L. G. (1956): Notas sobre la geologia petrolera de Mexico. — XX. Congr. geol. internac., 351—562. [271]
BENEO, E. (1956): Accumuli terziari da risedimentazione (olistostroma) nell'Apennino centrale et frane sottomarine. — Boll. Serv. Geol., Italia, **78**, 291—321. [82]
BENNACEF, A., BEUF, S., BIJU-DUVAL, B., DE CHARPAL, O., GARIEL, O. & ROGNON, P. (1971): Examples of cratonic sedimentation: Lower Paleozoic of Algerian Sahara. — Amer. Assoc. Petrol. Geol. Bull., **55**, 2225—2245. [107]
BENSON, L. V. & MATTHEWS, R. K. (1971): Electron microprobe studies of magnesium distribution in carbonate cements and recrystallized skeletal grainstones from the Pleistocene of Barbados, West Indies. — J. Sediment. Petrol., **41**, 1018—1025. [289]
BERG, G. (1944): Vergleichende Petrographie oolithischer Eisenerze. — Reichsamt Bodenforsch., Arch. Lagerstättenforsch., H. 76, 128 pp. [250, 333]
BERG, R. R. (1952): Feldspathized sandstone. — J. Sediment. Petrol., **22**, 221—223. [142]
BERGENBACK, R. E. & TERRIERE, R. T. (1953): Petrography and petrology of Scurry Reef, Scurry County, Texas. — Bull. Amer. Assoc. Petrol. Geol., **37**, 1014—1029. [272]
BERGER, W. (1971): Sedimentation of planktonic foraminifera. — Marine Geol., 11, 325—358. [206]
BERGQUIST, H. R. & COBBAN, W. A. (1957): Mollusks of the Cretaceous. — In: H. S. LADD (ed.), Paleoecology. — Geol. Soc. Amer. Memoir, **67**, 871—884. [241, 245]
BERKNER, L. V. & MARSHALL, L. C. (1965): On the origin and rise of oxygen-concentration in the earth's atmosphere. — J. Atm. Sci., May 1965, 225—261. [219]

Berner, R. A. (1967): Comparative dissolution characteristics of carbonate minerals in the presence and absence of aqueous magnesium ion. — Amer. J. Sci., 265, 45—70. [322]
— (1968 a): Calcium carbonate concretions formed by the decomposition of organic matter. — Science, 159, 195—197. [301]
— (1968 b): Rate of concretion growth. — Geochim. Cosmochim. Acta, 32, 477—483. [301, 302]
— (1971 a): Principles of chemical Sedimentology. — McGraw Hill, 240 pp. [301]
— (1971 b): Bacterial processes effecting the precipitation of calcium carbonate in sediments. — In O. P. Bricker (ed.): Carbonate cements. — The Johns Hopkins Univ. Stud. in Geol., 19, 247—251, Baltimore, Md. [302]
Bersier, A. (1959): Séquences détritiques et divagations fluviales. — Eclog. geol. Helvet., 51, 854—893. [361, 366, 380]
Berthois, L. & Portier, J. (1956): Recherches expérimentales sur le mode d'usure des graviers. — C. R. séanc. Acad. Sci., 243, 1778—1781. [65]
— — (1957 a): Recherches expérimentales sur le façonnement des graviers de quartz. — C. R. séanc. Acad. Sci., 244, 362—364. [65]
— — (1957 b): Recherches expérimentales sur le façonnement des grains de sable quartzeux. — C. R. séanc. Acad. Sci., 245, 1152—1154. [68]
Beuf, S., Biju-Duval, B., Charpal, O. De, Rognon, P., Gariel, O. & Bennacef, A. (1971): Les grès du Paléozoique inférieur au Sahara. Sédimentation et discontinuités — Evolution structurale d'un craton. — Publ. Inst. Franç. du Pétrole, 18, Ed. Technip, Paris, 464 pp. [107]
Biederman, E. W. (1962): Distinction of shoreline environments in New Jersey. — J. Sediment. Petrol., 32, 181—200. [69, 141]
Bien, G. S., Contois, D. E. & Thomas, W. H. (1958): Removal of soluble silica from fresh water entering the sea. — Geochim. Cosmochim. Acta, 14, 35—54. [329]
Bigarella, J. J., Mabesoone, J. M., Lins, C. J. & Mota, F. F. O. (1965): Paleogeographical features of the Serra Grande and Pimenteira formations (Parnaíba basin, Brazil). — Palaeogeogr., Palaeoclimat., Palaeoecol., 1, 259—296. [78]
Biju-Duval, B., Charpal, O. de & Merabet, O. (1966): Constance des directions des paléocourants dans les grès de base du Cambro-Ordovicien sur le pourtour du Hoggar. — C. R. Acad. Sci. Paris, Sér. D, 262, 48—50. [79]
Biscaye, P. E. (1964): Mineralogy and sedimentation of the Deep-Sea sediment fine fraction in the Atlantic ocean and adjacent seas and oceans. — Yale Univ., Dept. Geol., Geochem. Techn. Rept., 8, 86 pp. [318]
Bissell, H. J. (1959): Silica in sediments of the Upper Paleozoic of the Cordilleran area. — In: H. A. Ireland (ed.), Silica in sediments. — Soc. Econ. Paleont. Min. Spec. Publ., 7, 150—185. [293]
Bissell, H. J. & Chilingar, G. V. (1967): Classification of sedimentary carbonate rocks. — In: Chilingar, Bissell, Fairbridge (ed.): Carbonate rocks. Developm. in Sedimentol. 9 A. Elsevier Publ. Co., 87—168. [278]
Black, M. (1933): The algal sediments of Andros Island, Bahamas. — Phil. Trans. Roy. Soc. London, Ser. B, 222, 165—192. [223, 356]
Black, M. & Barnes, B. (1959): The structure of coccoliths from the English chalk. — Geol. Mag., 96, 322—327. [232]
Blackmon, P. D. & Todd, R. (1959): Mineralogy of some foraminifera as related to their classification and ecology. — J. Paleont., 33, 1—15. [228, 230]
Blackwelder, P. L. & Pilkey, O. H. (1972): Electron microscopy of quartz grain surface textures: The U.S. eastern Atlantic continental margin. — J. Sediment. Petrol., 42, 520—526. [69]
Blatt, H. (1959): Effect of size and genetic quartz type on sphericity and form of beach sediments, northern New Jersey. — J. Sediment. Petrol., 29, 197—206. [64]
— (1963): Selective destruction of undulatory quartz in sedimentary environments. — GSA etc. Joint Meet., Houston, Abstr., GSA Spec. Pap. No. 73, 118—119. [21]
— (1967): Original characteristics of clastic quartz grains. — J. Sediment. Petrol., 37, 401—424. [20]
Blatt, H. & Christie, J. M. (1963): Undulatory extinction in quartz of igneous and metamorphic rocks and its significance in provenance studies of sedimentary rocks. — J. Sediment. Petrol., 33, 559—579. [21]

BLATT, H., MIDDLETON, G. & MURRAY, R. (1972): Origin of sedimentary rocks. — Prentice Hall, Englewood Cl., N. J., 634 pp. [1, 76, 104, 206, 305, 313]
BLATT, H. & SUTHERLAND, B. (1969): Intrastratal solution and non-opaque heavy minerals in shales. — J. Sediment. Petrol., 39, 591—600. [41]
BLISSENBACH, E. (1954): Geology of alluvial fans in semiarid regions. — Bull. Geol. Soc. Amer., 65, 175—190. [57, 66]
— (1957): Die jungtertiäre Grobschotterschüttung im Osten des bayerischen Molassetroges. — Beih. Geol. Jb., H. 26, 9—48. [48]
BLUCK, B. J. (1965): The sedimentary history of some Triassic conglomerates in the Vale of Glamorgan, South Wales. — Sedimentology, 4, 225—245. [75]
BOCCALETTI, M. & MICHELI, P. (1968): Analisi statistica dell'orientamento dei granuli in una torbiditi della Marnoso-arenacea (Appennino settentrionale). — Boll. Soc. Geol. Ital., 87, 65—82. [72]
— — (1970): L'embriciamento dei granuli alla base di alcune torbiditi del macigno del Mugello e della Marnoso-areacea (Appennino settentrionale). — Atti Soc. Tosc. Sci. Nat. Mem., A, 76, 280—292. [70]
BOCK, W. D.: pers. commun. [200]
BÖGER, H. (1964): Palökologische Untersuchungen an Cyclothemen im Ruhrkarbon. — Paläont. Z., 38, 142—157. [375]
BØGGILD, O. B. (1912): The deposits of the sea-bottom. — Report of Danish Oceanographical Expeditions, 1908—1910, to the Mediterranean etc., Vol. 1, pt. 3, 255—269. [318]
— (1930): The shell structure of the mollusks. — Kgl. Danske Videnskab. Selsk. Skr., Naturvidensk. mathem. Afd., 9. Raekke, II. 2., 258—325. [208, 209, 235, 236, 241, 245—247, 349]
BOND, G. (1950a): The Lower Carboniferous reef limestones of northern England. — J. Geol., 58, 313—329. [271]
— (1950b): The Lower Carboniferous reef limestones of Cracoe, Yorkshire. — Quart. J. Geol. Soc., 105, 157 ff. [273]
BONHAM-CARTER, G. F. (1964): Preliminary report on some Pennsylvanian reefs from northwest Ellesmere Island. — Amer. Assoc. Petr. Geol. Bull., 48, 518 (Abstr.). [272]
BORDINE, B. W. (1965): Paleoecologic implications of strontium, calcium, and magnesium in Jurassic rocks near Thistle, Utah. — Brigham Young Univ. Geol. Stud., 12, 91—120. [379]
BOSELLINI, A. (1964): Sul significato genetico e ambientale di alcuni tipi di rocce calcare in base alle più recenti classificazioni. — Mem. Museo di Storia Nat., Trento, XV, Fasc. II, 5—58. [279]
— (1965): Analisi petrografica della "Dolomia principale" nel gruppo di Sella (regione Dolomitica). — Mem. Geopaleont. Univ. Ferrara, 1, II, 3, 49—109. [288, 355]
— (1967a): Erosione intercotidale presso la foce del Reno (Mare Adriatico). — Ann. Univ. Ferrara, N.S., Sez. IX, IV, 77—89. [23, 374]
— (1967b): La tematica deposizionale della Dolomia Principale (Dolomiti e Prealpi Venete). — Boll. Soc. Geol. Ital., 86, 133—169. [270]
BOTTINGA, Y., KUDO, A. & WEILL, D. (1966): Some observations on oscillatory zoning and crystallization of magmatic plagioclase. — Amer. Min., 51, 792—806. [25]
BOTWINKINA, L. N. (1962): Slosistost osadochnykh porod (Stratification of sedimentary rocks). — Akad. Nauk SSSR, Trudy Geol. Inst., 59, 542 pp. (Russ.). [72]
BOUMA, A. H. (1962): Sedimentology of some flysch deposits. — Elsevier, Amsterdam, 168 pp. [72, 128, 367, 369]
BOWSHER, A. L. (1957): Gastropods of the Paleozoic. — In: Paleoecology. — Geol. Soc. Amer. Mem., 67, 821—826. [245]
BOYD, D. R. & DYER, B. F. (1966): Frio barrier bar system of South Texas. — Bull. Amer. Assoc. Petrol. Geol., 50, 170—178. [121]
BOYD, F. R. (1961): Welded tuffs and flows in the rhyolite plateau of Yellowstone Park, Wyoming. — Geol. Soc. Amer. Bull., 72, 387—426. [176]
BOYER, B. W. (1972): Grain accretion and related phenomena in unconsolidated surface sediments of the Florida Reef Tract. — J. Sediment. Petrol., 42, 205—210. [282]
BOYER, F. (1949): Étude des alluvions de la Garonne en amont d'Agen. — Conf. séd. et quat., La Rochelle, 49—62. [68]
BRADLEY, J. (1973): Zoophycos and Umbellula (Pennatulacea): Their synthesis and identity. — Palaeogeogr., Palaeoclimat., Palaeoecol., 13, 103—128. [80]

BRADLEY, W. H. (1929): Algal reefs and oölites of the Green River Formation. — U.S. Geol. Surv., Prof. Pap., **154**, 203—233. [224]
— (1937): Non-glacial varves, with selected bibliography. — Nat. Res. Counc. Ann. Rept. App. A, Rept. of Committee on Geol. Time, 33—42. [363]
BRÄTTER, P., MÖLLER, P. & RÖSICK, U. (1971): Coprecipitation of trace elements in dolomite under sedimentary conditions (Abstr.). — Progr., VIII Internat. Sediment. Congr., Heidelberg, 11. [305]
BRAMER, H. (1965): Bestimmung der Oberflächenbeschaffenheit von Quarzkörnern mit dem Elektronenmikroskop. — Geologie, **9**, 1114—1117. [70]
BRAMLETTE, M. N. (1941): The stability of minerals in sandstone. — J. Sediment. Petrol., **11**, 32—36. [41]
— (1958): Significance of coccolithophorids in calcium carbonate deposition. — Bull. Geol. Soc. Amer. **69**, 121—126. [225]
BRAMLETTE, M. N. & MARTINI, E. (1964): The great change in calcareous nannoplankton fossils between the Maestrichtian and Danian. — Micropaleontol., **10**, 291—322. [225]
BRAMLETTE, M. N. & RIEDEL, W. R. (1954): Stratigraphic value of discoasters and some other microfossils related to recent coccolithophores. — J. Paleont., **28**, 385—403. [226]
V. BRAUN, E. (1953): Geologische und sedimentpetrographische Untersuchungen im Hochrheingebiet zwischen Zurzach und Eglisau. — Eclog. geol. Helvet., **46**, 143—170. [69]
BRAUN, M. & FRIEDMAN, G. M. (1969): Carbonate lithofacies and environments of the Tribes Hill Formation (Lower Ordovician) of the Mohawk Valley, New York. — J. Sediment. Petrol., **39**, 113—135. [204, 337]
BREDDIN, H. (1930): Die Milchquarzgänge des Rheinischen Schiefergebirges, eine Nebenerscheinung der Druckschieferung. — Geol. Rdsch., **21**, 367—388. [134]
BRENCHLEY, P. J. (1969): Origin of matrix in greywackes, Berwyn Hills, North Wales. — J. Sediment. Petrol., **39**, 1297—1301. [62]
BRETZ, J. H. & HORBERG, L. (1949): Caliche in southeastern New Mexico. — J. Geol., **57**, 491—511. [258]
BRICKER, O. P. (ed.) (1971): Carbonate cements. — Johns Hopkins Univ. Stud. in Geol., No. 19, Johns Hopkins Press, Baltimore and London, 376 pp. [282, 290]
BRIGGS, G. & CLINE, L. M. (1967): Paleocurrents and source areas of late Paleozoic sediments of the Ouachita Mountains, Southeastern Oklahoma. — J. Sediment. Petrol., **37**, 985—1000. [97]
BRIGGS, L. I. (1958): Evaporite facies. — J. Sediment. Petrol., **28**, 46—56. [314]
BRIGGS, L. I., MCCULLOCH, D. S. & MOSER, F. (1962): The hydraulic shape of sand particles. — J. Sediment. Petrol., **32**, 645—656. [39]
BRIGGS, L. I. & MIDDLETON, G. V. (1965): Hydrochemical principles of sediment structure formation. — In: MIDDLETON (ed.): Primary sedimentary structures and their hydrodynamic interpretation. — Soc. Econ. Paleont. and Mineral., Spec. Publ., **12**, 5—16. [59]
BRINKMANN, R. (1926): Tektonik und Sedimentation im deutschen Triasbecken. — Z. dt. geol. Ges., **78**, 52—74. [73]
— (1929): Statistisch-biostratigraphische Untersuchungen an mitteljurassischen Ammoniten. — Abh. Ges. Wiss. Göttingen, math.-phys. Kl. [N.F.], **13**, H. 3. [361, 374]
— (1930): Über die Schichtung und ihre Bedingungen. — Fortschr. Geol. Paläont., **XI**, H. 35, 187—219. [363]
— (1933): Über Kreuzschichtung im deutschen Buntsandsteinbecken. — Nachr. Ges. Wiss. Göttingen, math.-phys. Kl., **IV**, H. 32, 12 pp. [79]
— (1955): Gerichtete Gefüge in klastischen Sedimenten. — Geol. Rdsch., **43**, 562—568. [71]
BRODSKAJA cit. in RAUPACH 1952. [251]
BROECKER, W. S. & TAKAHASHI, T. (1966): Calcium carbonate precipitation on the Bahama banks. — J. Geophys. Res., **71**, 1575—1602. [196]
BRÖNNIMANN, P. (1955): Microfossils incertae sedis from the Upper Jurassic and Lower Cretaceous of Cuba. — Micropaleontol., **1**, 28—51. [232, 249]
BRÖNNIMANN, P. & NORTON, P. (1960): On the classification of fossil fecal pellets and description of new forms from Cuba, Guatemala and Libya. — Eclog. geol. Helvet., **53**, 832—842. [200]
BROGNIART, A. (1826): L'arkose, caractères minéralogiques et histoire géognostique de cette roche. — Ann. sci. nat., **8**, 113—163. [14]

Brown, C. E. & Thayer, T. P. (1963): Low-grade mineral facies in Upper Triassic and Lower Jurassic rocks of the Aldrich Mountains, Oregon. — J. Sediment. Petrol., 33. [144, 155]

Brown, C. N. (1956): The origin of caliche on the northeastern Llano Estracado, Texas. — J. Geol., 64, 1—15. [258]

Brückner, W. D. (1951): Lithologische Studien und zyklische Sedimentation in der helvetischen Zone der Schweizeralpen. — Geol. Rdsch., 39, 196—212. [371]

— (1953): Cyclic calcareous sedimentation as an index of climatic variations in the past. — J. Sediment. Petrol., 23, 235—237. [370, 371]

Brunnacker, K. (1965): Die Entstehung der Münchener Schotterfläche zwischen München und Moosburg. — Geol. Bavarica, 55, 341—359. [51]

v. Bubnoff, S. (1947): Rhythmen, Zyklen und Zeitrechnung in der Geologie. — Geol. Rdsch., 35, 6—22. [86, 363]

— (1950): Die Geschwindigkeit der Sedimentbildung und ihr endogener Antrieb. — Geotekt. Inst. Akad. Abh. z. Geotekt., Nr. 2, Akademie-Verlag, Berlin. [86]

Buchbinder, B. & Friedman, G. M. (1970): Selective dolomitization of micritic envelopes: A possible clue to original mineralogy. — J. Sediment. Petrol., 40, 514—517. [294, 313]

Bucher, W. H. (1919): On ripples and related sedimentary surface forms and their paleogeographic interpretation. — Amer. J. Sci., ser. 4, 47, 149—210, 241—269. [87]

Bullock, L. R. (1965): Paleoecology of the Twin Creek limestone in the Thistle, Utah area. — Brigham Young Univ. Geol. Stud., 12, 121—147. [379]

Burek, P. J. (1970): Magnetic reversals: Their application to stratigraphic problems. — Bull. Amer. Assoc. Petrol. Geol., 54, 1120—1139. [46]

Burne, R. V. (1970): The origin and significance of sand volcanoes in the Bude formation, Cornwall. — Sedimentology, 15, 211—228. [84]

Burrolet, P. F. & Byramjee, R. S. (1964): Shape and structure of Saharan Cambro-Ordovician sand bodies — paleocurrents and depositional environment. (Abstr.) — Bull. Amer. Assoc. Petrol. Geol., 48, 519—520. [79]

Burst, J. F. (1965): Subaqueously formed shrinkage cracks in clay. — J. Sediment. Petrol., 35, 348—353. [84]

Buryanova, E. Z. & Bogdanov, V. V. (1965): Distribution of the authigenic zeolites laumontite and heulandite in the sedimentary rocks of the Tarbagatai coal deposits. — Lithology and Mineral Resources, 2 (transl. from Russian, Consultants Bureau, 1967), 195—202. [144]

Busch, D. A. (1971): Genetic units in delta prospecting. — Amer. Assoc. Petrol. Geol. Bull., 55, 1137—1154. [113]

Button, A. (1973): Algal stromatolites of the early Proterozoic Wolkberg group, Transvaal sequence. — J. Sediment. Petrol., 43, 160—167. [219]

Byrne, J. V. & Maloney, N. J. (1965): Textural trends of continental margin sediments off the central coast of Oregon. — Ann. GSA & Assoc. Soc. Joint Meet., Kansas City, Progr., 23—24. [59, 60]

Cailleux, A. (1942): Les actions éoliens périglaciaires en Europe. — Mém. Soc. Géol. France, n. s. 21, No. 46, 1—176. [69]

— (1943): Distinction des sables marins et fluviatiles. — Bull. Soc. Géol. France, 5e Sér., 13, 125—138. [69]

— (1945): Distinction des galets marins et fluviatiles. — Bull. Soc. Géol. France, 5e Sér., 15, 375—404. [64, 70]

— (1952): Morphoskopische Analyse der Geschiebe und Sandkörner und ihre Bedeutung für die Paläoklimatologie. — Geol. Rdsch., 40, 11—19. [69]

— (1961): Application à la Géographie des Méthodes d'Étude des Sables et des Galets. — Centro Pesquisas Geograf. Brasil, Univ. de Brasil, Rio de Janeiro, 151 pp. [64]

Cailleux, A. & Tricart, J. (1959): Initiation à l'étude des sables et des galets. — T. I—III. Paris (C. D. U.). [69]

Cain, J. D. B. (1968): Aspects of the depositional environment and palaeoecology of crinoidal limestones. — Scott. J. Geol., 4, 191—208. [248]

Caine, N. (1972): Air photo analysis of Blockfield fabric in Talus Valley, Tasmania. — J. Sediment. Petrol., 42, 33—48. [71]

CALVERT, S. E. (1966): Accumulation of diatomaceous silica in the sediments of the Gulf of California. — Geol. Soc. Amer. Bull., 77, 569—594. [328]
— (1968): Silica balance in the ocean and diagenesis. — Nature, 219, 919—920. [336]
CAREY, W. C. & KELLER, M. D. (1957): Systematic changes in the beds of alluvial rivers. — Hydraulics Div. J., Amer. Soc. Civil Engineers Proc., Hy. 4, Pap. 1331, 24 pp. [75]
CARLISLE, D. (1963): Pillow breccias and their aquagene tuffs, Quadra Island, British Columbia. — J. Geol., 71, 48—71. [184]
CAROZZI, A. (1953a): Pétrographie des roches sédimentaires. — F. Rouge & Cie S. A., Lausanne, 250 pp. [333]
— (1953b): Données micrographiques sur le Crétacé supérieur helvétique. — Bull. Inst. Nat. Genevois, 76, 1—76. [370]
— (1955): Sédimentation récifale rhythmique dans le Jurassique supérieur du Grand-Salève (Haute-Savoie, France). — Geol. Rdsch., 43, 433—446. [271]
CAROZZI, A. V. (1960): Microscopic sedimentary petrography. — Wiley & Sons, New York, London, 485 pp. [134, 140, 142, 143, 201, 202, 250, 256, 326]
— (1961): Oolithes remaniées, boisées et régénérées dans le Mississippien des chaînes frontales, Alberta Central, Canada. — Arch. Sci., Genève, 14, fasc. 2, 281—296. [257]
— (1964): Complex oöids from Triassic lake deposit, Virginia. — Amer. J. Sci., 262, 231—241. [257]
CAROZZI, A. V. & LUNDWALL, W. R., JR. (1959): Microfacies study of a Middle Devonian bioherm, Columbus, Indiana. — J. Sediment. Petrol., 29. 343—353. [273]
CAROZZI, A. V. & TEXTORIS, D. A. (1963): Les Stromatactis des récifs Siluriens de l'Indiana sont des Bryozoaires. — Arch. Sci., Genève, Ed. Soc. Phys., Hist. Nat., 16, 188—192. [270]
CAROZZI, A. V. & ZADNIK, V. E. (1959): Microfacies of Wabash reef, Wabash, Indiana. — J. Sediment. Petrol., 29, 164—171. [273]
CARPENTER, G. C. & SCHMIDT, R. G. (1962): Insoluble residues in a portion of the Ordovician Cynthiana formation, North-central Kentucky. — J. Sediment. Petrol., 32, 423—434. [41]
CARRIGY, M. A. & MELLON, G. B. (1964): Authigenic clay mineral cements in Cretaceous and Tertiary sandstones of Alberta. — J. Sediment. Petrol., 34, 461—472. [143, 144]
CARRIKER, M. R. & SMITH, E. H. (1969): Comparative calcibiocavitology: Summary and conclusions. — In: CARRIKER et al. (ed.): Penetration of calcium carbonate substrates by lower plants and invertebrates (Sympos.). — Amer. Zoologist, 9, 1011—1020. [254]
CARROLL, D. (1953): Weatherability of zircon. — J. Sediment. Petrol., 23, 106—116. [39]
— (1970): Rock weathering. — Plenum Press, New York - London, 203 pp. [2]
CARSON, B. (1971): Holocene-Pleistocene sedimentation in Cascadia Basin, Northeast Pacific ocean (Abstr.) — Progr., VIII Internat. Sediment. Congr., Heidelberg, 16—17. [368]
CARTER, N. L., CHRISTIE, J. M. & GRIGGS, D. T. (1964): Experimental deformation and recrystallization of quartz. — J. Geol., 72, 687—733. [20]
CASEY, R. (1960): A Lower Cretaceous gastropod with fossilized intestines. — Palaeontol., 2, 270—276. [201]
CASSHYAP, S. M. (1970): Sedimentary cycles and environment of deposition of the Barakar coal measures of Lower Gondwana, India. — J. Sediment. Petrol., 40, 1302—1317. [380]
CASSHYAP, S. M. & QIDWAI (1971): Paleocurrent analysis of Lower Gondwana sedimentary rocks, Pench Valley coalfield, Madhya Pradesh (India). — Sediment. Geol., 5, 135—145. [106]
CAYEUX, L. (1929): Les roche sédimentaires de France, roches siliceuses. — Mém. carte géol. dét. France, 23, Paris, 774 pp. [129]
— (1931): Introduction à l'étude pétrographique des roches sédimentaires. — Mém. carte géol. dét. France, Texte: 524 pp., Atlas: 56 plates. [328, 329]
— (1932): Les accidents magnésiens du Bassin de Paris, envisagés dans leurs rapports avec les ruptures d'équilibre du fond des mers. — C. R., Acad. sci. (Paris), 194, 504—507. [319]
— (1935): Roches carbonatées. — In: CAYEUX, Les Roches Sédimentaires de France. — Masson et Cie., Paris, 463 pp. [197, 199, 228, 232, 236, 254, 329]
CHAMBERLAIN, T. K. (1964): Mass transport of sediment in the heads of Scripps submarine canyon, California. — In: R. L. MILLER (ed.), Pap. in marine Geol., SHEPARD comm. vol., New York, 42—64. [81, 92]

CHAPPEL, J. (1967): Recognizing fossil strand lines from grain-size analysis. — J. Sediment. Petrol., 37, 157—165. [121]
CHATTERJEE, N. D. (1966): On the widespread occurrence of oxidized chlorites in the Pennine zone of the western Italian Alps. — Beitr. Miner. Petrol., 12, 325—339. [26]
— (1973): Low-temperature compatibility relations of the assemblage quartz-paragonite and the thermodynamic status of the phase rectorite. — Contr. Miner. Petrol., 42, 259—271. [157]
CHAVE, K. E. (1954 a, b): Aspects of the biogeochemistry of magnesium. 1. Calcareous marine organisms; 2. Calcareous sediments and rocks. — J. Geol., 62, 266—283; 587—599. [208, 209, 211, 219]
— (1962): Factors influencing the mineralogy of carbonate sediments. — Limnol. Oceanogr., 7, 218—223. [322]
— (1964): Skeletal durability and preservation. — In: IMBRIE & NEWELL (ed.): Approaches to paleoecology, J. Wiley & Sons, New York, 377—387. [207, 321]
CHILINGAR, G. V. (1956): Dedolomitization: A review. — Bull. Amer. Assoc. Petrol. Geol., 40, 762—764. [321]
CHILINGAR, G. V. & BISSELL, H. J. (1961): Dolomitization by seepage refluxion (Discussion). — Bull. Amer. Assoc. Petrol. Geol., 45, 679—681. [308]
CHILINGAR, G. V., BISSELL, H. J. & WOLF, K. H. (1967): The diagenesis of carbonate rocks. — In: G. LARSEN & G. V. CHILINGAR (ed.), Diagenesis in Sediments. — Elsevier Publ. Co. [351]
CHIPPING, D. H. (1972): Sedimentary structure and environment of some thick sandstone beds. of turbidite type. — J. Sediment. Petrol., 42, 587—595. [82]
CHOQUETTE, P. W. (1955): A petrographic study of the „State College" siliceous oolite. — J. Geol., 63, 337—347. [332]
CHOQUETTE, P. W. & PRAY, L. C. (1970): Geologic nomenclature and classification of porosity in sedimentary carbonates. — Amer. Assoc. Petrol. Geol. Bull., 54, 207—250. [340, 341]
CHOQUETTE, P. W. & TRAUT, J. D. (1963): Pennsylvanian carbonate reservoirs, Ismay Field, Utah and Colorado. — In: R. O. BASS (ed.), Shelf Carbonates of the Paradox Basin. Four Corners Geol. Soc., Field Conf., 4th, 157—184. [271, 274, 314, 347]
CIFELLI, R., BOWEN, V. T. & SIEVER, R. (1966): Cemented foraminiferal oozes from the Midatlantic Ridge. — Nature, 209, No. 5018, 32—34. [282]
CLARK, A. M. (1957): Crinoids. — In J. W. HEDGPETH (ed.): Ecology. — Geol. Soc. Amer., Mem., 67, 1183—1185. [248]
CLARKE, F. W. (1924): Data of geochemistry. — U. S. Geol. Surv. Bull., 770, 841 pp. [192]
CLARKE, F. W. & WHEELER, W. C. (1922): The inorganic constituents of marine invertebrates. U. S. Geol. Surv. Prof. Pap., 124, 2. Ed. [208, 209]
CLAUS, G. (1936): Schwermineralien aus kristallinen Gesteinen des Gebietes zwischen Passau und Cham. — N. Jb. Miner., 71 A, 1—58. [31]
CLAYTON, R. N., FRIEDMAN, I., GRAF, D. L., MAYEDA, T. K., MEENTS, W. F. & SHIMP, N. F. (1966): The origin of saline formation waters. 1. Isotope composition. — J. Geophys. Res., 71, 3869—3882. [324]
CLIFTON, H. E. (1971): Orientation of empty pelecypod shells and shell fragments in quiet water. — J. Sediment. Petrol., 41, 671—682. [207]
— (1973): Pebble segregation and bed lenticularity in wave-worked versus alluvial gravel. — Sedimentology, 20, 173—187. [47]
CLIFTON, H. E. & BOGGS, S. JR. (1970): Concave-up pelecypod (Psephidia) shells in shallow marine sand, Elk River beds, southwestern Oregon. — J. Sediment. Petrol., 40, 888—897. [207]
CLOOS, H. (1938): Primäre Richtungen in Sedimenten der rheinischen Geosynkline. — Geol. Rdsch., 29, 357—367. [90, 91]
— (1941): Bau und Tätigkeit von Tuffschloten (Untersuchungen an dem Schwäbischen Vulkan). — Geol. Rdsch., 32, H. 6—8, 709—800. [182]
CLOUD, P. E., JR. (1942): Notes on Stromatolites. — Amer. J. Sci., 240, 363—379. [224]
— (1962 a): Behaviour of calcium carbonate in sea water. — Geochim. Cosmochim. Acta, 26, 867—884. [196]
— (1962 b): Environment of calcium carbonate deposition west of Andros Island, Bahamas. — U. S. Geol. Surv. Prof. Pap., 350, 1—138. [191, 196, 251, 356]

CLOUD, P. E., JR. & SEMIKHATOV (1969): Proterozoic stromatolite zonation. — Amer. J. Sci., 267, 1017—1061. [221]
COCH, N. K. & KRINSLEY, D. H. (1971): Comparison of stratigraphic and electron microscopic studies in Virginia Pleistocene coastal sediments. — J. Geol., 79, 426—437. [69]
COLACICCHI, R. (1959): Dicchi sedimentari del Flysch oligomiocenico della Sicilia Nord-Orientale. — Eclog. geol. Helvet., 51, 901—916. [84]
COLBURN, I. P. (1968): Grain fabrics in turbidite sandstone beds and their relationship to sole mark trends on the same beds. — J. Sediment. Petrol., 38, 146—158. [72]
COLE, W. S. (1957): Foraminifera of the Cenozoic. — In: H. S. LADD (ed.), Paleoecology. — Geol. Soc. Amer. Mem., 67, 2, 757—762. [228]
COLEMAN, J. M. (1969): Brahmaputra river: Channel processes and sedimentation. — Sediment. Geol., 3, 129—239. [85]
COLEMAN, J. M. & GAGLIANO, S. M. (1965): Sedimentary structures: Mississippi river deltaic plain. — In: G. V. MIDDLETON (ed.), Primary sedimentary structures and their hydrodynamic interpretation. — SEPM Spec. Publ., 12, 133—148. [80, 81, 114]
COLEMAN, J. M., GAGLIANO, S. M. & WEBB, J. E. (1964): Minor sedimentary structures in a prograding distributary. — Marine Geol., 1, 240—258. [103, 112]
COLLINSON, J. D. (1968): Deltaic sedimentation units in the Upper Carboniferous of Northern England. — Sedimentology, 10, 233—254. [110, 112]
CONOLLY, J. R. (1965): The occurrence of polycrystallinity and undulatory extinction in quartz in sandstones. — J. Sediment. Petrol., 35, 116—135. [21]
CONOLLY, J. R. & EWING, M. (1966): Modern graded beds and turbidity currents: Case history. — Bull. Amer. Assoc. Petrol. Geol., 50, 608—609. [56]
CONYBEARE, C. E. B. & CROOK, K. A. W. (1968): Manual of Sedimentary Structures. — Austral. Dept. nat. Devel., Bur. Min. Res., Geol., Geophys., Bull. No. 102, 327 pp.
[72, 80, 90, 91]
COOGAN, A. H. (1969): Recent and ancient carbonate cyclic sequences. — West Texas Geol. Soc. Sympos. on Cyclic Sedimentation, 1967, 5—16. [372]
COOK, D. O. (1970): The occurrence and geological work of rip currents off Southern California. — Marine Geol., 9, 173—186. [79]
COOKE, C. W. (1957a): Echinoids. — In: J. W. HEDGPETH (ed.) Ecology. — Geol. Soc. Amer. Memoir 67, 1, 1191—1192. [249]
— (1957b): Echinoids of the Post-Paleozoic. — In: H. S. LADD (ed.) Paleoecology. — Geol. Soc. Amer. Memoir 67, 2, 981—982. [249]
COOMBS, D. S., ELLIS, A. J., FYFE, W. S. & TAYLOR, A. M. (1959): The zeolite facies, with comments on the interpretation of hydrothermal syntheses. — Geochim. Cosmochim. Acta, 11, 53—107. [146, 155]
CORRENS, C. W. (1924): Beiträge zur Petrographie und Genesis der Lydite (Kieselschiefer). — Mitt. Abt. Erz-, Salz- u. Gesteinsmikrosk., preuß. geol. L. A. [329]
— (1937): Die Sedimente des äquatorialen Atlantischen Ozeans. — Wiss. Ergebn. dt. Atlant. Exped. Meteor 1925—1927, 3, Tl. 3, 298 pp. [318]
— (1939a): Die Sedimentgesteine. — In: BARTH, CORRENS & ESKOLA, Die Entstehung der Gesteine. — Springer, Berlin, 116—262. [130, 191, 236, 275, 297, 298, 301]
— (1939b): Pelagic sediments of the North Atlantic Ocean. — In: P. D. TRASK (ed.), Recent Marine Sediments. — Sympos., Amer. Assoc. Petrol. Geol., 373—395. [318]
— (1949): Einführung in die Mineralogie. — Springer, 414 pp. [337]
— (1950): Faktoren der Sedimentbildung, erläutert an Kalk- und Kieselsedimenten. — Dt. Hydrogr. Z., 3, 83 88. [196]
COTTER, E. (1966): Limestone diagenesis and dolomitization in Mississippian carbonate banks in Montana. — J. Sediment. Petrol., 36, 764—774. [281]
COX, L. R. (1960): The preservation of moulds of the intestine in fossil Nuculana (Lamellibranchia) from the Lias of England. — Palaeontology, 2, 262—269. [201]
CRONOBLE, W. R. & MANKIN, CH. J. (1963): Genetic significance of variations in the limestones of the Coffeyville and Hogshooter formations (Missourian), northeastern Oklahoma. — J. Sediment. Petrol., 33, 73—86. [272]
CROOK, K. A. W. (1960): Petrology of Parry group, upper Devonian — lower Carboniferous, Tamworth — Nundle district, New South Wales. — J. Sediment. Petrol., 30, 538—552.
[19]
CROSFIELD & JOHNSTON (1914) cit. in AGER 1963. [264]

CROWELL, J. C. (1955): Directional-current structures from the Prealpine flysch, Switzerland. — Geol. Soc. Amer. Bull., 66, 1351—1384. [128]
— (1957): Origin of pebbly mudstones. — Geol. Soc. Amer. Bull., 63, 993—1010. [47]
CROWELL, J. C. & FRAKES, L. A. (1972): Late Paleozoic glaciation: part V, Karroo Basin, South Africa. — Geol. Soc. Amer. Bull., 83, 2887—2912. [107]
Cullis, C. G. (1904): The mineralogical changes observed in the cores of the Funafuti borings. — In: W. J. SOLLAS et al. (ed.), The Atoll of Funafuti. — Rept. Coral Reef Comm., Roy. Soc., 392—420. [317]
CUMINGS, E. R. (1932): Reefs or bioherms? — Geol. Soc. Amer. Bull., 43, 331—352. [260]
CUMMINS, W. A. (1959): The Lower Ludlow Grits in Wales. — Liverpool, Manchester Geol. J., 2, 68—179. [19]
— (1962): The greywacke problem. — Liverpool Manchester Geol. J., 3, 51—72. [62, 128]
CURRAY, J. R. (1956): Dimensional grain orientation studies of recent coastal sands. — Bull. Amer. Assoc. Petrol. Geol., 40, 2440—2456. [72]
— (1960): Tracing sediment masses by grain size modes. — 21. Internat. Geol. Congr., Norden, Pt. 23, 119—130. [53, 56]
— (1964): Transgressions and regressions. — Pap. marine geol., SHEPARD commem. vol., R. L. MILLER (ed.), Macmillan Co., New York. [381]
— (1965): Late Quaternary history, Continental shelves of the United States. — In: WRIGHT, H. E., JR. & FREY, D. G. (eds.), The Quaternary of the United States. — Princeton, Univ. Press, 723—735. [87]
— (1969): Estuaries, lagoons, tidal flats, and deltas. — In: STANLEY, D. J. (ed.), The new concepts of continental margin sedimentation. — AGI short course lecture notes, Amer. Geol. Inst., Washington, lect. 3, 30 pp. [109, 111, 115, 116, 119, 120]
CURRAY, J. R., EMMEL, F. J., MOORE, D. G. & NORMAR, K. W. R. (1971): Migrating megachannels and sediment distribution, Bengal deep-sea fan (Abstr.). — Progr., VIII Internat. Sediment. Congr., Heidelberg, 19. [94]
CURTIS, D. M. (1970): Miocene deltaic sedimentation, Louisiana Gulf Coast. — In: MORGAN, J. P. (ed.), Deltaic sedimentation, modern and ancient. — Soc. Econ. Pal. Min., Spec. Publ., 15, 293—308. [111]
CURTIS, H. (1954): Mode of origin of pyroclastic debris in the Mehrten Formation of the Sierra Nevadas. — Calif. Univ., Dept. Geol. Sci. Bull., 29, 453—502. [183]
CURTIS, R., EVANS, G., KINSMAN, D. J. J. & SHEARMAN, D. J. (1963): Association of dolomite and anhydrite in the recent sediments of the Persian Gulf. — Nature, 197, 679—680. [315]

DAETWYLER, C. C. & KIDWELL, A. L. (1959): The Gulf of Batabano, a modern carbonate basin. — 5. World Petrol. Congr. New York, Sect. 1, 1—21. [251]
DAEVES, K. & BECKEL, A. (1948): Großzahl-Forschung und Häufigkeitsanalyse. — Verlag Chemie, Weinheim, 65 pp. (2. ed. 1958). [54]
DALEY, B. (1971): Diapiric and other deformational structures in an Oligocene argillaceous limestone. — Sediment. Geol., 6, 29—51. [84, 340]
DALY, R. A. (1910): Pleistocene glaciation and the coral reef problem. — Amer. J. Sci., ser. 4, 30, 297—308. [266]
DANGEARD, L. et al. (1964): Figures et structures observées au cours du tassement des vases sous l'eau. — C. R. Acad. Sci. Paris, 258, 5935—5938. [84]
DANIEL, E. J. (1954): Fractured reservoirs of Middle East. — Bull. Amer. Assoc. Petrol. Geol., 38, 774—815. [350]
DAPPLES, E. C. (1959): The behaviour of silica in diagenesis. — In: H. A. IRELAND (ed.), Silica in sediments. — Soc. Econ. Paleont. Miner. Spec. Publ., 7, 36—54. [149, 154]
— (1967 a): The diagenesis of sandstones. — In: G. LARSEN & G. V. CHILINGAR (ed.), Diagenesis in Sediments. — Elsevier Publ. Co., 91—125. [19]
— (1967 b): Silica as an agent in diagenesis. — In: G. LARSEN & G. V. CHILINGAR (ed.), Diagenesis in Sediments. — Elsevier Publ. Co., 323—342. [330]
DAPPLES, E. C. & ROMINGER, J. F. (1945): Orientation analysis of fine-grained clastic sediments: A report of progress. — J. Geol., 53, 246—261. [72]
D'ARGENIO, B. (1966): Le facies littorali mesozoiche nell'Appennino meridionale. — Boll. Soc. Natural. Napoli, 75, 493—552. [356]
DARWIN, C. (1837): On certain areas of elevation and subsidence in the Pacific and Indian oceans, as deduced from the study of coral formations. — Proc. Geol. Soc. London, 2, 552—554. [266]

DAUGHTRY, A. C., PERRY, D. & WILLIAMS, M. (1962): Magnesium isotopic distribution in dolomite. — Geochim. Cosmochim. Acta, 26, 857—866. [319]

DAVIES, D. K. (1968): Carbonate turbidites, Gulf of Mexico. — J. Sediment. Petrol., 38, 1100—1109. [360]

DAVIES, D. K., ETHRIDGE, F. G. & BERG, R. R. (1971): Recognition of barrier environments. — Amer. Assoc. Petrol. Geol. Bull., 55, 550—565. [119]

DAVIES, H. G. (1965): Convolute lamination and other structures from the lower coal measures of Yorkshire. — Sedimentology, 5, 305—325. [81]

DAVIS, R. A., JR. (1966): Willow River dolomite: Ordovician analogue of modern algal stromatolite environments. — J. Geol., 74, 908—923. [224]

DEBRENNE, F. (1964): Archaeocyatha. Contribution à l'étude des faunes cambriennes du Maroc, de Sardaigne et de France. — Notes et Mem. Serv. Geol. du Maroc, No. 179, Vol. 1 et 2. [232]

DEFFEYES, K. S., LUCIA, F. J. & WEYL, P. K. (1964): Dolomitization: Observations on the Island of Bonaire, Netherlands Antilles. — Science, 143, 678—679. [314, 315]

DEFLANDRE, G. (1952): Traité de Zoologie, Grassé: T. 1. — Masson et Cie., Paris, 439—470. [225]

DE GEER, G. (1912): A geochronology of the last 12 000 years. — Congr. Internat. Géol., Sess. 11, C. R., 241—258. [363]

DEGENS, E. T., KNETSCH, G. & REUTER, H. (1961): Ein geochemisches Buntsandstein-Profil vom Schwarzwald bis zur Rhön. — N. Jb. Geol. Paläont. Abh., 111, 181—233. [313]

DE GROOT, K. (1967): Experimental dedolomitization. — J. Sediment. Petrol., 37, 1216—1220. [321]

— (1969): The chemistry of submarine cement formation at Donat Hussain in the Persian Gulf. — Sedimentology, 12, 63—68. [284, 371]

DEININGER, R. W. (1964): Limestone-dolomite transition in the Ordovician Platteville formation in Wisconsin. — J. Sediment. Petrol., 34, 281—288. [314]

DEIRMENDJIAN, D. (1973): On volcanic and other particulate turbidity anomalies. — Adv. Geophys., 16, 267—296. [164]

DELMER, A. (1952): La sédimentation cyclique et notamment la sédimentation houillère considérée comme un phénomène d'oscillations de relaxation autoentretenues. — Troisième Congr. Avanc. études de stratigr. et géol. du Carbonif., Heerlen 1951, T. I, 135—139. [366]

DE MEIER, J. J. (1969): Fossil non-calcareous algae from insoluble residues of algal limestones. — Leidse Geol. Mededel., 44, 235—239. [224]

DENNEN, W. H. (1967): Trace elements in quartz as indicators of provenance. — Geol. Soc. Amer. Bull., 78, 125—130. [22]

DE RAAF, J. F. M. (1964): The occurrence of flute casts and pseudomorphs after salt crystals in the oligocene "Grès à ripplemarks" of the southern Pyrenees. — In: A. H. BOUMA & A. B. ROUWER (ed.), Turbidites. — Dev. in Sedimentol., 3, Elsevier, Amsterdam, 192—198. [91]

DE RAAF, J. F. M. & BOERSMA, J. R. (1971): Tidal deposits and their sedimentary structures. — Geol. en Mijnbouw, 50, 479—504. [116]

DE RAAF, J. F. M., READING, H. G. & WALKER, R. G. (1965): Cyclic sedimentation in the Lower Westphalian of North Devon, England. — Sedimentology, 4, 1—52. [380, 381]

DEUTSCH, R. (1972): Litho- und Biofazies des Korallenkalkes aus dem Oberen Jura im Gebiet von Valdegeña (Provinz Soria, Spanien). — Dipl.-Arbeit, Univ. Bochum, 39 pp. [234, 256]

DICKSON, J. A. D. (1966): Carbonate identification and genesis as revealed by staining. — J. Sediment. Petrol., 36, 491. [288]

DIETZ, V. (1973): Experiments on the influence of transport on shape and roundness of heavy minerals. — Contr. Sedimentol., 1, 103—125. Schweizerbart, Stuttgart. [40]

DILLER, J. S. (1890): Sandstone dikes. — Bull. Geol. Soc. Amer., 1, 411—442. [84]

DILLO, H. G. (1960): Sandwanderungen in Tideflüssen. — Mitt. Franzius Inst. f. Grund- u. Wasserbau, T. H. Hannover, 17. [77, 78]

DOBKINS, J. E. & FOLK, R. L. (1970): Shape development on Tahiti-Nui. — J. Sediment. Petrol., 40, 1167—1203. [64]

DODD, J. R. (1967): Magnesium and strontium in calcareous skeletons: A review. — J. Paleont., 41, 1313—1329. [208]

DODGE, C. F. (1965): Genesis of an upper Cretaceous offshore bar near Arlington, Texas. — J. Sediment. Petrol., **35**, 22—35. [72]
DOEGLAS, D. J. (1946): Interpretation of the results of mechanical analyses. — J. Sediment. Petrol., **16**, 19—40. [54]
— (1950): De interpretatie van Korrelgrootteanalysen I—V. — Verh. Geol. Mijnb. Gen., Geol. Ser., **15**, 247—328. [59]
— (1962): The structure of sedimentary deposits of braided rivers. — Sedimentology, **1**, 167—190. [70, 71]
— (1968): Grain-size indices, classification and environment. — Sedimentology, **10**, 83—100. [10, 12]
DONAHUE, J. (1965): Laboratory growth of pisolite grains. — J. Sediment. Petrol., **35**, 251—256. [251]
DONALDSON, A. C., MARTIN, R. H. & KANES, W. H. (1970): Holocene Guadalupe delta of Texas Gulf Coast. — In: MORGAN, J. P. (ed.), Deltaic sedimentation, modern and ancient. — Soc. Econ. Pal. Min., Spec. Publ., **15**, 107—137. [111, 113]
DONOVAN, R. N. & FOSTER, R. J. (1972): Subaqueous shrinkage cracks from the Caithness flagstone series (Middle Devonian) of Northeast Scotland. — J. Sediment. Petrol., **42**, 309—317. [84]
DOWNING, J. A. & COOKE, D. Y. (1955): Distribution of reefs of Woodbend group in Alberta, Canada. — Bull. Amer. Assoc. Petrol. Geol., **39**, 189—206. [273]
DREW, G. H. (1914): On the precipitation of calcium carbonate in the sea by marine bacteria, and on the action of denitrifying bacteria in tropical and temperate seas. — Carnegie Inst. Washington Publ., **182**, 7—45. [191]
DRONG, H. J. (1959): Zur Petrographie des Rotliegend-Eruptivs der Bohrung Weyhausen Z 1. — Geol. Rdsch., **48**, 55—65. [23]
— (1965): Die Schwerminerale des Dogger beta und ihre diagenetischen Veränderungen. — Vortr., Dt. Miner. Ges., Hannover. [41, 136]
DRUMMOND, J. M. (1964): An appraisal of fracture porosity. — Bull. Canad. Petr. Geol., **12**, 226—245. [346]
DRYDEN, A. L. & DRYDEN, C. (1946): Comparative rates of weathering of some common heavy minerals. — J. Sediment. Petrol., **16**, 91—96. [39]
DUFF, P. McL. D. & WALTON, E. K. (1962): Statistical basis for cyclothems: A quantitative study of the sedimentary succession in the East Pennine Coalfield. — Sedimentology, **1**, 235—255. [362, 376]
DUFF, P. McL. D., HALLAM, A. & WALTON, E. K. (1967): Cyclic Sedimentation. — Dev. in Sedimentology, **10**, Elsevier, Amsterdam, 280 pp. [362, 364, 365, 367, 374]
DUNBAR, C. O. & RODGERS, J. (1957): Principles of Stratigraphy. — Wiley & Sons, New York, 356 pp. [85, 109, 266, 329]
DUNCAN, J. R. & KULM, L. D. (1970): Mineralogy, provenance, and dispersal history of late Quaternary deep-sea sands in Cascadia basin and Blanco fracture zone off Oregon. — J. Sediment. Petrol., **40**, 874—887. [24, 62]
DUNHAM, R. J. (1962): Classification of carbonate rocks according to depositional texture. — In: W. E. HAM (ed.), Classification of carbonate rocks. — Amer. Assoc. Petrol. Geol. Mem., **1**, 108—121. [51, 276, 277]
— (1969 a): Early vadose silt in Townsend Mound (reef), New Mexico. — In: FRIEDMAN (ed.), Depositional environments in carbonate rocks. — Soc. Econ. Pal. Min., Spec. Publ., **14**, 139—181. [289]
— (1969 b): Vadose pisolite in the Capitan Reef (Permian), New Mexico and Texas. — In: G. M. FRIEDMAN (ed.), Depositional environments in carbonate rocks. — Soc. Econ. Pal. Min., Spec. Publ., **14**, 182—191. [222, 258]
— (1971): Meniscus cement. — In: O. P. BRICKER (ed.), Carbonate cements. — Johns Hopkins Stud. in Geol., **19**, 297—300, Johns Hopkins Press, Baltimore, Md. [289]
DUNNINGTON, H. V. (1954): Stylolite development post-dates rock induration. — J. Sediment. Petrol., **24**, 27—49. [297, 299]
— (1958): Generation, migration, accumulation, and dissipation of oil in Northern Iraq. — In: Habitat of Oil, ed. L. G. WEEKS. — Amer. Assoc. Petrol. Geol., 1194—1251. [348]
— (1967): Aspects of diagenesis and shape change in stylolitic limestone reservoirs. — 7th World Petrol. Congr., Mexico, Panel Disc. No. 3. [299]
DUNOYER DE SEGONZAC, G. (1969): Les minéraux argileux dans la diagenèse; passage au métamorphisme. — Mém. Serv. Carte géol. Als. Lorr., **29**, 320 pp. [146, 156]

— (1970): The transformation of clay minerals during diagenesis and low-grade metamorphism: A review. — Sedimentology, 15, 281—346. [146, 157]
DUNOYER DE SEGONZAC, G., FERRERO, J. & KUBLER, B. (1968): Sur la cristallinité de l'illite dans la diagenèse et l'anchimétamorphose. — Sedimentology, 10, 137—143. [156]
DUNOYER DE SEGONZAC, G., TARDY, Y. & MILLOT, G. (1969): Evolutions symétriques des micas au cours de l'alteration superficielle et de la diagenèse profonde. — Groupe franc., Argiles, 22, 51—67. [146]
DUPONT, E. (1881): Sur l'origine des calcaires dévoniens de la Belgique. — Bull. Acad. roy. Belg. ser. 3, 2, 264—280. [270]
DŻUŁYŃSKI, S., KSIAZKIEWICZ, M. & KUENEN, PH. H. (1959): Turbidites in flysch of the Polish Carpathian Mountains. — Bull. Geol. Soc. Amer., 70, 1089—1118. [123, 128, 369]
DŻUŁYŃSKI, S. & RADOMSKI, A. (1955): Origin of groove casts in the light of turbidity currents hypothesis. — Acta. Geol. Polon., 5, 47—66 (Engl. Summary 11—21). [96]
DŻUŁYŃSKI, S. & SANDERS, J. E. (1959): Bottom marks on firm lutite substratum underlying turbidite beds (Abstr.). — Bull. Geol. Soc. Amer., 70, 1594. [97]
DŻUŁYŃSKI, S. & SMITH, A. J. (1963): Convolute lamination, its origin, preservation, and directional significance. — J. Sediment. Petrol., 33, 616—627. [81]
DŻUŁYŃSKI, S. & WALTON, E. K. (1965): Sedimentary features of flysch and greywackes. — Developm. in Sedimentol., 7, Elsevier Publ. Co., Amsterdam, 274 pp.
[81, 84, 92, 94, 96, 97, 127, 369]

EARDLEY, A. J. (1938): Sediments of the Great Salt Lake, Utah. — Bull. Amer. Assoc. Petrol. Geol., 22, 1305—1411. [200, 220, 224, 252, 257]
EATON, G. P. (1964): Windborne volcanic ash. — J. Geol., 72, 1—35. [166]
ECKHARDT, F. J. & VON GAERTNER, H. R. (1955): Über Dolomite aus den Sedimenten des Ruhrkarbons. — Geol. Jb., 71, 427—432. [317]
— — (1962): Zur Entstehung und Umbildung der Kaolin-Kohlentonsteine. — Fortschr. Geol. Rheinl.-Westf., 3, 2, 623—640. [153]
EDELMAN, C. H. (1933): Petrologische provincies in het Nederlandse Kwartair. — Diss. Amsterdam. [43, 44]
EDELMAN, C. H. & DOEGLAS, D. J. (1934): Über Umwandlungserscheinungen an detritischem Staurolith und anderen Mineralien. — Tschermaks Miner. u. Petr. Mitt., 45, 225—234.
[41]
EDER, F. W. (1971): Riff-nahe detritische Kalke bei Balve im Rheinischen Schiefergebirge. — Göttinger Arb. Geol. Paläont., 10, 66 pp. [364]
EDIE, R. W. (1956): Origin and characteristics of evaporitic dolomite. — J. Alberta Soc. Petrol. Geol., 4, 16—23. [314]
— (1958): Mississippian sedimentation and oil fields in southeastern Saskatchewan. — Bull. Amer. Assoc. Petrol. Geol., 42, 94—126. [313]
— (1961): Devonian limestone reef reservoirs, Swan Hills oil field, Alberta. — Trans. Canad. Inst. Min. Metall., 54, 278—285. [268]
EDWARDS, W. N. & STUBBLEFIELD, C. J. (1948): Marine bands and other faunal marker horizons in relation to the sedimentary cycles of the middle coal measures of Nottinghamshire and Derbyshire. — Quart. J. Geol. Soc. London, 103, 4. [366]
EINSELE, G. (1963 a): "Convolute bedding" und ähnliche Sedimentstrukturen im rheinischen Oberdevon und anderen Ablagerungen. — N. Jb. Geol. Paläont., Abh., 116, 162—198.
[81, 91]
— (1963 b): Über Art und Richtung der Sedimentation im klastischen rheinischen Oberdevon (Famenne). — Abh. Hess. L. A. Bodenforsch. H. 43, 60 pp. [99, 108, 125, 369]
EINSELE, G., GENSER, H. & WERNER, F. (1967): Horizontal wachsende Riffplatten am Süd-Ausgang des Roten Meeres. — Senckenbergiana leth., 48 (Frankfurt), 359—379. [263]
EINSELE, G. & MOSEBACH, R. (1955): Zur Petrographie, Fossilerhaltung und Entstehung der Gesteine des Posidonienschiefers im Schwäbischen Jura. — N. Jb. Geol. Paläont., Abh., 101, 319—430. [301]
EKMAN, S. (1953): Zoogeography of the Sea. — London, 431 pp. [212]
ELIAS, G. K. (1963): Habitat of Pennsylvanian algal bioherms, Four Corners area. — In: R. O. BASS (ed.), Shelf Carbonates of the Paradox Basin. — Sympos. Field Conf., 4th, Four Corners Geol. Soc., 185—203. [274]
ELLIOTT, G. F. (1962): More microproblematica from the Middle East. — Micropaleontology, 8, 29—44. [200]

EMBRY, A. F., III & KLOVAN, J. E. (1972): Absolute water depth limits of late Devonian paleoecological zones. — Geol. Rdsch. 61, 2, 672—686. [264, 277]

EMERY, K. O. (1960): The sea off Southern California, a modern habitat of petroleum. — Wiley & Sons, New York, 366 pp. [86]

EMERY, K. O. & MILLIMAN, J. D. (1970): Quaternary Sediments of the Atlantic continental shelf. — Internat. Assoc. Quatern. Res., Quaternaria, 12, 3—18. [121]

EMERY, K. O. & RITTENBERG, S. C. (1952): Early diagenesis of California Basin sediments in relation to origin of oil. — Bull. Amer. Assoc. Petrol. Geol., 36, 735—806. [330]

EMERY, K. O., TRACEY, J. I., JR. & LADD, H. S. (1954): Geology of Bikini and nearby atolls. — Geol. Surv. Prof. Pap., 260-A, 265 pp. [228, 266, 272, 282, 285, 295]

ENAY, M. R. (1966): Les calcaires à coralliones de Saint-Germain-de-Joux (AIN). — Bull. Soc. Geol. France, Ser. 7, 7, 23—31. [263]

ENGEL, W. (1970): Die Nummuliten-Breccien im Flyschbecken von Ajdovscina in Slowenien als Beispiel karbonatischer Turbidite. — Verh. Geol. Bundesanst., Wien, H. 4, 570—582. [369]

ENGELHARDT, W. v. (1938): Über die Schwermineralsande der Ostseeküste zwischen Warnemünde und Darßer Ort und ihre Bildung durch die Brandung. — Z. angew. Mineral., 1, 30—59. [37, 38]

— (1960): Der Porenraum der Sedimente. — Springer, Berlin, Göttingen, Heidelberg, 207 pp. [131, 143, 158, 310, 338, 339, 345, 346]

— (1967): Interstitial solutions and diagenesis in sediments. — In: G. LARSEN & G. V. CHILINGAR (ed.), Diagenesis in Sediments. — Elsevier, Amsterdam, 503—521. [350]

— (1973): Die Bildung von Sedimenten und Sedimentgesteinen. — Sediment-Petrologie, Teil III. — Schweizerbart, Stuttgart, 378 pp. [27, 75, 89, 101, 145]

ENGELHARDT, W. v. & GAIDA, K. H. (1963): Concentration changes of pore solutions during the compaction of clay sediments. — J. Sediment. Petrol., 33, 919—930. [139]

ENGELHARDT, W. v. & PITTER, H. (1951): Über die Zusammenhänge zwischen Porosität, Permeabilität und Korngröße bei Sanden und Sandsteinen. — Heidelberg. Beitr. Miner. Petrogr., 2, 477—491. [158]

ENLOWS, H. E. & OLES, K. F. (1966): Authigenic silicates in marine Spencer formation at Corvallis, Oregon. — Bull. Amer. Assoc. Petrol. Geol., 50, 1918—1926. [27, 144]

EPSTEIN, S., GRAF, D. L. & DEGENS, E. T. (1963): Oxygen isotope studies on the origin of dolomites. — In: H. CRAIG, S. L. MILLER & G. J. WASSERBURG (eds.), Isotopic and cosmic chemistry. — North Holland Publ. Co., Amsterdam, 169—180. [305, 311, 318]

ERBEN, H. K., FLAJS, G. & SIEHL, A. (1968): Ammonoids: Early ontogeny of ultra-microscopical shell structure. — Nature, 219, 396—398. [246]

ERMANOVIES, I. F. (1967): Statistical application of plagioclase extinction in provenance studies. — J. Sediment. Petrol. 37, 683—687. [25]

ERNST, G. (1964): Zur Stratigraphie und Petrographie des Santon und Campan von Lägerdorf (Südwestholstein). — Z. dt. geol. Ges., 114, 575—582. [329]

ESCH, H. (1962): Zur Sedimentologie und Diagenese der Sandsteine und Schiefertone im Hangenden des Flözes P$_2$ (oberes Westfal B) in der Emscher-Mulde des Ruhrkarbons. — Fortschr. Geol. Rheinld.-Westf., 3, 647—666. [153, 154]

EVAMY, B. D. (1967): Dedolomitization and the development of rhombohedral pores in limestones. — J. Sediment. Petrol., 37, 1204—1215. [320]

— (1973): The precipitation of aragonite and its alteration to calcite on the Trucial Coast of the Persian Gulf. — In: PURSER, B. H. (ed.): The Persian Gulf. Springer, Berlin, Heidelberg, New York, 329—341. [285]

EVAMY, B. D. & SHEARMAN, D. J. (1962): The application of chemical staining techniques to the study of diagenesis in limestones. — Proc. Geol. Soc. London, 1599, 102. [282, 288]

— — (1965): The development of overgrowth from echinoderm fragments. — Sedimentology, 5, 211—233. [288, 289]

EVANS, G., MURRAY, J. W., BIGGS, H. E. J., BATE, R., BUSH, P. R. (1973): The oceanography, ecology, sedimentology and geomorphology of parts of the Trucial Coast barrier island complex, Persian Gulf. — In: PURSER, B. H. (ed.): The Persian Gulf. Springer, Berlin, Heidelberg, New York, 233—277. [285]

EVANS, G., SCHMIDT, V., BUSH, P. & NELSON, H. (1969): Stratigraphy and geologic history of the Sabkha, Abu Dhabi, Persian Gulf. — Sedimentology, 12, 145—159. [315, 325]

EVANS, G. & SHEARMAN, D. J. (1964): Recent celestine from the sediments of the Trucial Coast of the Persian Gulf. — Nature, 202, No. 4930, 385—386. [315]

Exon, N. (1972): Sedimentation in the outer Flensburg Fjord area (Baltic Sea) since the last glaciation. — Meyniana, 22, 5—62. [365]

Fabricius, F. H. (1961): Die Strukturen des „Rogenpyrits" (Kössener Schichten, Rät) als Beitrag zum Problem der „vererzten Bakterien". — Geol. Rdsch., 51, 647—657. [337]
— (1966): Beckensedimentation und Riffbildung an der Wende Trias/Jura in den bayerisch-tiroler Kalkalpen. — Internat. Sediment. Petrogr. Ser., Brill, Leiden, 143 pp.
[109, 241, 248, 249, 270, 275, 303, 329, 330, 355]
— (1972): Phytogenese mariner Ooide und „Grapestones" und strukturelle Abgrenzung der Ooide von natürlichen und künstlichen Sphäroiden. — Habil.-Schr., Techn. Univ. München, 190 pp. and 33 tables (to be publ. in English in "Contributions to Sedimentology", Schweizerbart, Stuttgart). [204, 252—257]
— thin sections [205, 207, 283, 291, 293, 331, 332]
Fabricius, F. H., Berdau, D. & Münnich, K.-O. (1970): Early Holocene oöids in modern littoral sands reworked from a coastal terrace, Southern Tunisia. — Science, 169, 757—760. [252]
Fairbridge, R. W. (1946): Submarine slumping and the location of oil bodies. — Bull. Amer. Assoc. Petrol. Geol., 30, 84—92. [84]
— (1950): Recent and pleistocene coral reefs of Australia. — J. Geol., 58, 330—401.
[266, 272, 295]
— (1957): The dolomite question. — In: R. J. Le Blanc & J. G. Breeding (ed.): Regional aspects of carbonate deposition. — Soc. Econ. Paleont. and Mineralogists, Spec. Publ. No. 5, 125—178. [311, 318]
— (1970): An ice age in the Sahara. — Geotimes 7/8, 18—20. [70, 267]
Falke, H. (1966): Zur Geochemie der Schichten der Kreuznacher Gruppe im Saar—Nahegebiet. — Geol. Rdsch., 55, 59—77. [57, 58]
Farinacci, A. (1964): Microrganismi dei calcari „Maiolica" e „Scaglia" osservati al microscopio elettronico (nannoconi e coccolithophoridi). — Boll. Soc. Paleont. Ital., 3, 172—181. [197, 232]
— (1965): Breccias and laminated dolomites of the Gavignano exposure. — Geol. Romana, IV, 129—144. [314]
— (1967): La serie giurassico-neocomiana di Monte Lacerone (Sabina). Nuove vedute sull' interpretazione paleogeografica delle aree di facies Umbro-Marchigiana. — Geol. Romana, VI, 421—480. [355]
Fenninger, A. (1968): Das Kalzitgefüge der sparitischen Kalke des Plassen (Tithonium, Nördliche Kalkalpen, Oberösterreich). — Sedimentology, 10, 273—291. [280]
Fenninger, A. & Flajs, G. (in press): Zur Mikrostruktur rezenter und fossiler Hydrozoa. — Biomineralisation, 7. [295]
Fenton, C. L. & Fenton, M. A. (1957): Paleoecology of the Precambrian of Northwestern North America. — In: H. S. Ladd (ed.), Paleoecology. — Geol. Soc. Amer. Mem., 67, 2, 103—116. [219]
Feray, D. E., Heuer, E. & Hewatt, W. G. (1962): Biological, genetic, and utilitarian aspects of limestone classification. — In: Ham (ed.), Classification of carbonate rocks. — Amer. Assoc. Petrol. Geol. Mem., 1, 20—32. [206]
Ferm, J. C. (1970): Alleghany deltaic deposits. — In: Morgan, J. P. (ed.), Deltaic sedimentation, modern and ancient. — Soc. Econ. Pal. Miner., Spec. Publ. 15, 246—255. [113]
Ferrero, J. & Kübler, B. (1964): Présence de dickite et de kaolinite dans les grès cambriens d'Hassi Messaoud. — Bull. Serv. Carte géol. Als. Lorr., 17, 247—261. [146, 157]
Fiege, K. (1936): Stratonomische Beobachtungen in der Grauwackenfazies des Harzer Kulms. — Stille-Festschr., Enke, Stuttgart, 44—64. [381]
— (1937): Untersuchungen über zyklische Sedimentation geosynklinaler und epikontinentaler Räume. — Abh. preuß. geol. L. A., N. F., H. 177, 218 pp. [381]
— (1938): Die Epirogenese des Unteren Muschelkalkes in Nordwestdeutschland, I. Teil. — Zbl. Miner., Geol. etc., Abt. B, 143—170. [314, 371]
— (1952): Sedimentationszyklen und Epirogenese. — Z. dt. geol. Ges., 103, 17—22. [361]
— (1960): Typologie und Entstehung der Sedimentationszyklen des Karbons, besonders der NW-europäischen Saumtiefe. — Quatrième Congr. Avanc. études de stratigr. et géol. du Carbonif., Heerlen 1958, T. I, 175—186. [381]
Fischbuch, N. (1962): Stromatoporoid zones of the Kaybob reef, Alberta. — J. Alberta Soc. Petrol. Geol., 10, 62—72. [268]

FISCHER, A. G. (1964): Sea-level oscillations in Triassic lagoonal limestones, Northern Alps. Abstr. — Ann. GSA & Assoc. Soc. Joint Meet., Miami Beach, Program, 61. [344, 84]
— (1965 a): The Lofer cyclothems of the Alpine Triassic. — In: D. F. MERRIAM (ed.), Symposium on cyclic sedimentation. — Kansas Geol. Surv. Bull., 169, 107—149.
[90, 204, 270, 356, 363, 374]
— electron micrograph [226]
FISCHER, A. G. & FINLEY, R., JR. (1949): Microstructure of some Pennsylvanian nautiloids. — Geol. Soc. Amer. Bull., 60, 1887. [246]
FISCHER, A. G. & GARRISON, R. E. (1967): Carbonate lithification on the sea floor. — J. Geol., 75, 488—496. [282, 288, 322, 359, 360]
FISCHER, A. G., GARRISON, R. E. & HONJO, S. (1964): Ultramicroscopic fossils and fabrics of limestones. — Ann. GSA & Assoc. Soc. Joint Meet., Miami Beach, Program, 61—62. [281]
FISCHER, A. G., HONJO, S. & GARRISON, R. E. (1967): Electron micrographs of limestones and their nannofossils. — Monogr. in Geol. and Paleontol. 1, Princeton Univ. Press, 141 pp. [197, 225]
FISHER, R. V. (1960): Criteria for recognition of laharic breccia, Southern Cascade Mountains, Washington. — Geol. Soc. Amer. Bull., 71, 127—132. [183]
— (1964): Maximum size, median diameter, and sorting of tephra. — J. Geophys. Res., 69, 341—355. [166, 168]
— (1966): Rocks composed of volcanic fragments and their classification. — Earth Sci. Rev., 1, 287—298. [160—162, 169, 181, 182]
— (1968): Puu Hou littoral cones, Hawaii. — Geol. Rdsch., 57, 837—864. [183]
— (1971): Features of coarse-grained, high-concentration fluids and their deposits. — J. Sediment. Petrol., 41, 916—927. [123]
FISHER, R. V. & WATERS, A. C. (1970): Base surge bed forms in maar volcanoes. — Amer. J. Sci., 268, 157—180. [90, 177]
FISHER, W. L. & MC GOWEN, J. H. (1967): Depositional systems in the Wilcox Group of Texas and their relationship to occurrence of oil and gas. — Gulf Coast Assoc. Geol. Soc. Trans., 17, 105—125. [115]
FISK, H. N. (1955): Sand facies of recent Mississippi delta deposits. — Proc. 4th World Petrol. Congr., Rome, Sec. I/c, 377—397. [109, 119, 376, 377]
— (1960): Recent Mississippi river sedimentation and peat accumulation. — Congr. Avan. Études Stratigr. Géol. Carbonif., C. R. 4, Heerlen 1958, 187—199. [109, 377]
— (1961): Bar-finger sands of Mississippi delta. — In: J. A. PETERSON & J. C. OSMOND (ed.), Geometry of sandstone bodies. — Amer. Assoc. Petrol. Geol., 29—52. [376, 377]
FISK, H. N. & MCFARLAN, E. Jr. (1955): Late Quaternary deposits of the Mississippi River. — Geol. Soc. Amer. Spec. Pap., 62, 279—302. [377]
FISKE, R. S. (1963): Subaqueous pyroclastic flows in the Ohanapecosh Formation, Washington. — Geol. Soc. Amer. Bull., 74, 391—406. [184]
— (1969): Recognition and significance of pumice in marine pyroclastic rocks. — Geol. Soc. Amer. Bull., 80, 1—8. [172, 185]
FISKE, R. S., HOPSON, C. A. & WATERS, A. C. (1963): Geology of Mount Rainier National Park Washington. —Geol. Surv. Prof. Pap., 444, 1—93. [183]
FISKE, R. S. & MATSUDA, T. (1964): Submarine equivalents of ash flows in the Tokiwa Formation, Japan. — Amer. J. Sci., 262, 76—106. [184—186]
FLAJS, G., pers. comm. —, micrograph. [237, 238]
FLÖRKE, O. (1955): Zur Frage des „Hoch"-Cristobalit in Opalen, Bentoniten und Gläsern. — N. Jb. Miner., Mh, 217—223. [328]
— (1972): Untersuchungen an amorphem und mikrokristallinem SiO_2. — Chemie d. Erde 22, 91—110. [328, 329]
— (1967): Die Modifikation des SiO_2. — Fortschr. Miner., 44, 181—230. [328]
FLÜGEL, E. (1966): Algen aus dem Perm der Karnischen Alpen. — Carinthia II, Sonderh. 25, Klagenfurt, 76 pp. [217, 227]
— (1967): Elektronenmikroskopische Untersuchungen an mikritischen Kalken. — Geol. Rdsch., 56, 341—358. [197]
— pers. comm. [220, 248]
FLÜGEL, E. & FLÜGEL-KAHLER, E. (1963): Mikrofazielle und geochemische Gliederung eines obertriadischen Riffes der nördlichen Kalkalpen (Sauwand bei Gußwerk, Steiermark, Österreich). — Mitt. Mus. f. Bergbau etc. am Landesmus. „Joanneum", Graz, H. 24, 128 pp. [228, 268, 270, 272]

FLÜGEL, E. & FRANZ, H. E. (1967): Über die lithologische Bedeutung von Coccolithen in Malmkalken des Flachwasserbereiches. — Eclog. geol. Helvet., **60**, 1—17. [225]
FLÜGEL, E., FRANZ, H. E. & OTT, W. F. (1968): Review on electron microscope studies of limestones. — In: MÜLLER & FRIEDMAN: Carbonate sedimentology in Central Europe. — Springer, 85—97. [296, 339]
FLÜGEL, E. & KIRCHMAYER, M. (1962): Zur Terminologie der Ooide, Onkoide und Pseudo-oide. — N. Jb. Geol. Pal., Mh., 113—123. [221]
FLÜGEL, E. & SY, E. (1959): Die Hydrozoen der Trias. — N. Jb. Geol. Paläont., Abh., **109**, 1—108. [234]
FLÜGEL, H. W. (pers. comm.) [123]
FLÜGEL, H. W. & PÖLSLER, P. (1965): Lithogenetische Analyse der Barmstein-Kalkbank B_2 nordwestlich von St. Koloman bei Hallein (Tithonium, Salzburg). — N. Jb. Geol. Paläont., Mh., 513—527. [282, 369]
FLÜGEL, H. W. & WEDEPOHL, K. H. (1967): Die Verteilung des Strontiums in oberjurassischen Karbonatgesteinen der Nördlichen Kalkalpen. — Contr. Miner. Petrol., **14**, 229—249. [268]
FOLK, R. L. (1951): Stages of textural maturity in sedimentary rocks. — J. Sediment. Petrol., **21**, 127—130. [14]
— (1959): Practical petrographic classification of limestones. — Bull. Amer. Assoc. Petrol. Geol., **43**, 1—38. [190, 203, 260, 275, 278, 293]
— (1960): Petrography and origin of the Tuscarora, Rose Hill, and Keefer Formations, Lower and Middle Silurian of eastern West Virginia. — J. Sediment. Petrol., **30**, 1—58. [134]
— (1962): Spectral subdivision of limestone types. — In: W. E. HAM (ed.), Classification of carbonate rocks. — Amer. Assoc. Petrol. Geol. Mem., **1**, 62—84. [191, 275, 277]
— (1965 a): On the earliest recognition of coprolites. — J. Sediment. Petrol., **35**, 272—273. [201]
— (1965 b): Some aspects of recrystallization in ancient limestones. — In: L. C. PRAY & R. C. MURRAY (ed.), Dolomitization and limestone diagenesis. — SEPM Spec. Publ., **13**, Tulsa, 14—48. [190, 191, 217, 280, 339]
— (1965 c): Petrology of sedimentary rocks. — Hemphill's, Austin, 159 pp. [25]
— (1966): A review of grain-size parameters. — Sedimentology, **6**, 73—93. [53]
— (1971) Longitudinal dunes of the northwestern edge of the Simpson Desert, Northern Territory, Australia. I. Geomorphology and grain size relationships. — Sedimentology **16**, 5—54. [101, 103]
— pers. comm. [323]
FOLK, R. L. & PITTMAN, J. S. (1971): Length-slow chalcedony, a new testament for vanished evaporites. — J. Sediment. Petrol., **41**, 1045—1058. [329]
FOLK, R. L. & WARD, W. (1957): Brazos River bar: a study in the significance of grain size parameters. — J. Sediment. Petrol., **27**, 3—26. [52, 55, 59]
FOLK, R. L. & WEAVER, C. E. (1952): A study of the texture and composition of chert. — Amer. J. Sci., **250**, 498—510. [328]
FONDEUR, C. (1964): Étude pétrographique détaillée d'un grès a structure en feuillets. — Rev. Inst. Franç. Pétrole, **19**, 901—920. [140, 149]
FONG, G. (1960): Geology of Devonian Beaverhill Lake formation, Swan Hills area, Alberta, Canada. — Bull. Amer. Assoc. Petrol. Geol., **44**, 195—209. [273, 348]
FORCE, L. M. (1969): Calcium carbonate size distribution on the west Florida shelf and experimental studies on the microarchitectural control of skeletal breakdown. — J. Sediment. Petrol., **39**, 902—934. [195]
FORCHE, F. (1935): Stratigraphie und Paläogeographie des Buntsandsteins im Umkreis der Vogesen. — Mitt. Geol. Staatsinst. Hamburg, H. 15, 15—55. [50, 56]
FORMAN, M. J. & SCHLANGER, S. O. (1957): Tertiary reef and associated limestone facies from Louisiana and Guam. — J. Geol., **65**, 611—627. [272]
FOTHERGILL, C. A. (1955): The cementation of oil reservoir sands and its origin. — Proc. 4th World Petrol. Congr., Sect. I, 300—312. [139, 149]
FRANKS, P. C. (1969): Nature, origin, and significance of cone-in-cone structures in the Kiowa formation (early Cretaceous), North-Central Kansas. — J. Sediment. Petrol., **39**, 1438—1454. [299, 300]
FREEMAN, T. (1962): Quiet water oolites from Laguna Madre, Texas. — J. Sediment. Petrol., **32**, 475—483. [257, 357]

— (1965): Post-lithification dolomite in the Joachim and Plattin formations (Ordovician), Northern Arkansas. — Ann. GSA & Assoc. Soc. Joint Meet., Kansas City, Program, 58. [320]
— (1971): Luminoscope detection and microprobe analysis of vadose calcite cement in an Ordovician biosparite (Abstr.). — Progr., VIII Internat. Sediment. Congr., Heidelberg, 116. [289]
FREISE, F. W. (1931): Untersuchungen über die Abnutzbarkeit von Mineralien bei der Verfrachtung in Wasser. — Miner. Petr. Mitt., 41, 1—7. [40]
FRENTZEN, K. (1932): Paläobiologisches über die Korallenvorkommen im oberen Weißen Jura bei Nattheim, O.-A.-Heidenheim. — Bad. Geol. Abh., 4, 43—57. [235, 263]
FREY, M. & HUNZIKER, C. (1973): Progressive niedriggradige Metamorphose glaukonitführender Horizonte in den helvetischen Alpen. — Contr. Miner. Petr., 39, 185—218. [156]
FREY, M. & NIGGLI, E. (1971): Illit-Kristallinität, Mineralfazien und Inkohlungsgrad. — Schweiz. Miner. Petr. Mitt., 51, 229—234. [155]
FREYBERG, B. v. (1957): Sedimentationszyklen und marine Transgressionen als tektonische Zeitmarken in der süddeutschen Trias. — Congr. geol. internat. XX Sess., Mexico, Sec. V, T. I., 201—208. [365]
FREYTET, P. (1965): Sédimentation microcyclothémique avec croûtes zonaires à algues dans le calcaire de Beauce de Chauffour-Étrechy (Seine et Oise). — Bull. Soc. Géol. France, sér. 7, 7, 309—313. [204]
FREYTET, P. & PLAZIAT, J.-C. (1965): Importance des constructions algaires dues à des Cyanophycées dans les formations continentales du Crétacé supérieur et de l'Éocène du Languedoc. — Bull. Soc. géol. de France, sér. 7, 7, 679—694. [222]
FRIEDMAN, G. M. (1961): Distinction between dune, beach, and river sands from their textural characteristics. — J. Sediment. Petrol., 31, 514—529. [59]
— (1962): On sorting, sorting coefficients, and the lognormality of the grain-size distribution of sandstones. — J. Geol., 70, 737—753. [53, 59]
— (1964): Early diagenesis and lithification in carbonate sediments. — J. Sediment. Petrol., 34, 777—813. [196, 204, 240, 275, 282, 288, 289, 292, 295, 315, 318, 321, 322, 349]
— (1965 a): Occurrence and stability relationships of aragonite, high-magnesian calcite, and low-magnesian calcite under deep-sea conditions. — Geol. Soc. Amer. Bull., 76, 1191—1196. [195, 322, 360]
— (1965 b): Terminology of crystallization textures and fabrics in sedimentary rocks. — J. Sediment. Petrol., 35, 643—655. [279, 296]
— (1965 c): On the origin of aragonite in the Dead Sea. — Isr. J. Earth Sci., 14, 79—85. [192]
— (1966): Occurrence and origin of Quaternary dolomite of Salt Flat, West Texas. — J. Sediment. Petrol., 36, 263—267. [316]
— (1967): Dynamic processes and statistical parameters compared for size frequency distribution of beach and river sands. — J. Sediment. Petrol., 37, 327—354. [56, 59, 61]
— (1968 a): Geology and geochemistry of reefs, carbonate sediments, and waters, Gulf of Aqaba (Elat), Red Sea. — J. Sediment. Petrol., 38, 895—919. [206]
— (1968 b): The fabric of carbonate cement and matrix and its dependence on the salinity of water. — In: G. MÜLLER & G. M. FRIEDMAN (ed.), Recent developments in carbonate sedimentology in Central Europe. — Springer, New York, 11—20. [284]
— (1972): Significance of Red Sea in problem of evaporites and basinal limestones. — Amer. Assoc. Petrol. Geol. Bull., 56, 1072—1086. [194]
— pers. comm. [201]
FRIEDMAN, G. M. & JOHNSON, K. G. (1966): The Devonian Catskill deltaic complex of New York, type example of a "Tectonic Delta Complex". — In: SHIRLEY (ed.), Deltas in their geologic framework. — Houston Geol. Soc., Spec. Publ., 171—188. [113]
FRIEDMAN, G. M. & SANDERS, J. E. (1967): Origin and occurrence of dolostones. — In: CHILINGAR, BISSELL & FAIRBRIDGE (ed.), Carbonate Rocks. — Developm. in Sedimentol., 9 A, Elsevier, 267—348. [308, 314, 316—318, 321]
FRIEDMAN, I. & HALL, W. E. (1963): Fractionation of O^{18}/O^{16} between coexisting calcite and dolomite. — J. Geol., 71, 238—243. [305]
FRIEDMAN, I. & SMITH, R. L. (1960): A new dating method using obsidian. Part I, The development of the method. — Amer. Antiquity, 25, 476—522. [187]
FRIEDMAN, I., SMITH, R. L. & LONG, W. D. (1966): Hydration of natural glass and formation of perlite. — Geol. Soc. Amer. Bull., 77, 323—328. [187]

FRIEDMAN, M. (1954): Miocene orthoquartzite from New Jersey. — J. Sediment. Petrol., **24**, 235—241. [140]
FRIEDMAN, S. A. (1954): Low temperature authigenic magnetite. — Econ. Geol., **49**, 101. [144]
FRITSCH, F. E. (1956): The structure and reproduction of the algae. — Cambridge Univ. Press, 2 Volumes. [215, 220]
FRITZ, G. K. (1958): Schwammstotzen, Tuberolithe und Schuttbreccien im Weißen Jura der Schwäbischen Alb. — Arb. Geol.-Pal. Inst. T.H. Stuttgart, N.F., Nr. 13, 118 pp. [228, 232, 271]
FRITZ, P. (1966): Zur Genese von Dolomit und zuckerkörnigem Kalk im Weißen Jura der Schwäbischen Alb (Württemberg). — Mikroskopische Untersuchungen und Isotopenanalysen. — Arb. Geol.-Pal. Inst. T.H. Stuttgart, N.F., Nr. 50, 104 pp. [318, 320, 321]
FRYE, J. C. & SWINEFORD, A. (1946): Silicified rock in the Ogallala Formation. — State Geol. Surv. Kansas Bull., **64**, 2, 33—76. [140]
FUDALI, R. F. & MELSON, W. G. (1972): Ejecta velocities, magma chamber pressure and kinetic energy associated with the 1968 eruption of Arenal volcano. — Bull. Volc., **35**, 383—401. [163, 164]
FÜCHTBAUER, H. (1948): Einige Beobachtungen an authigenen Albiten. — Schweiz. Miner. Petr. Mitt., **28**, 709—716. [333, 334]
— (1950): Die nichtkarbonatischen Bestandteile des Göttinger Muschelkalkes mit besonderer Berücksichtigung der Mineralneubildungen. — Heidelberg. Beitr. Miner. Petrogr., **2**, 235—254. [142, 334, 335]
— (1954): Eine sedimentpetrographische Grenze in der oberen Süßwassermolasse des Alpenvorlandes. — N. Jb. Geol. Paläont., Mh., 8, 337—347. [48]
— (1956): Zur Entstehung und Optik authigener Feldspäte. — N. Jb. Miner., Mh., 1, 9—23. [335, 336]
— (1958): Die petrographische Unterscheidung der Zechsteindolomite im Emsland durch ihren Säurerückstand. — Erdöl u. Kohle, **11**, 689—693. [293, 327]
— (1959): Zur Nomenklatur der Sedimentgesteine. — Erdöl u. Kohle, **12**, 605—613. [10, 61]
— (1961): Zur Quarzneubildung in Erdöllagerstätten. — Erdöl u. Kohle, **14**, 169—173. [148]
— (1963): Zum Einfluß des Ablagerungsmilieus auf die Farbe von Biotiten und Turmalinen. — Fortschr. Geol. Rheinl.-Westf., **10**, 331—336. [26, 106, 108]
— (1964a): Sedimentpetrographische Untersuchungen in der älteren Molasse nördlich der Alpen. — Eclog. geol. Helvet., **57**, 157—298. [7, 21, 23, 24, 29, 31, 33, 41, 61, 127, 128, 149, 197, 202, 272]
— (1964b): Fazies, Porosität und Gasinhalt der Karbonatgesteine des norddeutschen Zechsteins. — Z. dt. geol. Ges., **114**, 484—531. [271, 293, 309, 350, 363]
— (1967a): Die Sandsteine in der Molasse nördlich der Alpen. — Geol. Rdsch., **56**, 266—300. [19, 45, 49, 127, 382]
— (1967b): Der Einfluß des Ablagerungsmilieus auf die Sandstein-Diagenese im Mittleren Buntsandstein. — Sediment. Geol., **1**, 159—179. [25, 84, 106, 115, 134, 136, 137, 140, 144, 153, 313, 336]
— (1967c): Influence of different types of diagenesis on sandstone porosity. — 7th World Petrol. Congr., Mexico, Panel Disc., 3. Vol. 2, 353—369. [68, 132—134, 139, 148, 149, 153, 154, 157, 159]
— (1971): Cone-in-cone, a low-nucleation cement in marls. — In: O. P. BRICKER (ed.), Carbonate cements. — Johns Hopkins Stud. in Geol., **19**, 193—195, Johns Hopkins Press, Baltimore, Md. [272, 301]
— (1972): Influence of salinity on carbonate rocks in the Zechstein formation. — In: Geology of saline deposits. Proc. Hanover Symp., 1968 (Earth Sciences, 7) 23—31. — Unesco. [307, 314, 334, 336]
— (in press): Zur Diagenese nichtmariner Sandsteine. — Geol. Rdsch., **63**, 3. [148, 150]
FÜCHTBAUER, H. & ELROD, J. M. (1971): Different sources contributing to a beach sand, southeastern Bornholm (Denmark). — Sedimentology, **17**, 69—79. [68]
FÜCHTBAUER, H. & GOLDSCHMIDT, H. (1963): Beobachtungen zur Tonmineral-Diagenese. — Internat. Clay Conf., Stockholm. Pergamon Press, Vol. 1, 99—111. [294, 302]
— — (1964): Aragonitische Lumachellen im bituminösen Wealden des Emslandes. — Beitr. Miner. Petrogr., **10**, 184—197. [209, 245, 300, 322, 348, 374]

— — (1965): Beziehungen zwischen Calciumgehalt und Bildungsbedingungen der Dolomite. — Geol. Rdsch., 55, 29—40. [303, 304, 309, 310, 312, 315, 351]
FÜCHTBAUER, H. & REINECK, H.-E. (1963): Porosität und Verdichtung rezenter, mariner Sedimente. — Sedimentology, 2, 294—306. [86, 132]
FULLER, A. O. (1961): Size distribution characteristics of shallow marine sands from the Cape of Good Hope, South Africa. — J. Sediment. Petrol., 31, 256—261. [54]
FULLER, R. E. (1931): The aqueous chilling of basaltic lava on the Columbia River Plateau. Amer. J. Sci., 21, 281—300. [183]
FUNNELL, B. M. (1967): Foraminifera and radiolaria as depth indicators in the marine environment. — Marine Geol., 5, 333—347. [359]
FYFE, W. S. & BISCHOFF, J. L. (1965): The calcite-aragonite problem. — In: L. C. PRAY & R. C. MURRAY (ed.), Dolomitization and Limestone Diagenesis: a Symposium. — Soc. Econ. Paleont. Miner., Spec. Publ., 13, 3—13. [294]
FYFE, W. S., TURNER, F. J. & VERHOOGEN, J. (1958): Metamorphic reactions and metamorphic facies. — Geol. Soc. Amer. Mem., 73, 260 pp. [146]

GAARDER, K. R. & MARKALI, J. (1956): On the Coccolithophorid *Crystallolithus hyalinus* n. gen., n. sp. — Nytt. Mag. Bot., 5, 1—5. [225]
GARRELS, R. M. & CHRIST, C. L. (1965): Solutions, minerals, and equilibria. — Harper's Geosci. Ser., C. CRONEIS (ed.). — Harper & Row, New York, 450 pp. [145]
GARRISON, R. E. (1967): Pelagic limestones of the Oberalm beds (Upper Jurassic — Lower Cretaceous), Austrian Alps. — Bull. Canad. Petrol. Geol., 15, 21—49. [197]
GARRISON, R. E. & HONJO, S. (1964): Late Jurassic — Early Cretaceous pelagic sedimentation, western Salzburg province, Austria. — Abstr., Ann. GSA & Assoc. Soc. Joint Meet., Miami Beach, Program, 70. [366]
GARRISON, R. E., LUTERNAUER, J. L., GRILL, E. V., MAC DONALD, R. D. & MURRAY, J. W. (1969): Early diagenetic cementation of recent sands, Fraser River delta, British Columbia. — Sedimentology, 12, 27—46. [142, 143]
GASCHE, E. (1956): Über die Entstehung der Mumien und übrigen Kalkknollen aus dem Sequan des Berner Jura. — In: P. A. ZIEGLER, Geologische Beschreibung des Blattes Courtelary (Berner Jura) und zur Stratigraphie des Séquanien im zentralen Schweizer Jura. — Beitr. Geol. Karte d. Schweiz, N.F., 102. Lfg., 3 pp. [220, 221]
GASKELL, T. F. quoted in CHILINGAR et al. 1967. [351]
GASSER, W. (1967): Erste Resultate über die Verteilung von Schwermineralen in verschiedenen Flyschkomplexen der Schweiz. — Geol. Rdsch. 56, 300—308. [30]
GATES, O. (1959): Breccia pipes in the Shoshone Range, Nevada. — Econ. Geol., 54, 790—815. [182]
GAURI, K. L. & KALTERHERBERG, J. (1966): Sedimentstrukturen aus den niederrheinischen Braunkohlenschichten des Miozäns. — Sedimentology, 6, 115—133. [71]
GAY, P., JR. (1962): Origen, distribucion y movimiento de las arenas eolicas en el area de Yauca y Palpa. — Bol. Soc. Geol. Peru, 37, 37—58. [101]
GEBELEIN, C. D.: pers. comm. [223]
GEBELEIN, C. D. & HOFFMAN, P. (1969): Algal origin of dolomite in interlaminated limestone-dolomite sedimentary rocks. — In: O. P. BRICKER et al. (ed.), Carbonate Cements. — Bermuda Biol. Station, Spec. Publ., No. 3, 226—235. [222, 270, 308]
GEES, R. A. (1965): Moment measures in relation to the depositional environment of sands. — Eclog. geol. Helvet., 58, 209—213. [59]
— (1969): Surface textures of quartz sand grains from various depositional environments. — Beitr. elektronenmikroskop. Direktabb. Oberfl. 2, 283—297. [63]
GEIKIE, A. (1903): Textbook of geology, I, 4th ed. — McMillan & Co., New York. [319]
GEORGE, T. N. (1954): The Pre-Seminulan Main Limestone of the Avonian series in Breconshire. — Quart. J. Geol. Soc., 110, 283—322. [319]
— (1956): Carboniferous Main Limestone of the East Crop in South Wales. — Quart. J. Geol. Soc., 111, 309—322. [319]
— (1957): Limestones and dolomites. — Sci. Progr., 45, 95—103. [232]
GERMANN, K. (1966): Ablauf und Ausmaß diagenetischer Veränderungen im Wettersteinkalk (alpine Mitteltrias). — Diss. Univ. München, 122 pp. [288, 327]
GEVERS, T. W. (1930): Terrestrer Dolomit in der Etoscha-Pfanne, Südwestafrika. — Cbl. Miner. etc., Abt. B, 224—230. [316]

GEVIRTZ, J. L. & FRIEDMAN, G. M. (1966): Deep-sea carbonate sediments of the Red Sea and their implications on marine lithification. — J. Sediment. Petrol., 36, 143—151. [192, 281, 282]
GHENT, E. D. & HENDERSON, R. A. (1965): Significance of burrowing structures in the origin of convoluted laminae. — Nature, 207, 1286—1287. [81]
GIBBS, R. J. (1967): The geochemistry of the Amazon River system: Part I. The factors that control the salinity and the composition and concentration of the suspended solids. — Geol. Soc. Amer. Bull., 78, 10, 1203—1232. [7]
GIGNOUX, M. (1926): Géologie stratigraphique. — Masson, Paris, 588 pp. [241]
GILBERT, G. K. (1883): The topographic features of lake shores. — U.S. Geol. Surv. Ann. Rep., 5, 69—123. [111]
— (1890): Lake Bonneville. — U.S. Geol. Surv. Mon., 1. [111]
GILLBERG, M. (1965): A statistical study of till from Sweden. — Geol. Fören. Förhandl., 87, 84—108. [107]
GILMORE, C. W. (1926, 1927, 1928): Fossil footprints from the Grand Canyon. — Smithson. Misc. Coll., 77, no. 9, 41 pp., 80, no. 3, 78 pp., no. 8, 1—16. [103]
GINSBURG, R. N. (1953): Beachrock in South Florida. — J. Sediment. Petrol., 23, 89—92. [282]
— (1956): Environmental relationship of grain size and constituent particles in some South Florida carbonate sediments. — Bull. Amer. Assoc. Petrol. Geol., 40, 2384—2427. [195, 196, 217]
— (1957): Early diagenesis and lithification of shallow-water carbonate sediments in South Florida. — In: R. J. LE BLANC & J. G. BREEDING (ed.), Regional aspects of Carbonate deposition. — Soc. Econ. Pal. Miner. Spec. Publ., 5, 80—100. [195, 200, 289, 338]
— (1960): Ancient analogues of recent stomatolites. — Internat. Geol. Congr., Copenhagen, Part XXII, 26—35. [221]
GINSBURG, R. N. & LOWENSTAM, H. A. (1958): The influence of marine bottom communities on the depositional environment of sediments. — J. Geol., 66, 310—318. [193, 263]
GINSBURG, R. N. & SHINN, E. A. (1964): Distribution of the reef-building community in Florida and the Bahamas. — Bull. Amer. Assoc. Petrol. Geol., 48, (Abstr.) 527. [264]
GINSBURG, R. N., SHINN, E. A. & SCHROEDER, J. H. (1967): Submarine cementation and internal sedimentation within Bermuda reefs (Abstr.). — Geol. Soc. Amer., Progr. 1967, Ann. Meet., 78—79. [284]
GINSBURG, R. N., MARSZALEK, D. S. & SCHNEIDERMANN, N. (1971): Ultrastructure of carbonate cements in a Holocene algal reef of Bermuda. — J. Sediment. Petrol., 41, 472—482. [282, 285]
GLANGEAUD, L. (1957): Les éruptions sous-lacustres d'âge stampíen supérieure du plateau de Gergovie. — C. r. hebd. Séanc. Acad. Sci. Paris, 245, 338—340. [183]
GLOVER, E. D. & PRAY, L. C. (1971): High-magnesium calcite and aragonite cementation within modern subtidal sediment grains. — In: O. P. BRICKER (ed.), Carbonate cements. — Johns Hopkins Stud. in Geol., 19, 80—87, Johns Hopkins Press, Baltimore, Md. [285]
GLOVER, J. E. (1963): Studies in the diagenesis of some Western Australian sedimentary rocks. — J. Roy. Soc. Austral., 46, 33—56. [142]
GLYNN, P. W. (1971): Pacific coral reefs of Panama: Structure, distribution and predators (Abstr.). — Progr., VIII Internat. Sediment. Congr., Heidelberg, 36. [263]
GÖRLER, K. (1967): Ein Olisthostrom in der Molise-Zone (Süditalien) als Beispiel für Resedimentation aus Schlammströmen. — N. Jb. Geol. Paläont., Abh., 129, 65—82. [82]
GÖRLER, K. & REUTTER, K.-J. (1968): Entstehung und Merkmale der Olisthostrome. — Geol. Rdsch., 57, 484—514. [82]
GÖRZ, H. (1962): Zur Petrographie des Unterdevons im Westharz. — Beitr. Miner. Petrogr., 8, 232—266. [143]
GOLDICH, S. S. (1934): Authigenic feldspar in sandstones of southeastern Minnesota. — J. Sediment. Petrol., 4, 89—95. [142]
— (1938): A study in rock-weathering. — J. Geol., 46, 17—58. [24, 39]
GOLDSCHMIDT, H.: pers. comm. [328]
GOLDSMITH, J. R. (1953): A "simplexity principle" and its relation to "ease" of crystallization. — J. Geol., 61, 439—451. [305]
— (1960): Exsolution of dolomite from calcite. — J. Geol., 68, 103—109. [322]
GOLDSMITH, J. R. & GRAF, D. L. (1958): Structural and compositional variations in some natural dolomites. — J. Geol., 66, 678—693. [304, 305]

GOLDSMITH, J. R., GRAF, D. L. & JOENSUU, O. I. (1955): The occurrence of magnesian calcites in nature. — Geochim. Cosmochim. Acta, 7, 212—230. [303]
GOLDSMITH, J. R., GRAF, D. L., WITTERS, J. & NORTHROP, D. A. (1962): Studies in the system $CaCO_3$—$MgCO_3$—$FeCO_3$: 1. Phase relations; 2. A method for major-element spectrochemical analysis; 3. Compositions of some ferroan dolomites. — J. Geol., 70, 659—688. [321]
GOLDSMITH, J. R. & HEARD, H. C. (1961): Subsolidus phase relations in the system $CaCO_3$—$MgCO_3$. — J. Geol., 69, 45—74. [303]
GOREAU, T. F. (1959): The physiology of skeleton formation in corals. I. A method for measuring the rate of calcium deposition under different conditions. — Biol. Bull., 116 (1), 59—75. (With many references) [235, 358]
— (1963): Calcium carbonate deposition. — New York Acad. Sci., Ann., 160—165. [263]
— (1964): Fore-reef slope: structure, sediment, and community relationships. — Progr., 1964 Ann. Meet., Geol. Soc. Amer., Miami, 76. [269]
GOREAU, T. F. & HARTMAN, W. D. (1963): Boring sponges as controlling factors in the formation and maintenance of coral reefs. — In: Mechanics of hard tissue destruction. Amer. Assoc. Adv., Sci., Publ., 75, 25—54. [233]
GOULD, H. R. (1970): The Mississippi delta complex. — In: MORGAN, J. P. (ed.), Deltaic sedimentation modern and ancient. — Soc. Econ. Pal. Miner., Spec. Publ., 15, 3—30. [111, 377, 378]
GRABAU, A. W. (1913): Principles of Stratigraphy, vol. 1 & 2, New ed. 1960. — Dover Publications Inc., New York, 1185 pp. [190, 245, 249]
GRAF, D. L. (1960): Geochemistry of carbonate sediments and sedimentary carbonate rocks, part I, II, and IV B. — Illinois State Geol. Surv., Circ., 297, 39 pp., 298, 43 pp. [317]
GRAF, D. L., EARDLEY, A. J. & SHIMP, N. F. (1961): A preliminary report on magnesium carbonate formation in glacial Lake Bonneville. — J. Geol., 69, 219—223. [316]
GRAF, D. L. & GOLDSMITH, J. R. (1956): Some hydrothermal syntheses of dolomite and protodolomite. — J. Geol., 64, 173—186. [303, 304]
GRAF, D. L. & LAMAR, J. E. (1950): Petrology of Fredonia oolite in Southern Illinois. — Bull. Amer. Assoc. Petrol. Geol., 34, 2318—2336. [149, 281]
GRANT-MACKIE, J. A. & LOWRY, D. C. (1964): Upper Triassic rocks of Kiritehere, Southwest Auckland, New Zealand, pt. 1: Submarine slumping of Norian strata. — Sedimentology, 3, 296—317. [82]
GREENFIELD, L. J. (1963): Aragonite formation by marine bacteria. — Bull. Amer. Assoc. Petrol. Geol., 47, 358 (Abstr.). [193]
GREENMAN, N. N. (1951): The origin of the Randville dolomite of Dickinson and Iron Counties, Michigan. — Ph. D. thesis, Univ. Chicago (cit. in GRAF 1960). [314]
GREENSMITH, J. T. (1963): Clastic quartz, provenance and sedimentation. — Nature, 4865, 345—347. [21]
— (1965): Calciferous sandstone series sedimentation at the eastern end of the Midland Valley of Scotland. — J. Sediment. Petrol., 35, 223—242. [143]
GRÉGOIRE, CH. & MONTY, CL. (1963): Observations au microscope électronique sur le calcaire à pâte fine entrant dans la constitution de structures stromatolitiques du Viséen moyen de la Belgique. — Ann. Soc. Géol. Belg., 85, 10, 389—397. [224]
GREGORY, J. L. (1966): A Lower Oligocene delta in the subsurface of Southeastern Texas. — In: SHIRLEY, M. L. (ed.), Deltas in their geological framework. — Houston Geol. Soc., 213—227. [87]
GRIBNITZ, K.-H. (1952): Petrographische Untersuchungen an Nebengesteinen der Magerkohlenschichten (Namur ob. C) Westfalens, bei besonderer Berücksichtigung der Grauwacken. — N. Jb. Miner., Mh., 6, 174—190. (Abstr. in Z. dt. Geol. Ges., 103, 121—125, 1952.) [377]
GRIFFITHS, J. C. (1967): Scientific method in analysis of sediments. — Mc Graw Hill, 508 pp. [54]
GRIGGS, D. T., TURNER, F. J. & HEARD, H. C. (1960): Deformation of rocks at 500° C to 800° C. — In: Rock Deformation. — Geol. Soc. Amer. Mem., 79, 39—104. [301]
GRIMM, W.-D. (1964): Ausfällung von Kieselsäure in salinar beeinflußten Sedimenten. — Z. dt. Geol. Ges., 114, 590—619. [331]
— (1973): Stepwise heavy mineral weathering in the Residual Quartz Gravel, Bavarian Molasse (Germany). — Contr. Sedimentol., 1, 103—125, Schweizerbart, Stuttgart. [28]

GRIMSDALE, T. F. & VAN MORKHOVEN, F. P. C. M. (1955): The ratio between pelagic and benthonic foraminifera as a means of estimating depth of deposition of sedimentary rocks. — World Petrol. Congr. Proc., 4th., Rome 1/D (4), 473—491. [228, 359]
GRIPP, K. (1954): Kritik und Beitrag zur Frage der Entstehung der Kreide-Feuersteine. — Geol. Rdsch., **42**, 248—262. [329]
— (1956): Das Watt; Begriff, Begrenzung und fossile Vorkommen. Mit einem Anhang: Das Einzelwatt von Arcachon. — Senckenbergiana leth., **37**, 3/4, 149—181.
GROBA, E. & LUDWIG, G. (1956): Sedimentologische Untersuchungen zum Erkennen von Entmischungen der Sedimente an der Außenküste von SE-Rügen und im Greifswalder Bodden. — Geologie, **5**, 617—641. [39]
GROHNE, W. (1956): Die Geschichte des Jadebusens und seines Untergrundes. — Natur u. Volk, **86**, 225—233. [117]
GROISS, J. TH. (1964): Echinodermenreste in Schlämmproben aus dem Weißen Jura der Franken-Alb. — Geol. Bl. Nordost-Bayern, **14**, 2, 45—53. [249]
GROSS, G. (1964): Variations in the O^{18}/O^{16} and C^{13}/C^{12} ratios of diagenetically altered limestones in the Bermuda Islands. — J. Geol., **72**, 170—194. [321]
GRUMBT, E. (1966): Schichtungstypen, Marken und synsedimentäre Deformationsgefüge im Buntsandstein Südthüringens. — Ber. dt. Ges. geol. Wiss. A, **11**, 217—234, Berlin. [90]
GRY, H. (1938): Eine Methode zur Charakterisierung der Kornverteilung klastischer Sedimente. — Geol. Rdsch., **29**, 174—195. [54]
GUBLER, Y., BUGNICOURT, D., KÜBLER, B. & NYSSEN, R. (1966): Essai de nomenclature et caractérisation des principales structures sédimentaires. — Éd. Techniques, Paris, 291 pp. [72, 80, 92]
GWINNER, M. P. (1958): Schwammbänke, Riffe und submarines Relief im oberen weißen Jura der Schwäbischen Alb (Württemberg). — Geol. Rdsch., **47**, 402—418. [222]
— (1962): Geologie des Weißen Jura der Albhochfläche (Württemberg). — N. Jb. Geol. Paläont., Abh., **115**, 137—221. [271]
— (1971): Beobachtungen zur Sedimentation des Stubensandsteins (Mittlerer Keuper, km 4) im nördlichen Baden-Württemberg. — Z. dt. Geol. Ges., **122**, 209—212. [87]

HADDING, A. (1941): The pre-quaternary sedimentary rocks of Sweden. VI. Reef limestones. — Lunds Univ. Årsskr. N. F., Avd. 2, **37**, Nr. 10, 137 pp. [273]
— (1959): Silurian algal limestones of Gotland. — Lunds Univ. Årsskr. N. F., Avd. 2, **56**, Nr. 7, 26 pp. [273]
HÄNTZSCHEL, W. (1936): Die Schichtungs-Formen rezenter Flachmeer-Ablagerungen im Jade-Gebiet. — Senckenbergiana leth., **18**, 316—356. [74]
HÄNTZSCHEL, W., EL-BAZ, F. & AMSTUTZ, G. C. (1968): Coprolites. An annotated bibliography. — Geol. Soc. Amer. Mem. **108**, 132 pp. [201]
HÄNTZSCHEL, W. & REINECK, H.-E. (1968): Fazies-Untersuchungen im Hettangium von Helmstedt (Niedersachsen). — Geol. Staatsinst. Hamburg, Mitt., H. **37**, 5—39. [90]
HALLAM, A. (1964): Origin of the limestone-shale rhythm in the Blue Lias of England: a composite theory. — J. Geol., **72**, 157—169. [340, 364]
— (1967 a): Siderite- and calcite-bearing concretionary nodules in the Lias of Yorkshire. — Geol. Mag., **104**, 222—227. [301]
— (1967 b): Sedimentology and palaeogeographic significance of certain red limestones and associated beds in the Lias of the Alpine region. — Scott. J. Geol., **3**, 195—220. [108]
HALLAM, A. & O'HARA, M. J. (1962): Aragonitic fossils in the Lower Carboniferous of Scotland. — Nature, 4838, 273—274. [209, 246, 294]
HALLAM, A. & SWETT, K. (1966): Trace fossils from the Lower Cambrian Pipe Rock of the north-west Highlands. — Scott. J. Geol., **2**, 101—106. [80]
HAM, W. E. (1951): Dolomite in the Arbuckle limestone, Arbuckle Mountains, Oklahoma. — Geol. Soc. Amer. Bull., **62**, 1446—1447. [319]
HAMBLETON, A. W. (1962): Carbonate-rock fabrics of three Missourian stratigraphic sections in Socorro County, N. M. — J. Sediment. Petrol., **32**, 579—601. [195]
HAMBLIN, W. K. (1965): Internal structures of "homogeneous" sandstones. — Kansas State Geol. Surv. Bull., **175**, 1, 38 pp. [72]
HAMILTON, D. (1961): Algal growth in the Rhaetic Cothan Marble of Southern England. — Palaeontology, **4**, 3, 324—333. [224]
HANDIN, J. & HAGER, R. V. (1957): Experimental deformation of sedimentary rocks under confining pressure. — Bull. Amer. Assoc. Petrol. Geol., **41**, 1—50, (and **42**, 2897—2934). [346]

HARBAUGH, J. W. (1957): Mississippian bioherms in northeast Oklahoma. — Bull. Amer. Assoc. Petrol. Geol., **41**, 2530—2544. [271]
— (1967): Carbonate oil reservoir rocks. — In: CHILINGAR, BISSELL & FAIRBRIDGE (ed.), Carbonate Rocks. — Developm. in Sedimentol., **9 A**, Elsevier Publ. Co., 349—398. [352]
HARDER, H. (1965): Experimente zur „Ausfällung" der Kieselsäure. — Geochim. Cosmochim. Acta, **29**, 429—442. [329]
HARDER, H. & FLEHMIG, W. (1967): Bildung von Quarz aus verdünnten Lösungen bei niedrigen Temperaturen. — Naturwiss., **54**, 140. [138]
HARDIE, L. A. (1973): Recent cemented algal crusts from the carbonate tidal flats of Andros Island, Bahamas. — J. Sediment. Petrol. (in press) [223]
HARDIE, L. A. & GINSBURG, R. N. (in prepar.): Laminations and thin beds in the modern carbonate tidal flat sediments of northwest Andros Island, Bahamas. [223, 356, 365, 378]
HARMS, J. C. (1965): Sandstone dikes in relation to Laramide faults and stress distribution in the Southern Front Range, Colorado. — Geol. Soc. Amer. Bull., **76**, 981—1001. [84]
— (1969): Hydraulic significance of some sand ripples. — Geol. Doc. [88, 89]
HARMS, J. C. & FAHNESTOCK, R. K. (1965): Stratification, bed forms, and flow phenomena (with an example from the Rio Grande). — In: G. V. MIDDLETON (ed.), Primary sedimentary structures and their hydrodynamic interpretation. — SEPM Spec. Publ., **12**, 84—115. [73, 76, 78, 90, 104, 368]
HARRIS, S. A. (1958): Differentiation of various egyptian aeolian microenvironments by mechanical composition. — J. Sediment. Petrol., **28**, 164—174. [60]
HATCH, F. H., RASTALL, R. H. & BLACK, M. (1938): The petrology of the sedimentary rocks. 3rd ed. — Thomas Murby and Co., London, 383 pp. [319]
HATFIELD, C. B., ROHRBACHER, T. J. & FLOYD, J. C. (1968): Directional properties, paleoslope, and source of the Sylvania sandstone (Middle Devonian) of Southeastern Michigan and Northwestern Ohio. — J. Sediment. Petrol., **38**, 224—228. [79]
HATHAWAY, J. C. & ROBERTSON, E. C. (1961): Microtexture of artificially consolidated aragonitic mud. — Geol. Surv. Res., Art. No. 257, C 301—304. [295]
HAWLE, H., KRATOCHVIL, H., SCHMIED, H. & WIESENEDER, H. (1967): Reservoir geology of the carbonate oil and gas reservoir of the Vienna Basin. — 7th World Petrol. Congr., Mexico, Panel Disc. 3. [346]
HAY, R. L. (1966): Zeolites and zeolitic reactions in sedimentary rocks. — Geol. Soc. Amer. Spec. Pap., **85**, 130 pp. [142, 144, 155, 187]
HAY, R. L. & IIJIMA, A. (1968): Nature and origin of palagonite tuffs of the Honolulu Group on Oahu, Hawaii. — Stud. in Volcanology, COATS, HAY & ANDERSON (ed.). — Geol. Soc. Amer. Mem., **116**, 331—376. [188]
HAY, R. L. & MOIOLA, R. J. (1963): Authigenic silicate minerals in three desert lakes of eastern California. — Geol. Soc. Amer., Progr. Ann. Meet., p. 76 a. [144]
HAYES, J. B. (1964): Geodes and concretions from the Mississippian Warsaw Formation, Keokuk Region, Iowa, Illinois, Missouri. — J. Sediment. Petrol., **34**, 1, 123—133. [301]
HAYES, M. O. (1965): Role of mixing of grain-size modes in distinguishing between sedimentary environments, South Texas coast. — Ann. GSA & Assoc. Soc. Joint Meet., Kansas City, Progr., 73—74. [55]
— (1967): Hydrographic control of sediment patterns and environments on depositional coasts: A model for paleogeographic reconstruction. — 7th Internat. Sedimentol. Congr., preprint. [100, 121]
HEALD, M. T. (1950): Authigenesis in West Virginia sandstones. — J. Geol., **58**, 624—633. [149]
— (1952): Origin of chert in the Helderberg limestone of West Virginia. — Geol. Soc. Amer. Bull., **63**, 1261. [329]
— (1955): Stylolites in sandstones. — J. Geol., **63**, 101—114. [133, 134]
— (1956a): Cementation of Simpson and St. Peter sandstones in parts of Oklahoma, Arkansas, and Missouri. — J. Geol., **64**, 16—30. [134, 137, 138]
— (1956b): Cementation of Triassic arkoses in Connecticut and Massachusetts. — Bull. Geol. Soc. Amer., **67**, 1133—1154. [141]
— (1959): Significance of stylolites in permeable sandstones. — J. Sediment. Petrol., **29**, 251—253. [133]
HEALD, M. T. & ANDEREGG, R. C. (1960): Differential cementation in the Tuscarora sandstone. — J. Sediment. Petrol., **30**, 568—577. [139]

HEALD, M. T. & RENTON, J. J. (1966): Experimental study of sandstone cementation. — J. Sediment. Petrol., 36, 977—991. [149]
HECHT, F. E. (1935): Grundzüge der chemischen Fossilisation. — In: Erdölmuttersubstanz, Brennstoff-Geol., H. 10, Enke, 95—120. [200]
HECHT, F., HERING, O., KNOBLOCH, J., KUBELLA, K. & RÜHL, W. (1962): Stratigraphie, Speichergesteins-Ausbildung und Kohlenwasserstoff-Führung im Rotliegenden und Karbon der Tiefbohrung Hoya Z 1. — Fortschr. Geol. Rheinl.-Westf., 3, 1061—1074. [154]
HECKEL, P. H. (1972): Possible inorganic origin for stromatactis in calcilutite mounds in the Tully limestone, Devonian of New York. — J. Sediment. Petrol., 42, 7—18. [274]
HEDEMANN, H.-A. (1963): Die Gebirgstemperaturen in der Bohrung Münsterland 1 und die geothermische Tiefenstufe. — Fortschr. Geol. Rheinl.-Westf., 11, 403—418. [154]
HEEZEN, B. C. (1956): The origin of submarine canyons. — Sci. Amer. 195, 36—41. [92]
HEEZEN, B. C. & DRAKE, C. L. (1964): Grand Banks slump. — Amer. Assoc. Petrol. Geol. Bull., 48, 221—233. [92]
HEEZEN, B. C. & FISCHER, A. G. (1971): Initial reports of the Deep Sea Drilling Project, Vol. VI: 40. Regional problems, 1301—1305. [366]
HEEZEN, B. C., HOLLISTER, C. & RUDDIMAN, W. F. (1966): Shaping of the continental rise by deep geostrophic contour currents. — Science, 152, 502—508. [98, 123]
HEEZEN, B. C. & JOHNSON III, G. L. (1964): Submarine sand waves beneath the mediterranean undercurrent. — Amer. Geophys. Un. Trans., 45, 70. [123]
HEIKEN, G. (1972): Morphology and petrography of volcanic ash. — Geol. Soc. Amer. Bull., 83, 1961—1988. [162]
HEIM, A. (1916): Monographie der Churfirsten-Mattstock-Gruppe. III. Stratigraphie der Unteren Kreide und des Jura. Zur Lithogenesis. — Beitr. geol. Karte Schweiz, N. F., 20, 369—662. [220]
HELING, D. (1963): Zur Petrographie des Stubensandsteins. — Diss. Univ. Tübingen, 56 pp. [142, 151, 152]
— (1965): Zur Petrographie des Schilfsandsteins. — Beitr. Miner. Petrogr., 11, 272—296. [140, 142, 151]
HELLMERS, J. H. (1949): Der Vorgang der Verkieselung. — Abh. geol. L. A. Berlin, N. F., H. 218, 3—15. [140]
HELM, D. G. (1971): Succession and sedimentation of glacigenic deposits at Hendre, Anglesey. — Geol. J., 7, 271—298. [75]
HELMBOLD, R. (1952): Beitrag zur Petrographie der Tanner Grauwacken. — Beitr. Miner. Petrogr., 3, 253—288. [19, 22, 24]
HENNINGSEN, D. (1961): Untersuchungen über Stoffbestand und Paläogeographie der Gießener Grauwacke. — Geol. Rdsch., 51, 600—626. [19, 22]
HENSON, F. R. S. (1950): Cretaceous and Tertiary reef formations, and associated sediments in Middle East. — Bull. Amer. Assoc. Petrol. Geol., 34, 215—238. [263, 269, 270]
HERING, O. H. & ZIMMERLE, W. (1963): Simple method of distinguishing zircon, monazite, and xenotime. — J. Sediment. Petrol., 33, 472—473. [29, 30]
HERMAN, G. & BARKELL, C. A. (1957): Pennsylvanian stratigraphy and productive zones, Paradox salt basin. — Amer. Assoc. Petrol. Geol. Bull., 41, 861—881. [315]
HERRMANN, A. (1956): Der Zechstein am südwestlichen Harzrand (seine Stratigraphie, Fazies, Paläogeographie und Tektonik). — Geol. Jb., 72, 1—72. [272]
HERTWECK, G. (1972): Georgia coastal region, Sapelo Island, U.S.A.: Sedimentology and biology. V, Distribution and environmental significance of Lebensspuren and in-situ skeletal remains. — Senckenbergiana marit., 4, 125—167. [208]
HESEMANN, J. (1939): Diluvialstratigraphische Geschiebeuntersuchungen zwischen Elbe und Rhein. — Abh. naturwiss. Ver. Bremen, 31, 247—285. [47]
HESSE, R. (1965): Herkunft und Transport der Sedimente im bayerischen Flyschtrog. — Z. dt. Geol. Ges., 116, 403—426. [56, 123, 128]
HILLER, K. (1964): Über die Bank- und Schwammfazies des Weißen Jura der Schwäbischen Alb (Württemberg). — Arb. Geol.-Pal. Inst. T. H. Stuttgart, N. F., 40, 190 pp. [222, 232, 263, 271]
HILTERMANN, H. (1963): Erkennung fossiler Brackwassersedimente unter besonderer Berücksichtigung der Foraminiferen. — Fortschr. Geol. Rheinl.-Westf., 10, 49—52. [227]
— pers. comm. [232]
HINZE, C. & MEISCHNER, D. (1968): Gibt es rezente Rot-Sedimente in der Adria? — Marine Geol., 6, 53—71. [109]

HJULSTRÖM, F. (1939): Transportation of detritus by moving water. — In: TRASK (ed.): Recent Marine Sediments. — Amer. Assoc. Petrol. Geol., 5—31. [75]

HOBBS, C. R. B. (1957): Petrography and origin of dolomite-bearing carbonate rocks of Ordovician age in Virginia. — Virginia Polytechn. Inst. Bull., 50, 5, 1—128. [314]

HÖLDER, H. (1961): Das Gefüge eines Placunopsis-Riffs aus dem Hauptmuschelkalk. — Jber. u. Mitt. oberrh. geol. Ver., N. F., 43, 41—48. [245, 275]

HOEPPENER, R. (1956): Zum Problem der Bruchbildung, Schieferung und Faltung. — Geol. Rdsch., 45, 247—283. [134]

HÖTZL, H. (1966): Zur Kenntnis der Tressenstein-Kalke (Ober-Jura, Nördliche Kalkalpen). — N. Jb. Geol. etc., Abh. 123, 281—310. [204]

HÖVERMANN, J. & POSER, H. (1951): Morphometrische und morphologische Schotteranalysen. — Proc. 3rd. Internat. Congr. of Sedimentology, Groningen- Wageningen, 135—156. [66]

HOFFMAN, P., LOGAN, B. W. & GEBELEIN, C. W. (1971): Recent stromatolites and loferites, Shark Bay, Western Australia (Abstr.). — Progr., VIII Internat. Sediment. Congr. Heidelberg, 42. [224]

HOFFMEISTER, J. E. & MULTER, H. G. (1964): Growth-rate estimates of a Pleistocene coral reef of Florida. — Geol. Soc. Amer. Bull., 75, 353—357. [272]

HOFMANN, F. (1956): Ein experimentelles Verfahren zur Bestimmung der Kornform von Sanden. — Eclog. geol. Helvet., 49, 506—512. [67]

— (1960): Materialherkunft, Transport und Sedimentation im schweizerischen Molassebecken. — Jb. St.-Gall. naturwiss. Ges., 76, 1—28. [45]

HOFMANN, H. J. (1969): Attributes of stromatolites. — Geol. Surv. Canada, Pap. 39—69, 58 pp. [222]

HOHLT, R. B. (1948): The nature and origin of limestone porosity. — Quart. Colorado School of Mines, 43, 1—51. [348, 350]

HOLLAND, H. D., KIRSIPU, T. V., HUEBNER, J. S. & OXBURGH, U. M. (1964): On some aspects of the chemical evolution of cave waters. — J. Geol., 72, 36—67. [260]

HOLLENSHEAD, C. T. & PRITCHARD, R. L. (1961): Geometry of producing Mesaverde sandstones, San Juan Basin. — In: Geometry of sandstone bodies, J. A. PETERSON & J. C. OSMOND (ed.). — Amer. Assoc. Petrol. Geol., Tulsa, 98—118. [378]

HOLLMANN, R. (1962): Über Subsolution und die „Knollenkalke" des Calcare Ammonitico Rosso Superiore im Monte Baldo (Malm; Norditalien). — N. Jb. Geol. Paläont., Mh., 163—179. [303]

— (1964): Subsolutions-Fragmente (Zur Biostratonomie der Ammonoidea im Malm des Monte Baldo/Norditalien). — N. Jb. Geol. Paläont., Abh., 119, 22—82. [125, 303]

HONNOREZ, J. (1972): La Palagonitisation. — Publ. Vulkaninst. I. Friedlaender, 9, 1—131. [189]

HONJO, S. & FISCHER, A. G. (1964): Fossil coccoliths in limestone examined by electron microscopy. — Science, 144, No. 1620, 837—839. [197, 225]

HONJO, S., FISCHER, A. G. & GARRISON, R. (1965): Geopetal pyrite in fine-grained limestones. — J. Sediment. Petrol., 35, 480—488. [338]

HOPKINS, M. E. (1958): Geology and petrology of the Anvil Rock sandstone of Southern Illinois. — Illinois State Geol. Surv., Circ. 256, 49 pp. [122, 376]

HOPPE, G. (1951): Die akzessorischen Schwermineralien in Eruptivgesteinen am Beispiel des Ramberggranites und anderer Harzer Gesteine. — Geologica, Berlin, 9, 114 pp. [31]

— (1962): Petrogenetisch auswertbare morphologische Erscheinungen an akzessorischen Zirkonen. — N. Jb. Miner., Abh., 98, 35—50. [31]

— (1963): Die Verwendbarkeit morphologischer Erscheinungen an akzessorischen Zirkonen für petrogenetische Auswertungen. — Abh. dt. Akad. Wiss., Kl. Bergb., Hüttenw., Montangeol., 1, Berlin, 130 pp. [31]

HORN, D. (1965): Diagenese und Porosität des Dogger-beta-Hauptsandsteines in den Ölfeldern Plön-Ost und Preetz. — Erdöl u. Kohle, 18, 249—255. [139, 143]

HORN, D. R., DELACH, M. N. & HORN, B. M. (1969): Distribution of volcanic ash layers and turbidites in the north Pacific. — Geol. Soc. Amer. Bull., 80, 1715—1724. [164, 165]

HORN, D. R., EWING, M., DELACH, M. N. & HORN, B. M. (1971 a): Turbidites of the northeast Pacific. — Sedimentology, 16, 55—69. [124]

HORN, D. R., EWING, M., HORN, B. M. & DELACH, M. N. (1971 b): Turbidites of the Hatteras and Sohm abyssal plains, Western North Atlantic. — Marine Geol., 11, 287—323. [61]

HOROWITZ, A. S. & POTTER, P. E. (1971): Introductory petrography of fossils. — Springer, Berlin, Heidelberg, New York, 202 pp. [209, 212]

Hoskin, C. M. (1963): Recent carbonate sedimentation on Alacran Reef, Yucatan, Mexico. — Nat. Res. Counc. Publ., 1089. [196]
— (1966): Coral pinnacle sedimentation, Alacran reef lagoon, Mexico. — J. Sediment. Petrol. 36, 1058—1074. [282]
Houbolt, J. J. H. C. (1957): Surface sediments of the Persian Gulf near the Qatar peninsula. — Diss. Utrecht, 113 pp. [272]
Howard, W. V. & David, M. W. (1936): Development of porosity in limestones. — Bull. Amer. Assoc. Petrol. Geol., 20, 1389—1412. [349]
Hsü, K. J. (1959): Flute- and groove-casts in the pre-Alpine flysch, Switzerland. — Amer. J. Sci., 257, 529—536. [97]
— (1960): Paleocurrent structures and paleogeography of the Ultrahelvetic flysch basins, Switzerland. — Bull. Geol. Soc. Amer., 71, 577—610. [127]
Hsü, K. J. & Siegenthaler, Ch. (1969): Preliminary experiments on hydrodynamic movement induced by evaporation and their bearing on dolomite problem. — Sedimentology, 12, 11—25. [308, 315]
Huang, T. C., Watkins, N. D., Shaw, D. M. & Kennett, J. P. (1973): Atmospherically transported volcanic dust in south Pacific deep sea sedimentary cores at distances over 3 000 km from the eruptive source. — Earth Planet. Sci. Lett., 20, 119—124. [164]
Hubert, C. (1965): Stratigraphy of the Quebec complex in the l'Islet-Kamouraska area, Quebec. — Unpubl. Ph. D. thesis, McGill Univ., Montreal, 192 pp. [108]
Huckenholz, H. G. (1959): Sedimentpetrographische Untersuchungen an Gesteinen der Tanner Grauwacke. — Beitr. Miner. Petrogr., 6, 261—298. [19, 22, 24]
— (1963 a): Der gegenwärtige Stand in der Sandsteinklassifikation. — Fortschr. Miner., 40, 151—192. [13, 15, 24]
— (1963 b): Mineral composition and texture in greywackes from the Harz Mountains (Germany) and in arkoses from the Auvergne (France). — J. Sediment. Petrol., 33, 914—918. [13, 14, 25]
Hudson, J. D. (1962): Pseudo-pleochroitic calcite in recrystallized shell-limestones. — Geol. Mag., 99, 492—500. [209, 296]
— (1963): The recognition of salinity-controlled mollusc assemblages in the great Estuarine Series (Middle Jurassic) of the Inner Hebrides. — Palaeontology, 6, 2, 318—326. [241]
— (1964): The petrology of the sandstones of the Great Estuarine Series, and the palaeogeography of Scotland. — Proc. Geologists' Assoc., 75, 499—528. [41]
— (1967): Speculations on the depth relations of calcium carbonate solution in recent and ancient seas. — Marine Geol., 5, 473—480. [232, 288, 359, 360]
— (1968): The microstructure and mineralogy of the shell of a Jurassic mytilid (Bivalvia). — Palaeontology, 11, 2, 163—182. [241]
— (1970): Algal limestones with pseudomorphs after gypsum from the Middle Jurassic of Scotland. — Lethaia, 3, 11—40. [224]
Hudson, R. G. (1942): On the rhythmic succession of the Yoredale series in Wensleydale. — Yorkshire Geol. Soc. Proc., New Ser., 20, 125—135. [366]
Hübner, H. (1965): Permokarbonische glazigene und periglaziale Ablagerungen aus dem zentralen Teil des Kongobeckens. — Stockholm Contrib. Geol., 13, 5, 39—61. [363]
Hülsemann, J. (1955): Großrippeln und Schrägschichtungs-Gefüge im Nordsee-Watt und in der Molasse. — Senckenbergiana leth., 36, 359—388. [79]
Huffman, G. G. & Price, W. A. (1949): Clay dune formation near Corpus Christi, Texas. — J. Sediment Petrol., 19, 118—127. [103]
Hughes Clarke, M. W. & Keij, A. J. (1973): Organisms as producers of carbonate sediment and indicators of environment in the southern Persian Gulf. — In: Purser, B. H. (ed.), The Persian Gulf. Springer, Berlin, Heidelberg, New York, 33—56. [356, 357]
Humbert, F. L. (1968): Selection and wear of pebbles on gravel beaches. — Diss. Groningen, 144 pp. [64]
Hummel, P. (1960): Petrographie, Gliederung und Diagenese der Kalke im Oberen Weißen Jura der Schwäbischen Alb. — Arb. Geol.-Pal. Inst. T. H. Stuttgart, N. F., Nr. 26, 86 pp. [271]
Humphrey, W. E. (1953): La facies Urgoniana de Cretácico medio de la región de Tampico, by Federico Bonet (Review). — Bull. Amer. Assoc. Petrol. Geol., 37, 1086—1088. [271]
Hutton, C. O. (1950): Studies of heavy detrital minerals. — Geol. Soc. Amer. Bull., 61, 635—716. [27, 31, 32]

IIJIMA, A. & HAY, R. L. (1968): Analcime composition in tuffs of the Green River Formation of Wyoming. — Amer. Miner., 53, 184—200. [336]
IIJIMA, A. & UTADA, M. (1966): Zeolites in sedimentary rocks, with reference to the depositional environments and zonal distribution. — Sedimentology, 7, 327—357. [144, 155]
ILLIES, H. (1949 a): Über die erdgeschichtliche Bedeutung von Konkretionen. — Z. dt. Geol. Ges., 101, 95—98. [302]
— (1949 b): Zur Diagenese der südbaltischen Schreibkreide. — Geol. Fören. Stockh. Förh., 71, 41—50. [329, 330]
— (1949 c): Die Schrägschichtung in fluviatilen und litoralen Sedimenten, ihre Ursachen, Messung und Auswertung. — Mitt. Geol. Staatsinst. Hamburg, H. 19, 89—109. [77, 89]
ILLING, L. V. (1954): Bahaman calcareous sands. — Bull. Amer. Assoc. Petrol. Geol., 38, 1—95. [198, 200, 201, 204, 250, 251, 255, 281]
ILLING, L. V. & WELLS, A. J. (1964): Penecontemporary dolomite in the Persian Gulf. — Bull. Amer. Assoc. Petrol. Geol., 48, (Abstr.), 532. [315]
ILLING, L. V., WELLS, A. J. & TAYLOR, J. C. M. (1965): Penecontemporary dolomite in the Persian Gulf. — In: L. C. PRAY & R. C. MURRAY (ed.), Dolomitization and limestone diagenesis. — SEPM Spec. Publ., 13, 89—111. [315]
IMBRIE, J. (1957): Chemical facies in the Florence shale (Permian) of Nebraska, Kansas, and Oklahoma. — Amer. Assoc. Petrol. Geol. Ann. Convent. St. Louis (cit. in GRAF, 1960). [314]
IMBRIE, J. & BUCHANAN, H. (1965): Sedimentary structures in modern carbonate sands of the Bahamas. — In: G. V. MIDDLETON (ed.), Primary sedimentary structures and their hydrodynamic interpretation. — SEPM Spec. Publ., No. 12, 149—172. [76, 77]
IMBRIE, J. & VAN ANDEL, TJ. H. (1964): Vector analysis of heavy-mineral data. — Geol. Soc. Amer. Bull., 75, 1131—1156. [46]
IMBT, R. F. & MCCOLLUM, S. V. (1950): Todd Deep Field, Crockett County, Texas. — Bull. Amer. Assoc. Petrol. Geol., 34, 239—262. [271]
INGELS, J. J. C. (1963): Geometry, paleontology, and petrography of Thornton reef complex, Silurian of northeastern Illinois. — Bull. Amer. Assoc. Petrol. Geol., 47, 405—440. [273]
INGLE, J. C., JR. & SCHNACK, E. J. (1971): Sorting in the surf zone; new evidence from fluorescent tracer studies (Abstr.). — Progr., VIII Internat. Sediment. Congr., Heidelberg, 47. [57]
INMAN, D. L. (1952): Measures for describing the size distribution of sediments. — J. Sediment. Petrol., 22, 125—145. [52, 55]
Institut Français du Pétrole (Team «Géologie Sédimentaire») (1959): Relations entre mode de gisement et propriétés physicochimiques des dolomies. — Rev. Inst. franç. du pétrole, 14, 4/5, 474—518. [314, 319]
IPPEN, A. T. (ed.) (1966): Estuary and coastline hydrodynamics. — McGraw Hill, New York, etc., 744 pp. [115]
IRION, G. (1970): Mineralogisch-sedimentpetrographische und geochemische Untersuchungen am Tuz Gölü (Salzsee), Türkei. — Chemie d. Erde 29, 163—226. [192]
— pers. comm. [192]
IRION, G. & MÜLLER, G. (1968 b): Mineralogy, petrology and chemical composition of some calcareous tufa from the Schwäbische Alb, Germany. — In: FRIEDMANN, G. M. & MÜLLER, G. (ed.), Recent developments in carbonate sedimentology in Central Europe. Springer, Berlin - Heidelberg - New York, 157—169. [259]
ISPHORDING, WAYNE C. (1972): Analysis of variance applied to measures of central tendency and dispersion in sediments. — J. Sediment. Petrol., 42, 107—121. [53]
JAMES, N. P. (1972): Holocene and Pleistocene calcareous crust (caliche) profiles: Criteria for subaerial exposure. — J. Sediment. Petrol., 42, 817—836. [258]
JANKOWSKY, W. (1955): Schichtenfolge, Sedimentation und Tektonik im Unterdevon des Rheintales in der Gegend von Unkel-Remagen. — Geol. Rdsch., 44, 59—86. [84]
JAVOY, M. & FAYARD, M. (1964): Étude structurale d'un «front de dolomitisation». — C. R. Acad. Sci. Paris, 258, 3716—3719. [311]
JESSEN, W. (1957): Besondere sedimentologische Erkenntnisse aus dem Westfal B und C im Schacht Graf Bismarck 10. — Geol. Jb., 74, 400—446. [363, 375]
JESSEN, W., KREMP, G. & MICHELAU, P. (1952): Gesteinsrhythmen und Faunenzyklen des Ruhrkarbons und ihre Ursachen. — C. R. 3e Congr. avanc. étud. strat. et géol. carbon., Heerlen, t. I, 289—294. [380]

JIPA, D. (1966): Relationship between longitudinal and transversal currents in the Paleogene of the Tarcau valley (Eastern Carpathians). — Sedimentology, 7, 299—305. [97]
— (1968): Azimuthal relationship between cross-stratification and current markings in flysch deposits: Upper Precambrian of Central Dobrogea, Romania. — J. Sediment. Petrol., 38, 192—199. [96]
JODRY, R. L. (1955): Rapid method for determining magnesium-calcium ratios of well samples and its use in predicting structure and secondary porosity in calcareous formations. — Bull. Amer. Assoc. Petrol. Geol., 39, 493—511. [319]
JØRGENSEN, P. (1965): Mineralogical composition and weathering of some late Pleistocene marine clays from the Kongsvinger Area, Southern Norway. — Geol. Fören. Förhandl., 87, 62—83. [2]
JOHANSSON, C. E. (1965): Structural studies of sedimentary deposits. — Geol. Fören. Förhandl., 87, 3—61. [70]
JOHNSON, J. G. (1971): Timing and coordination of orogenic, epeirogenic, and eustatic events. — Geol. Soc. Amer. Bull., 82, 3263—3298. [366]
JOHNSON, J. H. (1951): An introduction to the study of organic limestones. — Quart. Colorado School of Mines, 46, 2, 185 pp. [215, 219, 227, 228, 236]
— (1961): Limestone-building algae and algal limestones. — Colorado School of Mines, 297 pp. [214, 218]
— (1963): Pennsylvanian and Permian algae. — Quart. Colorado School of Mines, 58, 3, 211 pp. [215]
JOHNSON, R. F. (1962): Geology and ore deposits of the Cachoeira do Campo, Don Bosco, and Ouro Branco quadrangles, Minas Gerais, Brazil. — U. S. Geol. Surv. Prof. Pap., 341—B, 39 pp. [215]
JOHNSON, R. H. (1920): The cementation process in sandstone. — Bull. Amer. Assoc. Petrol. Geol., 4, 33—35. [140]
JOHNSTON, W. P. & LOWELL, J. D. (1961): Geology and origin of mineralized breccia pipes in Copper Basin, Arizona. — Econ. Geol., 56, 916—940. [182]
JONES, D. J. (1956): Introduction to microfossils. — Harper Brs., New York, 406 pp. [228, 249]
JONES, J. G. (1969 a): Intraglacial volcanoes of the Laugarvatn Region, Southwest Iceland — I. — Q. J. Geol. Soc. Lond., 124. [183]
— (1969 b): A lacustrine volcano of Central France and the nature of peperites. — Proc. Geol. Ass., 80, 177—188. [183]
— (1970): Intraglacial volcanoes of the Laugarvatn Region, Southwest Iceland, II. — J. Geol., 78, 127—140. [183]
JOPLING, A. V. (1965 a): Laboratory study of the distribution of grain sizes in cross-bedded deposits. — In: G. V. MIDDLETON (ed.), Primary sedimentary structures and their hydrodynamic interpretation. — SEPM Spec. Publ., 12, 53—65. [74]
— (1965 b): Hydraulic factors controlling the shape of laminae in laboratory deltas. — J. Sediment. Petrol., 35, 777—791. [76]
JORDAN, R. & STAHL, W. (1970): Isotopische Paläotemperatur-Bestimmungen an jurassischen Ammoniten und grundsätzliche Voraussetzungen für diese Methode. — Geol. Jb., 89, 33—62. [295]
JORDAN, W. M. (1964): Prevalence of sand-dune types in the Sahara desert. — Ann. GSA & Assoc. Soc. Joint. Meet., Miami Beach, Program, 104—105. [101]
JUDSON, S. (1968): Erosion of the land, or What's happening to our continents? — Amer. Scientist, 56, 356—374. [7]
JÜNGST, H. (1934): Zur geologischen Bedeutung der Synärese. Ein Beitrag zur Entwässerung der Kolloide im werdenden Gestein. — Geol. Rdsch., 25, 312—325. [84]
JÜNTGEN, H. & KARWEIL, J. (1962): Künstliche Inkohlung von Steinkohlen. — Freiberg. Forschungsh., A 229, 27—36. [153]
JUX, U. (1957): Die Riffe Gotlands und ihre angrenzenden Sedimentationsräume. — Acta Univ. Stockh., Stockh. Contr. Geol. I, 4, 41—89. [273]
— (1960): Die devonischen Riffe im Rheinischen Schiefergebirge. — N. Jb. Geol. Paläont., Abh., 110, 186—258 u. 259—392. [271]

KABELAC, F. (1955): Beiträge zur Kenntnis und Entstehung des unteren Weißjuras am Ostrand des südlichen Oberrheingrabens. — Ber. Naturf. Ges. Freiburg (Br.), 45, 5—57. [271]

KAHLE, CH. F. (1965): Possible roles of clay minerals in the formation of dolomite. — J. Sediment. Petrol., 35, 448—453. [311]
KALEY, M. E. & HANSON, R. F. (1955): Laumontite and leonhardite cement in Miocene sandstone from a well in San Joaquin Valley, California. — Amer. Miner., 40, 923—925. [155]
KALKOWSKY, E. (1901): Die Verkieselung der Gesteine in der nördlichen Kalahari. — Sitz.-Ber. u. Abh. d. Nat. Ges. Isis, Dresden. [327]
— (1908): Oolith und Stromatolith im norddeutschen Buntsandstein. — Z. dt. Geol. Ges., 60, 68—125. [220, 223, 250, 255]
KALTERHERBERG, J. (1956): Über Anlagerungsgefüge in grobklastischen Sedimenten. — N. Jb. Geol. Paläont., Abh., 104, 30—57. [71]
— (1968): Beziehungen zwischen Korngrößenzusammensetzung und Porenvolumen in einigen vorbelasteten Schichten. — Fortschr. Geol. Rheinld.-Westf., 15, 167—180. [132]
KAMPTNER, E. (1952): Das mikroskopische Studium des Skeletts der Coccolithineen (Kalkflagellaten). Übersicht der Methoden und Ergebnisse. — Mikroskopie, 7, Wien. [225]
KARCZ, I. (1964): Grain growth fabrics in the Cambrian dolomites of Skye. — Nature, 204, No. 4963, 1080—1081. [293]
KARL, F. (1959): Vergleichende petrographische Studien an den Tonalitgraniten der Hohen Tauern und der Tonalit-Graniten einiger periadriatischer Intrusivmassive. Ein Beitrag zur Altersfrage der zentralen granitischen Massen in den Ostalpen. — Jb. geol. Bundesanst. Wien, 102, 1, 92 pp. [45]
KARPOVA, G. V. (1969): Clay mineral post-sedimentary ranks in terrigeneous rocks. — Sedimentology, 13, 5—20. [147, 157]
KASTNER, M. (1971): Authigenic feldspars in carbonate rocks. — Amer. Miner., 56, 1403—1442. [334, 336]
KASTNER, M. & WALDBAUM, D. R. (1968): Authigenic albite from Rhodes. — Amer. Miner., 53, 1579—1602. [333]
KATZ, A. (1968): Calcian dolomites and dedolomitization. — Nature, 217, No. 5127, 439—440. [320, 321]
KAUTSKY, G., QUENSEL, P., ÅHMAN, E., FRIETSCH, R. & GEIJER, P. (1960): Archean geology of Västerbotten and Norrbotten, Northern Sweden. — Guide to Excursions A 32 and C 26, Internat. Geol. Congr. XXI Session, Norden. [219]
KAY, M. (1955): Sediments and subsidence through time. — Geol. Soc. Amer., Spec. Pap., 62, 665—684. [86]
KAYE, C. A. (1959): Shoreline features and Quaternary shoreline changes in Puerto Rico. — U. S. Geol. Surv., Prof. Pap., 317-B. [274]
KAZAKOV, A. V., TIKHOMIROVA, M. M. & PLOTNIKOVA, V. I. (1959): The system of carbonate equilibria. — Internat. Geol. Rev., 1, 1—39. [316]
KELLER, J. & NINKOVICH, D. (1972): Tephra-Lagen in der Ägäis. — Z. dt. Geol. Ges., 123, 579—587. [164]
KELLER, W. D. (1952): Analcime in the Popo Agie member of the Chugwater formation. — J. Sediment. Petrol., 22, 70—82. [144]
KEMPER, E. (1966): Beobachtungen an obereozänen Riffen am Nordrand des Ergene-Beckens (Türkisch-Thrazien). — N. Jb. Geol. Paläont., Abh., 125, 540—554. [263, 270]
KENDALL, C. G. ST. C. & SKIPWITH, P. A. d'E. (1968): Recent algal mats of a Persian Gulf lagoon. — J. Sediment. Petrol., 38, 1040—1058. [223]
— — (1969): Holocene shallow-water carbonate and evaporite sediments of Khor al Bazam, Abu Dhabi, southwest Persian Gulf. — Bull. Amer. Assoc. Petrol. Geol., 53, 841—869. [204, 285]
KENNEDY, G. C. (1950): A portion of the system silica-water. — Econ. Geol., 45, 629—653. [133]
KENNEDY, W. J. (1969): The correlation of the Lower Chalk of south-east England. — Proc. Geologists' Assoc. Engl., 80, 459—551. [232]
KENNEDY, W. J. & HALL, A. (1967): The influence of organic matter on the preservation of aragonite in fossils. — Proc. Geol. Soc. London, 1643, 253—255. [294]
KENYON, N. H. & STRIDE, A. H. (1970): The tide-swept continental shelf sediments between the Shetland Isles and France. — Sedimentology, 14, 159—173. [127]
KEPPER, J. C. (1966): Primary dolostone patterns in the Utah-Nevada Middle Cambrian. — J. Sediment. Petrol., 36, 548—562. [314]

KHVOROVA, T. V. (1957): Essai d'analyse stadiale des roches carbonatées (sur l'exemple des dépôts carbonifères de la plate-forme russe). — In: STRACHOW: Méthodes d'étude des roches sédimentaires. Transl. in Ann. du service d'information géologique du Bureau de Recherch. geol., geophys. et minières. No. 35, 1958, **1**, 255—267. [257, 320, 321]

KIMURA, T. (1967): Thickness distribution of sandstone and shale beds of the southern part of the Shimanto group in central Japan. 7th Internat. Sedimentol. Congr., Preprint. [368]

KING, R. H. (1947): Sedimentation in Permian Castile sea. — Amer. Assoc. Petrol. Geol. Bull., **31**, 470—477. [308]

KINSMAN, D. J. J. (1964a): Reef coral tolerance of high temperatures and salinities. — Nature, **202**, No. 4939, 1280—1282. [235]

— (1964b): Dolomitization and evaporite development, including anhydrite, in lagoonal sediments, Persian Gulf. — Ann. GSA & Assoc. Soc. Joint Meet., Miami Beach, Progr., 108—109. [192, 315]

— (1965): Gypsum and anhydrite of recent age, Trucial Coast, Persian Gulf. — In: J. L. RAU (ed.), Second Symposium on Salt, 1, North Ohio Geol. Soc., Cleveland (Ohio), 302—326. [315]

— (1967): Huntite from a carbonate evaporite environment. — Amer. Miner., **52**, 1332—1340.

— (1969): Interpretation of Sr^{2+} concentration in carbonate minerals and rocks. — J. Sediment. Petrol., **39**, 486—508. [192, 204, 325]

(1971): Diagenetic history of limestones determined from Sr^{2+} distribution. — In: O. P. BRICKER (ed.), Carbonate Cements. — Johns Hopkins Stud. in Geol., **19**, Johns Hopkins Press, Baltimore, Md., 259—263. [325]

KIRCHMAYER, M. (1962): Zur Untersuchung rezenter Ooide. — N. Jb. Geol. Paläont. Abh., **114**, 245—272. [251]

— (1963): Höhlenperlen, Vorkommen, Definition sowie strukturelle Beziehung zu ähnlichen Sedimentsphäriten. — Anz. math.-naturw. Kl. Österr. Akad. Wiss. 1963, **10**, 1—7. [251]

KIRSCH, H. & HALLBAUER, D. (1960): Autigene Albite in Sandsteinen des Ruhrkarbons. — N. Jb. Miner., Mh., 11/12, 248—257. [142]

KIRTLEY, D. W. & TANNER, W. F. (1968): Sabellariid worms: builders of a major reef type. — J. Sediment. Petrol., **38**, 73—78. [275]

KISCH, H. J. (1966): Chlorite-illite tonstein in high-rank coals from Queensland, Australia: Notes on regional epigenetic grade and coal rank. — Amer. J. Sci., **264**, 386—397. [145, 154]

— (1969): Coal-Rank and burial-metamorphic mineral facies. — Advances in Organ. Geochem. 1968, Pergamon Press, Oxford, 1969, 407—425. [155, 156]

KITANO, Y. & HOOD, D. W. (1965): The influence of organic material on the polymorphic crystallization of calcium carbonate. — Geochim. Cosmochim. Acta, **29**, 29—41. [192]

KITANO, Y. & KANAMORI, N. (1966): Synthesis of magnesian calcite at low temperatures and pressures. — Geochem. J. (Japan), **1**, 1—10. [192]

KITTLEMAN, L. R. JR. (1964): Application of Rosin's distribution to size-frequency analysis of clastic rocks. — J. Sediment. Petrol., **34**, 483—502. [55]

— (1973): Mineralogy, correlation, and grain size distributions of Mazama tephra and other postglacial pyroclastic layers, Pacific Northwest.-Geol. Soc. Amer. Bull., **84**, 2957—2980. [164]

KLEIBER, K. (1937): Geologische Untersuchungen im Gebiet der Hohen Rone. — Eclog. geol. Helvet., **30**, 419—430. [47]

KLEIN, G. DE V. (1963): Analysis and review of sandstone classifications in the North American geological literature. — Geol. Soc. Amer. Bull., **74**, 555—576. [13]

— (1964): Diverse origins of graded bedding. — Ann. GSA & Assoc. Soc. Joint Meet., Miami Beach. Progr., 109. [73]

— (1965): Dynamic significance of primary structures in the Middle Jurassic Great Oolite series, Southern England. — In: G. V. MIDDLETON (ed.), Primary sedimentary structures and their hydrodynamic interpretation. SEPM Spec. Publ., **12**, 173—191. [73, 97]

— (1966): Dispersal and petrology of sandstones of Stanley-Jackfork boundary, Ouachita fold belt, Arkansas and Oklahoma. — Bull. Amer. Assoc. Petrol. Geol., **50**, 308—326. [98]

— (1967 a): Comparison of recent and ancient tidal flat and estuarine sediments. — In: LAUFF, G. H. (ed.), Estuaries. — Amer. Assoc. Advances Sci. Bull., **83**, 207—218. [116]
— (1967 b): Paleoccurent analysis in relation to modern marine sediment dispersal patterns. — Bull. Amer. Assoc. Petrol. Geol., **51**, 366—382. [98]
— (1970): Tidal origin of a Precambrian quartzite — the lower fine-grained quartzite (Middle Dalradian) of Islay, Scotland. — J. Sediment. Petrol., **40**, 973—985. [79, 117]
— (1971 a): A Cambrian tidal sand body — the Eriboll sandstone of Northwest Scotland: An ancient — recent analog. — J. Geol., **79**, 400—415. [116]
— (1971 b): A sedimentary model for determining paleotidal range. — Geol. Soc. Bull., **82**, 2585—2592. [116]
— (1972): Sedimentary model for determining paleotidal range: reply. — Geol. Soc. Amer. Bull., **83**, 539—546. [116]
KLEMENT, K. W. (1966): Studies on the ecological distribution of lime-secreting and sediment-trapping algae in reefs and associated environments. — N. Jb. Geol. Paläont., Abh., **125**, 363—381. [270]
KLIOUMIS, pers. comm. [255]
KLOVAN, J. E. (1964): Facies analysis of the Redwater reef complex, Alberta, Canada. — Bull. Canad. Petrol. Geol., **12**, 1—100. [268]
KLÜPFEL, W. (1916): Zur Kenntnis des Lothringer Bathonien. — Geol. Rdsch., **7**, 1—29. [208, 370, 371]
— (1917): Über die Sedimente der Flachsee im Lothringer Jura. — Geol. Rdsch., **7**, 97—109. [370]
KNETSCH, G. (1950): Beobachtungen in der libyschen Sahara. — Geol. Rdsch., **38**, 40—58. [258]
KNEUPER, G. (1957): Zur Petrographie der Sandsteine des flözführenden Ruhrkarbons. — Westf. Berggewerkschaftskasse, H. 12, (KUKUK-Festschr.) Bochum, 47—56. [153]
KNOBLAUCH, G. (1963): Sedimentpetrographische und geochemische Untersuchungen an Weißjurakalken der geschichteten Fazies im Gebiet von Urach und Neuffen. — Diss. Univ. Tübingen, 105 pp. [41]
KNOKE, R. (1966): Untersuchungen zur Diagenese an Kalkkonkretionen und umgebenden Tonschiefern. — Contr. Miner. Petrol., **12**, 139—167. [135]
KOCH, E. & BLISSENBACH, E. (1960): Die gefalteten oberkretazisch-tertiären Rotschichten im Mittel-Ucayali-Gebiet, Ostperu. — Geol. Jb., Beih., **43**, 103 pp. [68, 129, 218]
KOECHLIN, E. (1949): Ein Beitrag zur Kenntnis der Solenoporaceen und Chaetetiden des Berner Jura. — Eclog. geol. Helvet., **42**, 476—480. [218]
KÖGLER, F.-C. (1967): Geotechnical properties of recent marine sediments from the Arabian Sea, Gulf of Oman, and the Baltic Sea. — 7th Internat. Sedimentol. Congr., Preprint. [338]
KÖLBL, L. (1966): Geologische Studie über die Bildung der tortonen Zwischenhorizonte von Matzen und die Entstehung ihrer Lagerstätten. — Erdöl-Erdgas-Z., **82**, 45—65. [79]
KÖSTER, E. (1964): Granulometrische und morphometrische Meßmethoden an Mineralkörnern, Steinen und sonstigen Stoffen. — F. Enke, Stuttgart, 336 pp. [64, 65, 71]
KÖWING, K., KRAUS, L. & RÜCKERT, G. (1968): Erläuterungen zur Geologischen Karte von Bayern 1:25000, Bl. 7837 Markt Schwaben. — Bayer. Geol. Landesamt, München, 147 pp. [127]
KOLPACK, R. L. (1964): Abyssal sand ripples in the Drake passage, Antarctica. — Ann. GSA & Assoc. Soc. Joint. Meet., Miami Beach, Progr., 112. [123]
KONISHI, K. (1961): Studies of Paleozoic Codiaceae and allied algae. Pt. I: Codiaceae. — Sci. Rep. Kanazawa Univ., VII, 159—261. [217]
KOPSTEIN, F. P. H. W. (1954): Graded bedding of the Harlech Dome. — Diss. Univ. Groningen, 97 pp. [70, 72, 128]
KORNICKER, L. S. & PURDY, E. G. (1957): A bahaman faecal-pellet sediment. — J. Sediment. Petrol., **27**, 126—128. [200, 201]
KOSSOVSKAJA, A. G. & SHUTOV, V. D. (1955): Der Charakter der Veränderungen von detritischem Biotit im Prozeß der Epigenese. — Dokl. Akad. Nauk SSSR, N. Ser., **101**, 541—544 (russ.). [144, 145]
— — (1957): Minéraux authigenes principaux, et leur diagnose dans les lames minces. — In: Méthodes d'étude des roches sédimentaires. Transl. in Annales du serv. d'inform. géol., No. 35, 1958, 180—210. [155, 157]

— — (1958): Zonality in the structure of terrigene deposits in platform and geosynclinal regions. — Eclog. geol. Helvet., **51**, 656—666. [143, 145, 146, 154, 155, 327]
— — (1965): Facies of regional epi- and metagenesis. — Iswestija A. N. SSSR, ser. geol. 1963, no. 7, 3—18, Engl.: Internat. geol. Rev., **7**, 1965, 1157—1167. [146, 147, 154, 155]
— — (1970): Main aspects of the epigenesis problem. — Sedimentology, **15**, 11—40. [146]
KRÄMER, F. (1961): Sediment-Untersuchungen im Mittleren Buntsandstein (sm) Süd-Niedersachsens. — Diss. Frankfurt, 182 pp. [22]
KRAMER, J. R. (1959): Correction of some earlier data on calcite and dolomite in sea water. — J. Sediment. Petrol., **29**, 465—467. [305, 310]
KRAUSKOPF, K. B. (1959): The geochemistry of silica in sedimentary environments. — In: Silica in Sediments, H. A. IRELAND (ed.). — Soc. Econ. Paleont. Miner., Spec. Publ. No. 7, 4—19. [328]
— (1967): Introduction to Geochemistry. — McGraw-Hill, New York, 721 pp. [299]
KREBS, W. (1966): Der Bau des oberdevonischen Langenaubach-Breitscheider Riffes und seine weitere Entwicklung im Unterkarbon (Rheinisches Schiefergebirge). — Abh. senckenb. naturf. Ges., **511**, 1—105. [268, 269]
— (1969): Über Schwarzschiefer und bituminöse Kalke im mitteleuropäischen Variscikum. Erdöl u. Kohle, **22**, 2—6 and 62—67. [374]
KREJCI-GRAF, K. (1960): Zur Geologie der Makaronesen: 4. Krustenkalke. — Z. dt. geol. Ges., **112**, 36—61. [257]
KRINSLEY, D. (1965): Comparison of modern and fossil surface textures of sand grains using electron microscopy. — Amer. chem. Soc. Petrol. Res. Fund. 9th ann. Res. Rep., 54 (Abstr.). [70]
KRINSLEY, D. & MARGOLIS, S. (1969): A study of quartz sand grain surface textures with the scanning electron microscope. — Trans. New York Acad. Sci. Ser. II, **31**: Sect. geol. Sci., 467—477. [69]
KRINSLEY, D. & TAKAHASHI, T. (1962a): Applications of electron microscopy to geology. — N. York Acad. Sci. Trans. Ser. 2, **25**, 3—22. [70]
— — (1962b): Surface textures of sand grains — An application of electron microscopy: Glaciation. — Science, **138**, 3546, 1262—1264. [70]
KRUIT, C., BROUWER, J. & EALEY, P. (1972): A deep-water sand fan in the Eocene Bay of Biscay. — Nature Phys. Sci., **240**, 99, 59—61. [82]
KRUMBEIN, W. C. (1936): Application of logarithmic moments to size frequency distributions of sediments. — J. Sediment. Petrol., **6**, 35—47. [53]
— (1937): Sediments and exponential curves. — J. Geol., **45**, 577—601. [56]
— (1940): Flood gravel of San Gabriel Canyon, California. — Geol. Soc. Amer. Bull., **51**, 639—676. [71]
— (1941): Measurement and geologic significance of shape and roundness of sedimentary particles. — J. Sediment. Petrol., **11**, 64—72. [65, 66]
— (1965): The cyclothem as a response to sedimentary environment and tectonism. — In: D. F. MERRIAM (ed.), Symposium on cyclic sedimentation. — Kansas Geol. Surv. Bull., **169**, 239—247. [362]
KRUMBEIN, W. C. & SLOSS, L. L. (1963): Stratigraphy and sedimentation. — Freeman, San Francisco, 2nd ed., 660 pp. [65, 121, 125, 129]
KRUMBEIN, W. E. (1963): Über Riffbildung von Placunopsis ostracina im Muschelkalk von Tiefenstockheim bei Marktbreit in Unterfranken. — Abh. Naturwiss. Ver. Würzburg, **4**, 1—15. [245, 275]
— (1968): Zur Frage der biologischen Verwitterung: Einfluß der Mikroflora auf die Bausteinverwitterung und ihre Abhängigkeit von edaphischen Faktoren. — Z. Allg. Mikrobiol., **8**, 107—117. [2]
KRYNINE, P. D. (1935): Arkose deposits in the humid tropics. A study of sedimentation in southern Mexico. — Amer. J. Sci., ser. 5, **29**, 353—363. [24]
— (1946a): Microscopic morphology of quartz types. — Proc. 2nd. Panamer. Congr. Min. Eng. Geol., Petropolis, Vol. III, 35—49. [19]
— (1946b): The tourmaline group in sediments. — J. Geol., **54**, 65—87. [30]
— (1948): The megascopic study and field classification of sedimentary rocks. — J. Geol., **56**, 130—165. [99, 139]
— (1949):The origin of red beds. — New York Acad. Sci. Trans., ser. 2, **11**, 60—68. [108]
KSIAZKIEWICZ, M. (ed.) (1962): Geological Atlas of Poland. Fasc. 13: Cretaceous and Early Tertiary in the Polish External Carpathians. — Inst. Geol. Warsaw. [127]

KÜBLER, B. (1958): Calcites magnésiennes d'eau douce dans le Tertiaire supérieur du Jura neuchâtelois (canton de Neuchâtel), Suisse. — Eclog. geol. Helvet., **51**, 676—685. [195]
— (1962a): Étude pétrographique de l'Oehningien (Tortonien) du Locle (Suisse occidentale). — Beitr. Miner. Petrogr., **8**, 267—314. [195, 339]
— (1962b): Étude de l'Oehningien (Tortonien) du Locle (Neuchâtel — Suisse). — Bull. Soc. Neuchâtel Sci. Natur., **85**, 5—42. [326, 339]
— (1964): Les argiles, indicateurs de métamorphisme. — Rev. Inst. franç. Pétrole, **19**, 10, 1093—1113. [156]
KÜHN, R. (1948): Über einen rezenten Sandstein. — N. Jb. Miner., Mh., B, H. 9—12, 334—336. [142]
KÜHN-VELTEN, H. (1955): Subaquatische Rutschungen im höheren Oberdevon des Sauerlandes. — Geol. Rdsch., **44**, 3—25. [82]
KUENEN, PH. H. (1938): Density currents in connection with the problem of submarine canyons. — Geol. Mag., **75**, 241—249. [123]
— (1950): Marine Geology. — Wiley, New York, 568 pp. [8, 265, 266, 272]
— (1953): Graded bedding with observations on Lower Paleozoic rocks of Britain. — Verh. Koninkl. Ned. Akad. Wetensch. Amsterdam, Afd. Nat., **20**, 1—47. [81, 82, 368]
— (1956): Experimental abrasion of pebbles. 2. Rolling by current. — J. Geol., **64**, 336—368. [65, 66]
— (1960a): Experimental abrasion of sand grains. — Internat. Geol. Congr. XXI, Norden, Pt. X., 50—53. [68]
— (1960b): Experimental abrasion: 4. Eolian action. — J. Geol., **68**, 427—449. [68]
— (1964a): Experimental abrasion: 6. Surf action. — Sedimentology, **3**, 29—43. [64, 65]
— (1964b): Pivotability studies of sand by a shape-sorter. — Developm. in Sedimentol., **1**, VAN STRAATEN (ed.), Deltaic and Shallow Marine Depos., Elsevier, 207—215. [68]
— (1964c): Deep-sea sands and ancient turbidites. — In: A. H. BOUMA & A. BROUWER (ed.), Turbidites. — Developm. in Sedimentol., **3**, Elsevier, Amsterdam, 3—33. [92, 368]
— (1964d): The shell pavement below oceanic turbidites. — Marine Geol., **2**, 236—246. [367]
— (1965): Value of experiments in geology. — Geol. en Mijnb., **44**, 22—36. [84, 91]
— (1966a): Light thrown on general problems by the Roumanian results. — Sedimentology, **7**, 323—326. [97]
— (1966b): Matrix of turbidites: Experimental approach. — Sedimentology, **7**, 267—297. [63]
— (1967a): Geosynclinal sedimentation. — Geol. Rdsch., **56**, 1—19. [86]
— (1967b): Emplacement of flysch-type sand beds. — Sedimentology, **9**, 203—243. [98]
— pers. comm. [73]
KUENEN, PH. H. & HUMBERT, F. L. (1964): Bibliography of turbidity currents and turbidites. — In: A. H. BOUMA & A. BROUWER (ed.), Turbidites. — Developm. in Sedimentol., **3**, Elsevier, Amsterdam, 222—246. [92, 369]
— — (1969): Grain size of turbidite ripples. — Sedimentology, **13**, 253—261. [98]
KUENEN, PH. H. & MENRAD, H. W. (1952): Turbidity currents, graded and non-graded deposits. — J. Sediment. Petrol., **22**, 83—96. [368]
KUENEN, PH. H. & MIGLIORINI, C. I. (1950): Turbidity currents as a cause of graded bedding. — J. Geol., **58**, 91—127. [367]
KÜRSTEN, M. (1960): Zur Frage der Geröllorientierung in Flußläufen. — Geol. Rdsch., **49**, 498—501. [70]
KUKAL, Z. (1970): Geology of recent sediments. — Academic Press, London, New York, 490 pp. [365, 367, 379]
KULICK, J. (1960a): Zur Stratigraphie und Paläogeographie der Kulm-Sedimente im Eder-Gebiet des nordöstlichen Rheinischen Schiefergebirges. — Fortschr. Geol. Rheinld.-Westf., **3**, 243—288. [96, 369, 381]
— (1960b): Driftmarken im Kulm des Edersee-Gebietes. — Fortschr. Geol. Rheinld.-Westf., **3**, 289—296. [96]
KULKE, H. (1969): Petrographie und Diagenese des Stubensandsteins (mittlerer Keuper) aus Tiefbohrungen im Raum Memmingen (Bayern). — Contr. Mineral. Petrogr., **20**, 135—163. [152]
KUNO, H., ISHIKAWA, T., KATSUI, Y., YAGI, K., YAMASAKI, M. & TANEDA, S. (1964): Sorting of pumice and lithic fragments as a key to eruptive and emplacement mechanism. — Japan. J. Geol. Geograph., **35**, 223—238.

LACROIX, A. & BLONDEL, F. (1927): Sur l'existence dans le Sud de l'Annam d'une péperite résultant de l'intrusion d'une basalte dans un sédiment à diatomées. — Compt. Rend., Acad. Sci. Paris, **184**, 1145—1148. [183]
LADD, H. S. (1950): Recent reefs. — Bull. Amer. Assoc. Petrol. Geol., **34**, 203—214. [263, 266, 272]
LADD, H. S., INGERSON, E., TOWNSEND, R. C., RUSSELL, M. & STEPHENSON, H. K. (1953): Drilling on Eniwetok Atoll, Marshall Islands. — Amer. Assoc. Petrol. Geol. Bull., **37**, 2257—2280. [266, 295, 340]
LADD, H. S. & TRACEY, J. I. (1949): The problem of coral reefs. — Sci. Month., **69**, 297—305. [268]
LADD, H. S., TRACEY, J. I., JR., WELLS, J. W. & EMERY, K. O. (1950): Organic growth and sedimentation on an atoll. — J. Geol., **58**, 4, 410—425. [266]
LAIRD, M. G. (1968): Rotational slumps and slump scars in Silurian rocks, Western Ireland. — Sedimentology, **10**, 111—120. [123]
LAJOIE, J. (1968): Turbidites sans matrice: Produits de diagénèse. — Naturaliste canad., **95**, 1243—1255. [63, 146]
LALOU, CL. (1957): Studies on bacterial precipitation of carbonates in sea water. — J. Sediment. Petrol., **27**, 190—195. [191, 254]
— (1957): Étude expérimentale de la production de carbonates par les bactéries des vases de la baie de Villefranche-sur-Mer. — Ann. Inst. Oceanogr., Paris, **33**, 202—267. [309]
LAMB, H. H. (1971): Volcanic activity and climate. — Palaeogr. Palaeoclim., Palaeoec., **10**, 203—230. [164]
LAMING, D. J. C. (1964): Sedimentary structures and paleocurrents in the lower New Red Sandstone, Devonshire, England. — Bull. Amer. Assoc. Petrol. Geol., **48**, (Abstr.), 535. [84]
LAND, L. S. (1964): Eolian cross-bedding in the beach dune environment, Sapelo Island, Georgia. — J. Sediment. Petrol., **34**, 389—394. [76, 103]
— (1967): Diagenesis of skeletal carbonates. — J. Sediment. Petrol., **37**, 914—930. [289, 321, 322, 355]
— (1972): Contemporaneous dolomitization of Middle Pleistocene reefs by meteoric water, North Jamaica. — Abstr., AAPG & SEPM 46th Ann. Mtg., Denver, Amer. Assoc. Petrol. Geol. Bull., 635. [323]
LAND, L. S. & EPSTEIN, S. (1970): Late Pleistocene diagenesis and dolomitization, North Jamaica. — Sedimentology, **14**, 187—200. [311, 321, 325]
LAND, L. S., MACKENZIE, F. T. & GOULD, S. J. (1967): Pleistocene history of Bermuda. — Bull. Geol. Soc. Amer., **78**, 993—1006. [289]
LANG, H. B. (1964): Dolomit und zuckerkörniger Kalk im Weißen Jura der mittleren Schwäbischen Alb (Württemberg). — N. Jb. Geol. Paläont., Abh., **120**, 253—299. [297, 309, 320, 321]
LANGBEIN, R. (1973): Über die petrographischen Strukturen aczessorischen Anhydrits, sowie Geochemie und Mechanismen seiner Bildung. — Chemie d. Erde, **32**, 45—79. [325]
LANGENBERG, J. H. & DE ROEVER, W. P. (1955): Pumpellyite, a widespread detrital mineral in Quaternary deposits of the Netherlands. — Geol. en Mijnb., **17**, 163 ff. [30]
LAPORTE, L. F. (1967): Carbonate deposition near mean sea-level and resultant facies mosaic: Manlius formation (Lower Devonian) of New York State. — Amer. Assoc. Petrol. Geol. Bull., **51**, 73—101. [309, 313, 314, 355]
LAPORTE, L. F. & IMBRIE, J. (1965): Phases and facies in the interpretation of cyclic deposits. — In: D. F. MERRIAM (ed.), Symposium on cyclic sedimentation. — Kansas Geol. Surv. Bull., **169**, 249—263. [381]
LARSSON, W. (1937): Vulkanische Asche vom Ausbruch des chilenischen Vulkans Quizapu (1932) in Argentinien gesammelt. — Bull. Geol. Inst. Upsala, **26**, 27—52. [168]
LAUDON, L. R. & BOWSHER, A. L. (1941): Mississippian formations of Sacramento Mountains, New Mexico. — Bull. Amer. Assoc. Petrol. Geol., **25**, 2107—2160. [271]
LECOMPTE, M. (1958): Les recifs paleozoiques en Belgique. — Geol. Rdsch., **47**, 384—401. [263, 264, 271]
LE DANOIS, E. (1948): Les profondeurs de la mer. — Paris, 303 pp. [271—273]
LEES, A. (1964): The structure and origin of the Waulsortian (lower Carboniferous) "reefs" of West-Central Eire. — Philos. Trans. Roy. Soc. London Ser. B, No. 740, **247**, 483—531. [270, 272]
LEES, A. & BULLER, A. T. (1971): Temperate water shallow marine carbonate sediments and their ancient equivalents (Abstr.). — Progr. VIII Internat. Congr., Heidelberg, 58. [197]

LEIGHTON, M. W. & PENDEXTER, C. (1962): Carbonate rock types. — In: W. E. HAM, (ed.), Classification of Carbonate Rocks. A Symposium. — Amer. Assoc. Petrol. Geol. Mem., 1, 33—61. [275, 278]
LEINE, L. (1968): Rauhwackes in the Betic cordilleras, Spain. — Diss. Univ. Amsterdam, 112 pp. [352]
LEISCHNER, W. (1961): Zur Kenntnis der Mikrofauna und -flora der Salzburger Kalkalpen. — N. Jb. Geol. Paläont., Abh., 112, 1—47. [227]
LEITMEIER, H. (1910): Zur Kenntnis der Karbonate. Die Dimorphie des kohlensauren Kalkes. I. Teil. — N. Jb. Miner., 1, 49. [192]
LEMCKE, K. (1955): Die Fazies der Molasse der Bohrung Scherstetten 1. — Geol. bavar., 24, 12—21. [129]
— (1962): Beziehungen zwischen Molassesedimentation und Alpentektonik an der Wende Oligozän/Miozän. — Z. dt. Geol. Ges., 113, 280—281. [382]
— (1972): Die Lagerung der jüngsten Molasse im nördlichen Alpenvorland. — Bull. Ver. Schweiz. Petrol.-Geol. u. Ing., 39, 29—41. [5]
LEMCKE, K., ENGELHARDT, W. v. & FÜCHTBAUER, H. (1953): Geologische und sedimentpetrographische Untersuchungen im Westteil der ungefalteten Molasse des süddeutschen Alpenvorlandes. — Beih. Geol. Jb., 11, 110 pp. [39, 43]
LEMOINE, M. (1967): Brèches sédimentaires marines à la frontière entre les domaines briançonnais et piémontais dans les Alpes occidentales. — Geol. Rdsch., 56, 320—335. [127, 203]
LEONARDI, P. (1967): Le Dolomiti. Geologia dei monti tra Isarco e Piave. — Consiglio Naz. Ric.; Giunta Province di Trento. Vol. I, 522 pp., Vol. II, 467 pp. [266, 318, 320]
LEOPOLD, L. B., WOLMAN, M. G. & MILLER, J. P. (1964): Fluvial processes in geomorphology. — W. H. Freeman Co., San Francisco and London, 522 pp. [57, 104]
LERBEKMO, J. F. & PLATT, R. L. (1962): Promotion of pressure-solution of silica in sandstones. — J. Sediment. Petrol., 32, 514—519. [134]
LESSERTISSEUR, J. (1955): Traces fossils d'activité animale et leur signification paléobiologique. — Mém. Soc. Géol. France, N. S., 34, 74, 150 pp. [79]
LEVORSEN, A. I. (1956): Geology of Petroleum. — Freeman Co., 703 pp. [84, 271, 348, 350—352]
LEWIN, J. C. (1962): Calcification. — In: R. A. LEWIN, Physiology and biochemistry of algae. — Academie Press, New York, London, 457—465. [196]
LINDHOLM, R. C. & FINKELMAN, R. B. (1972): Calcite staining: Semiquantitative determination of ferrous iron. — J. Sediment. Petrol., 42, 239—245. [289]
LINDSTRÖM, M. (1963): Sedimentary folds and the development of limestone in an early Ordovician sea. — Sedimentology, 2, 243—276. [206, 288]
LINK, T. A. (1950): Theory of transgressive and regressive reef (bioherm) development and origin of oil. — Bull. Amer. Assoc. Petrol. Geol., 34, 263—294. [263, 265]
LIPMAN, P. W. (1965): Chemical comparison of glassy and crystalline volcanic rocks. — U. S. Geol. Surv. Bull., 1201-D, 1—24. [187]
LIPPMANN, F. (1955): Ton, Geoden und Minerale des Barrême von Hohenggelsen. — Geol. Rdsch., 43, 475—503. [301, 302]
— (1960): Versuche zur Aufklärung der Bildungsbedingungen von Calcit und Aragonit. — Fortschr. Miner., 38, 2, 156—161. [192]
— (1967): Die Synthese des Norsethit, $BaMg(CO_3)_2$ bei ca. 20° und 1 at. Ein Modell zur Dolomitisierung. — N. Jb. Miner., Mh., 23—29. [375]
— (1973): Sedimentary carbonate minerals. — Minerals, Rocks and Inorganic Materials, Vol. 6. Springer, Berlin, Heidelberg, New York. 228 pp. [192, 323]
— pers. comm. [323]
LIPPMANN, F. & SCHLENKER, B. (1970): Mineralogische Untersuchungen am Oberen Muschelkalk von Haigerloch (Hohenzollern). — N. Jb. Miner. Abh., 113, 68—90. [336]
LIRER, L., PESCATORE, T., BOOTH, B. & WALKER, G. P. L. (1973): Two plinian pumice fall deposits from Somma-Vesuvius, Italy. — Geol. Soc. Amer. Bull., 84, 759—772. [164]
LISITSYN, A. P. & PETELIN, V. P. (1967): Features of distribution and modification of $CaCO_3$ in bottom sediments of the Pacific Ocean. — Lithology and Mineral Resources 5 (transl. from Russ., Consultants Bureau), 565—578. [194]
LLOYD, R. M. (1971): Some observations on recent sediment alteration ("micritization") and the possible role of algae in submarine cementation. — In: O. P. BRICKER (ed.), Carbonate cements. — Johns Hopkins Stud. in Geol., 19, 72—79, Johns Hopkins Press, Baltimore, Md. [255, 288]

LOCHMAN, C. (1949): Paleoecology of the Cambrian in Montana and Wyoming. — Nat. Res. Council, Comm. on Treat. of Mar. Ecol., Paleoecol., Ann. Rept. 1948—1949, 31—71. [246]

LOGAN, B. W., REZAK, R. & GINSBURG, R. N. (1964): Classification and environmental significance of algal stromatolites. — J. Geol., 72, 68—83. [221—224, 275]

LOHSE, H.-H. (1957): Erfahrungen bei der röntgenographischen Identifizierung semisalinarer und nichtsalinarer Minerale der Salzlagerstätten. — Diss. Univ. Kiel. [337]

LOMBARD, A. (1953): Les rythmes sédimentaires et la sédimentation générale. — Rev. Inst. Franç. Pétrole, 8, 9—48. [85, 369]

— (1956): Géologie sédimentaire. Les séries marines. — Masson et Cie., Paris; Vaillant-Carmanne. S. A., Liège, 722 pp. [86, 361, 363, 366]

LONG, J. T. & SHARP, R. P. (1964): Barchan-dune movement in Imperial Valley, California. — Geol. Soc. Amer. Bull., 75, 149—156. [101]

LONGWELL, C. R., FLINT, R. F. & SANDERS, J. E. (1969): Physical Geology. — J. Wiley & Sons, New York, London, Sydney, 685 pp. [107]

LOPATIN, G. V. (1952): River deposits of USSR (in Russ.), 366 pp. — Moscow. [7]

LOREAU, J.-P. (1970): Ultrastructuree de la phase carbonatée des oolithes marines actuelles. — C. R. Acad. Sci. Paris, 271, 816—819, Pétrographie sédimentaire. [253]

LOREAU, J.-P. & PURSER, B. H. (1973): Distribution and ultrastructure of Holocene ooids in the Persian Gulf. — In: PURSER, B. H. (ed.): The Persian Gulf. — Springer, Berlin, Heidelberg, New York, 279—328. [252]

LORENZ, V. (1974): Vesiculated tuffs. — Sedimentology, in press. [180]

LOVE, L. G. & AMSTUTZ, G. C. (1966): Review of microscopic pyrite. (From the Devonian Chattanooga Shale and Rammelsberg Banderz.) — Fortschr. Miner., 43, 2, 273—309. [337]

LOVELL, J. P. B. (1972): Diagenetic origin of graywacke matrix minerals: Discussion of paper by WHETTEN & HAWKINS. — Sedimentology, 19, 141—143. [62]

LOWENSTAM, H. A. (1948): Marine pool, Madison county, Illinois, Silurian reef producer. — In: Structure of typical American Oilfields. Vol. 3. — Amer. Assoc. Petrol. Geol., 153—188. [273]

— (1949): Biostratigraphic Studies. Niagaran inter-reef formations, Northeastern Illinois. — Ill. State Mus. Sci. Pap., IV., 1—146. [248, 329]

— (1950): Niagaran reefs of the Great Lakes area. — J. Geol., 58, 4, 430—487. [261, 263, 273, 274]

— (1954): Factors affecting the aragonite: calcite ratios in carbonate-secreting marine organisms. — J. Geol., 62, 284—322. [208—210, 237, 241]

— (1961): Mineralogy, O^{18}/O^{16} ratios, and strontium and magnesium contents of recent and fossil brachiopods and their bearing on the history of the oceans. — J. Geol., 69, 241—260. [239]

— (1963): Biologic problems relating to the composition and diagenesis of sediments. — In: T. W. DONNELLY (ed.), The Earth Sciences, Problems and progress in current research. — Univ. Chicago Press, 137—195. [209, 239, 247, 322]

LOWENSTAM, H. A. & EPSTEIN, S. (1957): On the origin of sedimentary aragonite needles of the Great Bahama Bank. — J. Geol., 65, 364—375. [196]

LOWMAN, S. W. (1949): Sedimentary facies in Gulf Coast. — Bull. Amer. Assoc. Petrol. Geol., 33, 1939—1997. [377, 381]

LOWRY, W. D. & DE RUDDER, R. D. (1966): Stylolites in Antietam sandstone, Hellgate Canyon, Rockbridge County, Virginia. — Ann. GSA Southeastern Sect. Meet., Athens, Progr., 33. [134]

LUCAS, G. (1955): Oolithes marines actuelles et Calcaires oolithiques récents sur le rivage africain de la Méditerranée Orientale (Égypte et Sud Tunisien). — Bull. Stat. Oceanogr. Salammbo (Tunisie) No. 52, 19—38. [252]

LUCIA, F. J. (1961): Dedolomitization in the Tansill (Permian) formation. — Geol. Soc. Amer. Bull., 72, 1107—1110. [320]

— (1962): Diagenesis of a crinoidal sediment. — J. Sediment. Petrol., 32, 848—865. [313, 350]

— (1967): Recent sediments and diagenesis of south Bonaire, Netherlands Antilles. — J. Sediment. Petrol., 38, 845—858. [308, 315]

LUCIA, F. J. & MURRAY, R. C. (1967): Origin and distribution of porosity in crinoidal rocks. — 7th World Petrol. Congr., Mexico, Panel Disc. no. 3. [289, 350, 351]

LUDWIG, G. (1955): Neue Ergebnisse der Schwermineral- und Kornanalyse im Oberkarbon und Rotliegenden des südlichen und östlichen Harzvorlandes. — Beih. Z. Geol. Nr. 14, 76 pp. [33, 34, 41]
LUDWIG, V. (1968): Zur Lithologie des „Kulms" bei Erbendorf/Oberpfalz (Bayern). — N. Jb. Geol. Paläont. Mh., 407—412. [41]
LÜBBEN, H. (1969): Grundgegebenheiten für Planung und Ablauf der Förderung aus den emsländischen Valendis-Lagerstätten. — Erdöl u. Kohle, 22, 373—377. [159]
LÜTTIG, G. (1954): Alt- und mittelpleistozäne Eisrandlagen zwischen Harz und Weser. — Geol. Jb., 70, 43—125. [47, 70]
— (1958): Methodische Fragen der Geschiebeforschung. — Geol. Jb., 75, 361—418. [47]
— (1962 a): The shape of pebbles in the continental, fluviatile and marine facies. — Internat. Assoc. Sci. Hydrol., Publ. 59, 252—258. [64]
— (1962 b): Geröllmorphometrie des Zechsteinkonglomerats im Schacht Rossenray 1. — Fortschr. Geol. Rheinl.-Westf., 6, 385—390. [64]
— (1964 a): Zur Geröllmorphometrie von Transgressionskonglomeraten. — In: VAN STRAATEN (ed.), Developm. in Sedimentol., Vol. 1, Deltaic and Shallow Marine Sediments, Elsevier, Amsterdam, 253—256. [64, 65]
— (1964 b): Die Aufgaben des Geschiebeforschers und des Geschiebesammlers. — Lauenburg. Heimat, Ratzeburg, H. 45, 6—26. [47]
LYNTS, G. W. (1966): Relationship of sedimentsize distribution to ecological factors in Buttonwood Sound, Florida Bay. — J. Sediment. Petrol., 36, 66—74. [193]

MACGREGOR, A. G. (1952): Eruptive mechanisms: Mt. Pelée, the Soufrière of St. Vincent and the Valley of Ten Thousand Smokes. — Bull. Volc., 12, 49—74. [172]
MACKENZIE, D. B. (1972): Tidal sand flat deposits in Lower Cretaceous Dakota Group near Denver, Colorado. — Mountain Geol., 9, Nos. 2—3, 269—277. [117]
MACKIE, W. (1896): The sands and sandstones of Eastern Moray. — Trans. Edinburgh Geol. Soc., 7, 148—172. [20]
MÄGDEFRAU, K. (1937): Der Aufbau der thüringischen Zechstein-Riffe. — Natur u. Volk, Frankfurt, 67, 48—58. [263]
MAKEDONOV, A. V. (1957): Corrélations des coupes à l'aide des concrétions. (Sur l'exemple du gisement de Vorkuta.) — In: STRACHOW, Méthodes d'étude des roches sédimentaires. Transl. in Ann. du service d'information géol. du Bureau de Rech. géol., geophys. et min., No. 35, 1958, T. 2, 487—500, or in STRAKHOV, 1969. [301]
MANGER, G. E. (1963): Porosity and bulk density of sedimentary rocks. — Geol. Surv. Bull., 1144—E, 55 pp. [350]
MANNING, R. B. & KUMPF, H. E. (1959): Preliminary investigation of the fecal pellets of certain invertebrates of the South Florida area. — Bull. Mar. Sci. of the Gulf and Caribbean, 9, 291—309. [201]
MANTEN, A. A. (1962 a): Korallengestalten als Kennzeichen des Milieus. — Geol. Rdsch., 51, 665—671. [234]
— (1962 b): Some Middle Silurian reefs of Gotland. — Sedimentology, 1, 211—234. [273]
— (1966): Note on the formation of stylolites. — Geol. en Mijnb., 45, 269—274. [273, 297]
— (1971): Silurian reefs of Gotland. — Developm. in Sedimentology, 13, Elsevier, Amsterdam, 539 pp. [273]
MARCHETTI, M. P. (1960): The occurrence of slide and flowage materials (olistostromes) in the Tertiary series of Sicily. — Internat. Geol. Congr., Norden, Sect. V, 209—225. [82]
MARCHIG, V. (1970): Porenwässer in rezenten Sedimenten vor der indisch-pakistanischen Küste und Rückschlüsse auf frühdiagenetische Vorgänge. — Geol. Rdsch., 60, 275—293. [336]
MARGOLIS, S. (1965): Quartz sand surface texture as an indicator of shoreline energy levels. —Coast. Res. Notes, 11, 13—15. [70]
MARGOLIS, S. V. & KRINSLEY, D. H. (1971): Submicroscopic frosting on eolian and subaqueous quartz sand grains. — Geol. Soc. Amer. Bull., 82, 3395—3406. [70]
MARSAL, D. (1950): Über Windtransport von Kies in Wüstengebieten. II. Theoretische Überlegungen und experimentelle Untersuchungen. — N. Jb. Geol. Paläont., Mh., 295—304. [103]
— (1954): Die Darstellung und graphische Auswertung der Siebanalysen natürlicher Sande und Kiese nach dem RRS-Verfahren. — N. Jb. Miner., Abh., 87, 218—239. [55]

— (1959): Die Erdöl-Lagerstätten des Emslandes und ihre Produktionsprobleme. — Erdöl u. Kohle, **12**, 407—422. [348]
— (1967): Statistische Methoden für Erdwissenschaftler. — Schweizerbart, Stuttgart, 152 pp. [52, 53, 136, 362]
— figure. [4]
MARSAL, D. & PHILIPP, W. (1970): Compaction of sediments. A simple mathematical model for calculating the gravitational porosity-depth equilibrium-curve of shales. — Bull. Geol. Inst., Univ. Uppsala, N. Ser. II, **7**, 59—66. [6]
MARSCHNER, H. (1966): Mineralogisch-petrographische Untersuchungen an karbonatreichen Gesteinen aus dem Unteren Keuper des Weserberglandes. — Diss. Hamburg, 137 pp. [309]
MARTIN, G. P. R. & WEILER, H. (1963): Der Wealden in der Gegend von Barnstorf (Kreis Grafschaft Diepholz, Niedersachsen). — N. Jb. Geol. Paläont., Abh., **118**, 30—64. [374]
MARTINI, E. (1961): Nannoplankton aus dem Tertiär und der obersten Kreide von SW-Frankreich. — Senckenbergiana leth., **42**, 1—41. [225]
MASLOV, V. P. (1956): Fossil calcareous algae of the USSR (Russ.). — Akad. Nauk SSSR, Inst. Sci. geol., Trav., **160**, 1—301. [214]
— (1962): Algues rouges fossils d'URSS et leur rapport avec les facies. — Trudy geol. Inst. SSSR, **53**, 222 pp. (Russ.). [214, 224]
MASON, C. C. & FOLK, R. L. (1958): Differentiation of beach, dune, and aeolian flat environments by size analysis, Mustang Island, Texas. — J. Sediment. Petrol., **28**, 211—226. [59]
MATTAVELLI, L. (1966): Osservazioni petrografiche sulla sostituzione della dolomite con la calcite (dedolomitizzazione) in alcune facies carbonate italiane. — Atti Soc. Ital. Sci. Natur. e Mus. Civ. Storia Nat. Milano, CV, **3**, 294—316. [320]
MATTAVELLI, L. & TONNA, M. (1967): Osservazioni petrografiche su processi diagenetici in alcune facies carbonate mesozoichi italiane. — R. C. Soc. Miner. Ital., **23**, 245—273. [321]
MATTER, A. (1967): Tidal flat deposits in the Ordovician of Western Maryland. — J. Sediment. Petrol., **37**, 601—609. [356]
MATTHEWS, R. K. (1966): Genesis of recent lime mud in Southern British Honduras. — J. Sediment. Petrol., **36**, 428—454. [195]
— (1967): Diagenetic fabric in biosparites from the Pleistocene of Barbados, West Indies. — J. Sediment. Petrol., **37**, 1147—1153. [289]
MATTIAT, B. (1960): Beitrag zur Petrographie der Oberharzer Kulmgrauwacke. — Beitr. Miner. Petrogr., **7**, 242—280. [19, 22, 24]
MATTOX, R. B. (1955): Eolian shape sorting. — J. Sediment. Petrol., **25**, 111—114. [65]
MAUREL, P. (1962): Sur la présence d'albite dans le Permien supérieur des environs de Saint-Affrique (Aveyron) et de Lodève (Hérault). — C. R. Acad. Sci., Paris, **254**, 3003—3005. [142]
MAWSON, D. (1929): South Australian algal limestones in process of formation. — Geol. Soc. Lond. Quart. J., **85**, 613—621. [316]
MAXWELL, J. C. (1964): Influence of depth, temperature, and geologic age on porosity of quartzose sandstone. — Bull. Amer. Assoc. Petrol. Geol., **48**, 697—709. [132]
MAXWELL, W. G. H. (1962): Lithification of carbonate sediments in the Heron Island reef, Great Barrier Reef. — J. Geol. Soc. Austral., **8**, 2, 217—238. [199, 295]
— (1968): Atlas of the Great Barrier Reef. — Elsevier Publ. Co., Amsterdam, 258 pp. [266, 267]
MAXWELL, W. G. H., DAY, R. W. & FLEMING, P. J. G. (1961): Carbonate sedimentation on the Heron Island reef, Great Barrier Reef. — J. Sediment. Petrol., **31**, 215—230 [272]
MAYER, F. K. & WEINECK, E. (1932): Die Verbreitung des Kalziumkarbonates im Tierreich unter besonderer Berücksichtigung der Wirbellosen. — Jena. Z. Med. u. Naturwiss., **66**, 199—222. [208]
MAYER, G. (1956): Kotpillen als Füllmasse in Hoernesien und weitere Kotpillenvorkommen im Kraichgauer Hauptmuschelkalk. — N. Jb. Geol. Paläont., Mh., **12**, 531—535. [200]
MAYR, M. (1957): Geologische Untersuchungen in der ungefalteten Molasse im Bereich des unteren Inn. — Geol. Jahrb., Beih., H. **26**, 309—370. [81]
MAYRHOFER, H. (1955): Beiträge zur Kenntnis des alpinen Salzgebirges. — Z. dt. Geol. Ges., **105**, 752—775. [352]
MCBIRNEY, A. R. (1963): Factors governing the nature of submarine volcanism. — Bull. Volc., **26**, 455—469. [183]

McBirney, A. R. & Murase, T. (1970): Factors governing the formation of pyroclastic rocks. — Bull. Volc., 34, 372—384. [162]

McBride, E. F. (1962): Flysch and associated beds of the Martinsburg formation (Ordovician), Central Appalachians. — J. Sediment. Petrol., 32, 39—91. [72, 362]

McBride, E. F. & Kimberly, J. E. (1963): Sedimentology of Smithwick shale (Pennsylvanian), eastern Llano region, Texas. — Bull. Amer. Assoc. Petrol. Geol., 47, 1840—1854. [72]

McBride, E. F., Lindemann, W. L. & Freeman, P. S. (1968): Lithology and petrology of the Gueydan (Catahoula) formation in South Texas. — Bureau Econ. Geol. Univ. Texas, Rep. of Investig., 63, 122 pp. [84, 141]

McBride, E. F. & Yeakel, L. S. (1963): Relationship between parting lineation and rock fabric. — J. Sediment. Petrol., 33, 779—782. [90]

McCaleb, J. A. & Wayhan, D. A. (1969): Geologic reservoir analysis, Mississippian Madison formation, Elk Basin field, Wyoming-Montana. — Bull. Amer. Assoc. Petrol. Geol., 53, 2094—2113. [47]

McCallum, J. S. & Ginsburg, R. N. (1965): Formation of recent oolitic sands on Great Bahama Bank. — Trans. Amer. Geophys. Un., 46, 166. [251]

McGowen, J. H. & Garner, L. E. (1970): Physiographic features and stratification types of coarse-grained bars: Modern and ancient examples. — Sedimentology, 14, 77—111. [104, 106]

McIver, N. L. (1961): Upper Devonian marine sedimentation in the Central Appalachians. — Ph. D. Diss., Johns Hopkins Univ., 530 pp. [72]

McKee, E. D. (1944): Tracks that go uphill. — Plateau, 16, No. 4, 61—72. [103]
— (1957): Primary structures in some Recent sediments. — Bull. Amer. Assoc. Petrol. Geol., 41, 1704—1747. [76]
— (1962): Origin of the Nubian and similar sandstones. — Geol. Rdsch., 52, 551—587. [78, 82]
— (1965): Experiments on ripple lamination. — In: G. V. Middleton (ed.), Primary sedimentary structures and their hydrodynamic interpretation. — S. E. P. M. Spec. Publ., 12, 66—83. [76, 78]
— (1966 a): Significance of climbing-ripple structure. — U.S. Geol. Surv. Prof. Pap., 550, 94—103. [78, 81, 100]
— (1966 b): Structures of dunes at White Sands National Monument, New Mexico (and a comparison with structures of dunes from other selected areas). — Sedimentology, 7, 3—69. [101]

McKee, E. D. & Douglass, J. R. (1971): Growth and movement of dunes at White Sands National Monument, New Mexico. — Geol. Surv. Res., D108—D114. [101]

McKee, E. D. & Tibbits, G. C., Jr. (1964): Primary structures of a sief dune and associated deposits in Libya. — J. Sediment. Petrol., 34, 5—17. [101]

McKee, E. D. & Weir, G. W. (1953): Terminology for stratification and cross-stratification in sedimentary rocks. — Geol. Soc. Amer. Bull., 64, 381—390. [76, 77]

Meade, R. H. (1968): Relations between suspended matter and salinity in estuaries of the Atlantic seaboard, U.S.A. — Internat. Assoc. Sci. Hydrol., Gen. Assembly, Bern 1967, 4, (I.A.S.H. Publ. 78), 96—109. [115]
— (1969): Landward transport of bottom sediments in estuaries of the Atlantic coastal plain. — J. Sediment. Petrol., 39, 222—234. [115]

Meischner, K.-D. (1964): Allodapische Kalke, Turbidite in Riff-nahen Sedimentations-Becken. — In: A. H. Bouma & A. Brouwer (ed.), Developm. in Sedimentol., 3, Turbidites. — Elsevier, Amsterdam, 156—191. [125, 203, 282, 369]

Mellis, O. (1960): Gesteinsfragmente im roten Ton des Atlantischen Ozeans. — Medd. Oceanogr. Inst. Göteborg, 28, 3—18. [335]

Mellon, G. B. (1967): Stratigraphy and petrology of the Lower Cretaceous Blairmore and Mannville groups, Alberta foothills and plains. — Res. Counc. Alberta, Bull., 21, 270 pp. [113]

Mensink, H. (1960): Beispiele für die stratigraphische Kondensation, Schichtlücke und den Leitwert von Ammoniten aus dem Jura Spaniens im Vergleich zu NW-Europa. — Geol. Rdsch., 49, 70—82. [303]

Merriam, D. E. (ed.) (1964): Symposium on Cyclic Sedimentation. — Kansas State Geol. Surv. Bull., 169, vol. I, II, 636 pp. [367]

MESOLELLA, K. J. (1967): Zonation of uplifted Pleistocene coral reefs on Barbados, West Indies. — Science, 156, 638—640. [358]
MICHEELSEN, H. (1966): The structure of dark flint from Stevns, Denmark. — Medd. Dansk Geol. Foren. København, 16, 285—368. [329]
MICHEL, R. (1953): Contribution à l'étude pétrographique des pépérites et du volcanisme Tertiaire de la Grande Limagne. — Mém. Soc. Hist. nat. Auvergne, 5, 1—140. [183]
MIDDLETON, G. V. (1962): Size and sphericity of quartz grains in two turbidite formations. — J. Sediment. Petrol., 32, 725—742. [67]
— (1967): The orientation of concavo-convex particles deposited from experimental turbidity currents. — J. Sediment. Petrol., 37, 229—232. [98, 207]
— (1970): Experimental studies related to problems of flysch sedimentation. In: J. LAJOIE (ed.), Flysch sedimentology in North America. — Geol. Assoc. Canada Spec. Pap., 7, 253—272. [98]
— (1972): Albite of secondary origin in Charny sandstones, Quebec. — J. Sediment. Petrol., 42, 341—349. [25, 146]
MILLIMAN, J. D. (1966): Submarine lithification of carbonate sediments. — Science, 153, 994—997. [282, 322, 360]
— (1967): Carbonate sedimentation on Hogsty Reef, a Bahamian atoll. — J. Sediment. Petrol., 37, 658—676. [204]
— (1971): Carbonate lithification in the deep sea. — In: O. P. BRICKER (ed.), Carbonate cements. — Johns Hopkins Stud. in Geol., 19, 95—102, Johns Hopkins Press, Baltimore, Md. [194]
— (1974): Marine Carbonates. Recent Sedimentary Carbonates, part 1. — Springer Berlin, Heidelberg, New York, 375 pp. [212, 324]
— pers. comm. [122]
MILLIMAN, J. D. & MÜLLER, J. (1973): Precipitation and lithification of magnesian calcite in the deep-sea sediments of the eastern Mediterranean Sea. — Sedimentology, 20, 29—45. [192, 194]
MILLIMAN, J. D., ROSS, D. A. & KU, T. L. (1969): Precipitation and lithification of deep-sea carbonates in the Red Sea. — J. Sediment. Petrol., 39, 724—736. [192, 284]
MILLOT, G. (1960): Silice, silex, silicifications et croissance des cristaux. — Bull. Serv. Carte géol. Als. Lorr., 13, 4, 129—146. [141, 327—329]
— (1970): Geology of Clays. — Springer New York, Heidelberg, Berlin, 429 pp. [143]
MILNE, J. (1897): Sub-oceanic changes. — Geol. J., 10, 129—146, 259—285. [81]
MILNER, H. B. (1962): Sedimentary Petrography. — G. Allen & Unwin Ltd., London, 715 pp. [19, 27, 31]
MILTON, CH., CHAO, E. C. T., FAHEY, J. J. & MROSE, M. E. (1960): Silicate mineralogy of the Green River formation of Wyoming, Utah, and Colorado. — 21. Internat. geol. Congr., Norden., Pt. XXI, 171—184. [334]
MINATO, M., ISHII, M. & HONJO, S. (1967): Electron microscopic study of oil bearing calcilutite, Khafji oil field, Neutral Zone. — 7th World Petrol. Congr., Panel Disc. no. 3. [197]
MITTERER, R. M. (1972): Calcified proteins in the sedimentary environment. — Advances in Organ. Geochem., 1971, Proc. V. Internat. Meet. Organ. Geochem., Hannover, Germany; Pergamon Press, 441—451. [211, 254]
MIZUTANI, S. (1957): Permian sandstones in the Mugi area, Gifu Prefecture, Japan. — J. Earth Sci., Nagoya Univ., 5, 135—151. [19, 25]
— (1959): Clastic plagioclase in Permian graywacke from the Mugi area, Gifu Prefecture, central Japan. — J. Earth Sci., Nagoya Univ., 7, 108—136. [24, 25]
— (1967): Kinetic aspects of diagenesis of silica in sediments. — J. Earth Sci., Nagoya Univ., 15, 99—111. [330]
— (1970): Silica minerals in the early stage of diagenesis. — Sedimentology, 15, 419—436. [330]
MOBERLY, R., JR. (1970): Microprobe study of diagenesis in calcareous algae. — Sedimentology, 14, 113—123. [219]
MOIOLA, R. J. & WEISER, D. (1968): Textural parameters: An evaluation. — J. Sediment. Petrol., 38, 45—53. [59]
MOJSISOVICS, E. v. (1879): Die Dolomitriffe von Südtirol und Venetien. Beitrag zur Bildungsgeschichte der Alpen. — Hölder, Wien, 552 pp. [263]

MONAGHAN, P. H. & LYTLE, M. L. (1956): The origin of calcareous ooliths. — J. Sediment. Petrol., 26, 111—118. [254]
MONROE, J. N. (1951): Woodbine sandstone dikes of Northern McLennan County, Texas. — In: F. E. LOZO (ed.), The Woodbine and adjacent strata of the Waco area of Central Texas. A Symposium. — Fondren Sci. Ser., No. 4, 93—100. [84]
MONTY, C. L. V. (1963): Biostromes stromatolithiques dans le Viséen moyen de la Belgique. — C. R., 256, 2, 5603—5606. [271]
— (1967): Distribution and structure of recent stromatolitic algal mats, Eastern Andros Island, Bahamas. — Ann. Soc. Géol. Belg., 90, 55—99. [198, 215, 222—224, 288, 356]
— (1971): An autoecological approach of intertidal and deep water stromatolites. — Ann. Soc. Géol. Belg., 94, 265—276. [215, 220]
MOORE, D. G. (1966): Deltaic sedimentation. — Earth-Sci. Rev., 1, 87—104. [199]
MOORE, D. G., CURRAY, J. R. & RAITT, R. W. (1971): Structure and history of the Bengal deep-sea fan and geosyncline (Abstr.). — Progr., VIII Internat. Sediment. Congr., Heidelberg, 69. [94]
MOORE, H. B. (1933): The faecal pellets of the Anomura. — Proc. Roy. Soc. Edinburgh, 52, 296—308. [200]
— (1939): Faecal pellets in relation to marine deposits. — In: Recent marine sediments, ed. by P. D. TRASK, 516—524. [201]
MOORE, J. G. (1966): Rate of palagonitization of submarine basalt adjacent to Hawaii. — U. S. Geol. Surv. Prof. Pap., 550—D, D163—D171. [188]
— (1967): Base surge in recent volcanic eruptions. — Bull. Volcanol., 30, 337—363. [177, 178]
— (1970a): Pillow lava in a historic lava flow from Hualalai volcano, Hawaii. — J. Geol., 78, 239—243. [188]
— (1970b): Water content of basalt erupted on the ocean floor. — Contr. Miner. Petrogr., 28, 272—279. [162, 184]
MOORE, J. G. & FISKE, R. S. (1969): Volcanic substructure inferred from dregde samples and ocean bottom photographs. — Geol. Soc. Amer. Bull., 80, 1191—1202. [183]
MOORE, J. G. & MELSON, W. G. (1969): Nuées ardentes of the 1968 eruption of Mayon volcano, Philippines. — Bull. Volcanol., 33, 600—620. [182]
MOORE, J. G., NAKAMURA, K. & ALCARAZ, A. (1966): The 1965 eruption of Taal Volcano. — Science, 151, 955—960. [177]
MOORE, J. G. & PECK, D. L. (1962): Accretionary lapilli in volcanic rocks of the western continental United States. — J. Geol., 70, 182—193. [163]
MOORE, J. G., PHILLIPS, R. L., GRIGG, R. W., PETERSON, D. W. & SWANSON, D. A. (1973): Flow of lava into the sea, 1969—1971, Kilauea Volcano, Hawaii. — Geol. Soc. Amer. Bull., 84, 537—546. [183]
MOORE, R. C. (1949): Meaning of facies. — In: Sedimentary facies in geologic history. — Geol. Soc. Amer. Mem., 39, 1—34. [378]
— (1950): Late Paleozoic cyclic sedimentation in Central United States. — 18th Internat. Geol. Congr., London, Rept., Pt. IV, 5—16. [366]
MOOS, A. v. (1935): Sedimentpetrographische Untersuchungen an Molassesandsteinen. — Schweiz. Miner. Petr. Mitt., 15, 170—265. [31]
MORET, L. (1940): Rôle probable des Holothuries dans la genèse de certains sédiments calcaires. — C. R. somm. Soc. géol. France, 5, 10, 11—12. [201]
MORGENSTEIN, M. E. (1972): The sideromelane-palagonite transition in authigenic marine sediments. — Geol. Soc. Amer. 68th Meet. Cord. Sec. (Abstr.), 4, 203. [188, 189]
MORLOT, A. v. (1847): Über den Dolomit und seine künstliche Darstellung aus Kalkstein. — Haidinger Naturwiss. Abh., 1, 305. [320]
MOSEBACH, R. (1954): Auswertung und Darstellung von Kornanalysen und Anwendung ihrer Ergebnisse auf petrologische Fragen. — Geologie, 3, 413—440. [52]
MOSS, A. J. (1962): The physical nature of common sandy and pebbly deposits, Pt. I, Amer. J. Sci., 260, 337—373. [56]
— (1966): Origin, shaping and significance of quartz sand grains. — J. Geol. Soc. Austral., 13, 1, 97—136. [68]
MÜLLER, A. H. (1950): Stratonomische Untersuchungen im oberen Muschelkalk des Thüringer Beckens. — Geologica, 4, Akademie-Verlag, Berlin, 74 pp. [239]
— (1956): Die Knollenfeuersteine der Schreibkreide, eine frühdiagenetische Bildung. — Ber. Geol. Ges., 1, 136—146. [329, 330]

— (1957): Lehrbuch der Paläozoologie, Bd. I. Allgemeine Grundlagen. — VEB Gustav Fischer, Jena, 322 pp. [70]
— (1958): Lehrbuch der Paläozoologie, Bd. II. Invertebraten, Tl. 1 Protozoa — Mollusca 1. — VEB Gustav Fischer, Jena, 566 pp. [232, 233, 236, 240]
— (1964): Die präkambrische Lebewelt. Erscheinungen und Probleme. — Biol. Rdsch., 2, 53—67. [219]
MÜLLER, GERMAN (1952): Vorkommen und Entstehung der Karbonate, insbesondere der Eisenkarbonate, in den Steinkohlenflözen des Ruhrgebietes. — Diss. Univ. Bonn, 92 pp. [142]
— (1962): Zur Geochemie des Strontiums in ozeanen Evaporiten unter besonderer Berücksichtigung der sedimentären Coelestinlagerstätte von Hemmelte-West (Süd-Oldenburg). — Geologie, 11, Beih. 35, 1—90. [326]
— (1964): Frühdiagenetische allochthone Zementation mariner Küsten-Sande durch evaporitische Calcit-Ausscheidung im Gebiet der Kanarischen Inseln. — Beitr. Miner. Petrogr., 10, 125—131. [289]
— (1966a): Die Sedimentbildung im Bodensee. — Naturwiss., 53, 237—247. [195, 197—199, 221, 355]
— (1966b): Grain size, carbonate content, and carbonate mineralogy of recent sediments of the Indian Ocean off the eastern coast of Somalia. — Naturwiss., 53, 547—550. [195]
— (1967): Methods in Sedimentary Petrology. — In: v. ENGELHARDT, FÜCHTBAUER & MÜLLER, Sedimentary Petrology, pt. I, Schweizerbart, Stuttgart, Hafner Publ., New York, 283 pp. [67, 72, 158]
— (1970a): High-magnesian calcite and protodolomite in Lake Balaton (Hungary) sediments. — Nature, 226, 749—750. [193, 195, 317]
— (1970b): Petrology of the Cliff Limestone (Holocene), North Bimini, Bahamas. — N. Jb. Miner. Mh., 11, 507—523. [287]
— (1971): Gravitational cement: An indicator for the vadose zone of the subaerial environment. — In: O. P. BRICKER (ed.), Carbonate cements. — Johns Hopkins Stud. in Geol., 19, 301—302, Johns Hopkins Press, Baltimore, Md. [285]
— pers. comm. [313]
MÜLLER, G. & BLASCHKE, R. (1969a): Zur Entstehung des Tiefsee-Kalkschlammes im Schwarzen Meer. — Naturwiss., 56, 561—562. [197]
— — (1969b): Zur Entstehung des Posidonienschiefers (Lias ζ). — Naturwiss., 56, 635. [197]
— — (1971): Coccoliths: Important rock-forming elements in bituminous shales of Central Europe. — Sedimentology, 17, 119—124. [197, 225]
MÜLLER, G. & IRION, G. (1969): Subaerial cementation and subsequent dolomitization of lacustrine carbonate muds and sands from Paleo-Tuz Gölü ("Salt Lake"), Turkey. — Sedimentology, 12, 193—204. [316]
MÜLLER, G., IRION, G. & FÖRSTNER, U. (1972): Formation and diagenesis of inorganic Ca-Mg carbonates in the lacustrine environment. — Naturwiss., 59, 158—164. [191, 194, 307, 353—355]
MÜLLER, G. & MÜLLER, J. (1967): Mineralogisch-sedimentpetrographische und chemische Untersuchungen an einem Bank-Sediment (Cross-Bank) der Florida Bay, USA. — N. Jb. Miner., Abh., 106, 257—286. [195]
MÜLLER, G. & TIETZ, G. (1966): Recent dolomitization of Quaternary biocalcarenites from Fuerteventura (Canary Islands). — Contr. Miner. Petrol., 13, 89—96. [317]
MÜLLER, H. (1958): Die Petrographie der Röt-Muschelkalkgrenzschichten bei Steudnitz nördlich Jena. — Chemie d. Erde, 19, 4, 391—435. [320, 334, 335]
MÜLLER, J. & MILLIMAN, J. D. (1971): Precipitation of magnesian calcite in Eastern Mediterranean deep-sea sediments (Abstr.). — Progr., VIII Internat. Sediment. Congr. Heidelberg, 70. [194]
MÜNZBERGER, E. (1958): Die Coccolithen der Rügenschen Schreibkreide. — Dipl.-Arb. Univ. Greifswald, 127 pp. [225]
MUKHOPADHYAY, A. & CHANDA, S. K. (1972): Silica diagenesis in the banded hematite jasper and bedded chert associated with the Iron Ore Group of Jamda-Koira Valley, Orissa, India. — Sediment. Geol., 8, 113—135. [331]
MULLINEAUX, D. R. & CRANDELL, D. R. (1962): Recent Lahars from Mount St. Helens, Washington. — Geol. Soc. Amer. Bull., 73, 855—870. [176, 180]

MULTER, H. G. & HOFFMEISTER, J. E. (1968): Subaerial laminated crusts of the Florida Keys. — Geol. Soc. Amer. Bull., 79, 183—192. [258]
MURAI, I. (1961): A study of the textural characteristics of pyroclastic flow deposits in Japan. — Earthquake Res. Inst. Bull., 39, 133—254. [175]
MURAVYOV, V. I. (1970): Formation of carbonate cement in clastic rocks. — Sedimentology, 15, 139—145. [142, 281]
MURRAY, A. N. (1930): Limestone oil reservoirs of the northeastern United States and of Ontario, Canada. — Econ. Geol., 25, 459 f. [348]
MURRAY, J. W. (1954): The deposition of calcite and aragonite in caves. — J. Geol., 62, 481—492. [192, 260]
MURRAY, R. C. (1957): Hydrocarbon fluid inclusions in quartz. — Bull. Amer. Assoc. Petrol. Geol., 41, 950—956. [20]
— (1960): Origin of porosity in carbonate rocks. — J. Sediment. Petrol., 30, 59—84. [346—348, 351]
— (1964 a): Origin and diagenesis of gypsum and anhydrite. — J. Sediment. Petrol., 34, 512—523. [325]
— (1964 b): Preservation of primary structures and fabrics in dolomite. — In: IMBRIE & NEWELL (ed.), Approaches to paleoecology. — Wiley & Sons, 388—403. [313, 320, 325]
— pers. comm. [290]
MURRAY, R. C. & LUCIA, F. J. (1967): Cause and control of dolomite distribution by rock selectivity. — Geol. Soc. Amer. Bull., 78, 21—36. [320]
MUTTI, E. & RICCI LUCCHI, F. (1972): Le torbiditi dell' Appennino settentrionale: introduzione all' analisi di facies. — Mem. Soc. Geol. Ital., 11, 161—199. [369]

NADSON, G. (1927): Les algues perforantes, leur distribution et leur rôle dans la nature. — C. R., 184, 1015—1017. [356]
NÄGELE, E. (1962): Zur Petrographie und Entstehung des Albsteins. — N. Jb. Geol. Paläont., Abh., 115, 44—120. [258, 259, 317]
NAGELSCHMIDT, G., GORDON, R. L. & GRIFFIN, O. G. (1952): The determination of quartz by X-ray. — Nature, 169, no. 4300, 539—540. [69]
NAGTEGAAL, P. J. C. (1969): Microtextures in recent and fossil caliche. — Leidse geol. Meded., 42, 131—142. [258]
— (1973): Adhesion-ripple and barchan-dune sands of the Recent Namib (SW Africa) and Permian Rotliegend (NW Europe) deserts. — Madoqua, ser. II, 2, 5—19. [89, 101]
NAKAMURA, K. (1964): Volcano-stratigraphic study of Oshima Volcano, Izu. — Earthquake Res. Inst. Bull., 42, 649—728. [164]
NATLAND, M. L. (1957): Paleoecology of west coast Tertiary sediments. — In: H. S. LADD (ed.), Paleoecology. — Geol. Soc. Amer. Mem., 67, 543—572. [228, 241]
NATLAND, M. L. & KUENEN, PH. H. (1951): Sedimentary history of the Ventura Basin, Calif., and the action of turbidity currents. — Soc. Econ. Paleont. Miner., Spec. Publ., 2, 76—107. [128]
NECHAEV, S. V. (1959): The origin of dolomites and dolomitized limestones along the southwestern border of the Donets Basin. — Dokl. Akad. Nauk SSSR, Earth Sci. sec., 124, 8991. Transl.: Amer. Geol. Inst., 1960. [319]
NEEV, D. (1963): Recent precipitation of calcium salts in the Dead Sea. — Res. Counc. Israel Bull., 11 G, 153—154. [192]
NEEV, D. & EMERY, K. O. (1967): The Dead Sea. Depositional processes and environments of evaporites. — Geol. Surv. Israel Bull., 41, 147 pp. [363]
NEHER, J. & ROHRER, E. (1958): Dolomitbildung unter Mitwirkung von Bacterien. — Eclog. geol. Helvet., 51, 213—215. [309]
NELSON, B. W. (1970): Hydrography, sediment dispersal, and historical development of the Po river delta, Italy. — In: MORGAN, J. P. (ed.), Deltaic sedimentation, modern and ancient. — Soc. Econ. Pal. Miner., Spec. Publ., 15, 152—184. [112, 113]
NELSON, H. F. (1959): Deposition and alteration of the Edwards limestone, Central Texas. — In: Symposium on Edwards limestone in Central Texas. — Univ. of Texas, Austin, Publ. No. 5905, 21—95. [272]
NELSON, H. F., BROWN, C. W. & BRINEMAN, J. H. (1962): Skeletal limestone classification. — In: W. E. HAM (ed.), Classification of carbonate rocks. — Amer. Assoc. Petrol., Geol., Mem., 1, 224—252. [261, 275]

NELSON, H. W. & NIGGLI, E. (1950): Röntgenologisch onderzoek van de ondoorzichtige zware fractie van enkele nederlandse zanden. — Proc. Kon. Nederl. Akad. Wetensch., 53, 1240—1246. [32]
NESTLER, H. (1965): Die Rekonstruktion des Lebensraumes der Rügener Schreibkreide-Fauna (Unter-Maastricht) mit Hilfe der Paläoökologie. — Geologie, Beih. 49, 147 pp. [332]
NEUGEBAUER, J. (1972): Warum ist Schreibkreide weich? — Vortrag, Dt. Geol. Ges., Braunschweig, 27. [339]
NEUHAUS, A. (1940): Über die Erzführung des Kupfermergels der Haaseler und der Gröditzer Mulde in Schlesien (nebst Beitrag zur Frage der „vererzten Bakterien"). — Z. angew. Miner., 2, 304—343. [337]
NEUMAN, A. C. (1966): Observations on coastal erosion in Bermuda and measurements of the boring rate of the sponge, Cliona lampa. — Limnol. & Oceanogr., 11, 92—108. [206]
NEUMANN, A. G., GEBELEIN, C. D. & SCOFFING, T. P. (1970): The compositiom, structure and erodability of subtidal mats, Abaco, Bahamas. — J. Sediment. Petrol., 40, 274—297. [223]
NEUMANN, R. (1963): Die Auswertung von Korngrößenverteilungen durch Häufigkeitsanalyse. — N. Jb. Geol. Paläont., Mh., 9, 492—501. [54]
NEWELL, N. D., PURDY, E. G. & IMBRIE, J. (1960): Bahamian oölitic sand. — J. Geol., 68, 481—497. [251, 253—255]
NEWELL, N. D. & RIGBY, J. K. (1957): Geological studies on the Great Bahama Bank. — In: R. J. LEBLANC & J. G. BREEDING (ed.), Regional aspects of carbonate deposition. — Soc. Econ. Pal. Miner. Spec. Publ., 5, 15—79. [195]
NEWELL, N. D., RIGBY, J. K., FISCHER, A. G., WHITEMAN, A. J., HICKOX, J. E. & BRADLEY, J. S. (1953): The Permian Reef Complex of the Guadalupe Mountains Region, Texas and New Mexico. — Freeman, San Francisco, 236 pp. [232, 263, 275, 293, 329, 330, 339]
NEWTON, R. S. (1968): Internal structure of wave-formed ripple marks in the nearshore zone. — Sedimentology, 11, 275—292. [89]
NEWTON, R. S. & WERNER, F. (1972): Transitional-size ripple marks in Kiel Bay (Baltic Sea). — Meyniana, 22, 89—94. [103]
NICKEL, E. (1973): Experimental dissolution of light and heavy minerals in comparison with weathering and intrastratal solution. — Contr. Sedimentol., 1, Schweizerbart, Stuttgart, 1—68. [40, 42]
NIEHOFF, W. (1958): Die primär gerichteten Sedimentstrukturen, insbesondere die Schrägschichtung im Koblenzquarzit am Mittelrhein. — Geol. Rdsch., 47, 252—321. [76, 77, 369]
NINKOVICH, D. & HEEZEN, B. C. (1965): Santorini tephra. — In: WHITTARD, W. F. & BRADSHAW, R. (ed.), Submarine geology and geophysics (Colston Pap. no. 17); 413—452, London, Butterworths. [164, 174]
NISSEN, H.-U. (1963): Röntgengefügeanalyse am Kalzit von Echinodermskeletten. — N. Jb. Geol. Paläont., Abh., 117, LOTZE-Festbd., 230—234. [248]
NOBLE, D. C. (1967): Sodium, Potassium, and ferrous iron contents of some secondarily hydrated natural silicic glasses. — Amer. Miner., 52, 285—286. [187]
NOËL, D. (1965): Sur les coccolithes du Jurassique Européen et d'Afrique du Nord. — Éd. Centre Nat. Rech. Sci., Paris, 209 pp., 29 tab. [225]
— (1970): Coccolithes Crétacés. La crais Campanienne du bassin de Paris. — Ed. Centre Nat. Rech. Sci., Paris, 129 pp., 48 tab. [225]

OBERHAUSER, R. (1960): Foraminiferen und Mikrofossilien „incertae sedis" der ladinischen und karnischen Stufe der Trias aus den Ostalpen und aus Persien. — Jb. Geol. B. A. Wien, Sonderbd. 5, 5—46. [227]
OERTEL, G. & CURTIS, C. D. (1972): Clay-ironstone concretion preserving fabrics due to progressive compaction. — Geol. Soc. Amer. Bull., 83, 2597—2606. [302]
OKAMOTO, G., OKURA, T. & GOTO, K. (1957): Properties of silica in water. — Geochim. Cosmochim. Acta, 12, 123—132. [134]
OLDERSHAW, A. E. & SCOFFIN, T. P. (1967): The source of ferroan and non-ferroan calcite cements in the Halkin and Wenlock limestones. — Geol. J., 5, 2, 309—319. [288]
ONIONS, D. & MIDDLETON, G. V. (1968): Dimensional grain orientation of Ordovician turbidite greywackes. — J. Sediment. Petrol., 38, 164—174. [72]
OOMKENS, E. (1966): Environmental significance of sand dikes. — Sedimentology, 7, 145—148. [84]

— (1967): Depositional sequences and sand distribution in a deltaic complex. — Geol. en Mijnb., 46, 265—278. [111, 113]
— (1970): Depositional sequences and sand distribution in the postglacial Rhône delta complex. — In: MORGAN, J. P. (ed.), Deltaic sedimentation, modern and ancient. — Soc. Econ. Pal. Miner., Spec. Publ., 15, 198—212. [113, 382]
OPPENHEIMER, C. H. & MASTER, I. M. (1963): Transition of silicate and carbonate crystal structure by photosynthesis and metabolism. — Geol. Soc. Amer. Ann. Meet., Progr., 125 A. [309]
ORME, G. R. & BROWN, W. W. M. (1963): Diagenetic fabrics in the Avonian limestone of Derbyshire and North Wales. — Proc. Yorksh. Geol. Soc., 34, 1, 51—66. [206, 293]
OSBURN, R. C. (1957): Marine Bryozoa. — In: J. W. HEDGPETH (ed.), Ecology. — Geol. Soc. Amer. Mem. 67, 1109—1112. [236]
OTT, E. (1966): Zwei neue Kalkalgen aus den Cassinaer Schichten Südtirols (Oberladin, mittlere Trias). — Mitt. Bayer. Staatssamml. Paläont. hist. Geol., 6, 155—166. [220]
— (1967 a): Dasycladaceen (Kalkalgen) aus der nordalpinen Obertrias. — Mitt. Bayer. Staatssamml. Paläont. hist. Geol., 7, 205—226. [270]
— (1967 b): Segmentierte Kalkschwämme (Sphinctozoa) aus der alpinen Mitteltrias und ihre Bedeutung als Riffbildner im Wettersteinkalk. — Bayer. Akad. Wiss., Math. Naturwiss. Kl., H. 131, 1—96. [232, 263, 268]
— pers. comm. [217]
OTTE, C., JR. & PARKS, J. M., JR. (1963): Fabric studies of Virgil and Wolfcamp bioherms, New Mexico. — J. Geol., 71, 380—396. [270, 271]
OTTE, O. (1972): Schichtfolgen, Fazies und Gebirgsbau des Mesozoikums der Vorarlberger Kalkalpen südlich des Großen Walstertales (Österreich). — Diss., Freie Univ. Berlin, 196 pp. [370]

PACKHAM, G. H. & CROOK, K. A. W. (1960): The principle of diagenetic facies and some of its implications. — J. Geol., 68, 392—407. [146, 153, 155]
PAGE, N. J. & CAROZZI, A. V. (1962): Étude du remplacement diagénétique du quartz détritique par les carbonates dans des dolomies Cambriennes. — Arch. Sci., Genève, 14, 461—491. [149]
PANIN, N. (1965): Coexistence de traces de pas de vertébrés et des mécanoglyphes dans la molasse miocène des Carpates orientales. — Rev. Roum. Géol. Géophys. Géogr., Sér. de Géol., Bucarest, 9, 141—163. [91]
PANTIN, H. M. (1958): Rate of formation of a diagenetic calcareous concretion. — J. Sediment. Petrol., 28, 366—371. [302]
PAPENFUSS, K.-H. (1963): Das Schlotkonglomerat des Bürzeln bei Eningen u. d. Achalm (Schwäbische Alb). — Jh. geol. Landesamt Baden-Württ., 6, 461—505. [182]
PAPP, A. (1963): Das Verhalten neogener Molluskenfaunen bei verschiedenen Salzgehalten. — Fortschr. Geol. Rheinl.-Westf., 10, 35—47. [241]
PAREA, G. C. (1965): Evoluzione della parte settentrionale della geosynclinate appennica dell'Albiano all' Eocene Superiore. — Atti Mem. Acc. Naz. Sci. Lett. Arti Modena, 7, 5—97. [369]
PARK, W. C. & SCHOT, E. H. (1968): Stylolites: Their nature and origin. — J. Sediment. Petrol., 38, 175—191. [297]
PARKE, M. & ADAMS, I. (1960): The motile (*Crystallolithus hyalinus* GAARDER & MARKALI) and nonmotile phases in the life history of *Coccolithus pelagicus* (WALLICH) SCHILLER. — Mar. Biol. Assoc. U. K. J., 39, 263—274. [225]
PARKER, R. H. (1959): Macro-invertebrate assemblages of Central Texas coastal bays and Laguna Madre. — Bull. Amer. Assoc. Petrol. Geol., 43, 2100—2166. [241]
PARKINSON, D. (1964): Problematic fabrics in the Carboniferous reef limestone of Dovedale. — Mercian Geol., Nottingham, 1, 49—59. [274]
PARSONS, W. H. (1969): Criteria for the recognition of volcanic breccias: review. — Geol. Soc. Amer. Mem., 115, 263—304. [181, 183]
PASSEGA, R. (1957): Texture as a characteristic of clastic deposition. — Bull. Amer. Assoc. Petrol. Geol., 41, 1952—1984. [57]
PASSEGA, R. & BYRAMJEE, R. (1969): Grain size image of clastic deposits. — Sedimentology, 13, 233—252. [57]
PASSERINI, P. (1966): Gradazione inversa alla base degli strati del Macigno. — Boll. Soc. Geol. Ital., 85, 157—165. [123]

PATRIQUIN, D. G. (1972): Carbonate mud production by epibionts on Thalassia: An estimate based on leaf growth rate data. — J. Sediment. Petrol., 42, 687—689. [193]
PATZELT, W. J. (1964): Lithologische und paläogeographische Untersuchungen im Unteren Keuper Süddeutschlands. — Erlanger Geol. Abh., H. 52, 30 pp. [21]
PAULUS, B., BROCKERT, M., HINSCH, W. & ZIMMERLE, W. (1964): Der tiefere Untergrund unter besonderer Berücksichtigung des von den Bohrungen Landsham und Pliening 101—105 erschlossenen Tertiärs. — Erläut. Geol. Kte. Bayern 1 : 25 000 Bl. 7736 Ismaning, Bayer. Geol. Landesamt, München, 9—53, 93—96. [127]
PEARN, W. C. (1965): Finding the ideal cyclothem. — In: Symposium on cyclic sedimentation. — Kansas Geol. Surv. Bull., 169, 399—413. [362]
PEARSE, A. S. & GUNTER, G. (1957): Salinity. — In: J. W. HEDGPETH (ed.), Ecology. — Geol. Soc. Amer. Mem., 67, 129—158. [192]
PELLETIER, B. R. (1958): Pocono paleocurrents in Pennsylvania and Maryland. — Bull. Geol. Soc. Amer., 69, 1033—1064. [48, 50, 56]
— (1965): Paleocurrents in the Triassic of Northeastern British Columbia. — In: G. V. MIDDLETON (ed.), Primary sedimentary structures and their hydrodynamic interpretation. — SEPM Spec. Publ., 12, 233—245. [78]
PELTO, C. R. (1956): A study of chalcedony. — Amer. J. Sci., 254, 32—50. [328]
PERKINS, R. D. & HALSEY, S. D. (1971): Geologic significance of microboring fungi and algae in Carolina shelf sediments. — J. Sediment. Petrol., 41, 843—853. [288]
PERRENOUD, J.-P. (1952): Etude du feldspath potassique contenu dans le «Pontiskalk» (Trias, Valais). — Schweiz. Miner. Petrogr. Mitt., 32, 179—184. [335]
PERRET, F. A. (1937): The eruption of Mt. Pelée 1929—1932. — Carnegie Inst. Washington Publ., 549, 1—126. [175]
PETERSON, D. W. (1970): Ash-flow deposits — their character, origin and significance. — J. Geol. Educ., 18, 66—76. [176]
PETERSON, J. A. & OHLEN, H. R. (1963): Pennsylvanian shelf carbonates, Paradox Basin. — Four Corners Geol. Soc. Sympos.: R. O. BASS (ed.), Shelf Carbonates of the Paradox Basin, 65—79. [271, 316]
PETERSON, M. N. A. (1966): Calcite: rates of dissolution in a vertical profile in the central Pacific. — Science, 154, 1542—1544. [360]
PETTIJOHN, F. J. (1941): Persistence of heavy minerals and geologic age. — J. Geol., 49, 610—625. [40]
— (1957): Sedimentary Rocks. — Harper & Broth., New York, 718 pp. [14, 20, 24, 27, 40, 42, 47, 50, 56, 57, 64—68, 71, 124, 125, 127, 128, 139, 149, 161, 201, 298, 301, 329, 369]
— (1960): Some contributions of sedimentology to tectonic analysis. — XXI. Internat. Geol. Congr., Norden, Part XVIII, 446—454. [48]
— (1963): Chemical composition of sandstones — excluding carbonate and volcanic sands: data of geochemistry. — U.S. Geol. Surv. Prof. Pap., 440—S, 21 pp. [15, 25]
— J. Sediment. Petrol., 13, 69—78. [64, 65]
PETTIJOHN, F. J. & LUNDAHL, A. C. (1943): Shape and roundness of Lake Erie beach sands.
PETTIJOHN, F. J. & POTTER, P. E. (1964): Atlas and glossary of sedimentary structures. — Springer, Berlin, Göttingen, Heidelberg, New York, 370 pp. [72, 77, 88, 90—92, 94]
PETTIJOHN, F. J., POTTER, P. E. & SIEVER, R. (1965): Geology of sand and sandstone. — Indiana Univ., Bloomington, 205 pp. [75, 100, 109, 120]
— — — (1972): Sand and Sandstone. — Springer, New York, Heidelberg, Berlin, 618 pp. [17, 25, 60, 68, 74, 78, 84, 89, 100, 119, 122, 135, 138, 143]
PFANNENSTIEL, M.: pers. comm. [298]
PFLAUMANN, U. (1967): Zur Ökologie des bayerischen Flysches auf Grund der Mikrofossilführung. — Geol. Rdsch., 56, 200—227. [124]
PHILCOX, M. E. (1971): Stromatactis in Upper Devonian bioherms, Alberta, Canada (Abstr.). — Progr., VIII Internat. Sediment. Congr., Heidelberg, 77. [270]
PHILIPP, W. (1961): Struktur- und Lagerstättengeschichte des Erdölfeldes Eldingen. — Z. dt. Geol. Ges., 112, 414—482. [363]
PHILIPP, W., DRONG, H. J., FÜCHTBAUER, H., HADDENHORST, H.-G. & JANKOWSKY, W. (1963): The history of migration in the Gifhorn trough (NW-Germany). — Sixth World Petrol. Congr., Sect. I, Pap. 19, PD 2, 457—481. [41, 86, 148]
PHLEGER, F. B., JR. (1951): Displaced foraminifera faunas in turbidity currents. — Soc. Econ. Pal., Miner. Spec. Publ., 2, 66—75. [228]
— (1960): Ecology and distribution of recent foraminifera. — Baltimore, 297 pp. [220]

— (1964): Foraminiferal ecology and marine geology. — Mar. Geol., 1, 16—43. [228, 359]
PHOENIX, D. A. (1956): Relation of carnotite deposits to permeable rocks in the Morrison formation, Mesa County, Colorado. — Internat. Conf. Peaceful Uses Atom. Energ. Proc., U. N., New York, 6, 321—325. [144]
PIA, J. v. (1926): Pflanzen als Gesteinsbildner. — Borntraeger, Berlin, 355 pp. [220]
— (1928): Die Anpassungsformen der Kalkalgen. — Palaeobiologica, 1, 211—224 (Wien). [220]
— (1933): Die rezenten Kalksteine. — Z. Krist. etc., B. Miner., Petrogr. Mitt., Erg. Bd., 1—418. [259, 260]
PICCOLI, G. (1966): Subaqueous and subaerial basic volcanic eruptions in the paleogene of the Lessinian Alps (Southern Alps, NE-Italy). — Bull. Volc., 29, 253—270. [183]
PICHLER, H. (1963): Geologische Untersuchungen im Gebiet zwischen Roßfeld und Markt Schellenberg im Berchtesgadener Land. —Beih. Geol. Jb., 48, 129—204. [288]
PICKEL, W. (1937): Stratigraphie und Sedimentanalyse des Kulms an der Edertalsperre. — Z. dt. Geol. Ges., 89, Abh., A., 233—280. [56]
PIERCE, J. W. (1968): Sediment budget along a barrier island chain. — Sediment. Geol., 3, 5—16. [120]
PILLER, H. (1951): Über den Schwermineralgehalt von anstehendem und verwittertem Brokkengranit nördlich St. Andreasberg. — Heidelb. Beitr. Miner. Petrogr., 2, 523—537. [31, 39]
PIMM, A. C, GARRISON, R. E. & BOYCE, R. E. (1971): Sedimentology synthesis: Lithology, chemistry and physical properties of sediments in the north-western Pacific Ocean. — In: FISCHER, A. G. et al. (ed.), Initial reports of the Deep Sea Drilling Project, Vol. VI, Washington (U. S. Governm. Print Off.), 1131—1252. [330, 331]
PIRSSON, L. V. (1915): The microscopical characters of volcanic tuffs — a study for students. — Amer. J. Sci., 40, 191—211. [162]
PITTMAN, E. D. (1963): Use of zoned plagioclase as an indicator of provenance — J. Sediment. Petrol., 33, 380—386. [24, 25]
— (1970): Plagioclase feldspar as an indicator of provenance in sedimentary rocks. — J. Sediment. Petrol., 40, 591—598. [24]
— (1972): Diagenesis of quartz in sandstones as revealed by scanning electron microscopy. — J. Sediment. Petrol., 42, 507—519. [138]
PLESSMANN, W. (1961): Strömungsmarken in klastischen Sedimenten und ihre geologische Auswertung. Untersuchungsergebnisse im Oberharzer Kulm und im westalpinen Flyschbecken von San Remo. — Geol. Jb., 78, 503—566. [70, 94, 96, 369]
— (1964): Gesteinslösung, ein Hauptfaktor beim Schieferungsprozeß. — Geol. Mitt., Aachen, 4, 69—82. [134]
— (1972): Horizontal-Stylolithen im französisch-schweizerischen Tafel- und Faltenjura und ihre Einpassung in den regionalen Rahmen. — Geol. Rdsch., 61, 332—347. [297]
PLUMLEY, W. J. (1948): Black Hills terrace gravels: a study in sediment transport. — J. Geol., 56, 526—577. [48, 57, 66]
PLUMLEY, W. J., RISLEY, G. A., GRAVES, R. W. JR. & KALEY, M. E. (1962): Energy index for limestone interpretation and classification. — In: W. E. HAM (ed.), Classification of carbonate rocks. — Amer. Assoc. Petrol. Geol. Mem., 1, 84—107. [278]
POBEGUIN, TH. (1954): Contribution à l'étude des carbonates de calcium. Précipitation du calcaire par les végétaux. Comparaison avec le monde animal. — Ann. des Soc. Nat. Bot., Ser. 11, 29—109. [215, 236]
POKORNY, V. (1958): Grundzüge der zoologischen Mikropaläontologie, I, 582 pp; II, 453 pp. — VEB Dtsch. Verlag d. Wiss., Berlin. [232, 247]
POLDERVAART, A. (1950): Statistical studies of zircon as a criterion in granitization. — Nature, 165, 574—575. [31]
— (1955): Chemistry of the earth's crust. — Geol. Soc. Amer. Spec. Pap., 62, 119—144. [8]
PORTER, J. J. (1962): Electron microscopy of sand surface texture. — J. Sediment. Petrol., 32, 124—135. [70]
POSER, H. & HÖVERMANN, J. (1951): Untersuchungen zur pleistozänen Harzvergletscherung. Abh. Braunschw. Wiss. Ges., 3, 61—115. [71]
— — (1952): Beiträge zur morphometrischen und morphologischen Schotteranalyse. — Abh. Braunschw. Wiss. Ges., 4, 12—36. [71]
POSTMA, H. (1967): Sediment transport and sedimentation in the estuarine environment. — In: LAUFF, G. H. (ed.), Estuaries. — Amer. Assoc. Advances, Sci., Publ. 38, 158—179. [3]

POTTER, P. E. (1955): The petrology and origin of the Lafayette gravel, pt. I, mineralogy and petrology. — J. Geol., 63, 1—38. [56]
— (1963): Late paleozoic sandstones of the Illinois basin. — Ill. Geol. Surv. Rep. Invest., 217, 92 pp. [100]
— (1967):Sand bodies and sedimentary environments. A Review. — Bull. Amer. Assoc. Petrol. Geol., 51, 337—365. [100]
POTTER, P. E. & BLAKELY, R. F. (1967)· Generation of a synthetic vertical profile of a fluvial sandstone body. — Soc. Petrol. Engin. J., 243—251. [78, 100]
POTTER, P. E. & GLASS, H. D. (1958): Petrology and sedimentation of the Pennsylvanian sediments in Southern Illinois: A vertical profile. — Ill. State Surv. Rep. Invest., 204, 60 pp. [39]
POTTER, P. E. & MAST, R. F. (1963): Sedimentary structures, sand shape fabrics, and permeability. I. — J. Geol., 71, 441—471. [72]
POTTER, P. E. & PETTIJOHN, F. J. (1963): Paleocurrents and basin analysis. — Springer, Berlin, Göttingen, Heidelberg, 296 pp. [65, 71, 72, 77, 78, 79, 81, 82, 84, 89, 90, 92, 94]
POTTER, P. E. & PRYOR, W. A. (1961): Disposal centers of Paleozoic and later clastics of the Upper Mississippi Valley and adjacent areas. — Bull. Geol. Soc. Amer., 72, 1195—1250.
POWERS, R. W. (1962): Arabian Upper Jurassic carbonate reservoir rocks. — In: W. E. HAM (ed.), Classification of carbonate rocks. — Amer. Assoc. Petrol. Geol. Mem., 1, 122—192. [215, 347, 351]
PRAY, L. C. (1958): Fenestrate bryozoan core facies, Mississippian bioherms, southwestern United States. — J. Sediment. Petrol., 28, 261—273. [270, 271]
— (1964): Limestone clastic dikes in Mississippian bioherms, New Mexico. — Geol. Soc. Amer. Ann. Meet., Miami, Progr., 154—155. [84, 288]
PRAY, L. C. & HOROWITZ, A. S. (1959): Mississippian bioherms, southwestern United States. — AAPG-SEPM Joint Ann. Meet., p. 79. [270]
PRESTON, F. W. & HENDERSON, J. H. (1965): Fourier series characterization of cyclic sediments for stratigraphic correlation. — In: D. F. MERRIAM (ed.), Symposium on cyclic sedimentation. — Kansas Geol. Surv. Bull., 169, 415—425. [362]
PROKOPOVICH, N. (1952): The origin of stylolites. — J. Sediment. Petrol., 22, 212—220. [298]
PROZOROVICH, G. E. (1970): Determination of the time of oil and gas accumulation by epigenesis studies. — Sedimentology, 15, 41—52. [144, 147]
— (1971): Sedimentary and epigenetical trends aiding the hydrocarbon exploration in W. Siberia. — Sedimentology, 17, 233—239. [144]
PRYOR, W. A. (1961): Sand trends and paleoslope in Illinois basin and Mississippi embayment. — In: J. A. PETERSON & J. E. OSMOND (ed.), Geometry of sandstone bodies. — Amer. Assoc. Petrol. Geol., 119—133. [100]
PÜMPIN, V. F. (1965): Riffsedimentologische Untersuchungen im Rauracien von St. Ursanne und Umgebung (Zentraler Schweizer Jura). — Eclog. geol. Helvet., 58, 799—876. [221]
PURDY, E. G. (1963): Recent calcium carbonate facies of the Great Bahama Bank. 2. Sedimentary facies. — J. Geol., 71, 472—497. [196, 200, 204, 206, 251, 285]
— (1968): Carbonate diagenesis: an environmental survey. — Geol. Romana, 7, 183—228. [285]
PURDY, E. G. & IMBRIE, J. (1964): Carbonate sediments, Great Bahama Bank. — Geol. Soc. Amer. Convent. Miami, Guidebook Field Trip2, 1—58. [197, 251, 257]
PURSER, B. H. (1969): Syn-sedimentary marine lithification of Middle Jurassic limestones of the Paris Basin, France. — Sedimentology, 12, 205—230. [288, 370]
— (1971): The relationship between syn-sedimentary lithification and subsequent pressure solution in the Middle Jurassic Limestones of the Paris Basin (Abstr.). — Progr., VIII Internat. Sedimentol. Congr., Heidelberg, 79. [288]
— (1973): Sedimentation around bathymetric highs in the southern Persian Gulf. — In: PURSER, B. H. (ed.), The Persian Gulf. — Springer, Berlin, Heidelberg, New York, 157—177. [284]
PURSER, B. H. & EVANS, G. (1973): Regional sedimentation along the Trucial Coast, SE Persian Gulf. — In: PURSER, B. H. (ed.), The Persian Gulf. — Springer, Berlin, Heidelberg, New York, 211—231. [253]
PURSER, B. H. & LOREAU, J.-P. (1973): Aragonitic, supratidal encrustations on the Trucial Coast, Persian Gulf. — In: PURSER, B. H. (ed.), The Persian Gulf. — Springer, Berlin, Heidelberg, New York, 343—376. [260]
PUSTOWALOW, L. V. (1940): Petrography of Sedimentary Rocks (Russ.). — Moscow. [279]

QUESTER, H. (1964): Petrographie des erdgashöffigen Hauptdolomits im Zechstein 2 zwischen Weser und Ems. — Z. dt. Geol. Ges., 114, 461—483. [221, 325, 345, 347]

RAAM, A. (1968): Petrology and diagenesis of Broughton sandstone (Permian), Kiama district, New South Wales. — J. Sediment. Petrol., 38, 319—331. [155]

RABIEN, A. (1956): Zur Stratigraphie und Fazies des Ober-Devons in der Waldecker Hauptmulde. — Abh. hess. Landesamt Bodenforsch., 16, 1—83. [81]

RAD, W. v. (1970): Comparison between "magnetic" and sedimentary fabric in graded and cross-laminated sand layers. Southern California. — Geol. Rdsch., 60, 331—354. [72]

RADOMSKI, A. (1961): On some sedimentological problems of the Swiss flysch series. — Eclog. geol. Helvet., 54, 451—459. [369]

RAMSDEN, R. M. (1952): Stylolites and oil migration. — Bull. Amer. Assoc. Petrol. Geol., 36, 2185—2186. [299]

RANKAMA, K. & SAHAMA, T. G. (1950): Geochemistry. — Univ. of Chicago Press, 912 pp. [192]

RAPSON-MCGUGAN, J. E. (1970): The diagenesis and depositional environment of the Permian Ranger Canyon and Mowitch formations, Ishbel group, from the Southern Canadian Rock Mountains. — Sedimentology, 15, 363—417. [141]

RAUPACH, F. v. (1952): Die rezente Sedimentation im Schwarzen Meer, im Kaspi und im Aral und ihre Gesetzmäßigkeiten. — Geologie, 1, 78—132. [241, 316]

READ, W. A. & DEAN, J. M. (1967): A quantitative study of a sequence of coal-bearing cycles in the Namurian of Central Scotland, 1. — Sedimentology, 9, 137—156. [362, 376]

REES, A. I. (1965): The use of anisotropy of magnetic susceptibility in the estimation of sedimentary fabric. — Sedimentology, 4, 257—271. [72]

REIFF, W. (1955): Über den pleistozänen Sauerwasserkalk von Stuttgart-Münster-Bad Cannstatt. — Jber. u. Mitt. oberrhein. geol. Ver., N. F., 37, 56—91. [259]

REIMER, T. O. (1971): Volcanic quartz as indicator mineral in graywackes. — Sedimentology, 17, 125—128. [21]

REINECK, H.-E. (1955): Haftrippeln und Haftwarzen, Ablagerungsformen von Flugsand. — Senckenbergiana leth., 36, 347—357. [89]

— (1958 a): Wühlbau-Gefüge in Abhängigkeit von Sediment-Umlagerungen. — Senckenbergiana leth., 39, 1—23. [80, 89]

— (1958 b): Über Gefüge von orientierten Grundproben aus der Nordsee. — Senckenbergiana leth., 39, 25—41. [71]

— (1958 c): Longitudinale Schrägschicht im Watt. — Geol. Rdsch., 47, 73—82. [79, 116]

— (1960 a): Über die Entstehung von Linsen- und Flaserschichten. — Abh. dt. Akad. Wiss. Berlin, III, H. 1, 369—374. [74]

— (1960 b): Über Zeitlücken in rezenten Flachsee-Sedimenten. — Geol. Rdsch., 49, 149—161. [85—87]

— (1961): Sedimentbewegungen an Kleinrippeln im Watt. — Senckenbergiana leth., 42, 51—67. [74]

— (1962): Schichtungsarten in Wattenböden. — Pflanzenernähr., Düng., Bodenkde., 99, 154—159. [74, 118]

— (1963): Sedimentgefüge im Bereich der südlichen Nordsee. — Abh. senckenb. naturf. Ges., Frankfurt, 505, 138 pp. [73, 74, 76—80, 89, 115, 119, 121]

— (1967 a): Parameter von Schichtung und Bioturbation. — Geol. Rdsch., 56, 420—438. [112]

— (1967 b): Layered sediments of tidal flats, beaches, and shelf bottoms of the North Sea. — In: Estuaries, G. H. LAUFF (ed.). — Amer. Assoc. Advances of Sci. (AAAS) Washington, D. C., 191—206. [112]

— (1969): Die Entstehung von Runzelmarken. — Natur u. Museum, 99, 386—388. [90]

— (1971): Das Watt, Ablagerungs- und Lebensraum. — Senckenberg-Buch 50, W. KRAMER, Frankfurt, 142 pp. [118]

— figure [116]

REINECK, H.-E. & SINGH, I. B. (1971): Genesis of laminated sand and graded rhythmites in storm sand layers of shelf mud. — Sedimentology, 18, 123—128. [74, 90, 365]

REINECK, H.-E., SINGH, I. B. & WUNDERLICH, F. (1971): Einteilung der Rippeln und anderer mariner Sandkörper. — Senckenbergiana Marit., 3, 93—101. [78, 118]

REINECK, H.-E. & WUNDERLICH, F. (1968 a): Classification and origin of flaser and lenticular bedding. — Sedimentology, 11, 99—104. [74]

— — (1968 b): Zur Unterscheidung von asymmetrischen Oszillationsrippeln und Strömungsrippeln. — Senckenbergiana leth., 49, 321—345. [89]
REMANE, A. (1963): Biologische Kriterien zur Unterscheidung von Süß- und Salzwassersedimenten. — Fortschr. Geol. Rheinl.-Westf., 10, 9—34. [241]
REULING, H. T. (1931): Dolomit-Studien im Devon der Eifel. — Senckenbergiana 13, 271—298. [319]
— (1934): Der Sitz der Dolomitisierung: Versuch einer neuen Auswertung der Bohrergebnisse von Funafuti. — Abh. senckenb. naturf. Ges., 428, 44 pp. [317]
REVELLE, R. & FAIRBRIDGE, R. (1957): Carbonates and Carbon Dioxide. — In: J. W. HEDGPETH (ed.), Geol. Soc. Amer. Mem., 67, 1, 239—296. [209, 236]
REZAK, R., KAN, D. L. & BUCHBINDER, B. (1971): Laboratory experiments in calcium carbonate cementation (Abstr.). — Progr., VIII. Internat. Sediment. Congr., Heidelberg, 82—83. [284]
RICCI LUCCHI, F. (1969): Considerazioni sulla formazione di alcune impronte da corrente. — Giorn. Geol. (2), 36, 363—438. [96]
— (1970): Sedimentografia. Atlante fotografico delle strutture primarie dei sedimenti. — Zanichelli Ed. Bologna, 288 pp. [72, 96]
RICCI LUCCHI, F. & DALLA CASA, G. (1970): Surface textures of desert quartz grains. A new attempt to explain the origin of desert frosting. — Giorn. Geol. (2), 36, 751—776. [69]
RICH, M. (1965): "Calcispheres" from the Duperow formation (Upper Devonian) in western North Dakota. — J. Paleont., 39, 143—145. [227]
RICHTER, D. K. (1971): Fazies- und Diagenesehinweise durch Einschlüsse in authigenen Quarzen. — N. Jb. Geol. Paläont., Mh., 604—622. [314]
— (1972a): Eine subrezente spätdiagenetische Dolomitisierung mit prätertiären Dolomiten als Keime (Bucht von Volos, Griechenland). — N. Jb. Geol. Paläont. Mh. 8, 490—506. [305]
— (1972b): Authigenic quartz preserving skeletal material. — Sedimentology, 19, 211—218. [235, 236]
— (1974): Origin and diagenesis of Devonian and Permotriassic dolomites in the Eifel Mts. (Germany). — Contr. to Sedimentol., 2, Schweizerbart, Stuttgart, 101 pp. [311, 314, 319—321, 335]
RICHTER, K. (1932): Die Bewegungsrichtung des Inlandeises, rekonstruiert aus den Kritzen und Längsachsen der Geschiebe. — Z. Geschiebeforsch., 8, 62—66. [71]
— (1933): Gefüge und Zusammensetzung des norddeutschen Jungmoränengebietes. — Abh. Geol.-Pal. Inst. Greifswald, 11, 1—63. [47]
— (1958): Bildungsbedingungen pleistozäner Sedimente Niedersachsens aufgrund morphometrischer Geschiebe- und Geröllanalysen. — Z. dt. Geol. Ges., 110, 400—435. [47, 64]
RICHTER, R. (1919): Vom Bau und Leben der Trilobiten. I. Das Schwimmen. — Senckenbergiana leth., 1, 213—238. [246]
— (1926): Flachseebeobachtungen zur Paläontologie und Geologie XV. Die Großrippeln unter Gezeitenströmungen im Wattenmeer und die Rippeln im Pirnaer Turon. — Senckenbergiana leth., 8, 297—315. [77]
— (1942): Die Einkippungsregel. — Senckenbergiana, 215—244. [207]
RICHTER-BERNBURG, G. (1955): Über salinare Sedimentation. — Z. dt. Geol. Ges., 105, 593—645. [314, 337, 363, 364]
— (1957): Isochrone Warven im Anhydrit des Zechstein 2. — Geol. Jb., 74, 601—610. [363]
— (1960): Zeitmessung geologischer Vorgänge nach Warven-Korrelationen im Zechstein. — Geol. Rdsch., 49, 132—148. [364]
RIEDEL, W. R., LADD, H. S. TRACEY, J. I., JR. & BRAMLETTE, M. N. (1961): Preliminary drilling phase of Mohole project, II. Summary of coring operations. — Bull. Amer. Assoc. Petrol. Geol., 45, 1793—1798. [319]
RIGBY, J. K. (1953): Some transverse stylolites. — J. Sediment. Petrol., 23, 265—271. [297]
RIMŠAITE, J. (1957): Über die Eigenschaften der Glimmer in den Sanden und Sandsteinen. — Beitr. Miner. Petrogr., 6, 1—51. [26]
RITTENHOUSE, G. (1943): Transportation and deposition of heavy minerals. — Bull. Geol. Soc. Amer., 54, 1725—1780. [24]
— (1961): Problems and principles of sandstone-body classification. — In: J. A. PETERSON & J. C. OSMOND (ed.), Geometry of sandstone bodies. — Amer. Assoc. Petrol. Geol., 3—12. [100]

Rittmann, A. (1962): Volcanoes and their activity. — New York, John Wiley & Sons, Inc., 305 pp. [183]
Robertson, E. C. (1964): Experimental consolidation of carbonate mud. — Bull. Amer. Assoc. Petrol. Geol., 48, 544 (Abstr.). [294, 339]
Robinson, R. B. (1966): Classification of reservoir rocks by surface texture. — Bull. Amer. Assoc. Petrol. Geol., 50, 547—559. [346]
— (1967): Diagenesis and porosity development in Recent and Pleistocene oolites from Southern Florida and the Bahamas. — J. Sediment. Petrol., 37, 355—364. [349]
Roll, A. (1934): Form, Bau und Entstehung der Schwammstotzen im süddeutschen Malm. — Paläont. Z., 16, 197—246. [263, 271, 274]
Rona, P. A. (1973): Relations between rates of sediment accumulation on continental shelves, sea-floor spreading, and eustacy inferred from the Central North Atlantic. — Geol. Soc. Amer. Bull., 84, 2851—2872. [9, 366]
Ronov, A. B. (1949): Geschichte der Sedimentbildung sowie der epirogenen Bewegungen im europäischen Teil der UdSSR. — Trudi Geophys. In-tan, Akad. Nauk SSSR, H. 3 (Russ.). [86]
— (1956): The chemical composition and the conditions of formation of Paleozoic carbonate layers of the Russian Platform, based on data of lithologic-geochemical maps (Russ.). — Akad. Nauk SSSR, Trudy Geol. Inst., no. 4, 256—343. [314]
— (1968): Probable changes in the composition of sea water during the course of geological time. — Sedimentology, 10, 25—43. [8, 9]
Rosenfeld, M. A. (1949): Some aspects of porosity and cementation. — Prod. month., 13, 39—42. [135]
Ross, C. S. & Smith, R. L. (1955): Water and other volatiles in volcanic glasses. — Am. Miner., 40, 1071—1089. [186, 187]
— — (1961): Ash-flow tuffs: Their origin, geologic relations, and identification. — U. S. Geol. Surv. Prof. Pap., 366, 81 pp. [175, 176]
Ross, J. V. (1957): Combination twinning in plagioclase feldspar. — Amer. J. Sci., 255, 650—655. [25]
Rothpletz, A. (1892): On the formation of oolite. — Amer. Geologist, 10, 279—282. [253, 255]
Rottgardt, D. (1952): Mikropaläontologisch wichtige Bestandteile recenter brackischer Sedimente an den Küsten Schleswig-Holsteins. — Meyniana (Kiel), 1, 169—228. [227]
Rubey, W. W. (1951): The geologic history of sea water. — Geol. Soc. Amer. Bull., 62, 1111—1147. [9]
Rubey, W. W. & Hubbert, M. K. (1959): Role of fluid pressure in mechanics of overthrust faulting. II. Overthrust belt in geosynclinal area of Western Wyoming in light of fluid-pressure hypothesis. — Bull. geol. Soc. Amer., 70, 167—206. [80]
Ruchin, L. B. (1958): Grundzüge der Lithologie. Lehre von den Sedimentgesteinen. — Akademie Verlag, Berlin, 806 pp. (transl. from Russ. A. Schüller). [67, 70, 71, 86, 317]
Rudolf, W. F. (1959): Zur Dolomitisierung und Petrogenese im unteren Hauptmuschelkalk Württembergs. — Diss. Univ. Tübingen. [311, 313]
Rücklin, H. (1938): Strömungs-Marken im Unteren Muschelkalk des Saarlandes. — Senckenbergiana leth., 20, 94—114. [90]
— (1955): Das Holzer Konglomerat im Saarkarbon. — Geol. Jb., 70, 435—510. [50, 51]
Rupp, A. (1967): Origin, structure, and environmental significance of Recent and fossil calcispheres. — Geol. Soc. Amer., Spec. Pap., 101, 186 (Abstr.). [227]
Rusnak, G. A. (1957): The orientation of sand grains under condition of "unidirectional" fluid flow. 1. Theory and experiment. — J. Geol., 65, 384—409. [71]
— (1960 a): Some observations of recent oolites. — J. Sediment. Petrol., 30, 471—480. [251]
— (1960 b): Sediments of Laguna Madre, Texas. — In: Shepard, F. P., Phleger, F. B. & Van Andel, Tj. (ed.), Northwest Gulf of Mexico. — Amer. Assoc. Petrol. Geol., 153—196. [118]
— (1964): Late Pleistocene marine sedimentation rates. — Inst. Marine Sci., Publ., Univ. Miami, Florida. [86]
Rusnak, G. & Nesteroff, W. (1964): Modern turbidites: terrigenous abyssal plain versus bioclastic basin. — In: Miller, R. L. (ed.), Papers in marine geology, Shepard Comm.-Vol. — New York, Macmillan, 488—507. [360, 368, 369]
Russell, R. D. (1936): The size distribution of minerals in Mississippi river sands. — J. Sediment. Petrol., 6, 125—142. [40]

— (1937): Mineral composition of Mississippi River sands. — Bull. geol. Soc. Amer., **48**, 1307—1348. [24]
— (1939): Effects of transportation on sedimentary particles. — In: P. D. TRASK (ed.), Recent Marine Sediments. — Amer. Assoc. Petrol. Geol., 32—47. [57]
RUSSELL, R. D. & TAYLOR, R. E. (1937): Roundness and shape of Mississippi River sands. — J. Geol., **45**, 225—267. [67]
RUSSELL, R. J. (1962): Origin of beach rock. — Z. Geomorph., N. F., **6**, 1—16. [281]
— (1968): Where most grains of very coarse sand and fine gravel are deposited. — Sedimentology, **11**, 31—38. [56]
RUTSCH, R. F. (1968): Herkunft und Bedeutung des Begriffs „Nagelfluh". — Mitt. Naturf. Ges. Bern, N.F., **25**, 69—79. [47]
RUTTE, E. (1955): Süßwasserkalke und Kalkalgenbildungen in der chattischen Unteren Süßwassermolasse von Hoppetenzell nördlich Stockach/Baden. — Geol. Jb., **69**, 517—536. [221]
— (1958): Kalkkrusten in Spanien. — N. Jb. Geol. Paläont., Abh., **106**, 52—138. [258]
— (1961): Kalkkrusten im östlichen Mittelmeergebiet. — Z. dt. Geol. Ges., **112**, 81—90. [258]
RUTTEN, M. G. (1952): Rhythm in sedimentation and erosion. — 3ième Congr. Avanc. Etudes de Stratigr. et de Géol. du Carbonif., C. R., t. 2, 529—537. [366]
— (1958): Detailuntersuchungen an gotländischen Riffen. — Geol. Rdsch., **47**, 359—384. [261, 273]
— (1962): The geological aspects of the origin of life on earth. — Elsevier, Amsterdam, 146 pp. [219]
RYAN, W. B. F. (1972): Stratigraphy of late Quaternary sediments in the Eastern Mediterranean. — In: STANLEY, D. J. (ed.), The Mediterranean Sea: A natural sedimentation laboratory. — Dowden, Hutchinson & Ross, Inc., 149—169. [86]

SAEMUNDSSON, K. (1967): Vulkanismus und Tektonik des Hengill-Gebietes in Südwest-Island. — Acta Nat. Island., **11**, 1—105. [183]
SAHU, B. K. (1964): Significance of the size-distribution statistics in the interpretation of depositional environments. — Res. Bull., N. S., Panjab Univ., **15**, III—IV, 213—219. [59]
SANDER, B. (1936): Beiträge zur Kenntnis der Anlagerungsgefüge. — Tscherm. Miner. Petr. Mitt., **48**, 27—139, 141—209. [203, 204, 277, 279, 281, 288, 314, 361]
— (1948): Einführung in die Gefügekunde der geologischen Körper. 1. Allgemeine Gefügekunde und Arbeiten im Bereich Handstück bis Profil. — Springer, Wien, Innsbruck, 215 pp. [371]
— (1950): Einführung in die Gefügekunde der geologischen Körper. II. Korngefüge. — Springer, Wien, Innsbruck, 409 pp. [277, 281]
SANDERS, J. E. (1965): Primary sedimentary structures formed by turbidity currents and related resedimentation mechanisms. — In: G. V. MIDDLETON (ed.), Primary sedimentary structures and their hydrodynamic interpretation. — SEPM Spec. Publ., **12**, 192—219. [81, 97, 123]
SANDERS, J. E. & FRIEDMAN, G. M. (1967): Origin and occurrence of limestones. — In: CHILINGAR et al. (ed.), Carbonate rocks. — Developm. in Sedimentol., 9 A, Elsevier Publ. Co. Amsterdam, 169—265. [203, 259, 260]
SAPOZHNIKOV, D. G. (1951): Sovemennye osadki i geologiya ozero Balkhash. — Akad. Nauk SSSR. Inst. Geol. Nauk Trudy, **132**, Geol. Ser. no. 53, 207 pp. [317]
SARIN, D. D. (1962): Cyclic sedimentation of primary dolomite and limestone. — J. Sediment. Petrol., **32**, 451—471. [314]
SARKISYAN, S. G. (1949): Petrographic—mineralogical investigation of the Upper Permian and Triassic variegated deposits of the pre-Urals region. — Acad. Nauk USSR, Moskau, 191 pp. (Russ.) [45]
— (1958): Upper Permian continental molasse of the pre-Urals region. — Eclog. geol. Helvet., **51**, 1043—1051. [45]
SARNTHEIN, M. (1965): Sedimentologische Profilreihen aus den mitteltriadischen Karbonatgesteinen der Kalkalpen nördlich und südlich von Innsbruck. — Verh. Geol. Bundesanst. Wien, H. 1/2, 119—162. [374]
— (1967): Versuch einer Rekonstruktion der mitteltriadischen Paläogeographie um Innsbruck, Österreich. — Geol. Rdsch., **56**, 116—127. [263]

SARNTHEIN, M. & WALGER, E. (1973): Classification of modern marl sediments in the Persian Gulf by factor analysis. — In: PURSER, B. H. (ed.), The Persian Gulf. Springer, Berlin, Heidelberg, New York, 81—97. [197]

SCHAD, A. (1964): Feingliederung des Miozäns und die Deutung der nacholigozänen Bewegungen im Mittleren Rheingraben. — Abh. geol. Landesamt Baden-Württ., 5, 1—56. [125]

SCHÄFER, W. (1954): Dehnungsrisse unter Wasser im meerischen Sediment. — Senckenbergiana leth., 35, 87—99. [80]
— (1956): Der kritische Raum und die kritische Situation in der tierischen Sozietät. — Aufs. Reden Senckenb. naturf. Ges., Frankfurt. [80]
— (1962): Aktuo-Paläontologie nach Studien in der Nordsee. — W. Kramer, Frankfurt/Main, 666 pp. [80]
— (1963): Biozönose und Biofazies im marinen Bereich. — Aufs. Reden Senckenb. naturf. Ges., 11, 1—37. [80]

SCHEIDEGGER, A. E. & POTTER, P. E. (1965): Textural studies of graded bedding. Observation and theory. — Sedimentology, 5, 289—304. [369]
— — (1967): Bed thickness and grain size: Crossbedding. — Sedimentology, 8, 39—44. [75]
— — (1968): Textural studies of grading: volcanic ash falls. — Sedimentology, 11, 163—170. [166]

SCHELLMANN, W. (1959): Experimentelle Untersuchungen über die sedimentäre Bildung von Goethit und Hämatit. — Chemie d. Erde, 20, 105—135. [144]

SCHERP, A. (1963): Die Petrographie der paläozoischen Sandsteine in der Bohrung Münsterland 1 und ihre Diagenese in Abhängigkeit von der Teufe. — Fortschr. Geol. Rheinl.-Westf., 11, 251—282. [145, 153—155, 157]

SCHIEMENZ, S. (1960): Fazies und Paläogeographie der Subalpinen Molasse zwischen Bodensee und Isar. — Beih. geol. Jb., 38, 119 pp. [47, 50, 51, 71, 73, 203, 366, 380]

SCHINDEWOLF, O. H. (1928): Über Farbstreifen bei *Amaltheus (Paltopleuroceras) spinatus* (BRUG.). — Paläont. Z., 10, 136—143. [245]

SCHLANGER, S. O. (1957): Dolomite growth in coralline algae. — J. Sediment. Petrol., 27, 181—186. [313]

SCHLEE, J. (1957): Upland gravels of southern Maryland. — Bull. Geol. Soc. Amer., 68, 1371—1410. [56]
— (1963): Sandstone pipes of the Laguna area, New Mexico. — J. Sediment. Petrol., 33, 112—123. [91]

SCHLEGELMILCH, V. (1968): Rotfärbungen im Thüringer Schiefergebirge. — Geologie, 17, 136—155. [108]

SCHMALZ, R. F. (1965): Brucite in carbonate secreted by the red alga *Goniolithon* sp. — Science, 149, 993—996. [219]

SCHMALZ, R. F. & CHAVE, K. E. (1963): Calcium carbonate: factors affecting saturation in ocean waters off Bermuda. — Science, 139, 1206—1207. [322]

SCHMASSMANN, H. J. (1954): Stratigraphie des mittleren Doggers der Nordschweiz. — Tätigkeitsber. naturf. Ges. Baselland, 14, 13—180 (Diss. Basel). [370]

SCHMEER, D. (1962): Zur Sedimentpetrographie der Oberen Süßwassermolasse. — In: K. BRUNNACKER, Erläuterungen zur Geol. Karte von Bayern 1:25 000, Bl. 7536 Freising Nord, 12—22. [317]

SCHMIDEGG, O. (1928): Über geregelte Wachstumsgefüge. — Jb. Geol. Bundesanst. Wien, 78, 1—52. [254, 288]

SCHMIDT, H. (1935a): Einführung in die Paläontologie. — Enke, Stuttgart, 253 pp. [206]
— (1935b): Die bionomische Einteilung der fossilen Meeresböden. — Fortschr. Geol. Paläont., 12, 38, 154 pp. [243]

SCHMIDT, M. (1928): Die Lebewelt unserer Trias. — Hohenlohesche Buchhandlung, Öhringen, 1928, 461 pp., Nachtrag 1938, 143 pp. [220]

SCHMIDT, V. (1961): Petrographische und fazielle Untersuchungen an Karbonatgesteinen des Oberkimmeridge und des Oberen Malm 1 in Südoldenburg. — Diss. Kiel, 287 pp. [250, 257, 295, 297, 313, 325—327, 351]
— (1965): Facies, diagenesis, and related reservoir properties in the Gigas beds (Upper Jurassic), Northwestern Germany. — In: L. C. PRAY & R. C. MURRAY (ed.), Dolomitization and limestone diagenesis. — SEPM Spec. Publ., 13, 124—168. [295, 301, 311, 313, 327, 333, 351, 353]

SCHMIDT, W. F. (1963): Untersuchungen über das Pliozän der Insel Cypern. 1. Ein Wurmröhren-(Serpuliden-)Bryozoenriff-Horizont südlich Leucossia (Nicosia, Cypern). — Ann. Géol. des Pays Hellén., 14, 109—132. [275]

SCHMINCKE, H.-U. (1967a): Graded lahars in the type section of the Ellensburg Formation, South Central Washington. — J. Sediment. Petrol., 37, 438—448. [180, 181]

— (1967b): Fused tuff and péperites in south-central Washington. — Geol. Soc. Amer. Bull., 78, 319—330. [183]

SCHMINCKE, H.-U., FISHER, R. V. & WATERS, A. C. (1973): Antidune and chute and pool structures in base surge deposits from the Laacher See area (Germany). — Sedimentology, 20, 553—574. [90, 177, 179]

SCHNEIDER, H. E. (1970): Problems of quartz grain morphoscopy. — Sedimentology, 14, 325—335. [70]

SCHNEIDER, H. E. & CAILLEUX, A. (1959): Signification géomorphologique des formes des grains de sables des Etats-Unis. — Z. Geomorph., N.F., 3, 114—125. [69]

SCHNEIDER, H.-J. (1954): Die sedimentäre Bildung von Flußspat im Oberen Wettersteinkalk der nördlichen Kalkalpen. — Abh. Bayer. Akad. Wiss., math.-naturw. Kl., N. F., 66, 1—37. [327]

SCHNEIDERHÖHN, H. (1923): Chalkographische Untersuchung des Mansfelder Kupferschiefers. — N. Jb. Miner., Geol., Paläont., B.-Bd. 47, 1—38. [337]

SCHNEIDERHÖHN, P. (1954): Eine vergleichende Studie über Methoden zur quantitativen Bestimmung von Abrundung und Form an Sandkörnern (im Hinblick auf die Verwendbarkeit an Dünnschliffen). — Heidelb. Beitr. Miner. Petrogr., 4, 172—191. [64]

SCHNITZER, W. A. (1957): Die Quarzkornfarbe als Hilfsmittel für die stratigraphische und paläogeographische Erforschung sandiger Sedimente. — Erlanger Geol. Abh., H. 23, 13 pp. [22]

— (1963): Zur Methodik der Quarzkornfarben-Untersuchung in feinkörnigen sandigen Sedimenten. — Geol. Bl. NO-Bayern, 13, 1—11. [22]

SCHÖNER, H. (1960): Über die Verteilung und Neubildung der nichtkarbonatischen Mineralkomponenten der Oberkreide aus der Umgebung von Hannover. — Beitr. Miner. Petrogr., 7, 70—103. [334, 335, 337]

SCHÖNE-WARNEFELD, G. & DAHM, H. (1962): Tutenmergel im Ruhrkarbon. — Fortschr. Geol. Rheinl.-Westf., 3, 2, 643—646. [299]

SCHÖTTLE, M. & MÜLLER, G. (1968): Recent carbonate sedimentation in the Gnadensee (Lake Constance), Germany. — In: MÜLLER & FRIEDMAN (ed.), Recent developments in carbonate sedimentology in Central Europe. — Springer, Berlin, Heidelberg, New York, 148—156. [222]

SCHOKLITSCH, A. (1930): Der Wasserbau, Bd. 1. — Springer, Wien, 484 pp. [56]

SCHOLLE, P. A. & KLING, S. A. (1972): Southern British Honduras: Lagoonal coccolith ooze. — J. Sediment. Petrol., 42, 195—204. [359]

SCHOPF, TH. J. & MANHEIM, F. T. (1967): Chemical composition of Ectoprocta (Bryozoa). — J. Paleont., 41, 1197—1225. [237]

SCHOT, E. H. & PARK, W. C. (1968): Note on the formation of stylolites. — Geol. en Mijnb., 47, 112—113. [299]

SCHRAMM, M. W. (1963): Oölites and algal aggregates of the West Spring Creek formation (Ordovician), Arbuckle Mountains, Oklahoma. — Oklah. Geol. Notes, 23, 6, 152—162. [221]

SCHROEDER, J. H. (1972a): Calcified filaments of an endolithic alga in Recent Bermuda reefs. — N. Jb. Geol. Paläont. Mh., 1, 16—33. [285]

— (1972b): Fabrics and sequences of submarine carbonate cements in Holocene Bermuda cup reefs. — Geol. Rdsch., 61, 2, 708—730. [285, 290]

— (1973): Submarine and vadose cements in Pleistocene Bermuda reef rock. — Sediment. Geol., 10, 179—204. [295]

SCHROEDER, J. H., DWORNIK, E. J. & PAPIKE, J. J. (1969): Primary protodolomite in echinoid skeletons. — Bull. Geol. Soc. Amer., 80, 1613—1616. [211]

SCHROLL, E. & WIEDEN, P. (1960): Eine rezente Bildung von Dolomit im Schlamm des Neusiedler Sees. — Tschermaks miner. petrogr. Mitt., 7, 286—289. [317]

SCHÜLLER, M. (1967): Petrographie und Feinstratigraphie des Unteren Muschelkalkes in Südniedersachsen und Nordhessen. — Sediment. Geol., 1, 353—401. [371]

SCHUMANN, H. (1941): Zur Korngestalt der Quarze in Sanden. — Chemie d. Erde, 14, 131—151. [64]

SCHURAVLEVA, Z. A. (1964): Riphean and Lower Cambrian oncolithes and catagraphes of Siberia and their stratigraphic importance. — Akad. Nauk SSSR, Geolog. Instit., Trudy, 114, 1—73 (Russ.). [220, 224]
SCHWARZ, A. (1928): Die Natur des culmischen Kieselschiefers. — Abh. senckenb. naturf. Ges., 41, Lfg. 4, 191—241. [332]
SCHWARZACHER, W. (1947): Über die sedimentäre Rhythmik des Dachsteinkalkes von Lofer. — Verh. geol. Bundesanst. Wien, 175—188. [314]
— (1948): Sedimentpetrographische Untersuchungen kalkalpiner Gesteine. Hallstätter Kalk von Hallstatt und Ischl. — Jb. geol. Bundesanst. Wien, Jg. 1946, 91, 1—48. [227, 272]
— (1951): Grain orientation in sands and sandstones. — J. Sediment. Petrol., 21, 162—172. [72]
— (1953): Cross-bedding and grain size in the Lower Cretaceous sands of East Anglia. — Geol. Mag., 90, 322—330. [78]
— (1954): Die Großrhythmik des Dachsteinkalkes von Lofer. — Tschermaks Miner. Petr. Mitt., 4, 44—54. [374]
— (1961): Petrology and structure of some Lower Carboniferous reefs in northwestern Ireland. — Bull. Amer. Assoc. Petrol. Geol., 45, 1481—1503. [71, 270, 271]
SCHWARZMANN, S. (1956): Über die Lichtbrechung und die Achsenwinkel von Hochtemperaturplagioklasen und ihre Entstehungsbedingungen. — Heidelb. Beitr. Miner. Petr., 5, 105—112. [334]
SCHWERTMANN, U. (1966): Die Bildung von Goethit und Hämatit in Böden und Sedimenten. — Proc. Int. Clay Conf. Jerusalem, 1, 159—165. [144]
SCOTT, G. (1940): Paleoecological factors controlling the distribution and mode of life of Cretaceous ammonoids in the Texas area. — J. Paleont., 14, 199—233 (cit. from BERGQUIST & COBBAN 1957). [245]
SCOTT, K. M. (1966): Sedimentology and dispersal pattern of a Cretaceous flysch sequence, Patagonian Andes, Southern Chile. — Bull. Amer. Assoc. Petrol. Geol., 50, 72—107. [72, 98, 362]
SCRUTON, P. C. (1955): Sediments of the eastern Mississippi delta. — In: Finding ancient shorelines. — SEPM Spec. Publ., No. 3, 21—51. [85]
Sedimentary Petrology Seminar (1964): Gravel fabric in Wolf Run. — Sedimentology, 4, 273—283. [71]
Sedimentation Seminar (1966): Cross-bedding in the Salem limestone of Central Indiana. — Sedimentology, 6, 95—114. [79]
SEIBOLD, E. (1952): Chemische Untersuchungen zur Bankung im unteren Malm Schwabens. — N. Jb. Geol. Paläont., Abh., 95, 337—370. [371]
— (1958): Jahreslagen in Sedimenten der mittleren Adria. — Geol. Rdsch., 47, 100—117. [363]
— (1963): Geological investigation of near-shore sand-transport. — In: Progress in Oceanography, vol. 1 (ed. M. SEARS), Pergamon Press, Oxford, 1—70. [39, 55, 57, 61, 67, 71, 72, 79, 301, 302]
— (1964): Chemische Bestandteile der marinen Sedimente. — In: R. BRINKMANN (Red.), Lehrbuch der Allgemeinen Geologie Bd. I. — Enke, Stuttgart, 331—356. [86]
— (1964a): Organogene Bestandteile der marinen Sedimente. — In: BRINKMANN (Red.), Lehrbuch der allgemeinen Geologie Bd. I. — Enke, Stuttgart, 357—406. [85, 86, 225, 239, 245, 248]
— (1964b): Beobachtungen zur Schichtung in Sedimenten am Westrand der Great Bahama Bank. — N. Jb. Geol. Paläont., Abh., 120, 233—252. [90, 251]
— (1970): Nebenmeere im humiden und ariden Klimabereich. — Geol. Rdsch., 60, 73—105. [118]
SEIBOLD, E., DILL, R. F. & WALGER, E. (1961): Tauchbeobachtungen und petrographische Untersuchungen zur Sedimentumlagerung in der Kieler Außenförde. — Meyniana, 11, 82—96. [55, 59]
SEIBOLD, E. & WIEGERT, R. (1960): Untersuchungen des zeitlichen Ablaufs der Sedimentation im Malo Jezero (Mljet, Adria) auf Periodizitäten. — Z. Geophys., 26, 87—104. [362]
SEILACHER, A. (1959): Zur ökologischen Charakteristik von Flysch und Molasse. — Eclog. Geol. Helvet., 51, 1062—1078. [124]
— (1960): Strömungsanzeichen im Hunsrückschiefer. — Notizbl. hess. Landesamt Bodenf. Wiesbaden, 88, 88—106. [71]

— (1962): Paleontological studies on turbidite sedimentation and erosion. — J. Geol., **70**, 227—234. [99, 369]
— (1963): Lebensspuren und Salinitätsfazies. — Fortschr. Geol. Rheinl.-Westf., **10**, 81—94. [108]
— (1964): Biogenic sedimentary structures. — In: IMBRIE, J. & NEWELL, N. (ed.), Approaches to paleoecology. — J. Wiley & Sons, New York, 296—316. [80, 87, 90, 98, 99, 122, 124]
— (1967 a): Tektonischer, sedimentologischer oder biologischer Flysch? — Geol. Rdsch., **56**, 189—200. [127]
— (1967 b): Bathymetry of trace fossils. — Marine Geol., **5**. 413—428. [117]
SELLEY, R. C. (1970): Ancient sedimentary environments. — Chapman & Hall Ltd., London, 237 pp. [100]
SELLWOOD, B. W. (1971): The genesis of some sideritic beds in the Yorkshire Lias (England). — J. Sediment. Petrol., **41**, 854—858. [372]
SHARMA, G. D. (1964): Genesis of silica cement in sandstones and its replacement by carbonates. — Bull. Amer. Assoc. Petrol. Geol., **48** (Abstr.), 546. [149]
SHARMA, G. D. & BURRELL, D. C. (1970): Sedimentary environment and sediments of Cook Inlet, Alaska. — Bull. Amer. Assoc. Petrol. Geol., **54**, 647—654. [115]
SHARP, W. E. & FAN, POW-FOENG (1963): A sorting index. — J. Geol., **71**, 76—83. [55]
SHATFORD, R. A. (1956): Redwater, p. 44—49, in Anonym ("Alberta Soc. of Petrol. Geologists"). Geographical and geological distribution of oil and gas in Canada. — XX. Congr. geol. internac., Mexico, Sympos. yacimentos de petrol. y gas, **3**, 9—110. [273]
SHAUB, B. M. (1958): Some apparently misunderstood aspects concerning stylolites. — J. Sediment. Petrol., **28**, 376—378. [298]
SHAW, A. B. (1964): Time in Stratigraphy. — Mc Graw Hill, New York, 363 pp. [122]
SHEARMAN, D. J. (1963): Demonstration of recent anhydrite, gypsum, dolomite, and halite from the coastal flats of the Arabian shore of the Persian Gulf. — Proc. Geol. Soc. London, 1607, 63—64. [315]
— (1966): Origin of marine evaporites by diagenesis. — Trans. Inst. Miner. Metall., Sec. B, **75**, 208—215. [325]
SHEARMAN, D. J., KHOURI, J. & TAHA, S. (1961): On the replacement of dolomite by calcite in some Mesozoic limestones from the French Jura. — Proc. Geol. Assoc., **72**, 1, 1—12. [320]
SHEARMAN, D. J. & SKIPWITH, P. A. D'E. (1965): Organic matter in recent and ancient limestones and its role in their diagenesis. — Nature, **208**, 1310—1311. [285]
SHELTON, J. W. (1967): Stratigraphic models and general criteria for recognition of alluvial, barrier-bar, and turbidity-current sand deposits. — Bull. Amer. Assoc. Petrol., **51**, 2441—2461. [100]
SHELTON, J. W. & MACK, D. E. (1970): Grain orientation in determination of paleocurrents. — Amer. Assoc. Petrol. Geol., **54**, 1108—1119. [72]
SHEPARD, F. P. (1964): Criteria in modern sediments useful in recognizing ancient sedimentary environments. — In: VAN STRAATEN (ed.), Deltaic and shallow marine deposits. — Developm. in Sedimentol., **1**. — Elsevier, Amsterdam, 1—25. [112]
SHEPARD, F. P. & LANKFORD, R. R. (1959): Sedimentary facies from shallow borings in lower Mississippi delta. — Bull. Amer. Assoc. Petrol. Geol., **43**, 2051—2067. [376]
SHEPARD, F. P. & MOORE, D. G. (1955): Sediment zones bordering the barrier islands of Central Texas coast. — In: Finding ancient shorelines. — SEPM Spec. Publ., Nr. 3, 78—98. [113]
SHEPARD, F. P. & YOUNG, R. (1961): Distinguishing between beach and dune sands. — J. Sediment. Petrol., **31**, 196—214. [65, 68]
SHEPPARD, R. (1971): Zeolites in sedimentary deposits of the United States — A review. — Advances Chem. Ser., **101**, 279—310. [187]
SHEPPARD, R. A. & GUDE, A. J. 3d (1969): Diagenesis of tuffs in the Barstow Formation, Mud Hills, San Bernardino County, California. — U. S. Geol. Surv. Prof. Pap., **634**, 35 pp. [336]
SHIER, D. E. (1965): Vermetid reefs and coastal development in Southwest Florida. — Thesis, Florida State Univ.; Diss. Abstr. 26, No. 3, 1594—1595. [275]
SHINN, E. A. (1964): Recent dolomite, Sugarloaf Key. — Geol. Soc. Amer. Convent. Miami, Guidebook Field Trip 1, 62—67. [315]

— (1968a): Practical significance of birdseye structures in carbonate rocks. — J. Sediment. Petrol., 38, 215—223. [355]
— (1968b): Burrowing in recent lime sediments of Florida and the Bahamas. — J. Paleont., 42, 879—894. [274]
— (1969): Submarine lithification of Holocene carbonate sediments in the Persian Gulf. — Sedimentology, 12, 109—144. [198—200, 204, 282, 284, 285, 293, 295, 339, 370, 371]
SHINN, E. A. & GINSBURG, R. N. (1964): Formation of recent dolomite in Florida and the Bahamas. — Bull. Amer. Assoc. Petrol. Geol., 48 (Abstr.), 547. [315]
SHINN, E. A., GINSBURG, R. N. & LLOYD, R. M. (1965): Recent supratidal dolomite from Andros Island, Bahamas. — In: L. C. PRAY & R. C. MURRAY (ed.), Dolomitization and limestone diagenesis. — SEPM Spec. Pap., 13, 112—123. [315]
SHOJI, R. & FOLK, R. L. (1964): Surface morphology of some limestone types as revealed by electron microscope. — J. Sediment. Petrol., 34, 144—155. [251]
SHROCK, R. R. (1948): Sequence in layered rocks. — McGraw-Hill Book Co., 507 pp. [96]
SIBLEY, D. F. & MURRAY, R. C. (1972): Marine diagenesis of carbonate sediment, Bonaire, Netherlands Antilles. — J. Sediment. Petrol., 42, 168—178. [290]
SIEGEL, F. R. (1960): The effect of strontium on the aragonite-calcite ratios of Pleistocene corals. — J. Sediment. Petrol., 30, 297—304. [295]
— (1965): Aspects of calcium carbonate deposition in Great Onyx Cave, Kentucky. — Sedimentology, 4, 285—299. [260]
SIEHL, A. (1968): Raster-elektronenmikroskopische Aufnahmen von Turon-Kalken. — In: LANGHEINRICH, G. & PLESSMANN, W., Zur Entstehungsweise von Schieferungsflächen in Kalksteinen. — Geol. Mitt., 8, Aachen, 134—139. [227]
SIEVER, R. (1959): Petrology and geochemistry of silica cementation in some Pennsylvanian sandstones. — In: H. A. IRELAND (ed.), Silica in Sediments. — Soc. Econ. Paleont. Spec. Publ., No. 7, 55—79. [133, 139, 149]
— (1962): Silica solubility, 0°—200° C, and the diagenesis of siliceous sediments. — J. Geol., 70, 127—150. [69, 133, 328, 329, 332, 333]
— (1968): Sedimentological consequences of a steady-state ocean-atmosphere. — Sedimentology, 11, 5—29. [2, 8]
SIEVER, R., BECK, K. & BERNER, R. (1965): Composition of interstitial waters of modern sediments. — J. Geol., 73, 39—73. [336]
SIMMONS, G. & BELL, P. (1963): Calcite-aragonite equilibrium. — Science, 139, 1197—1198. [294]
SIMONS, D. B. & RICHARDSON, E. V. (1966): Resistance to flow in alluvial channels. — Geol. Surv. Prof. Pap., 422—J, 1—61. [73]
SIMONS, F. S. (1956): A note on Pur-Pur dune, Virú valley, Peru. — J. Geol., 64, 517—521. [101]
SINDOWSKI, K.-H. (1948): Fennoskandia als Sediment-Liefergebiet. — Naturwiss. Ver. Schlesw.-Holstein, Rundschreiben Juni 1948, 4 p. [44]
— (1956): Korngrößen- und Kornform-Auslese beim Sandtransport durch Wind (nach Messungen auf Norderney). — Geol. Jb., 71, 517—526. [65]
— (1957): Die synoptische Methode des Kornkurven-Vergleiches zur Ausdeutung fossiler Sedimentationsräume. — Geol. Jb., 73, 235—275. [59, 60]
SIPPEL, R. F. (1968): Sandstone petrology, evidence from luminescence petrography. — J. Sediment. Petrol., 38, 530—554. [138]
SKINNER, H. C. W., SKINNER, B. J. & RUBIN, M. (1963): Age and accumulation rate of dolomite-bearing carbonate sediments in South Australia. — Science, 139, 335—336. [316]
SKIPPER, K. (1971): Antidune cross-stratification in a turbidite sequence, Cloridorm formation, Gaspé, Quebec. — Sedimentology, 17, 51—68. [90, 98]
SKOLNICK, H. (1965): The quartzite problem. — J. Sediment. Petrol., 35, 12—21. [134, 141]
SLOSS, L. L., DAPPLES, E. C. & KRUMBEIN, W. C. (1960): Lithofacies maps. An atlas of the United States and Southern Canada. — J. Wiley & Sons, New York, 108 pp. [129]
SLOSS, L. L. & FERRAY, D. E. (1948): Microstylolites in sandstone. — J. Sediment. Petrol., 18, 3—13. [139]
SLOSS, L. L., KRUMBEIN, W. C. & DAPPLES, E. C. (1949): Integrated facies analysis. — Geol. Soc. Amer. Mem., 39, 91—124. [124]
SMITH, D. B. (1958): Observations on the Magnesian Limestone reefs of North-Eastern Durham. — Bull. Geol. Surv. Great Brit., 15, 71—84. [272]

SMITH, F. D. (1955): Planktonic foraminifera as indicators of depositional environment. — Micropaleontology, 1, 147—151. [359]
SMITH, H. T. U. (1965): Wind-formed pebble ripples in Antarctica. — Ann. GSA & Assoc. Soc. Joint Meet., Kansas City, Progr., 157. [60, 75]
SMITH, N. D. (1971): Transverse bars and braiding in the Lower Platte river, Nebraska. — Geol. Soc. Amer. Bull., 82, 3407—3420. [78]
SMITH, R. L. (1960a): Ash flows. — Bull. Geol. Soc. Amer., 71, 795—842. [172, 173, 176]
— (1960b): Zones and zonal variations in welded ash flows. — U. S. Geol. Surv. Prof. Pap., 354—F, 149—159. [176]
SMYERS, N. B. & PETERSON, G. L. (1971): Sandstone dikes and sills in the Moreno Shale, Panochettills, California. — Geol. Soc. Amer. Bull., 82, 3201—3208. [84]
SNEED, E. D. & FOLK, R. L. (1958): Pebbles in the Lower Colorado River, Texas, a study in particle morphogenesis. — J. Geol., 66, 114—150. [64]
SNYDER, F. G. & ODELL, J. W. (1958): Sedimentary breccias in the southeast Missouri lead district. — Bull. Geol. Soc. Amer., 69, 899—926. [339]
SORBY, H. C. (1856): On the physical geography of the Old Red Sandstone of the central district of Scotland. — New Philos. J., New Ser. 3, Edinburgh, 112—122. [90]
SOROTCHINSKY, C. (1954): Formation des minéraux granitiques dans les calcaires. — 19. Congr. Géol. Internat., Alger 1952, Sect. XIII, 552—569. [337]
SPARKS, R. S. J. & WALKER, G. P. L. (1973): The ground surge deposit: a third type of pyroclastic rock. — Nature, 241, 62—64. [180]
SPECK, J. (1953): Geröllstudien in der subalpinen Molasse am Zugersee und Versuch einer paläogeographischen Auswertung. — Diss. Zürich. [47]
SPENCER, C. W. (1964): Unconsolidated Miocene dolomite in northern peninsular Florida. — Ann. GSA & Assoc. Soc. Joint Meet., Miami Beach, Progr., 192—193. [351]
SPENCER, MARIA (1967): Bahamas deep test. — Bull. Amer. Assoc. Petrol. Geol., 51, 263—268. [353]
SPOTTS, J. H. (1964): Grain orientation and imbrication in Miocene turbidity current sandstones, California. — J. Sediment. Petrol., 34, 229—253. [72, 96]
SPOTTS, J. H. & WESER, O. E. (1964): Directional properties of a Miocene turbidite, California. — In: A. H. BOUMA & A. BROUWER (ed.), Turbidites. — Developm. in Sedimentol., 3, Elsevier, Amsterdam, 199—221. [72]
SQUIRES, D. F. (1964): Fossil coral thickets in Wairarapa, New Zealand. — J. Paleont., 38, 904—915. [271]
STADLER, G. (1963): Die Mineralführung der Klüfte in der Bohrung Münsterland 1. — Fortschr. Geol. Rheinl.-Westf., 11, 293—304. [155]
STANLEY, D. J. (1967): Comparing patterns of sedimentation in some modern and ancient submarine canyons. — Earth and Planet. Sci. lett., 3, 371—380, N-Holland Publ. Co., Amsterdam. [123]
STANLEY, S. M. (1966): Paleoecology and diagenesis of Key Largo limestone, Florida. — Bull. Amer. Assoc. Petrol. Geol., 50, 1927—1947. [295]
STANTON, R. J. (1966): The solution brecciation process. — Geol. Soc. Amer. Bull., 77, 843—848. [352]
STAUFFER, K. W. (1962): Quantitative petrographic study of Paleozoic carbonate rocks, Caballo Mountains, New Mexico. — J. Sediment. Petrol., 32, 357—396. [280]
STAUFFER, P. H. (1967): Grain-flow deposits and their implications, Santa Ynez Mountains, California. — J. Sediment. Petrol., 37, 487—508. [123]
STEARN, C. (1966): The microstructure of stromatoporoids. — Palaeontology, 9, 74—124. [235]
STEHLI, F. G. (1956): Shell mineralogy in paleozoic invertebrates. — Science, 123, 1031—1032. [209, 236, 237, 240, 245—247, 294]
STEHLI, F. G. & CREATH, W. B. (1964): Foraminiferal ratios and regional environments. — Bull. Amer. Assoc. Petrol. Geol., 48, 1810—1827. [359]
STEHLI, F. G. & HOWER, J. (1961): Mineralogy and early diagenesis of carbonate sediments. — J. Sediment. Petrol., 31, 358—371. [195, 321]
STEPHAN, W. (1965): Zur faziellen und zyklischen Gliederung der chattischen Brackwasser-Molasse in Oberbayern. — Geol. Bavarica, 55, 239—257. [375]
STERNBERG, E. T., FISCHER, A. G. & HOLLAND, H. D. (1959): Strontium contents of calcites from the Steinplatte reef complex, Austria. — Bull. Geol. Soc. Amer., 70, 1681 (Abstr.). [268]

STETSON, T. R., SQUIRES, D. F. & PRATT, R. M. (1962): Coral banks occurring in deep water on the Blake Plateau — Amer. Mus. Nov., New York, 2114, 1—39. [271]

STEWART, A. D. (1962): Greywacke sedimentation in the Torridonian of Colonsay and Oronsay. — Geol. Mag., 99, No. 5, 399—419. [82]

STIEFEL, J. (1957): Ein Beitrag zur Gliederung der oberen Süßwassermolasse in Niederbayern. — Beih. Geol. Jb., H. 26, 201—259. [48, 49, 56]

STIEGLITZ, R. D. (1972): Scanning electron microscopy of the fine fraction of recent carbonate sediments from Bimini, Bahamas. — J. Sediment. Petrol., 42, 211—216. [195, 197, 206]

STIRN, A. (1964): Kalktuffvorkommen und Kalktufftypen der Schwäbischen Alb. — Abh. Karst- u. Höhlenkde., E, H. 1, 91 pp. [222, 259]

STOCKDALE, P. B. (1926): The stratigraphic significance of solution in rocks. — J. Geol., 34, 399—414. [297]

STOCKMAN, K. W., GINSBURG, E. N. & SHINN, E. A. (1967): The production of lime mud by algae in South Florida. — J. Sediment. Petrol., 37, 633—648. [196]

STODDART, D. R. & CANN, J. R. (1965): Nature and origin of beach rock. — J. Sediment. Petrol., 35, 243—247. [281]

STOKES, W. L. (1961): Fluvial and eolian sandstone bodies in Colorado Plateau. — In: J. A. PETERSON & J. L. OSMOND (ed.), Geometry of sandstone bodies. — Amer. Assoc. Petrol. Geol., 151—178. [100, 106]

STORZ, M. (1931): Die sekundäre authigene Kieselsäure in ihrer petrogenetisch-geologischen Bedeutung. II. Teil: Die Einwirkung der sekundären authigenen Kieselsäure auf vorhandene Gesteine (Einkieselung und Verkieselung). — Monogr. z. Geol. u. Paläont., Ser. II, H. 5, Berlin, 139—479. [141, 328]

STOUT, J. L. (1964): Pore geometry as related to carbonate stratigraphic traps. — Bull. Amer. Assoc. Petrol. Geol., 48, 329—337. [345]

STOUT, W. (1931): Pennsylvanian cycles in Ohio. — Ill. Geol. Surv. Bull., 60, 195—216. [366]

STRADNER, H. (1964): Die Ergebnisse der Aufschlußarbeiten der ÖMV AG. in der Molassezone Niederösterreichs in den Jahren 1957—1963. Ergebnisse der Nannofossil-Untersuchungen (Teil III). — Erdöl-Z., 80, 133—139. [225]

STRAKHOV, N. M. (1969): Principles of Lithogenesis. Vol. 2. — Oliver & Boyd, Edinburgh (Transl. by J. P. FITZSIMMONS, ed. by S. I. TOMKEIEFF & J. E. HEMINGWAY), 609 pp. [14, 195]

STRAKHOV, N. M. & ZWETKOV, A. I. (1946): Paragenese der Karbonate in den Ablagerungen der Salzwasserlagunen. — Soc. Nat. Moscow (Russ.). [316]

STRAUCH, F. (1966): Sedimentgänge von Tjörnes (Nord-Island) und ihre geologische Bedeutung. — N. Jb. Geol. Paläont., Abh., 124, 259—288. [84]

STRUVE, W. (1961): Das Eifeler Korallen-Meer. — Aufschluß, Sonderh. 10, 81—107. [271]

STUMPFL, E. (1958): Erzmikroskopische Untersuchungen an Schwermineralien in Sanden. — Geol. Jb., 73, 685—724. [33]

STURM, M. & MATTER, A. (1971): Sediment distribution and mechanisms of clastic sedimentation in a freshwater basin (Lake Thun, Switzerland) (Abstr.). — Progr., VIII Internat. Sediment. Congr., Heidelberg, 98. [97]

SUESS, E. (1968): Calcium carbonate interaction with organic compounds. — Diss. Lehigh Univ., Marine Sci. Center, USA, 153 pp. [285]

— (1970): Interaction of organic compounds with calcium carbonate. I. Association phenomena and geochemical implications. — Geochim. Cosmochim. Acta, 34, 157—168. [196]

SUJKOWSKI, ZB. (1930): Étude pétrographique du Crétacé de Pologne. La série de Lublin et sa comparaison avec la craie blanche. — Bull. Serv. Géol. Pol., 6, 627.

SUJKOWSKI, ZB. L. (1958): Diagenesis. — Amer. Assoc. Petrol. Geol. Bull., 42, 2692—2717. [329]

SULLWOLD, H. H., JR. (1961): Turbidites in oil exploration. — In: Geometry of sandstone bodies, ed. by J. A. PETERSON & J. C. OSMOND. — Amer. Assoc. Petrol. Geol., Tulsa, 63—81. [123]

SUNDBORG, Å. (1956): The river Klarälven, a study of fluvial processes. — Geogr. Ann., 38, 127—316. [75]

SUTTON, R. G. & LEWIS, TH. L. (1966): Regional patterns of cross-laminae and convolutions in a single bed. — J. Sediment. Petrol., 36, 225—229. [82]

SWANSON, D. A. (1966): Tieton Volcano, a Miocene eruptive center in the southern Cascade Mountains, Washington. — Geol. Soc. Amer. Bull., 77, 1293—1314. [183]
SWETT, K. (1968): Authigenic feldspars and cherts resulting from dolomitization of illitic limestones: a hypothesis. — J. Sediment. Petrol., 38, 128—135. [336]
SWIFT, D. J. P. (1969a): Inner Shelf sedimentation: Processes and products. — In: STANLEY, D. J. (ed.), The new concepts of continental margin sedimentation. — AGI short course lect. notes, no. 4, 46 pp. [120, 122]
— (1969b): Outer Shelf sedimentation: Processes and products. — Ibid., no. 5, 25 pp. [122]
— (1969c): Evolution of the shelf surface, and the relevance of modern shelf studies to the rock record. — Ibid., no. 7, 19 pp. [122]
SWINCHATT, J. P. (1965): Significance of constituent composition, texture, and skeletal breakdown in some recent carbonate sediments. — J. Sediment. Petrol., 35, 71—90. [195]
SWINEFORD, A. & FRYE, J. C. (1951): Petrography of the Peoria loess in Kansas. — J. Geol., 59, 306—322. [56]

TAFT, W. H. (1967): Modern carbonate sediments. — In: CHILINGAR, BISSELL & FAIRBRIDGE (ed.), Carbonate rocks. — Developm. in Sedimentol., 9 A, Elsevier Publ. Co., 29—50. [281]
TAFT, W. H. & HARBAUGH, J. W. (1964): Modern carbonate sediments of southern Florida, Bahamas, and Espíritu Santo Island, Baja California: a comparison of their mineralogy and chemistry. — Stanford Univ. Publ., Univ. Ser., Geol. Sci., 8 (2), 1—133. [195, 282]
TANNER, W. F. (1959): Sample components obtained by the method of differences. — J. Sediment. Petrol., 29, 408—411. [54]
— (1967): Ripple mark indices and their uses. — Sedimentology, 9, 89—104. [88]
— (1971): Numerical estimates of ancient waves, water depth and fetch. — Sedimentology, 16, 71—88. [103]
TATARSKY, V. B. (1949): About the occurrence of rocks in which dolomite is replaced by calcite. — Dokl. Akad. Nauk SSSR, 69, 849—851 (Russ.). [321]
TAYAMA, R. (1935): Table reefs, a particular type of coral reef. — Proc. Imperial Acad. Japan, 11, 268—270. [266]
TAYLOR, J. C. M. & ILLING, L. V. (1969): Holocene intertidal calcium carbonate cementation at Qatar, Persian Gulf. — Sedimentology, 12, 69—108.
[198, 281, 282, 284, 285, 295, 339, 357, 370]
TAYLOR, J. H. (1964): Some aspects of diagenesis. — Advanc. Sci., London, 22, 417—436. [133, 281, 302]
TAYLOR, J. M. (1950): Pore space reduction in sandstones. — Bull. Amer. Assoc. Petrol. Geol., 34, 701—716. [136, 137]
TEICHERT, C. (1958): Cold- and deep-water coral banks. — Bull. Amer. Assoc. Petrol. Geol., 42, 1064—1082. [236, 264, 271]
— (1970): Runzelmarken (winkle marks). — J. Sediment. Petrol., 40, 1056—1057. [90]
TEICHMÜLLER, R. (1955a): Über Küstenmoore der Gegenwart und die Moore des Ruhrkarbons. — Geol. Jb., 71, 197—220. [376]
— (1955b): Sedimentation und Setzung im Ruhrkarbon. — N. Jb. Geol. Paläont., Mh., 145—168. [378]
— (1962): Zusammenfassende Bemerkungen über die Diagenese des Ruhrkarbons und ihre Ursachen. — Fortschr. Geol. Rheinl.-Westf., 3, 2, 725—734. [154]
— (1964): Zur Stratigraphie und Inkohlung des jüngsten Oberkarbons (Silesium) in Nordwestdeutschland. — 5. Congr. Internat. Stratigr. Géol. Carbonifère, Paris, C. R., 813—820. [108]
TEMMLER, H. (1966): Über die Nusplinger Fazies des Weißen Jura der Schwäbischen Alb (Württemberg). — Z. dt. Geol. Ges., 116, 891—907. [274, 369]
TEN HAAF, E. (1959): Graded beds of the Northern Appenines. — Thesis, Rijksuniv. Groningen, 102 pp. [96]
TEODOROVICH, G. I. (1946): On the genesis of the dolomite of sedimentary deposits (Russ.). — Dokl. Akad. Nauk SSSR, 53, 817—820. [317]
— (1961): On the origin of sedimentary dolomites. — Internat. Geol. Rev., 3, 5, 373—384. [316]
TERMIER, H. & TERMIER, G. (1963): Erosion and Sedimentation. — Van Nostrand Co., London etc., 433 pp. [219, 275]
THOMAS, C. M. (1965): Origin of pisolites. — Bull. Amer. Assoc. Petrol. Geol., 49, Abstr., 360. [222, 258]

THOMAS, G. E. (1962): Grouping of carbonate rocks into textural and porosity units for mapping purposes. — In: W. E. HAM (ed.), Classification of carbonate rocks. — Amer. Assoc. Petrol. Geol. Mem., **1**, 193—223. [348, 352]
THOMSON, A. (1959): Pressure solution and porosity. — In: H. A. IRELAND (ed.), Silica in sediments. — S.E.P.M. Spec. Publ., No. 7, 92—110. [133, 134]
THORARINSSON, S. (1954): The eruption of Hekla, 1947—1948. The tephra-fall from Hekla on March 29, 1947. — Mus. Nat. Hist. Soc. Sci. Island., Reykjavik, 68 pp. [56, 160, 164]
THORARINSSON, S. & SIGVALDASON, G. E. (1972): The Hekla eruption of 1970. — Bull. Volcanol., **36**, 269—288. [166, 167]
THOREZ, J. (1964): Sur la présence de granocroissances et de granodécroissances dans la sédimentation du Famennien supérieur au bord nord du synclinorium de Dinant (Belgique). — Sedimentology, **3**, 226—232. [369]
THORSLUND, P. (1960): Notes on the geology and stratigraphy of Dalarna. — XXI. Internat. geol. Congr., Norden, Guide to Excurs., No. A 23 and C 18, 23—27. [272]
TILL, R. (1971): Are there geochemical criteria for differentiating reef and nonreef carbonates? — Bull. Amer. Assoc. Petrol. Geol., **55**, 523—530. [268]
TOLLMANN, A. (1967): Das Längen-Breiten-Verhältnis der geosynklinalen Sedimenttröge. — Geol. Rdsch., **56**, 78—94. [125]
TOMKEIEFF, S. I. (1927): Proc. Geol. Assoc., **38**, 518—547. [302]
TOPKAYA, M. (1950): Recherches sur les silicates authigènes dans les roches sédimentaires. — Bull. Lab. de Géol., Min., Geophys. et du Mus. Géol., Lausanne, **97**, 1—132. [333, 337]
TRASK, P. D. (1932): Origin and environment of source sediments of petroleum. — Houston, Gulf Publ. Co., 323 pp. [52]
TREFETHEN, J. M. & DOW, R. L. (1960): Some features of modern beach sediments. — J. Sediment. Petrol., **30**, 589—602. [47]
TRICART, J. (1951): Études sur le façonnement des galets marins. — Proc. 3. Internat. Congr. Sedimentol., Groningen-Wageningen, 245—255. [64]
TRICHET, J. (1971): Recent aragonite deposition on algal substrates during blue-green algae decomposition. Role of organic substances (Abstr.). — Progr., VIII Internat. Sediment. Congr., Heidelberg, 103. [211, 284]
TRÖGER, W. E. (1967): Optische Bestimmung der gesteinsbildenden Minerale. Teil 2. Textband. — E. Schweizerbart, Stuttgart, 822 pp. (2nd ed. 1969). [26, 147]
— (1971): Optische Bestimmung der gesteinsbildenden Minerale. Teil 1. Bestimmungstabellen. — E. Schweizerbart, Stuttgart, 188 pp. (4th ed. 1971). [26]
TROELL, A. R. (1962): Lower Mississippian bioherms of southwestern Missouri and northwestern Arkansas. — J. Sediment. Petrol., **32**, 629—644. [271]
TRÜMPY, R. (1960): Paleotectonic evolution of the central and western Alps. — Bull. Geol. Soc. Amer., **71**, 843—908. [125, 129, 203]
— (1965): Zur geosynklinalen Vorgeschichte der Schweizer Alpen. — Umschau, 1965, H. 18, 573—577. [129]
TRÜMPY, R. & RYF, W. (1965): Erläuterungen zur Exkursion in die Glarner Alpen. — Symposium sul „Verrucano". — Istit. Geol. e Paleont. dell'Univ. Pisa, 22 pp. [47]
TRURNIT, P. (1968): Pressure solution phenomena in detrital rocks. — Sediment. Geol., **2**, 89—114. [298]
TURNER, F. J. (1951): Observations on twinning of plagioclase in metamorphic rocks. — Amer. Mineralogist, **36**, 581—589. [25]
TWENHOFEL, W. H. (1950a): Principles of sedimentation. — McGraw-Hill, 673 pp. [232, 239, 243, 256]
— (1950b): Coral and other organic reefs in geologic column. — Bull. Amer. Assoc. Petrol. Geol., **34**, 182—202. [269, 271, 272]
— (1961): Treatise on sedimentation, 2nd edition (= 1932), republished by Dover Publ. Inc., New York, 926 pp. [115]

UNRUG, R. (1963): Istebna beds — a fluxoturbidite formation in the Carpathian flysch. — Ann. Soc. Géol. Pologne, **33**, 49—92. [127]
UREY, H. C. (1951): The origin and development of the earth and other terrestrial planets. — Geochim. Cosmochim. Acta, **1**, 209—277. [9]
USDOWSKI, H.-E. (1962): Die Entstehung der kalkoolithischen Fazies des norddeutschen Unteren Buntsandsteins. — Beitr. Miner. Petrogr., **8**, 141—179. [250, 257, 281, 282]
— (1963a): Der Rogenstein des norddeutschen Unteren Buntsandsteins, ein Kalkoolith des marinen Faziesbereiches. — Fortschr. Geol. Rheinl.-Westf., **10**, 337—342. **[254]**

— (1963b): Die Genese der Tutenmergel oder Nagelkalke (cone-in-cone). — Beitr. Miner. Petrogr., 9, 95—110. [299]
— (1964): Dolomit im System Ca^{2+}—Mg^{2+}—CO_3^{2-}—Cl_2^{2-}—H_2O. — Naturwiss., 51, 357.
— (1967): Die Genese von Dolomit in Sedimenten. — Miner. u. Petrogr. in Einzeldarst., Vol. 4, Springer, Berlin etc., 95 pp. [309]
UTADA, M. (1971): Zeolitic zoning of the Neogene pyroclastic rocks in Japan. — Sci. Pap. Coll. Gen. Ed. Tokyo, 21, 189—221. [187]

VALETON, I. (1953): Petrographie des süddeutschen Hauptbuntsandsteins. — Heidelb. Beitr. Miner. Petrogr., 3, 335—379. [25, 153]
— (1955): Beziehungen zwischen petrographischer Beschaffenheit, Gestalt und Rundungsgrad einiger Flußgerölle. — Petermanns Geograph. Mitt. 1955, I, 13—17. [51, 63—65]
— (1960): Vulkanische Tuffiteinlagerung in der nordwestdeutschen Oberkreide. — Mitt. Geol. Staatsinst. Hamburg, 29, 26—41. [329]
VAN ANDEL, TJ. H. (1950): Provenance, transport and deposition of Rhine sediments. — (Diss.) H. Veenman & Zonen, Wageningen, 129 pp. [29, 38, 40, 44]
— (1952): Zur Frage der Schwermineralverwitterung in Sedimenten. — II. Fazielle Bedingungen und stratigraphische Bedeutung der Schwermineralverwitterung. — Erdöl u. Kohle, 5, 100—104. [45]
— (1959): Reflections on the interpretation of heavy mineral analyses. — J. Sediment. Petrol., 29, 153—163. [38]
— (1967): The Orinoco delta. — J. Sediment. Petrol., 37, 297—310. [111, 116, 118]
VAN ANDEL, TJ. H. & POOLE, D. M. (1960): Sources of recent sediments in the northern Gulf of Mexico. — J. Sediment. Petrol., 30, 91—122. [39, 44]
VAN ANDEL, TJ. H. & POSTMA, H. (1954): Recent sediments of the Gulf of Paria. Reports of the Orinico Shelf Expedition. — Verh. Kon. Nederl. Akad. Wetensch., Afd. Natuurk. 1, 20, No. 5, North-Holland, Amsterdam, 1, 245 pp. [59, 68]
VAN ANDEL, TJ. H. & SACHS, P. L. (1964): Sedimentation in the Gulf of Paria during the Holocene transgression; a subsurface acoustic reflection study. — Sears Foundation: J. Marine Res., 22, 30—50. [381]
VAN ANDEL, TJ. H., WIGGERS, A. J. & MAARLEVELD, G. (1954): Roundness and shape of marine gravels from Urk (Netherlands), a comparison of several methods of investigation. — J. Sediment. Petrol., 24, 100—116. [66]
VAN BEMMELEN, R. W. (1949): The geology of Indonesia, General Geology. — The Hague, Govt. Print. Office, 1, 732 pp. [180]
VAN DER KNAAP, W. & EIJPE, R. (1968): Some experiments on the genesis of turbidity currents. — Sedimentology, 11, 115—124. [93, 367]
VAN DER LINDEN, W. J. M. (1963): Sedimentary structures and facies interpretation of some Molasse deposits. — Geol. Ultraiectina, Utrecht, No. 12, 42 pp. [78]
VAN DER PLAS, L. (1966): The identification of detrital feldspars. — Elsevier Publ. Co., Amsterdam, 305 pp. [24]
VAN HOUTEN, F. B. (1962): Cyclic sedimentation and the origin of analcime-rich Upper Triassic Lockatong formation, west-central New Jersey and adjacent Pennsylvania. — Amer. J. Sci., 260, 561—576. [144]
— (1965a): Cyclic lacustrine sedimentation, Upper Triassic Lockatong formation, Central New Jersey and adjacent Pennsylvania. — In: D. F. MERRIAM (ed.), Symposium on cyclic sedimentation. — Kansas Geol. Surv. Bull., 169, 497—531. [84, 103, 363, 379]
— (1965b): Composition of Triassic Lockatong and associated formations of Newark Group, Central New Jersey and adjacent Pennsylvania. — Amer. J. Sci., 263, 825—863. [142]
— pers. comm. [80, 107]
VAN SICLEN, D. C. (1965): Depositional topography in relation to cyclic sedimentation. — In: D. F. MERRIAM (ed.), Symposium on cyclic sedimentation. — Kansas Geol. Surv. Bull., 169, 533—539. [365]
VAN STRAATEN, L. M. J. U. (1948): Note on the occurrence of authigene feldspar in nonmetamorphic sediments. — Amer. J. Sci., 246, 569—572. [333]
— (1953): Rhythmic patterns on Dutch North Sea beaches. — Geol. en Mijnb., N. S., 2, 15. Jg., 31—43. [79, 89]
— (1954): Composition and structure of recent marine sediments in the Netherlands. — Leidse geol. Meded., 19, 110 pp. [80, 84, 115—117]

— (1959): Minor structures of some recent littoral and neritic sediments. — Geol. en Mijnb., N. S., **21**, 197—216. [119, 121]
— (1960): Some recent advances in the study of deltaic sedimentation. — Liverpool & Manchester Geol. J., Centenary Issure, **2**, 3, 411—442. [109]
— (1965): Coastal barrier deposits in South- and North-Holland. — Meded. Geol. Sticht., N. S., **17**, 41—75. [119]
VATAN, A. (1949): La sédimentation détritique dans la zone subalpine et le Jura méridional au Crétacé et au Tertiaire. — C. R. somm. Séanc. Soc. géol. France, **6**, 102—104. [45]
— (1950): General aspects of sedimentation in the geological basins of France. — J. Sediment. Petrol., **20**, 65—73. [45, 46]
— (1962): Les grès et leur milieu. — C. R. séanc. Acad. Sci., **254**, 2026—2028. [145]
VELLA, P. (1964): Foraminifera and other fossils from late Tertiary deep-water coral thickets, Wairarapa, New Zealand. — J. Paleont., **38**, 916—928. [271]
VENZLAFF, H. (1965): Zur Stratigraphie und Tektonik der Kulmstufe III westlich des Oberharzer Diabaszuges. — Geol. Jb., **82**, 243—270. [123]
VERHOOGEN, J. (1951): The chemical potential of a stressed solid. — Amer. Geophys. Union Trans., **32**, 251—258. [133]
VINE, J. D. & TOURTELOT, E. B. (1973): Geochemistry of Lower Eocene sandstones in the Rocky Mountain region. — U. S. Geol. Surv. Prof. Pap., **789**, 36 pp. [17]
VISHER, G. S. (1965): Use of vertical profile in environmental reconstruction. — Bull. Amer. Assoc. Petrol. Geol., **49**, 41—61. [103]
— (1967): Grain size distributions and depositional processes. — 7th Internat. Sedimentol. Congr., Preprint. [56]
VITANAGE, P. W. (1954): Sandstone dikes in the South Platte area, Colorado. — J. Geol., **62**, 493—500. [84]
VÖLK, H. R. (1966): Aggradational directions and biofacies in the youngest Postorogenic deposits of southeastern Spain. A contribution to the determination of the age of the east mediterranean coast of Spain. — Palaeogeogr., Palaeoclimatol., Palaeoecol., **2**, 313—331. [112]
VOGLER, H. (1956): Die Unterkarbonkonglomerate des Frankenwaldes und ihre paläogeographische Deutung. — Geol. Bavarica, **27**, 232—272. [47]
VOIGT, E. (1929): Die Lithogenese der Flach- und Tiefwassersedimente des jüngeren Oberkreidemeeres. — Jb. Hall. Verb. z. Erforsch. d. mitteldeutsch. Bodenschätze, N. F., **8**, 1—138. [200, 232]
— (1956): Der Nachweis des Phytals durch Epizoen als Kriterium der Tiefe vorzeitlicher Meere. — Geol. Rdsch., **45**, 97—119. [215]
— (1959): Die ökologische Bedeutung der Hartgründe („Hardgrounds") in der oberen Kreide. — Paläont. Z., **33**, 129—147. [288, 371]
— (1962): Frühdiagenetische Deformation der turonen Plänerkalke bei Halle/Westf. — Mitt. Geol. Staatsinst. Hambg., **31**, 146—275. [82]
— (1963): Über Randtröge vor Schollenrändern und ihre Bedeutung im Gebiet der Mitteleuropäischen Senke und angrenzender Gebiete. — Z. dt. Geol. Ges., **114**, 378—418. [125]
VOLL, G. (1960): New work on petrofabrics. — Liverpool and Manch. Geol. J., **2**, 503—567. [292]
VOLLMAYR, TH. (1958): Erläuterungen zur Geologischen Karte von Bayern 1 : 25 000, Bl. 8426 Oberstaufen. 55 pp. [380]
VON DER BORCH, C. C. (1965): Th distribution and preliminary geochemistry of modern carbonate sediments of the Coorong area, South Australia. — Geochim. Cosmochim. Acta, **29**, 781—799. [192]
VOORT, H. B. (1963): Zum Flyschproblem in den Westpyrenäen. — Geol. Rdsch., **53**, 220—233. [98]

WAAGE, K. M. (1965): Origin of repeated fossiliferous concretion layers in the Fox Hills formation. — In: D. F. MERRIAM (ed.), Symposium on cyclic sedimentation. — Kansas Geol. Surv. Bull., **169**, 541—563. [367]
WADELL, H. (1935): Volume, shape and roundness of quartz particles. — J. Geol., **43**, 250—280. [67]
WAGNER, C. W. & VAN DER TOGT, C. (1973): Holocene sediment types and their distribution in the southern Persian Gulf. — In: PURSER, B. H. (ed.), The Persian Gulf, Springer, Berlin, Heidelberg, New York, 123—155. [352, 358]

WAGNER, GEORG (1931 a): Einführung in die Erd- und Landschaftsgeschichte. — Hohenlohe'sche Buchhandlung, Öhringen, 664 pp. [300]
— (1931 b): Stylolithen und Drucksuturen. — Geol. paläont. Abh., 11, 2, 101—128. [298]
WAGNER, G. H. (1964): Kleintektonische Untersuchungen im Gebiet des Nördlinger Rieses. — Geol. Jb., 81, 519—600. [297]
WALDSCHMIDT, W. A. (1941): Cementing materials in sandstones and their influence on the migration of oil. — Bull. Amer. Assoc. Petrol. Geol., 25, 1839—1879. [138, 149]
WALGER, E. (1962): Die Korngrößenverteilung von Einzellagen sandiger Sedimente und ihre genetische Bedeutung. — Geol. Rdsch., 51, 494—507. [54—56]
— (1965): Zur Darstellung von Korngrößenverteilungen. — Geol. Rdsch., 54, 976—1002. [52]
— (1966): Untersuchungen zum Vorgang der Transportsonderung von Mineralen am Beispiel von Strandsanden der westlichen Ostsee. — Meyniana, 16, 55—106. [37, 38]
WALKER, G. P. L. (1971): Grain-size characteristics of pyroclastic deposits. — J. Geol., 79, 696—714. [166, 169, 177]
WALKER, G. P. L. & CROASDALE, R. (1971): Two Plinian-type eruptions in the Azores. — J. Geol. Soc., 127, 17—55. [170]
— — (1972): Characteristics of some basaltic pyroclastics. — Bull. Volc., 35, 303—317. [170, 184]
WALKER, G. P. L., WILSON, L. & BOWELL, E. L. G. (1971): Explosive volcanic eruptions — I. The rate of fall of pyroclasts. — Geophys. J. R. Astr. Soc., 22, 377—383. [164]
WALKER, R. G. (1966): Deep channels in turbidite-bearing formations. — Bull. Amer. Assoc. Petrol. Geol., 50, 1899—1917. [123]
— (1967): Turbidite sedimentary structures and their relationship to proximal and distal depositional environments. — J. Sediment. Petrol., 37, 25—43. [123, 369]
— (1970): Review of the geometry and facies organization of turbidites and turbidite-bearing basins. — Geol. Assoc. Canada, Spec. Pap., 7, 219—251. [51, 123, 369]
WALKER, T. R. (1960): Carbonate replacement of detrital silicate minerals as a source of authigenic silica in sedimentary rocks. — Bull. Geol. Soc. Amer., 71, 145—152. [141]
— (1962): Reversible nature of chert-carbonate replacement in sedimentary rocks. — Bull. Geol. Soc. Amer., 73, 237—242. [332]
— (1967 a): Formation of red beds in modern and ancient deserts. — Bull. Geol. Soc. Amer., 78, 353—368. [40, 108, 144]
— (1967 b): Color of recent sediments in tropical Mexico: A contribution to the origin of Red Beds. — Bull. Geol. Soc. Amer., 78, 917—920. [108, 109]
— (1968): Formation of red beds in modern an ancient deserts: Reply. — Bull. Geol. Soc. Amer., 79, 281—282. [108]
WALKER, T. R. & HARMS, J. C. (1972): Eolian origin of flagstone beds, Lyons Sandstone (Permian), type area, Boulder County, Colorado. — Mountain Geol., 9, Nos. 2—3, 279—288. [100, 101, 103]
WALTON, E. K. (1967): The sequence of internal structures in turbidites. — Scott. J. Geol., 3, 306—317. [94]
WALTON, W. R. (1964): Recent foraminiferal ecology and paleoecology. — In: J. IMBRIE & N. NEWELL (ed.), Approaches to paleoecology. — Wiley, New York, 151—237. [359]
WANLESS, H. R. (1965): Local and regional factors in Pennsylvanian cyclic sedimentation. — In: D. F. MERRIAM (ed.), Sympos. on cyclic sedimentation. — Kansas Geol. Surv. Bull., 169, 593—606. [362, 366]
WANLESS, H. R. & SHEPARD, F. P. (1936): Sea level and climatic changes related to late Paleozoic cycles. — Bull. Geol. Soc. Amer., 47, 1117—1206. [365]
WANLESS, H. R. & WELLER, J. M. (1932): Correlation and extent of Pennsylvanian cyclothems. — Bull. Geol. Soc. Amer., 43, 1003—1016 [361, 366, 375]
WARING, W. W. & LAYER, D. B. (1950): Devonian dolomitized reef D 3 reservoir, Leduc field, Alberta, Canada. — Bull. Amer. Assoc. Petrol. Geol., 34, 295—312. [273, 352]
WARREN, A. (1972): Observations on dunes and bi-modal sands in the Ténéré desert. — Sedimentology, 19, 37—44. [56]
WATERS, A. C. & FISHER, R. V. (1971): Base surges and their deposits: Capelinhos and Taal volcanoes. — J. Geophys. Res., 76, 5596—5614. [177]
WEAVER, CH. E. (1960): Possible uses of clay minerals in search for oil. — Clays and clay minerals. 8th Nat. Conf., 214—227. [156]

WEBER, J. N. & KAUFMAN, J. W. (1965): Brucite in the calcareous alga Goniolithon. — Science, 149, 996—997. [219]

WEBER, K. (1972a): Notes on determination of illite crystallinity. — N. Jb. Min., Mh., 267—276. [156]

— (1972b): Kristallinität des Illits in Tonschiefern und andere Kriterien schwacher Metamorphose im nordöstlichen Rheinischen Schiefergebirge. — N. Jb. Geol. Pal. Abh., 141, 333—363. [156]

— pers. comm. [156]

WEDEPOHL, K. H. (1963): Einige Überlegungen zur Geschichte des Meerwassers. — Fortschr. Geol. Rheinl.-Westf., 10, 129—150. [9]

— (1967): Geochemie. — Samml. Göschen, Bd. 1224—1224b, W. de Gruyter, Berlin, 221 pp. [8, 9]

— pers. comm. [260]

WEEKS, L. G. (1957): Origin of carbonate concretions in shales, Magdalena Valley, Colombia. — Bull. Geol. Soc. Amer., 68, 95—102. [302]

WEGEHAUPT, H. (1962): Zur Petrographie und Geochemie des höheren Westfal D von Westerholt. — Fortschr. Geol. Rheinl.-Westf., 3, 2, 445—496. [268]

WEGNER, TH. (1926): Geologie Westfalens. — Schöningh, Paderborn, 500 pp. [232]

WEILER, H. (1957): Untersuchungen zur Frage der Kalk-Mergel-Sedimentation im Jura Schwabens. — Diss. Tübingen. [297]

WEIMER, R. C. (1970): Rates of deltaic sedimentation and intrabasin deformation, Upper Cretaceous of Rocky Mountain region. — In: MORGAN, J. P. (ed.), Deltaic sedimentation, modern and ancient. — Soc. Econ. Pal. Miner. Spec. Publ., 15, 271—292. [87, 113]

WEISS, M. P. (1954): Feldspathized shales from Minnesota. — J. Sediment. Petrol., 24, 270—274. [142]

WELLER, J. M. (1931): The conception of cyclical sedimentation during the Pennsylvanian period. — Illinois Geol. Surv. Bull., 60, 163—177. [375]

— (1956): Argument for diastrophic control of late Paleozoic cyclothems. — Bull. Amer. Assoc. Petrol. Geol., 40, 17—50. [366]

— (1960): Stratigraphic principles and practice. — Harper & Brs., New York, 725 pp. [257]

WELLS, A. J. (1962): Recent dolomite in the Persian Gulf. — Nature, 194, No. 4825, 274—275. [315]

WELLS, A. J. & ILLING, L. V. (1964): Present-day precipitation of calcium carbonate in the Persian Gulf. — In: Developm. in Sedimentol., 1. VAN STRAATEN (ed.), Deltaic and shallow marine deposits. — Elsevier, 429—435. [192]

WELLS, J. W. (1942): Supposed color-marking in Ordovician trilobites from Ohio. — Amer. J. Sci., 240, 710—713. [247]

— (1957a): Coral Reefs. — In: J. W. HEDGPETH (ed.), Ecology. — Geol. Soc. Amer. Mem., 67, 1, 609—631. [233, 264, 266]

— (1957b): Corals. — In: J. W. HEDGPETH (ed.), Ecology. — Geol. Soc. Amer. Mem., 67, 1, 1087—1104. [233, 235, 359]

WENDT, A. (1965): Der Finefrausandstein. — Sedimentation und Epirogenese im Ruhrkarbon. — Forsch. Ber. Nordrhein-Westf. Nr. 1396, Westdeutscher Verlag, Köln, Opladen, 48 pp. [203]

WENDT, J. (1965): Synsedimentäre Bruchtektonik im Jura Westsiziliens. — N. Jb. Paläont., Mh., 286—311. [203]

WENGERD, S. A. (1951): Reef limestone of Hermosa formation, San Juan Canyon, Utah. — Bull. Amer. Assoc. Petrol. Geol., 35, 1038—1051. [271]

WENTWORTH, C. K. (1919): A laboratory and field study of cobble abrasion. — J. Geol., 27, 507—522. [64]

— (1922): A scale of grade and class terms for classifying sediments. — J. Geol., 30, 377—392. [12]

WENTWORTH, C. K. & WILLIAMS, H. (1932): The classification and terminology of the pyroclastic rocks. — Natl. Res. Council Bull., 89, 19—53. [160]

WENTZLAU, D. (1960): Stratinomische, stratigraphische und tektonische Untersuchungen in den Mittleren Siegener Schichten südöstlich des Siegener Schuppensattels auf den Blättern Freudenberg und Siegen. — Abh. Hess. Landesamt Bodenforsch., H. 29, 157—250. [381]

WERMUND, E. G. (1965): Cross-bedding in the Meridian sand. — Sedimentology, 5, 69—79. [76]

WERNER, E. (1961): Zu Verkittungsvorgängen an Psammiten. — Geol. Rdsch., **51**, 507—517. [139]
WETZEL, O. (1933): Die in organischer Substanz erhaltenen Mikrofossilien des baltischen Kreidefeuersteins. — Palaeontographica, **77**, 141—186; **78**, 1—110. [329]
WETZEL, W. (1923): Sedimentpetrographie. — Fortschr. Miner. Krist. Petrogr., **8**, 101—198. [200]
— (1957): Selektive Verkieselung. — N. Jb. Geol. Paläont., Abh., **105**, 1—10. [329]
WEYL, P. K. (1959): Pressure solution and the force of crystallization — a phenomenological theory. — J. Geophys. Res., **64**, 2001—2025. [134, 299]
— (1960): Porosity through dolomitization: conservation-of-mass requirements. — J. Sediment. Petrol., **30**, 85—90. [311]
WEYL, R. (1951): Schwermineralverwitterung in schleswig-holsteinischen Böden. — Schr. Naturwiss. Ver. Schlesw.-Holstein, **25**, 157—165. [39]
WEYL, R. & WERNER, H. (1951): Schwermineraluntersuchungen im Jungtertiär und Altquartär Schleswig-Holsteins. — Proc. 3. Internat. Congr. Sedimentol., Groningen-Wageningen, 293—303. [41]
WHETTEN, J. T. & HAWKINS, J. W. J. (1970): Diagenetic origin of graywacke matrix minerals. — Sedimentology, **15**, 347—361. [62, 128]
WHITE, J. F. & CORWIN, J. F. (1961): Synthesis and origin of chalcedony. — Amer. Miner., **46**, 112—119. [330]
WHITE, W. A. (1961): Colloid phenomena in sedimentation of argillaceous rocks. — J. Sediment. Petrol., **31**, 560—570. [84]
WHITNEY, M. I. & DIETRICH, R. V. (1973): Ventifact sculpture by windblown dust. — Geol. Soc. Amer. Bull., **84**, 2561—2582. [103]
WIEDEN, P. (1964): Neubildung von Dolomit im Schlamm des Neusiedlersees. — Fortschr. Miner., **41**, 179. [317]
WIESENEDER, H. (1953): Zur Diagenese klastischer Sedimente im Wiener Becken. — Tschermaks miner. petrogr. Mitt., **3**, 142—153. [42, 45]
— (1962): Sedimentologische und sedimentpetrographische Beobachtungen im Profil Pazin-Poljice. — Verh. geol. Bundesanst. Wien, H. 2, 235—238. [71]
— (1967): Zur Petrologie der ostalpinen Flyschzone. — Geol. Rdsch., **56**, 227—241. [128]
WILCOX, R. E. (1965): Volcanic-ash chronology, p. 807—815. — In: WRIGHT, H. E. & FREY, D. G. (ed.), The Quaternary of the United States, Princeton, N. J., Princeton Univ. Press, 922 pp. [164]
WILLIAMS, A. (1956): The calcareous shell of the Brachiopoda and its importance to their classification. — Biol. Rev., **31**, 3, 243—287. [240]
WILLIAMS, G. D. (1966): Origin of shale-pebble conglomerate. — Bull. Amer. Assoc. Petrol. Geol., **50**, 573—577. [47]
WILLIAMS, G. E. (1966): Planar cross-stratification formed by the lateral migration of shallow streams. — J. Sediment. Petrol., **36**, 742—746. [79]
WILLIAMSON, W. C. (1880): On the organization of the fossil plants of the Coal-Measures. — Part 10: Roy. Soc. London, Philos. Trans., **17**, 493—539. [227]
WILSON, I. G. (1972): Aeolian bedforms — their development and origins. — Sedimentology, **19**, 173—210. [102, 103]
WILSON, J. L. (1969): Microfacies and sedimentary structures in "Deeper Water" lime mudstones. — In: G. M. FRIEDMAN (ed.), Depositional environments in carbonate rocks. — Soc. Econ. Pal. Miner. Spec. Publ., **14**, 3—19. [356, 359]
WILSON, R. C. L. (1967): Particle nomenclature in carbonate sediments. — N. Jb. Geol. Paläont., Mh., 498—510. [204]
— (1968): Carbonate facies variation within the Osmington oolite series in southern England. — Palaeogeogr., Palaeoclimatol., Palaeoecol., **4**, 89—123. [257, 285, 353]
WINKELMOLEN, A. M. (1969): The rollability apparatus. — Sedimentology, **13**, 291—305. [68]
WINKLER, H. G. F. (1970): Abolition of metamorphic facies, introduction of four divisions of metamorphic stage, and of a classification based on isogrades in common rocks. — N. Jb. Miner., Mh., **5**, 189—248. [155, 157]
WINLAND, H. D. (1968): The role of high Mg calcite in the preservation of micrite envelopes and textural features of aragonite sediments. — J. Sediment. Petrol., **38**, 1320—1325. [285, 294]

WISE, S. W. (1973): Calcareous nannofossils from cores recovered during Leg 18, Deep Sea Drilling Project: Biostratigraphy and observations of diagenesis. — Initial Reports, D.S.D.P., Vol XVIII, 569—615. [293]

WISE, S. W. JR. & KELTS, K. R. (1971): Submarine lithification of Middle Tertiary chalks in the South Atlantic Ocean basin (Abstr.). — Progr., VIII Internat. Sediment. Congr. Heidelberg, 110. [282]

WITTHUHN, W. (1968): Schalensubstanz und Schalenstruktur der Gattung *Bolivina* ORB. (Foram.) aus dem Mittleren Lias Nordwestdeutschlands. — Beih. Ber. naturhist. Ges., 5, KELLER-Festschr., Hannover, 445—455. [230]

WOLBURG, J. (1961): Sedimentationszyklen und Stratigraphie des Buntsandsteins in NW-Deutschland. — Geotekt. Forsch., 14, 7—74. [381]

WOLDSTEDT, P. (1955): Norddeutschland und angrenzende Gebiete im Eiszeitalter. — K. F. Koehler, Stuttgart, 467 pp. [47]

WOLETZ, G. (1963): Charakteristische Abfolgen der Schwermineralgehalte in Kreide- und Alttertiär-Schichten der nördlichen Ostalpen. — Jb. geol. Bundesanst. Wien, 106, 89—119. [30, 45]

WOLF, K. H. (1962): The importance of calcareous algae in limestone genesis and sedimentation. — N. Jb. Geol. Paläont., Mh., 245—261. [279]

— (1963): Syngenetic to epigenetic processes, paleoecology, and classification of limestones; in particular reference to Devonian algal limestones of central New South Wales. — Thesis, Univ. of Sydney. [281, 282]

— (1965 a): Petrogenesis and palaeoenvironment of Devonian algal limestones of New South Wales. — Sedimentology, 4, 113—178. [282]

— (1965 b): "Grain diminution" of algal colonies to micrite. — J. Sediment. Petrol., 35, 420—427. [219]

— (1965 c): Littoral environment indicated by open-space structures in algal limestones. — Palaeogeogr., Palaeoclimat., Palaeoecol., 1, 183—223. [270]

WOLF, M. (in print): Inkohlungsunterschiede in Gesteinen des nördlichen Rheinischen Schiefergebirges und ihre geotektonische Ursache. — Fortschr. Geol. Rheinl.-Westfalen. [156]

WOLFENDEN, E. B. (1958): Paleoecology of the Carboniferous reef complex and shelf limestones in northwest Derbyshire, England. — Bull. Geol. Soc. Amer., 69, 871—898. [273]

WOOD, A. (1949): The structure of the wall of the test in the foraminifera; its value in classification. — Quart. J. Geol. Soc. London, 104, 229—255. [228]

WOOD, G. V. & WOLFE, M. J. (1969): Sabkha cycles in the Arab/Darb formation off the Trucial Coast of Arabia. — Sedimentology, 12, 165—191. [353, 356]

WOODLAND, B. G. (1964): The nature and origin of cone-in-cone structure. — Fieldiana: Geology, 13, 185—305, Chicago Nat. Hist. Mus. [299—301]

WRIGHT, A. E. & BOWES, D. R. (1963): Classification of volcanic breccia: a discussion. — Geol. Soc. Amer. Bull., 74, 79—86. [181, 182]

WRIGHT, L. D. & COLEMAN, J. M. (1973): Variations in morphology of major river deltas as function of ocean wave and river discharge regimes. — Amer. Assoc. Petrol. Geol. Bull., 57, 370—398. [115]

WRIGHT, M. D. (1959): The formation of cross-bedding by a meandering or braided stream. — J. Sediment. Petrol., 29, 610—615. [79]

WURSTER, P. (1964): Geologie des Schilfsandsteins. — Mitt. Geol. Staatsinst. Hamburg, H. 33, 140 pp. [77, 115]

YANAT'EVA, O. K. (1957): About polytherm of solubility of ($CaCO_3$ + $MgSO_4$ ⇌ $CaSO_4$ + $MgCO_3$) − H_2O system. — Dokl. Akad. Nauk SSSR, 12, no. 6 (Russ.). [305]

YAPAUDJIAN, L. (1972): Une approche actualiste en géologie sédimentaire (Quelques données d'interprétation des sequences de plate-forme). — Mém. B.R.G.M., Fr., 77, 715—744. [378]

YASUO, M. & SUGIMURA, Y. (1961): Ionium-thorium chronology of deep-sea sediments of the western North Pacific Ocean. — Science, 133, 1823—1824. [86]

YONGE, C. M. (1957): Symbiosis. — In: J. W. HEDGPETH (ed.), Ecology. — Geol. Soc. Amer. Mem., 67, 1, 429—442. [235]

YOUNG, K. (1959): Edwards limestone fossils as depth indicators. — In: Sympos. on Edwards limestone in central Texas. — Univ. Texas Bull., 5905, 97—104. [263]

YURKOVA, R. M. (1970): Comparison of postsedimentary alteration of oil-, gas- and water-bearing rocks. — Sedimentology, 15, 53—68. [41]

ZADNIK, V. E. & CAROZZI, A. V. (1963): Sédimentation cyclique dans les dolomies du cambrien supérieur de Warren County, New Jersey, USA. — Bull. Inst. Nat. Genevois, **62**, 1—55. [203]

ZALMANSON, E. S. (1951): Sediment formation in Lake Balkash (Russ.). — Bull. Moskov. Obshchestva Ispytatd. Prirody, Otdel Geol., **26**, 41—59. [317]

ZANKL, H. (1964): Neue Untersuchungen an Triasriffen in den Alpen. — Z. dt. Geol. Ges., **114**, 697—698. [288, 358]

— (1967): Die Karbonatsedimente der Obertrias in den nördlichen Kalkalpen. — Geol. Rdsch., **56**, 128—139. [125]

— (1969): Structural and textural evidence of early lithification in fine-grained carbonate rocks. — Sedimentology, **12**, 241—256. [340]

ZAPFE, H. (1936): Die Erhaltungsmöglichkeit des Aragonit im Fossilisationsprozeß, untersucht mit Hilfe des Reagens von FEIGL und LEITMEIER. — Anz. Akad. Wiss. Wien, **73**, 11, 110—111. [294]

ZELLER, E. J. & WRAY, J. (1956): Factors influencing precipitation of calcium carbonate. — Bull. Amer. Assoc. Petrol. Geol., **40**, 140—152. [209]

ZEN, E-AN (1959): Mineralogy and petrography of marine bottom sediment samples off the coast of Peru and Chile. — J. Sediment. Petrol., **29**, 513—539. [318]

ZIEGENHARDT, W. (1962): Sedimentologische und fazielle Untersuchungen am eeminterglazialen Travertin von Taubach bei Weimar. — Geologie, **11**, 1029—1051. [203]

— (1966): Frühdiagenetische Deformationen im Schaumkalk (Unterer Muschelkalk) des Meßtischblattes Plauen (Thüringen). — Geologie, **15**, 159—165. [204]

ZIEGLER, B. (1967): Ammoniten-Ökologie am Beispiel des Oberjura. — Geol. Rdsch., **56**, 439—464. [241, 357, 359]

ZIEGLER, M. A. (1962): Beiträge zur Kenntnis des unteren Malm im zentralen Schweizer Jura. — Diss. Zürich, Buchdruckerei Winterthur AG, 55 pp. [272]

ZIMDARS, J. (1958): Über Korn-Oberflächen von Sanden. Eine kritische Betrachtung der morphoskopischen Quarzkornanalyse. — Diss. Univ. Tübingen, 92 pp. [68—70]

ZIMMERLE, W. (1963): Zur Petrographie und Diagenese des Dogger-beta-Hauptsandsteins im Erdölfeld Plön-Ost. — Erdöl u. Kohle, **16**, 9—16. [25, 29, 149]

ZIMMERLE, W. & BONHAM, L. C. (1962): Rapid methods for dimensional grain orientation measurements. — J. Sediment. Petrol., **32**, 751—763. [72]

ZINGG, TH. (1935): Beitrag zur Schotteranalyse. — Schweiz. miner. petr. Mitt., **15**, 39—140. [63]

ZUMPE, H. H. (1964): The detection of phosphatization in calcareous sediments — a fluorescence method. — J. Sediment. Petrol., **34**, 691—692. [201]

Subject Index

Main references are printed in bold. Page numbers in italics refer to figures. The suffix "f." refers to the next page, "ff." to the following pages.

Reference is not made to the different heavy minerals and to the key to their identification (p. 31 f.), to the key to identification of skeletal grains (p. 212 ff.), and to the table of biostromes and bioherms (p. 271 ff.).

Abrasion
 of pebbles 48 f.
 of sand grains 65 ff.
Absolute time in sedimentology 363
Accretionary lapilli *161*, **163**, *163*, 170, 187
Achnelith 170
Acropora 235 f., 358
Actualism 122, 352
Agglutinate lapilli 180
Agglutinating foraminifera 228
Aggrading crystallization *217*, 292
Albite, authigenic
 in sandstones 141 f., *142*, *150*, 152 f.
 in carbonate rocks **333** ff., *334*
Algal mats **222** ff., 355 f., 365
Alkali basaltic magma 162, 184
Alloclastic volcanic breccia 181
Allodapic limestones 125
Alluvial plain 103 f.
Alluvial fan 103 f.
Alteration 130, 146
Alternations limestone-marl 364, 371 f.
Amazon River 7 f.
Ammonites 245 f.
Analcime 144, 147, *150*, 152 f., 187 f., 379
Anatase formation in sandstones 144
Anchimetamorphism 155 ff.
Andesite 180 f.
Anhedral crystals 279
Anhydrite formation 325 f., *327*, 355, 373
Anisomyaria 243 f.
Anthozoa 235 f., *237*, 265, 267
Antidunes 73, **90**, 106, *179*, 180
Aptycha 246, 359
Aquagene tuff 183
Archaeocyathidae 232
Arkose **14** ff., *16*
Arroyo 104
Arthropoda 246 f.
Ash, volcanic 162
Ash fall 164
Ash flow *176*

Asteroidea 249
Atlantic Coast, USA
 barriers 120
 estuaries *110*, 115
Atmospheric turbidity 164
Atolls 266 f.
Australia, Coorong 316
Authigenesis 130, 142
Authigenic silicates
 in carbonate rocks 333 ff.
Autoclastic volcanic breccia 181
Average grain size 52

Back-reef 270, 358
Bacteria, their role in
 carbonate precipitation 191 ff.
 dolomitization 309
 pyrite formation 337
 weathering 2
Bafflestone 277
Bahamas
 cementation *283*, *284*, *286*, *287*
 corals 235
 dolomite 315
 grapestone 204
 mud 196
 oöids *250*, 251, *257*
 pellets *250*, 251, *252*
 tidal flats 223
 well Andros 1 353
Bahamites 198
Balance, sedimentary 6 f.
Baltic Sea
 cross bedding 79
Barchans 100
Barfinger sands 100, *376*
Barite formation 143
Barrier bars 100
Barrière en creux *129*
Barrier islands 119, 120
Base surge deposits 160, 163, 177, *178*, *179*
Beaches 119 ff.

Beach ridges 119
Beach sand
 grain surface 69
 roundness 68
 sorting 60
Bed roughness 73
Bedding plane 85 ff.
 upper surface 87
 lower surface 91
Belemnites 245 f.
Bermuda
 cementation 282, 284 f.
 cup reef 263
Biancone 231
Bimodal cross-bedding 79, 117
Bindstone 277
Biocalcarenite 206, *207*, 358
Biocalcirudite 206, *207*
Biocalcisiltite 206, *225*
Biocoenoses 206, *208*, 260
Bioerosion 195, 208, 241
Bioherms **261** ff.
 defined 260
 dimensions *274*
 facies 262
 morphological 261 f.
 occurrences 271 ff.
 quiet-water bioherms 261 ff., *269*, *270*, *274*
 reef bioherms 261 ff., *262*, *265*, *267*
Biostromes 261 ff.
 defined 260
 occurrences 271
Birds eye structures 270, *308*, *341*
 (= fenestrae) 342, 355 f.
Black Sea
 coccolith layers 197
 varves 363
Blastoidea 248
Blocks, pyroclastic **160**, *163*, *164*
Blue-green algae *219* ff.
 clotted texture 198
 coccoid 224
 color absorption 214 f.
 in oöids 255
 lumps 204
 micro- 356
 s. oncolites
 s. stromatolites
Bombs, volcanic **160**, *163*
Bomb sag 170, *172*
Borings, defined 254
Bottomset beds *110*, 111
Boulder clays 47
Bouma-sequence 367 ff.
Bounce casts 96
Brachiopoda **239**, *239*
Brecciation, early diagenetic 82, 201 ff., *204*, *205*, 314, 355 f.
Brush casts 96
Bryozoa 236, *238*, 358

Burrows 80, 90
 defined 254

Calcarenite **201** f., *202*
 defined 190
Calcareous algae 214 ff.
Calcilutite
 defined 190
 origin 191 ff.
Calcirudite, defined 190
Calcisiltite, defined 190
Calcisphaera 226, 227
Calcispongea 232, *233*
Calcrete 259
Caliche 257, *322*
Calpionella 232
Capillary pressure curves of porous rocks *342—345*
Carbonaceous sandstones, diagenesis 153 ff.
Carbonate
 cycles 353, *372*, *373*, *374*, *379*
 neoformation in sandstones 142
Carbonization 155 f.
Caribbean Sea
 Batabano (oöids) 251
 Bonaire (dolomitization) 314 f.
 coral zonation in reefs 358
Caverns 343
Celestite formation 326
Cement
 A **281** ff., *282*, *283*
 B 205, *283*, **288** ff., *292*
 cryptocrystalline 281, 357
 defined 12
 gravitational 285, *286*, *287*
 meniscus *286*, *287*, **289**
 overgrowths 289, *332*
 whisker *286*, *287*
Cementation of calculitites 338 ff.
 deep marine sediments 359
 limestones 276, **279** ff.
 marls 299 ff., 301 ff.
 sandstones 130, **135** ff., *135*, *137*, *140*, *142*, *148*, *150*
 shallow marine sediments 284, 357
Cephalopoda **245** f., 359
Chabazite 187, 188
Chalcedony **328** f.
 in sandstones *140*, 141
Chalk 197, **225**, **232**, 339, 359
Chambers (pores) 341
Channels (pores) 343 f.
Channel sands 99 f., *105*, 106
Charophyta 218, *218*
Chemical analyses, sandstones 13
Cheniers 119
Chert 329, 360
Chlorite formation in sandstones 143, 147, 150 ff., *150*, 157
Chloritic oöids 26, *255*, 257
Chute and pool 73, *179*, 180

Classification of
  carbonate rocks 257 ff.
  sandstones 13 ff., *13, 18*
Clastic wedges 121, 125, 362, 366, 376
Clay, influencing
  cementation of limestones 340
  compaction 340
  crystal enlargement 293
  stylolite formation 298 f.
Clay mineral formation in sandstones 143
Clinoptilolite 187
Cliona (sponge) 233
Clotted limestones **197** ff., *198*, 201, 355
Coalification 155 f.
Coccoliths *196*, 197, **225**, *225*, 359
Codiaceae 217
Coelenterates 233 f.
Collenia 223
Colorado River 7
Color of quartz grains 22
Columbia River 7
Commensalism, defined 208
Compaction
  of clastic sediments 80 ff., 130 ff.
  of limestones 269, 297 ff., 338 ff.
  stream 5
Compensation depth in oceans 360
Competition as constructive principle
  in bioherms 358
Composition, primary
  of carbonate rocks 208 ff.
  of sandstones 13 ff.
Concretions 301 ff.
Cone-in-cone 299 ff., *300*
Conglomerates and breccias 46 ff., *63* f., 65 ff., 201 ff.
Conglomeratic mudstones 47
Contact strength 136
Contour currents 98, **123**
Convex-up position of shells 207
Convolute lamination 80 ff., *81*
Cooling unit 176
Coprolites, defined 201
Coquina *242, 244, 294, 296*
Corallinaceae 218 f.
Corals **235** f., 264 ff.
  ahermatypic 235
  hermatypic 235, 264 ff., *265, 267*, 358
Coriolis force
  in estuaries: Cook Inlet, Alaska 115
  on continental slopes 123
Crinoidea 248 f., *249*
Cross-bedding **74** ff., *75, 77*, 355
  longitudinal 79
Crustose limestones 257 ff.
Cruziana facies 90
Cryptozoon 223 f.
Crystal enlargement 206, **290** ff., *291*, 339
  defined 280
Crystal tuff *161*

Cubichnia 90
Cummulative curves (grain size) 35, 54
Current lineation 90, *91*
Currents, marine
  contour 98, **123**
  longshore 120
  rip 120
  tidal 116, 120
  turbidity 92, 94, 123
  wave 120
Cycles, causes 364 ff.
Cycles, defined
  composite 362
  modal 362
  superimposed 362
  theoretical 362
Cycles, examples
  carbonate 371 ff., 378
  cementation 370 f.
  coal 362, **375** ff., 380
  delta 378, 382
  detrital 116, 369, 380
  estuarine 116
  eustatic 365
  fluvial 106, 380
  limestone-marl 370 ff.
  limnic 379
  reef 372 f.
  sabkha 379
  tidal (clastic) 117 f.
  turbidite 367 ff.
Cyclothems
  asymmetrical 361, 370, *375, 376*
  symmetrical 361, 370, *378*

Dacite 185
Dasycladaceae **215**, *215, 216*, 357
Dead Sea
  deep blue-green algae 224
  evaporitic rhythms 363
  inorganic carbonate precipitation 192
Decementation of sandstones 143
Dedolomitization 320 ff.
Deeper water environments 359 f.
Deep sea
  calcite cement 282, 284, 288
  chert 330, 366
  contour currents 98, **123**
  dolomite 318 f.
  turbidites 91 ff., 123 f., 125, *126*, 127 f., 360, 367 ff.
Deep Spring Lake dolomitization 316
Deltas 109 ff., *110, 111*
  cycles (destructive-constructive) 378
  Mississippi 376 ff., *376, 377*
Density differentiation 169
Desiccation cracks
  in carbonate sediments 204
  in clastic sediments 82, 84
Detrital limestones 201 f.
Detritus, defined 3

Diagenesis, defined 4
  hampered by hydrocarbons 41 f., 132, 139, 144, 148, 242, 294, 299, 351
  of carbonate rocks 279 ff.
  of sandstones 130 ff.
  of shales 4, 156
  of volcanic glass 186 ff.
Diatoms 328
Dickite formation 146, 157
Dielectric anisotropy 72
Discharge of rivers 6 f.
Discoaster 226
Dish structures 82
Disintegration of skeletons 195 f.
Dispersion factor (grain size) 55
Dissolution in sandstones 133
Dolomite-arenite 202
Dolomites 203, 205, 343, 344, 345, 347, 349
Dolomitization 303 ff.
  ascendent 314, 318
  descendent 309, 318
  early diagenetic 305 f., 314 ff.
  homoaxial 313
  in lakes 314 f., 316 f., 318, 354
  late diagenetic 309, 318 ff.
  porosity 351 f.
Domichnia 80
Draa (star dune) 102
Drag coefficient (volcanic bomb) 163, 164
Dripstone 257, 260
Drumlins, defined 107
Dunes 73, 78, 100 ff.
  longitudinal 101, 102
  parabolic 101
  pyramidal 101
  star dunes 101, 102
  transverse 75, 100

Earthquakes, influence in turbidity currents 92 f., 367
Echinodermata 248 f.
Echinoidea 248 f., 332
Ejection (volcanic bomb)
  angle 163
  velocity 163, 164
Enfacial junction 269, 280
Entophysalis 223, 285, 288
Environmental indicators
  in carbonate rocks 191 ff., 352 ff.
  in sandstones 56 ff., 100 ff.
Eolian differentiation 166
Eolianites, defined 355
Eolian sand
  cross-bedding 76
  grain surface 69
  roundness 68
  sorting 59, 60
Epigenesis 146 f.
Epirelief of bedding planes 87 ff.
Erionite 187

Erosion 3, 7
Eruption column 175, 176
Eruptive cloud 162
Esker, defined 107
Estuaries 110, 115 f.
Etoscha pan (S.-Africa) dolomitization 316
Euhedral crystals 279
Eustatic
  cycles 365
  sea level variations 121 f., 365 f., 368, 376, 378
Evaporation-dolomitization 307 ff.
Explosion breccia 182

Facies maps 126, 129
Fall unit (pyroclastic) 164
Fanglomerates, defined 47
Feldspar authigenic
  in carbonate rocks 333 ff., 334, 335
  in sandstones 141 f., 142, 150, 151 ff.
Feldspar, detrital 24 f.
Fenestrae (pores) 269, 270, 308, 341, 342, 356
Filtration in sandstones 139
Fining-downward sequences 361, 369, 378, 381 ff.
Fining-upward sequences 106, 116, 361, 369, 378, 381
Fissure-type eruptions 175
Flaser bedding 74, 117
Flint 329
Floatstone 277
Flood (plain) deposits 105, 106
Florida Bay & Keys
  bacteria 193
  dolomite 315
  pellets 200
  whitings 194
Flow regime
  lower 73 f., 104, 368
  upper 73, 104, 368
Flow unit (pyroclastic) 173, 176
Flow velocity
  contour currents 123
  required for transport 3, 60, 78
  turbidites 98, 123
Fluorite formation 327
Flute casts 93, 94 f., 97 f.
Fluxoturbidites 123, 369
"Fly paper" (algal mats) 365
Flysch (Alpine and Carpathian) 44 f., 126, 128
"flysch" 127
  Cambrian (Great Britain) 128
  Cretaceous (Chile) 98
  Devonian, Ordovician (Pennsylvania) 128
  Lower Carboniferous (Germany) 16, 19, 22 f., 25, 369
  Plio-Pleistocene (California) 128
  Tertiary (Spain) 95
  Upper Devonian (Germany) 81, 128
Fodinichnia 80

Foraminifera 227 ff.
  as depth indicators 228, **359**
  encrusting 268
  hyaline *224,* 227, 228, 229
  imperforate 228
  occurrence 357 f.
  perforate 230
  porcellaneous 228, *229,* 230
  rotaloid *229*
Foreset beds *110,* 111
Fore-reef *268,* 269
Fourier analysis of cyclic sequences 362
Fractures in carbonate rocks 344, 346
Framestone 277
Fusulinidae 230

Ganister, defined 19
Gas, magmatic 162, 184
Gastropoda *243,* **244** f., *244,* 358
Geopetal fabric *269, 277, 308*
Geostrophic currents 123
Gilbert-type delta 111
Girvanella 220
Glacial
  cycles 363 ff.
  deposits 107
  drift 107
Glaciofluvial deposits 107
Glass shard *161,* **162**, 164
Globigerina 197, 228
Goniolithon 219
Gonnardite 188
Graded bedding
  flysch-type 73, 123, *368*
  inverse 123
  matrix-free 73, 106, 380
Grain 280
  contacts 136, 147
  growth 280
  shape 63 ff., *63*
  size (sandstones) 51 ff., *51*
  size and environment 56, 58 ff.
  support 276
Grainstone 277
Grapestone 204, 256, 357
Graywackes 14 ff., *16,* 128
Great Barrier Reef 267
Great Salt Lake
  blue-green algae 220
  oöids 252, 257
  pellets 200, 220, 224
Greece, orogenic tectofacies *129*
Green algae **215**, *215, 216, 217*
  color absorption 214
  in reefs 263, 266 f.
  in the shallow sea 357
  lime mud production 196
  spores **217**
Grit, defined 19
Groove casts *93, 95,* **96**, 97 f.
Ground surge deposit 180

Growth framework 341 f.
Gulf Coast (USA)
  ancient 115, 121, 125, 381
  barriers, modern *120*
  heavy minerals 44
  sedimentation rates 87
  shallow sea 359
  (see also Mississippi delta)
Gulf of Batabano (Cuba; oöids) 251
Gulf Stream 123

Halimeda 196, **217**, *217*
  in reefs 263, 266 f.
Halite formation (sandstones) 144
Halmyrolysis 130
Hardground 288, 371
Hawaiian
  deposits 170
  eruptions 169
Heavy minerals 26 ff., *28*
  correlation *43*
  grain size 33 ff.
  identification 27 ff.
  stability 39 ff.
  use 42 ff.
Hematite formation (sandstones) 144
Hexactinellidea 233
Holothuroidea 249
Homomyaria 244
Homopycnic inflow (deltas) 111
Horizontal bedding 73
Hyaloclastites 183
Hybrid sandstones, defined 17
Hydration (volcanic glass) 187
Hydrozoa 233 ff.
Hyperpycnic inflow (deltas) *110,* 111
Hypidiomorphous fabric 279
Hypopycnic inflow (deltas) *110,* 112
Hyporelief of bedding planes 91 ff.

Icebergs transporting pebbles 367
Identification keys
  heavy minerals 31 f.
  skeletal grains 212 ff.
Idiomorphous fabric 279
Ignimbrite *174,* **177**
Illite crystallinity 156
Impact casts 96
Inclusions in quartz 20
Injection structures 82 f., *83,* **84**
Inoceramus 232
Intercrystalline pores 343, **344**, 346, *347*
Internal sediment *269, 277, 308*
Interparticle pores 341, *342,* 345, *347*
Interstitial water, flow of 5 f.
Intertidal environments *116* ff., 223, 356
Intraclasts 201, *203,* 203 ff., 314, 355 f.
Intraparticle pores 341 f., *343,* 345 f., *347*
Intrastratal solution 40, 131
Intrusive breccia 182
Inverse graded bedding 123

Subject Index

Isopach maps (pyroclastic deposits) 166, *167*, *168*
Isotopic composition, carbonates *324*
Itacolumite, defined 19

Joints in Limestones 350 f.

Kames, defined 107
Kaolinite formation 138, 145, 156
Key to identification of
  heavy minerals 31 f.
  skeletal grains 212 ff.
Kongo River 7
Kurtosis, defined 53

Lag concentrates *55*, 118
Laguna Madre (Texas)
  oöids 251
  sedimentation rate 118
Lahar *171*, **180**, *181, 182*
Lakes
  Aral (oöids) 251
  Balaton 193, 317
  Balkash (dolomitization) 317
  Bonneville (delta) 111
  carbonate sedimentation *354*
  Constance 221 f., 355
  environments 103, 353 ff.
  Kara Bogas (dolomitization) 316
  Neusiedl (dolomitization) 317
  Thun (Turbidites) 97
  Tuz Gölü (dolomitization) 317
Lamellibranchiata 243 ff.
Laminated sediments
  deltas 113
  intertidal 116, 223
  marginal seas 118
  shallow sea 74, 365
  storm layers 223, 365, 378
  supratidal 222 f.
  turbidites 367 ff.
  varves *307*, 363 f., *364*
Laminites 369, 378
Lapilli 160
  breccia 162
  tuff **160**, *161*, *171*
Lapillistone 160, *161*
Laumontite formation (sandstone) 155
Lava flow breccia 182
Lebensspuren (= trace fossils) 208
Light penetration in the sea 214 f., 224, 356
Limestone-marl alternations 364, 371 f.
Lithic arenites, defined 17
Lithification, defined 4
Lithistidea 233
Lithothamnion **219**, *224*, 267
Load casts 91, *92*
Loferites 224
Lognormal grain size distribution 54
Lombardia 249
Lower flow regime **73**, 74
Luminescence of quartz grains 22

Lumps 204
Lydites, defined 125
Lyngbya 223

Maar *178*, 180
Magma 162
Magnetic susceptibility 72
Magnetite formation (sandstones) 144
Main River, pebbles 63
Mantle bedding *165*
Marginal basins 125
Marginal seas 118 f.
Marine sands 121 ff.
Massif Central (France) 45 f.
Mass-mortality (cyclic) 367
Matrix
  defined 12
  in limestones **275 ff.**
  in sandstones **61 ff.**, 128
Matrix grains, defined 206
Maturity of sandstones 14
Median diameter 52 f.
Mediterranean Sea
  cementation 285
  deep sea sediments 194
Megacycles
  asymmetrical 381 f.
  symmetrical 380 f.
Melobesieae 219
Metagenesis 146 f.
Metamorphism (very low stage) of sandstones 155
Metaquartzites, defined 141
Meteoric water, influencing
  cementation 289
  dolomitization 320 f., 323
  isotope exchange 325
  Mg-loss of Mg-calcites 321 f.
Micrite, defined 191
Micritization *207*, *247*, *255*, 281, **285**, 288, *292*, *293*, *294*, 357, 360
Microspar 190, **191**, 290 f.
Microstalactitic cement 285, *287*
Miliolidea 228, *230*, 357
Milleporina 233
Mineral composition, skeletons 208 ff.
Minus-cement-porosity 135
Mississippi
  River 1, 7, *57*
  delta *111*, 112, 114, *376*, *377*
Molasse (Alpine) 7, *16*, 26, *41*, *43*, 44 f., *48*, *49*, *50*, *56*, *61*, *62*, *126*, 149, 382
"molasse" 128
  Devonian (Pennsylvania) 129
  Mississippian (Pennsylvania) 50
  Pennsylvanian (Pennsylvania) *48*
  Permian (USSR, Ural) 45
  Tertiary (Greece) *129*
  Tertiary (Peru) 33, 129
  Upper Carboniferous (Germany) *16*, *150*, 153 ff.

Molds 342, *343*, 345f., *347, 349*
Mollusca 241 ff.
Moment measures (grain size) 53
Montastrea 265, 358
Montmorillonite 2, *16*, 143, 188
Mordenite 187
Mottled structures 79 f.
Mucilage 204, *254*, 255 f., 285, 357
Mud
    cloud *172*
    cracks 82 f.
    support 276
    volcanoes 84
Mudstone 277
Mummy Limestone 221

Nagelfluh 47, 203
Nannoconus *231*, 232
Nannofossils *196*, 225 f., *225, 226, 231*, 359 f.
Nannoplankton mud 197
Natrolite 188
Natural levees 105
Nearshore Modern Sand Prism 121 f.
Neoformation 130 f., 135
Neomorphism, defined 280, 290 ff.
Nereites facies 98, 124
Nomenclature principles
    carbonate rocks 275 ff.
    grain size 10 ff., *12*
    sandstone composition 10 ff., *11*
North Sea
    cross-bedding 79
    heavy minerals *43* f.
    tidal flats 117 f.
Nucleation rate, carbonate cement 290
Nuée ardente 175
Nuée ardente deposit *182*
Nummulites 224, **230**

Obliteration of sedimentary textures by
    anhydritization 325 f.
    crystal enlargement *291*, 292 f.
    dissolution 133, *215, 322, 343, 349*
    dolomitization 313
    micritization 285, *292*
    other replacements *142*, 144 ff., 234
    silicification *242*, 330, *331*
    stylolitization 297
    transformation 295
Obsidian 186
Oil and gas pools 271 ff.
    Ain Zalah (Iraque, Cretaceous) 350
    Andrews South (Texas, Devonian) 349
    Asmari (Iran, Tertiary) 350
    Dalum (Germany, Cretaceous) *242, 243*, 244, 348
    Ekofisk (Norway, Cretaceous) 339
    Elkton (Canada, Mississippian) 351 f.
    Emsland area (Germany, Cretaceous) 159
    Ghawar (Saudi-Arabia, Jurassic) 347
    Golden Lane (Mexico, Cretaceous) 348
    Kirkuk (Iraque, Tertiary) 348, 350
    Leduc (Canada, Devonian) 352
    Magnolia (Arkansas, Jurassic) 347
    Mara (Venezuela, Cretaceous) 350
    Redwater (Canada, Devonian) 348
    Rehden (Germany, Permian) *347*
    Scurry-Snyder (Texas, Pennsylvanian) 350
    Swan Hills (Canada, Devonian) 348
    Umm Shaif (Abu Dhabi M.A., Jurassic) 353
Oil saturation of sandstones 158 f.
Oligostegina 227
Olisthostrome, defined 82
Oncoid, oncolite, defined 220 f., *199, 203, 219, 220, 221*, 270, 313, *343, 345, 349*, 357
Oöid, oölite 201, 204, *205*, 250 ff., *250, 252—256, 322*, 358
Oölitic micrite 257, 355
Oömoldic pores (= dissolved oöids) *322*, 349
Opal 188, 328 f.
Ophiuroidea 249
Orbitoidae 230
Ordering degree of dolomite *304*, 305
Organic compounds, carbonate sediments 192, 211, 285
Orientation
    of pebbles 70 f.
    of sand grains 71 f.
Orinoco delta 111, 116
Orthoquartzite 17
Ostracoda *247*, 358
Overbank deposits 105
Overburden pressure 4 f.

Packstone 277
Palagonite 170, 188, *188*
Paragonite formation, sandstones 157
Parasitism, defined 208
Pascichnia 98
Peléan pyroclastic flow 175
Pelecypoda *242*, 243 ff.
    pelagic 359
Pelletal limestones 199 ff., *200*
Pellets 199 ff., 357
Penicillus 196, 217, 267
Peperites 183
Permeability **158**, *159*, 342 ff., *347*
Persian Gulf
    carbonate cycles 353, 379
    cementation 284 f.
    corals 235
    dolomite 315
    marine tufa 260
    oöids 251 f., *254*
    restricted environment 357 f.
    shallow sea 356 f.
    sites of carbonate deposition 353
    south coast *253*, 356 ff.
Philippsite 187 ff.

Phi scale *12*, 52
Photosynthesis (its role in carbonate precipitation) 191
Phreatic zone 289
Phreatomagmatic eruptions 180, **184**
Phyllosilicates in sandstones 25 f.
Pillow
    breccia 184
    lava 183
Pingo, defined 107
Pisolites **221**, 250
Pivotability of grains 68
Plate tectonics and geosynclinal concept 125
Plinian
    deposits *170, 171*
    eruptions **169**
Po delta *112*
Point bar 105
Porites 235
Porosity of carbonate rocks
    calcilutites 338 ff.
    particle limestones and dolomites 340 ff.
    pore types 340 f., *341—345*
    primary *341, 342*, 347
    secondary *341, 343—347*, 346, *349*
Porosity of sandstones, relation with depth 130 ff., *132*, 147, 155, 157
    relation with grain size and permeability 158 f., *159*
Porosity of shales 4
Porostomata 220
Potassium feldspar, authigenic
    in carbonate rocks 335 ff., *335*
    in sandstones 141 f., *150*, 151 ff.
Pottsville fm. 48
Pressure
    hydrostatic 5
    overhydrostatic 5
Pressure solution (pressolution) 131, 133 ff., *134*, 140
Protodolomite, defined 303
Pseudomorph 130, 144
Pseudosparite 191
Pteropoda 245
Pumice **162**, 164, *171*
Pumice flow deposit *173, 174*
Pyrite, authigenic 337 f., *338*
Pyroclastic
    breccia 160, *161*
    fall deposits 162, *177*
    flow *171*, **172**, *173, 174*
    flow deposits 177
    rocks 160
Pyrophyllite formation in sandstones 156

Quartzine (length-slow chalcedony) 328
Quartzites
    cemented 141
    pressolved 141
Quartzose sandstone, defined 17

Quartz overgrowths 147 f., *148*
Quinqueloculina 228, *230*

Radiographs 72
Radiolaria 328, 360
Rain prints 90
Rate of
    accumulation 85
    sediment formation 85 ff., 363, 372
Rauhwackes (cellular dolomites) 321, **352**
Recrystallization 280, 292
Recycled sandstones 14
Red algae **218** f., *224*
    color absorption 214 f.
    encrusting 358
    in reefs 263, 267
    specific surface 313
Red beds *83*, **108** f., 152 f.
Red Sea 192
Reef cycles 372 f.
Reefs 261 ff., *262, 265, 267*, 358
Regressions 119
Regressive sequences 113, 117, 119, 122, *373, 374, 378, 379*, 382
Repichnia 90
Replacement 130, 131, **144** ff.
Restricted environment 357 f.
Rhine River, heavy minerals 38, 44
Rhipidolite formation in sandstones 157
Rhone delta 113
Rhone River, heavy minerals 38 f.
Rhyolitic ash *161*, 186
Rhythmites
    alternation 361
    diagenetically accentuated 364
    superposition 361
Rhythms, defined 361
    causes 364
Rip currents 79, 120
Ripple marks 87 ff.
Ripples
    accretion 76
    adhesion 89
    avalanche 76
    combined-flow 89
    complex 87
    current 74, 87 ff., *88*
    linguoid 76, *88*, 89
    longitudinal 87, 89
    rhomboid *88*, 89
    wave (oscillation) *88*, 89
River sand
    grain surface 69
    roundness 68
    sorting *59*, 61
Rivers
    braided 104
    meandering 104
Rock fragments **22** ff.
    unstable 23 f., 128
Rogenstein 223

Rosin-Rammler-Sperling grain size
  distribution 55
Rotaliidae 230
Roundness
  pebbles 65 ff.
  sand grains 67
Rugosa 236
Rudistids 241, 244
Rudstone 277

Sand
  bodies 99 ff.
  dikes 83 f.
  volcanoes 84
  waves 89
Sandstones, thin sections *16, 138, 140, 142*
Sandur 107
Saccocoma 248 f.
Sahara, dunes 101
Sapelo Island, USA *120*
Schizophyta 214 f., **219**
Schizothrix **222** ff., 365
Scleractinia 236
Scolithus 80
Scoria 162
Scytohema 222 f.
Sea floor spreading, influence on
  transgressions 366
Sea urchins **248** f. *332,* 356
Seaweed (sea grass) *193*
Sedimentary cover, continents and oceans 8
Sedimentary structures
  carbonate rocks 222 ff., 257 ff., 260 ff.,
    314, **352** ff.
  conglomerates 106
  sandstones 72 ff.
Sedimentation
  carbonate sediments 191 ff.
  sands 3, 100 ff.
  pebbles 73, 106, 123
Sedimentation rates
  rate of accumulation 85
  rate of sediment formation 85 ff., 363,
    372
Seif (longitudinal dune) 101
Sequences
  composite 362
  negative 361
  positive 361
  regressive 113, 117, *373* f., 378
  transgressive 113, 118, 374
Serpulidae 240, *240*
Settling velocity (pyroclastic particles) 169
Shallow marine
  environments 119 ff., 356 ff.
  sand (sorting) *59, 60*
Shape
  pebbles 64
  sand grains 64 f.
Shards (s. glass shards)
Sheet cracks 355

Shelf Relict Sand Blanket 122
Shell structures, mollusks 241 f.
Shelter pores 343, 346
Shrinkage cracks 82 f., 355
Siderastrea *237,* 358
Sideromelane 178, 188
Silex 329
Silicification *242,* **327** ff., 330, *331, 332*
Silicispongea 232, *234*
Skeletal limestones **206** ff., *207, 215—249*
Skewness 53, *58*
Slumping 80 f., *126*
Soft pebble conglomerate *203*
Soils 2
Solenoporaceae 218
Somphospongia 220
Sorting
  pebbles 48 f.
  sandstones 53 ff., *55, 61*
Source index, sandstones 14
Sparagmite defined 19
Specific surface of grains 67, **158**
Speleothems 260
Sphaeractinoidea 235
Sphaerocodium 220
Spheroid, spherulite 250 f.
Sphericity index 64
Spine-like intergrowth 157
Spirorbinae 240
Sponges 232 f., *233, 234*
Spongiomorphidea 235
Spongiostromata 220
Spreite (trace fossil) 99, 117
Staining of carbonate rocks 288 f.
Starved sedimentation 125
Steam explosions 169, **177**, 184
Stilpnomelane formation in sandstones 156
Storm layers
  carbonate 223, 365
  sand 74
Stratification 72 ff.
  cross-bedding **74** ff., *75, 77*
  horizontal **73** f., *123, 165,* 222 ff., *307,*
    359, 363, 364
Streams
  bed-load 106
  mixed-load 106
Stromatactis **270**, 273, *308*
Stromatolites *219,* **220**, 222, 270, *308,* 313,
  356, *364*
Stromatopora 234 f., 264, 358
Strombolian
  deposit 170
  eruptions 169
St. Vincent pyroclastic flow 175
Styliolina 245
Styliolites
  carbonate rocks **269** ff., *297,* 351
  sandstones 133, *134*
Subaqueous pyroclastic flow 184
Subarkoses *16,* 17

Subgraywackes *16*, 17
Subhedral crystals 279
Subsidence rates
    different from sediment formation rates 85 ff., 125 ff., 363
    equal to sediment formation rates 85 ff., 128 f., *377*
    influence on cyclicity 366
Subsolution 125, 303
Substitution, defined 130
Substratum influence on biocoenoses 207, 241
Sulphate formation in sandstones 143
Supratidal
    carbonate sedimentation 355 f.
    environments 222 f., 355 f.
Surface of grains 69
Surface roughness (volcanic bomb) 163
Surface tension (magma) 162
Surtseyan
    deposits 170
    eruptions 184
Symbiosis 208, 235
Syneresis cracks 82, 84, 379
Syntaxial
    enlargement 280
    rim cementation 280

Tabulata 236
Taphocoenosis 206, 208
Teepee structures 285
Tensile strength (magma) 162
Tentaculites 245
Tephra **160**, 162
Tephrochronology 164
Terra rossa 108
Terrigeneous lime mud 197
Thermal shock 184
Tholeiitic basaltic magma 162, 184
Thomsonite 188
Thrombolite 220, 223
Tidal sedimentation
    carbonate sediments 222 f., **355** f.
    clastic sediments 116 ff.
    cycles 117 f., *378*
    flasers 117
Tigillites 80
Tillites 47, **107**
Tintinnines 232
Tool marks 94
Topset beds 109, *110*
Tourmaline formation in carbonate rocks *335*, 337
Trace fossils 99
    in the interior of beds **79** f., 122
    on the lower surface 98 f., 124
    on the upper surface 90, 103, 122
    pre- and postdepositional 124
Transformation, defined 280
    aragonite-calcite **294** f., 339

Transgressive sequences 113, 115, 118 f., 122, 355, 365, 374, 377, *378*
Transport
    bed load 7
    dissolved material 7
    grains 3, 56 ff., 64 ff., 71 ff., 87 ff., 100 ff., 203 ff., 206 f., 248, 250 ff., 269 f.
    heavy minerals 40
    mud 61 ff., 193 ff., 289, 360
    pebbles 48 ff., 63 ff., 70 f., 106
    suspended load 3, 7
Travertine 259, *259*
Trilobita 246, *246*
Trivoli sandstone, Pennsylvanian *105*
Tufa 257, **259**, *259*
Tuff 160, *161*
Tuff breccia 160, *161*
Tuffite 172
Turbidites **91** ff., 123, **367** ff., *368*
    distal 369
    grain orientation 72
    proximal 369
Turbidity currents 92

Uintacrinus 248
Undulose extinction, quartz 20 f.
Uniformitarian principle 352 f.
Upper flow regime 73
Uranium precipitation in sandstones 144

Vadose
    silt 289
    zone 289
Varves
    evaporites *307*, 363, *364*
    glacial clays 363
Verrucano, defined 47
Vesicles 162
Vesiculated tuff 180
Viscosity (magma) 162, 182
Vitric tuff *161*, *163*
Volatiles (magmatic) 162
Volcanic
    breccia 181
    dust 164
    explosion 162
    glass, alteration 186
    mudflow 180
Vugs 343, *345*, *346*

Wackestone 61, 277
Weathering
    chemical 2
    clay minerals 2
    feldspars 2
    heavy minerals 39 f.
    mechanical 1
    role of bacteria 2
Welded tuff *174*, *175*, **177**
Welding temperature 176
Whitings 192, *194*, 196, 354

Worms 240 f., *240*
Wrinkle marks 90

Xenomorphous fabric 279

Yorked basins 125

Zeolites
    pyroclastic rocks 187
    sandstones 144
Zoantharia 236
Zoophycos facies 80
Zooxanthellae 235

QE
471
E523
v.2

JUN 15 1976